Books are to be returned on or before
the last date below.

LIBREX —

EMC '91: NON-FERROUS METALLURGY—PRESENT AND FUTURE

Papers presented at the First European Metals Conference, 'EMC '91: Non-Ferrous Metallurgy—Present and Future', organized by Benelux Metallurgie, die Gesellschaft Deutscher Metallhütten- und Bergleute, and the Institution of Mining and Metallurgy, and held in Brussels, Belgium, 15–20 September 1991.

      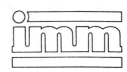

ORGANIZING COMMITTEE

Jean Vereecken (Chairman) Belgium
Herbert Aly Germany
Michael Anthony UK
Michael Jones UK
Helmut Maczek Germany
Robert Maes Belgium
Luc Segers Belgium

PROGRAMME COMMITTEE

Jussi Asteljoki Finland
Helena Bastos Portugal
Pierre Chabry France
Douglas Flett UK
Renato Guerriero Italy
Reinhard Hähn Germany
Claude Job France
Theo Lehner Sweden
Carlos Nuñez Spain
Harald Øye Norway
Michael Stefanakis Greece
Ser van der Ven The Netherlands

EMC '91: NON-FERROUS METALLURGY—
PRESENT AND FUTURE

ELSEVIER APPLIED SCIENCE
LONDON and NEW YORK

ELSEVIER SCIENCE PUBLISHERS LTD
Crown House, Linton Road, Barking, Essex IG11 8JU, England

Sole Distributor in the USA and Canada
ELSEVIER SCIENCE PUBLISHING CO., INC.
655 Avenue of the Americas, New York, NY 10010, USA

WITH 154 TABLES AND 396 ILLUSTRATIONS

© 1991 INSTITUTION OF MINING AND METALLURGY

British Library Cataloguing in Publication Data

European Metals Conference (1st: 1991: Brussels,
Belgium)
EMC '91: non-ferrous metallurgy.
I. Title
669

ISBN 1-85166-715-6

Library of Congress CIP data applied for

Printed in Great Britain at The Alden Press, Oxford

Foreword

This volume contains the papers that will be presented at 'EMC '91'—the European Metals Conference—to be held in Brussels, Belgium, from 15 to 20 September 1991, and organized by Benelux Metallurgie, GDMB (Gesellschaft Deutscher Metallhütten- und Bergleute) and IMM (the Institution of Mining and Metallurgy).

'EMC '91' is the first of an intended major series organized at the European level with the aim of bringing together all those who are involved with the extraction and processing of non-ferrous metals—European metallurgists and their international colleagues—to provide them with the opportunity to exchange views on the state and evolution of their industry.

The programme covers all the different aspects of the metallurgy of non-ferrous metals from mining to fabricated products. Particular attention is being paid to the European non-ferrous industry with respect to changes in demand, the technology used, pressures on the environment and the competitive position of manufacturers.

The contributions of the plenary lecturers (copies of which will appear in the IMM journal *Minerals Industry International* in 1991–92) and the many authors are gratefully acknowledged. Thanks are also due to the referees of the papers, the sponsors, the companies that have allowed registrants to visit their operations, the chairmen of the technical sessions and the staffs of the organizing bodies for their efficient administrative work.

Jean Vereecken
Chairman, Organizing Committee

July 1991

Contents

Nickel

Copper

Recycling

Minor Metals

Environment

Mineral Processing

Design of non-cyanide technology for flotation of lead–zinc ores: energy prerequisites, implementation and results

V. Panayotov
University of Mining and Geology, Sofia, Bulgaria

SYNOPSIS

A technological determination for flotation of lead-zinc ores is offered, which affords an opportunity to reduce or to stop the addition of common used depressants (NaCN, $ZnSO_4$). Laboratory and plant tests have been made. Results obtained show the perspective of cyanide - free technology.

Some problems by realization, as well as the ways of their elimination are revealed.

INTRODUCTION

A very important problem in modern mineral processing is how to increase the efficiency of the processes and particularly the rate of recovery of noble and valuable components from the ever impoverished and complex ores, at the same time removing or minimizing the harmful effect of toxic or costly reagents such as sodium cyanide and copper sulphate usually applied in these processes.

This problem has not only a technological but also ecological and social aspect so that its solution is of particular importance.

The present paper is devoted to the above-mentioned problem. We consider principles determining the energy effect on the mineral phase (in contrast with the effect of the aforementioned reagents), laboratory experiments and industrial implementation of the noncyanide technology for floating lead-zinc ores, accompanied with the activation of zinc minerals without copper sulphate.

The basic means of achieving this technological design are presented. The results of several years of implementation in a Bulgarian plant con-

nected with considerable increase in the recovery of noble elements when eliminating toxic reagents at minimum costs, are given. We have applied some relationships which could make the technology versatile and allow for its control and automation.

METHODS AND MEANS

Over the last few years some aspects of the energy effect on mineral processing have been discussed in detail /1,2/. These studies made it possible to consider certain new principles of introducing modern methods in mineral processing /3/ and developing an energy classification of minerals having semiconductor properties thus allowing for the assessment of their natural flotability and degree of energy effect on the mineral surface in order to obtain the best selective separation from the remaining associated mineral components /4/.

One of the most appropriate impacts which may be used in the flotation process is the electrochemical attack by which it is possible to reach both depression /5/ and activation /6/ of the mineral (in this case zinc) components. The activation and depression of the mineral components was carried out for an arbitrary type of minerals in a specially designed flotation chamber /7/, and the theoretical explanation of some problems concerned with the activation and depression of pyrite (which is largely valid for other minerals) is given /8/.

Using the basic relationships of solid physics, when the mineral semiconductor particle gets into contact with a charged metal surface - Fig. 1, we can say that in this contact there are

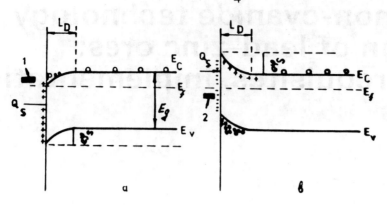

Fig. 1 Bending of energy zones on the mineral surface under the influence of a contact with a charged metal surface

1a - enriched zone bending, 1b - impoverished zone bending, Ec - bottom of conduction band, Ev - top of valence band, Ef - Fermi level, Eg - width of forbidden zone

two principal moments: either electrons are injected in the mineral particle or it is "enriched" in these basic carriers (in considering n-type minerals) - Fig. 1a, or conversely, when in this contact electrons are lost then the result is an "impoverished" bending of the energy zones - Fig. 1b.

The bending of the zones on the mineral surface is given by the relationships:

$$\frac{dV}{dx}\bigg|_{x=0} = \frac{dV}{dx}\bigg|_{V=V_S} = \frac{Q}{\mathcal{X}\,\mathcal{E}_0}$$

where: V_S - surface barrier which characterizes the bending of the energy zones

Q_S - density of the surface charge $[C/cm^2]$

If we consider the contact of the mineral particle with a metal electrode, then it is obvious that the depth of penetration of the metal electric field in the semiconductor mineral, determined by the difference in the work functions, should be presented most generally with the expression:

$$L_D = \left[\frac{\mathcal{E}\cdot(\mathcal{X}_2 - \mathcal{X}_1)}{2\pi e^2 n_0}\right]^{1/2}$$

where it is evident that the penetration depth of the field is so much greater as the difference in the work functions is stronger and as the concentration of the charge carriers in the semiconductor mineral is lower, $n_0\,[cm^{-3}]$, for a distance $x \geq L_D$ in the semiconductor there being no field. The consideration of other effects chan-

ging the energy configuration of the mineral surface and their relation to the flotation results as well as the devices for measuring these changes, the quantitative assessment and prediction of the technological results are given in other papers /9,10/.

We shall discuss briefly the problem raised above. In order to assess quantitatively and qualitatively the degree of energy effect on the mineral surface (in this case reached by electrochemical impact) it is necessary to have information about the energy state of the mineral phase. For this purpose energy diagrams of the minerals are plotted. In these diagrams all possible states in nature are reflected. In Fig. 2 is presented the joint energy diagram of the most common minerals in dressing lead-zinc ores: ZnS, FeS_2, PbS, $CuFeS_2$, in which the concentration of the free charge carriers $n\,[cm^{-3}]$ Q_S - mineral surface charge $[C/cm^2]$; N - number of possible fixed particles of the collector $[cm^{-2}]$; G - the same concentration expressed in $[mg/cm^2]$, which permits the easier determination of the reagent-collector concentration and its monitoring. The isolines from 1 to 7 show, respectively, the degree of bending of the surface zones. For $e\,\psi_S/\kappa T = 1$ straightened zones are observed and for $e\,\psi_S/\kappa T = 7$ inversion is observed. These are boundary cases and between these are all possible states of zone bending which can be caused by outside impacts, admixtures, defects, etc., i.e. these are all the

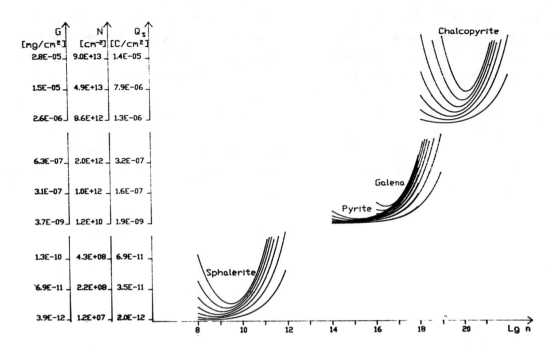

Fig. 2 General energy diagram of the minerals

possible states of the mineral surface in nature. Comprising the entire concentration interval of the charge carriers $n\left[cm^{-3}\right]$ and taking into account the energy diagrams of each mineral – Fig. 2, it is possible, by means of a specially designed apparatus, to examine the concentration of the charge carriers and the degree of zone bending /1/, in order to determine the transient energy state of mineral, to compare its flotability with that of the other, accompanying mineral components and to predict an impact, in this case electrochemical, which should have a donor or acceptor character relating to the mineral surface, depressing or activating the respective mineral. It is also possible to make a quantitative assessment of the necessary energy impact in view of obtaining particular flotation properties, by using the so-called coefficient of energy flotation insufficiency /9/. As far as the optimum selection of flotation reagents depending on the energy state of the mineral phase fed for flotation is concerned, our latest investigations /11/ give a tentative direction in this respect.

When testing the hypotheses for purposeful depression and activation of the zinc minerals in the process of flotation without $NaCN$ and $CuSO_4$ we had in mind the following factors:

a) the bending of the energy zones on the mineral surface actually determines its degree of activation and depression under certain condi-

tions;

b) along with charging the mineral surface there is also a possibility for isomorphic replacement of ions close in size to those of the crystal lattice.

c) if we consider the process of activation of zinc minerals by means of the classical activator copper sulphate, the process in this case is usually given by the reaction which does not exclude this mechanism:

$$ZnS + Cu^{2+} \rightleftharpoons CuS + Zn^{2+}$$

If for this process we apply the law of mass action and deduce the reaction rate constant, which also determines the activation rate, we shall obtain:

$$K = \frac{[CuS] \cdot [Zn^{2+}]}{[ZnS] \cdot [Cu^{2+}]}$$

where: A - activation; D - depression;
The acceleration of the process of transmitting electrons from the mineral to the liquid phase speeds up the activation process. We start from the basic equation for each semiconductor mineral,

$$n.p = n_i^2 = const$$

and from this equation for the concentration of electrons we obtain:

$$n = \frac{const}{p}$$

i.e. in order to speed up the activation process it is necessary to cause an impact creating free vacancies (gaps) in the semiconductor mineral. This can be done either by doping with donor admixtures in the crystal lattice, e.g. with sulphur or by treating the surface with acceptors, such as oxygen and ozone (9). Previous papers have reported on the establishment of the positions of Fermi level in the semiconductor mineral so that it should have the highest and the lowest flotability (3); These are respectively:

$$E_F^A = E_{g/2} + 4kT$$

$$E_F^D = E_c + 4kT$$

In measuring the initial Fermi level, which is performed as shown (9) by measuring the contact potential difference (cpd), it is possible to find Fermi levels and then, by appropriate impacts, to approach these levels of Fermi which guarantee activation of depression of the mineral component under given conditions. Another method which could be realized faster at the presence of the energy diagrams takes into account the conditions of "degeneration" of the electron gas in the semiconductor, namely $n_i/n < 1$, and the condition for which the Fermi level transcends the forbidden area and vacancies are generated in the mineral, which guarantee the transmission of electrons from the liquid to the solid phase or activation $n_i/n > 1$. In the first case this means reaching the "top" in the Fermi level, and in the second case, it is unnecessary to lower more considerably the Fermi level. For each mineral we can find the Fermi level at which it is most successfully activated – these are (by cpd) for galena CPD=-240mV, for sphalerite CPD=-260mV and for chalcopyrite CPD=-160 mV. Accordingly all minerals are depressed when being near the Fermi level up to the zone of conductivity 3 – 4 kT which is approximation to the limit of "degeneration". The change in the free energy of the electron gas when "raising" and "lowering" the Fermi level is given with the following relationship /3,9/ by which the qualitative assessment of the processes of depression and activation is given:

$$\Delta U = \frac{3}{2} \cdot k \cdot T \cdot N_c \cdot exp\left(-\frac{E_{f_1}}{kT}\right) - exp\left(-\frac{E_{f_2}}{kT}\right)$$

where: U – change in the free energy of the electron gas (eV/cm^3); N_c – reduced density of states; E_{f1}, E_{f2} – Fermi levels before and after the given process, respectively (eV).

EXPERIMENTAL RESULTS

The laboratory tests for controlling the depression and activation of zinc minerals by using an electrochemical impact (without sodium cyanide and copper sulphate) were proved – Fig. 3, where: I – dependence of the degree of depression by electrochemical impact without sodium cyanide on the duration of influence, II – dependence of the degree of activation of zinc minerals by electrochemical impact (without copper sulphate) on the time of processing. With a dotted line 1 is shown the maximum degree of activation of zinc minerals without electrochemical impact (only with copper sulphate). With dotted lines 2 and 3 respectively, are given the minimum and maximum admixture of zinc in a Pb concentrate obtained under laboratory conditions /3,11/. As can be seen, for an optimum time of eletrochemical depression the lower boundary of admixture is approached (in industrial conditions we were able to decrease this admixture down to 2,5-3 %, the latter being an important reserve) and for an optimum time of activation Δ_2 we reached improvement of the indices of zinc mineral recovery which under industrial conditions was brought up to maintaining the indices obtained by the classical regime – see the Table. For testing the principles described here an industrial system for activation and depression of the mineral components was designed, which can be used in the flotation tanks – Fig. 4 or in the flotation machines – Fig. 5 (in our case we used Denver 300 industrial machines). The principle of selection of the material for preparing the electrode systems, their design as laboratoly and industrial versions, as well as some more detailed laboratory and industrial results are given in some previous papers /3,5,7/. When inplementing the industrial system in a Bulgarian processing plant we experimented a complete noncyanide flotation of lead-zinc ores and a partially noncyanide one (in which 10 % of the initial amount of sodium cyanide, being 74 g per tonne of ore, was fed). The same applies to the activator copper sulphate.

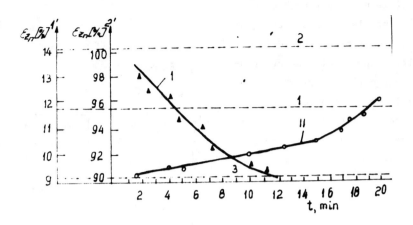

Fig. 3 Results of depression and activation of sphalerite without sodium cyanide and copper sulphate but with electrochemical impact

Table

Industrial technological results			
Regime	Pb recov. %	Zn recov. %	Ag recov. %
classical regime with NaCN CuSO$_4$	92,61	85,81	75,19
flotation without NaCN CuSO$_4$	93,38	85,57	77,00

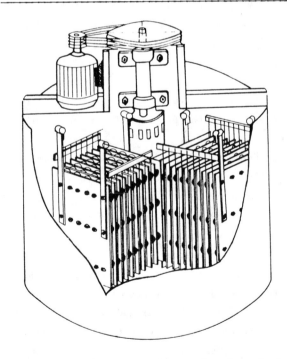

Fig. 4 Industrial system for activation of zinc minerals located in agitation tanks

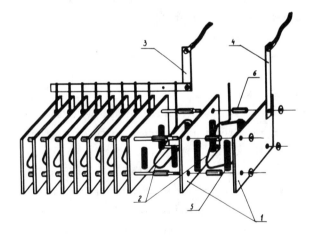

Fig. 5 A separate electrochemical module introduced into practice, for depression of zinc minerals, located in flotation machines where: 1,2 – working and auxiliary electrodes, respectively, 3,4 – power lines, 5 – studs regulating the distance between electrodes, 6 – rubber limiters

The relationship between the strength of current passed to the working electrode and the quality of the concentrates obtained was studied. Fig. 6 and Fig. 7 show respectively the processing of these data. The equations derived can be very useful in controlling these processes. Similar relationships are obtained also when controlling the admixtures of Pb and Zn in the corresponding concentrates.

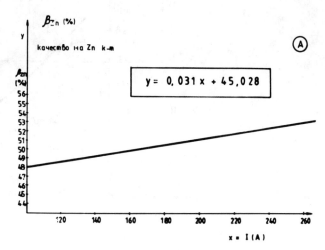

Fig. 7 Dependence of the quality of the zinc concentrate under industrial conditions on the strength of current of the working electrode in (A) in the process of activation of zinc minerals without copper sulphate

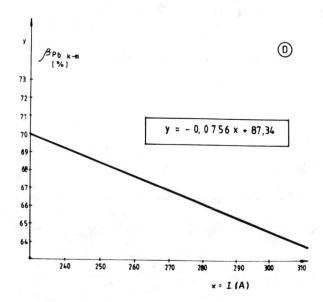

Fig. 6 Dependence of the quality of the Pb concentrate under industrial conditions on the strength of current in (A) of the working electrode in the process of depression of zinc minerals without sodium cyanide

CONCLUSION

A method has been designed for energy assessment of minerals fed for flotation in terms of their flotability.

The energy diagrams plotted make it possible to estimate the place of each mineral in relation to the accompanying minerals as well as the type of impact necessary to induce in order to carry out the most successful separation between them. By means of electrochemical models an effect is obtained on the mineral surface thus enabling the particles to reach a degree of depression or activation under the definite conditions of flotation. The relationships obtained make it possible to start controlling and monitoring the dressing processes on the basis of the energy principle. This is partially done in /12/.

The implementation of the method for depression of zinc minerals without NaCN in a Bulgarian processing plant led to increasing the silver recovery by 2,5 %, of Pb by approximately 1 %, the admixture of Zn in the lead concentrates being decreased by 1,5 %.

Apart from achieving technological results, the method has a considerable ecological and social effect related mostly to elimination of cyanide ions in the waters and atmosphere of the plant.

The paper presents a basis for a transition to realizing a nonreagent depression and activation of other mineral components, an analysis being necessary of their energy state and assessment of the degree of type of impact (cathode or anode processing).

For assessment of almost all minerals used in mineral processing for presenting a prediction picture of their behaviour and possible technological results depending on the properties of each mineral, as well as for controlling the degree of impact, a special computer system is being designed.

References

1. Carta M. at M. Improvement in electric separation and flotation by modification of energy levels in surface layers, Tenth International Mineral processing (IMPC), London, 1973.
2. Hoberg H., Schneider F., Investigations into
the improvement of flotability of minerals by
means of radiation, XI IMPC, Cagliary, 1975.
3. Panayotov V., Panayotova M., New Technological decisions, 3rd International Symposium on
Benefication and Agglomeration - ISBA 91, India
Institute of metals, Bhubaneswar, January 1991.
4. Panayotov V., Energic classification of semiconductor minerals, ISBA 91, January 1991.
5. Panayotov V., et al, A device for depression
of mineral components, auth.cert. Bulgaria, VOZ
V5/28 N$^{\underline{o}}$ 39337.
6. Banerji B.K., A new technique for activation
of sphalerite, proc. Australas, Inst. min and
Metall, N$^{\underline{o}}$ 245, Masch, 1973.
7. Panayotov V., et al, Electrochemical Flotation
Chambers, Auth. cert. N$^{\underline{o}}$77198 Bulgaria.
8. Janetski N.D., Woodbush S.I. and Woods R., An
electrochemical investigations of pyrite flotation and Depression, International Journal of Mineral processing 4 1977, p. 227-239.
9. Gaidarjiev S., Panayotov V., An energy approach. The prognosis of technological flotation
parameters, XV, IMPC, Cannes, France, vol.II,
1985, p. 255-264.
10. Gaidarjiev S., Panayotov V., The apparatus and
Methodology for energi-technological monitoring
of flotation, XIX, Symposium processing control,
Katowice, 1990, 10-12.
11. Panayotov V., Panayotova M., Studies of surface potential of Frothers Aimed of their optimal selection in Flotation, III International
mineral processing Symposium, Istambul, Turkey,
11-13, IX, 1990.
12. Tsekov Tsv., Panayotov V., at all, An Automated system for controlling a low cyanide technology of Dressing lead-zinc ores, International
congres in mining, Ostrawa, 1990.

Recovery of gold and silver from plumbojarosite-containing hematite tailings by alkaline pretreatment and cyanidation

J. Viñals
A. Roca
M. Cruells
C. Núñez
Department of Chemical Engineering and Metallurgy, Faculty of Chemistry, University of Barcelona, Barcelona, Spain

SYNOPSIS

The tailings from the sulphatizing roasting of Spanish complex pyrites consist of an impregnation of ferric sulphate, argentian plumbojarosite and minor amounts of anglesite and elemental sulphur in a porous hematite matrix. The gold and silver contents are 1-4 g/t and 50-150 g/t, respectively. A two-step alkaline pretreatment and subsequent cyanidation was investigated.

The effects of prior conditioning with $CaCO_3$, the NaCN concentration, particle size, temperature and basic aspects of the alkaline decomposition of the plumbojarosite were looked at. Extensive batch testing showed that 85-90% Ag and 90-95% Au can be recovered by: (1) a first pretreatment with $CaCO_3$ at pH 5-5.5; (2) a second pretreatment with CaO at 60ºC, pH 11-12, for 60 min; and (3) cyanidation at 60ºC for 30 min. The reagent consumption was 25-60 kg/t $CaCO_3$, 25-35 kg/t CaO and 0.2-1.5 kg/t NaCN.

During the second half of the 19th century, large amounts of complex cupriferous pyrites were beneficiated in southwest Spain by the process known as "artificial cementation"[1]. More than 10^6 t of tailings from this treatment are currently available, with a mean chemical and mineralogical composition as shown in Table 1.

In accordance with the mineralogy of these tailings -particularly because of the presence of large amounts of ferric sulphate and the binding of approximately 70% of the silver in the plumbojarosite lattice -conventional cyanidation results in low silver recovery and high reagent consumption. The problem of low silver recovery in jarositic products has also been reported for the residues of the pressure leaching of complex sulphides. Decomposition in hot alkaline media with or without simultaneous sulphidization has recently been reported as a suitable pretreatment[2,3]

In 1984-85, the Tharsis Sulphur and Copper Co. contracted the University of Barcelona to investigate treatment processes for these old tailings . Two possible routes were examined based on characterization studies[4]: 1) the recovery of Au, Ag and Pb by a chloride-hydrometallurgy[5,6] and 2) the recovery of Au and Ag by a two-step alkaline pretreatment and subsequent cyanidation process. The alkaline route was chosen by the Tharsis Co. for capital cost reasons, since there is a Merrill-Crowe plant in operation in Tharsis for gossan processing.

This paper summarizes the response of these old tailings to the alkaline pretreatment and cyanidation process, and describes basic aspects of plumbojarosite decomposition in $Ca(OH)_2$ solutions.

EXPERIMENTAL

Hematite tailings

The materials used correspond basically to samples 1 and 2 described in a previous paper[6]. The composition of sample 1 (Au 3.1 g/t, Ag 52 g/t, ferric

11

Table 1

Element	Mean %	Range %	Mineral	Mean %
Au	2.6 g/t	1.0-4.5 g/t	Hematite	73
Ag	98 g/t	52-164 g/t	Plumbojarosite*	8
Pb	2.4	0.55-4.4	Ferric sulphate**	5
Fe	51.3	30.9-63.4	Anglesite	1
S	2.0	0.97-3.9	Elemental sulphur	0.2
Cu	0.07	0.03-0.21	Quartz	10
Zn	0.05	0.03-0.14		

* as $Pb_{0.5}Fe_3(SO_4)_2(OH)_6$; ** as $Fe_2(SO_4)_3.9H_2O$

sulphate 3.7%, plumbojarosite 6.9%, S_{elem} 0.05%) is roughly an average, whereas sample 2 (Au 1.3 g/t, Ag 136 g/t, ferric sulphate 9.8%, plumbojarosite 11%, S_{elem} 0.9%) is representative of zones in the deposits especially rich in ferric sulphate and plumbojarosite.

Tests of alkaline pretreatment were made with 500 g solid (80% < 100 µm) and 1 l of solution in a conventional stirred reactor equipped with pH-register. For the cyanidation tests the pulp was transferred to a pachuca reactor inmersed in a thermostatic bath. The process was followed by solution sampling and analysis of Au and Ag by Atomic Absorption Spectrometry (AAS) and CN^- by volumetric analysis. The solid residues were analyzed by fire assay and selected samples were examined by Scanning Electron Microscopy (SEM) in conjunction with X-Ray Energy Dispersive Analysis (EDS).

Synthetic lead-jarosite

Synthetic samples of lead-jarosite were used to determine certain aspects of its decomposition in $Ca(OH)_2$ solutions, which are difficult to observe directly from hematite tailings. The synthesis was carried out by a procedure similar to that described by Dutrizac[7,8] For the firts synthesis, 1 l 0.3M Fe^{III} (as $Fe_2(SO_4)_3.nH_2O$) and 0.03M H_2SO_4 was reacted with 30 g $PbSO_4$ at 100ºC for 24 hours in a stirred reactor. The product obtained was washed four times with 1 l 10% NH_4CH_3COO at 25ºC for one hour to remove the ex-

cess $PbSO_4$. A second synthesis was performed in the same way but with the addition of 5 g of lead-jarosite seed from the first synthesis.

The XRD characterization of the resulting product indicated only jarosite phase (Fig.5). The chemical composition (PbO 12.8%, SO_3 30.5%, Fe_2O_3 45.2%) was that of a plumbojarosite-hydroniumjarosite solid solution whose approximate formula is $Pb_{0.3}(H_3O)_{0.4}Fe_3(SO_4)_2(OH)_6$, similar to the products obtained by Dutrizac under like conditions. Most of the material was distributed in aggregates of 5-10 µm, made up of rhombohedral crystals measuring 0.5-2 µm.

In the experiments for the decomposition of synthetic samples in $Ca(OH)_2$ media dilute pulps were used to prevent contamination of the residues by formation of gypsum. Typically, 2.0 g of sample were treated with 1 l of $Ca(OH)_2$ solutions of different concentrations at differents temperatures. The degree of decomposition was determined by analyzing the sulphates in the solid residue and by X-ray diffraction. The residues were also characterized by SEM and EDS.

RESULTS AND DISCUSSION

Pretreatment with $CaCO_3$

Because of the presence of ferric sulphate, direct conditioning of the tailings with CaO presents serious difficulties for stabilizing the pH of

13

the pulp during the conventional cyanidation process. A gelatinous precipitate is generated which coats the CaO particles requiring 50-100 kg/t CaO ; that is an excess of alkali over the theoretical consumption. Furthermore, the precipitate tends to clog the pores of the hematite particles and part of the ferric sulphate remains unreacted, which leads to alterations of the pH in the cyanidation process and an increase in the cyanide consumption (NaCN 2-3 kg/t).

A first pretreatment with $CaCO_3$ (< 50 μm) was employed for neutralizing the ferric sulphate, and not only for economic reasons, according to the following process:
$$Fe_2(SO_4)_3 + 3CaCO_3 + 9H_2O \longrightarrow 2Fe(OH)_3 + 2CaSO_4 \cdot 2H_2O + 3CO_2$$

The precipitate generated on the $CaCO_3$ surface was dispersed as a results of a simultaneous generation of CO_2, with the consequent advantages for stabilizing the pH of the pulps. The treatment made it possible to achieve a pH of 5-5.5. The initial pH of the pulp was 1.5-2.3 and the $CaCO_3$ consumption was 25-60 kg/t. This first pretreatment was used throughout the entire research involving tailings.

Preliminary cyanidation tests

The effects of the cyanide concentration, particle size and temperature on the cyanidation rate of tailings pretreated with $CaCO_3$ were examined. Prior to cyanidation, the pulp was conditioned with an excess of CaO in relation to the plumbojarosite content (CaO 25-35 kg/t). Under these conditions the pH attained in the pulp (11-12) was practically stable.

Typical results on the effect of the cyanide concentration are given in Fig.1. The extraction rate of gold increased as the cyanide concentration was increased. However, the cyanide concentration over the range studied did not affect the silver extraction, whose values were around 30% at ambient temperature.

The effect of particle size was studied by analyzing the gold and silver in fractions of the

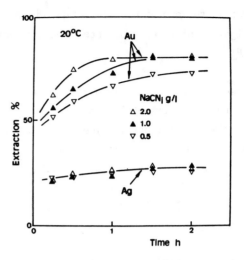

Fig.1. Effect of CN⁻ concentration for tailings pretreated with $CaCO_3$ and CaO at 20ºC (sample 1)

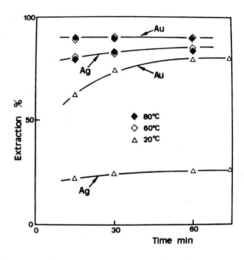

Fig.2. Temperature effect on the cyanidation of tailings pretreated with $CaCO_3$ and CaO at 20ºC. (sample 1, $NaCN_i$ 2 g/l)

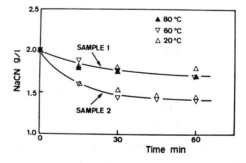

Fig.3. Evolution of the CN⁻ concentration

cyanidation residue with different particle sizes . No significant effect was noted in fractions of < 100 μm.

The effect of temperature (Fig.2) indicated that this is the main parameter causing a dramatic increase in silver extraction. The effect was considered to be associated with the extent of alkaline decomposition of the plumbojarosite. The rate and degree of gold recovery also improved considerabily with high temperature cyanidation. Nonetheless, the relation between this effect and the gold mineralogy in these tailings is not known. It is possible that there is a certain association between sub-micron particles of gold the plumbojarosite aggregates.

The change in cyanide concentration with time at different temperatures is shown in Fig.3. No significant differences were observed in the 20-80ºC range. The NaCN consumptions were 0.5 kg/t and 1 kg/t for samples 1 and 2, respectively. The larger consumption for sample 2 was most likely due to the higher elemental sulphur content, which may generate small amounts of sulphidizing species by reactions of the following type:

$$4S + 4OH^- \dashrightarrow S_2O_3^{2-} + 2HS^- + H_2O$$

which may consume cyanide by the formation of thiocyanates. This question has been investigated in other jarositic materials containing larger amounts of elemental sulphur[2,3]. We did not look at this aspect in this study since the cyanide consumptions obtained were considered acceptable.

Decomposition of plumbojarosite in Ca(OH)$_2$ solutions

Synthetic lead jarosite

The results obtained with synthetic samples confirmed the noteworthy effect of the temperature on the plumbojarosite decomposition in Ca(OH)$_2$ solutions (Fig.4). On the other hand, the decomposition rate showed little sensitivity to the Ca(OH)$_2$ concentration over the range studied (0.1-1.2 g/l).

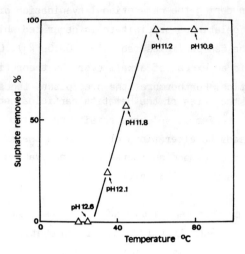

Fig.4. Temperature effect on the decomposition of lead-jarosite (Ca(OH)$_2$ 0.8 g/l, 1 h)

In the temperature range of 30-80ºC the process can de represented stoichiometrically as follows:

$$Pb_{0.5}Fe_3(SO_4)_2(OH)_6 \text{ (s)} + 4OH^- \text{(aq)} \dashrightarrow$$
$$2SO_4^{2-} \text{(aq)} + 0.5Pb(OH)_2 \text{ (s)} + 3Fe(OH)_3 \text{ (s)}$$

That is, the decomposition is characterized by the removal of the sulphate ions from the lattice and their diffusion to the bulk solution. At the same time, a weakening of the reflection intensity of the reticular planes of the plumbojarosite is observed. Fig.5 shows the diffractograms obtained. The solids resulting from the decomposi- are amorphous, although the morphology of the original plumbojarosite aggregates is left relatively undisturbed in the decomposition process (Fig.6).

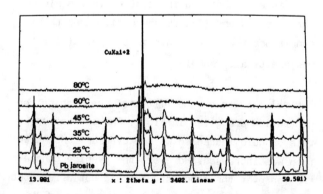

Fig.5. X-ray diffraction patterns of lead-jarosite and decomposition solids at different temperatures (Ca(OH)$_2$ 0.8 g/l, 1 h)

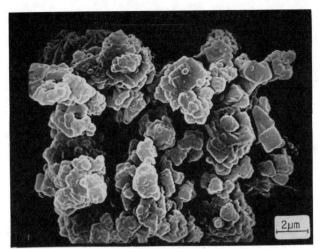

Fig.6. Secondary electron micrograph of decomposition solids (Ca(OH)$_2$ 0.8 g/l, 60ºC, 1 h)

Plumbojarosite-containing hematite tailings

The decomposition of the plumbojarosite associated to the hematite tailings was studied by ambient-temperature cyanidation of the tailings previously conditioned with CaO at different temperatures and residence times (Figs. 8 and 9). The results were similar to those of the synthetic samples. The decomposition process was fast (< 1 h) in the region of 60ºC.

Furthermore, the cyanidation rate of the decomposed plumbojarosite fraction was also high and practically independent of the degree and temperature of decomposition (Fig. 10).

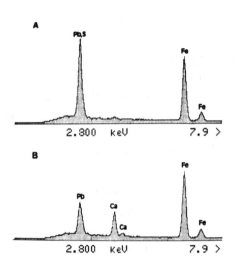

Fig.7. X-ray spectra of: A) Nucleus of lead-jarosite. B) Halo of decomposition (Ca(OH)$_2$ 0.8 g/l, 45ºC, 1 h).

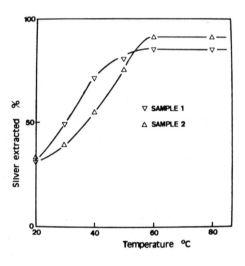

Fig.8. Temperature effect on the decomposition of plumbojarosite associated to hematite tailings (pretreatment time 30 min, cyanidation time 30 min, NaCN$_i$ 2 g/l)

The EDS examination of polished sections of partially decomposed aggregates indicated the presence of a reaction front. The X-ray spectra of the decomposed areas (Fig.7) indicated the removal of the sulphate ions. Furthermore, it was felt that the presence of Ca in these areas was because the solid decomposition products were porous enough to permit the diffusion of ions from the aqueous phase. This would explain the ease with which the decomposition solids were cyanided even though the silver was originally found in dilute solid solution in the plumbojarosite.

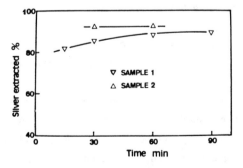

Fig.9. Effect of the pretreatment time at 60ºC (cyanidation time 30 min, NaCN$_i$ 2 g/l)

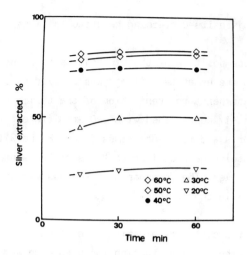

Fig.10. Cyanidation rates for tailings pretreated at different temperatures (sample 1, pretreatment time 15 min, $NaCN_i$ 2 g/l)

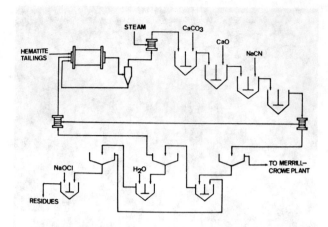

Fig.11. Basic flowsheet

DESCRIPTION OF THE PROCESS

A parameter-optimizing campaign to design an engineering project to process 125,000 t/y of tailings was undertaken. Samples from the entire range of compositions were tested to establish common operating conditions. The results obtained showed that 85-90% Ag and 90-95% Au can be recovered by: (1) a first pretreatment of the tailings milled to <100 μm with $CaCO_3$ at pH 5-5.5; (2) a second pretreatment with CaO at 60ºC and pH 11-12 for 60 min and (3) cyanidation at 60ºC for 30 min. The reagent consumption was 25-60 kg/t $CaCO_3$, 25-35 kg/t CaO and 0.2-1.5 kg/t NaCN. The mean composition of the leaching solutions (Au 1.2 mg/l, Ag 42 mg/l, Cu 29 mg/l, Zn 15 mg/l, As 5 mg/l, Pb 1 mg/l, Ca 0.6 g/l, SO_4 1.5 g/l NaCN 0.8 g/l) were similar to those treated in the Merrill-Crove plant operating in Tharsis for the gossan ores processing.

The basic flowsheet for the project is shown in Fig. 11. The tailings were reduced to the specified size in the milling circuit; a heat-exange system maintained the pulp at operating temperature and was send to the conditioning and cyanidation tanks. The cyanided pulp went through a thickening circuit and was washed by countercurrent repulping. The pregnant solution was sent to the Merrill-Crowe plant and the residues were treated with NaOCl to eliminate traces of cyanide.

CONCLUSIONS

The decomposition of plumbojarosite in $Ca(OH)_2$ solutions is characterized by the removal of the sulphate ions from the jarosite lattice. An amorphous product which retains the shape of the original aggregate is formed. The reaction is extremely slow at 25-35ºC, becoming effective at temperatures of >50ºC. The ions in aqueous solution can easly diffuse through the decomposition products, resulting in high cyanidation rates for the silver originally present as dilute solid solution in plumbojarosite.

The application of a two-step alkaline pretreatment and cyanidation process makes it possible to recover 85-90% Ag and 90-95% Au from the old tailings of sulphatizing roasting of Spanish complex pyrites. The process is characterized by : (1) a first pretreatment with $CaCO_3$ at pH 5-5.5 ; (2) a second pretreatment with CaO at 60ºC, pH 11-12; and (3) cyanidation at 60ºC. The reagent consumption is 25-60 kg/t $CaCO_3$, 25-35 kg/t CaO and 0.2-1.5 kg/t NaCN.

AKNOWLEDGEMENTS

The authors wish to thank the Tharsis Sulphur and Copper Co. for the financial support of this research and the permission to publish this paper. Thanks are also due to the Servei de Micoscòpia Electrònica (Universitat de Barcelona) for their assistance in the SEM and EDS studies.

References

1. Piritas de Huelva: Su historia, minería y aprovechamiento. I. Pinedo. Madrid: Summa, 1963, 1003 pp.

2. Hydrothermal sulfidation of plumbojarosite for recovery of lead and silver. K. Lei and N. Gallagher. In: Proceedings of EPD Congres'90. Warrendale: The Minerals, Metals & Materials Society, 1990. p. 13-20.

3. Silver and gold recovery from pressure leach residue. R. Berezowsky, J. Stiksma, D. Kerfoot and B. Krysa. In: Lead-Zinc'90. Warrendale: The Minerals, Metals & Materials Society, 1990. p. 135-150.

4. Viñals J., Roca A., Cruells M. and Núñez C. Tratamiento de morrongos procedentes de tostación sulfatante. Unpublished report, University of Barcelona, 1986.

5. Viñals J. and Núñez C. Dissolution kinetics of argentian plumbojarosite from old tailings of sulfatizing roasting pyrites by $HCl-CaCl_2$ leaching. Metallurgical Transactions B, vol. 19B, 1988 . p. 365-373.

6. Viñals J., Núñez C. and Carrasco J. Leaching of gold, silver and lead from plumbojarosite-containing hematite tailings in $HCl-CaCl_2$ media. Hydrometallurgy, vol 26, 1991. In press.

7. Dutrizac J., Dinardo O. and Kaiman S. Factors affecting lead jarosite formation. Hydrometallurgy, vol. 5, 1980. p. 305-324.

8. Formation and characterization of argentojarosite and plumbojarosite and their relevance to metallurgical processing. J. Dutrizac and J. Jambor. In: Applied Mineralogy. Warrendale: AIME, 1984. p. 507-530.

Grinding of the 'Jales de Santa Julia de la Compañía Real del Monte y Pachuca', SA de CV, Mexico

F. Patiño
Instituto de Ciencias de la Tierra, Universidad Autónoma de Hidalgo, Mexico
J. Ramírez
Instituto de Investigaciones Metalúrgicas, Universidad Michoacana de San Nicolás de Hidalgo, Mexico

SYNOPSIS

A study of a scheme about the optimum grinding process of the "Jales de Santa Julia de la Compañía Real del Monte y Pachuca", México, is presented. The study consists of an experimental work of grinding at laboratory scale. The results were applied to a digital simulator. The configuration of two closed grinding circuits in series (direct primary circuit and inverted secondary circuit) was defined with the application of the following selection criterion: (a) upper grinding to 48.3% - 400 mesh, (b) optimum utilization water of pulp dilution, (c) cut correct size of minimum classification, and (d) to keep with the capacity limitations recommended by hydrocyclone manufacturers. Nominal capacity investigated was of 150 metric ton/hour by grinding line.

This paper considers the study of a grinding process of "Jales de Santa Julia", which contain residual values of gold and silver. High percent of the "Jales" were produced by cyanidation of quarziferous ores. At present, the "Jales" production are of two classes: (a) cyanide tails of Pb/Ag flotation concentrates and (b) tails of this flotation circuit that have not been cyanided.

Silver recoveries obtained by cyanidation of the ground ores as well as of the flotation concentrates were not high in any case. For this reason, the "Jales" re-grinding should be necessary for the recovery of the metallic values contained.

The practice for many years in the industrial plant of cyanidation has been to mill the quarziferous ores only 60-65% -75μm, whereas the optimum size is 75-80% -75μm.

The objective of this paper is to define an optimum grinding scheme using a mathematical modelling of unitary operations involved. The re-grinding is the requisite of these "jales" beneficiation. Recent metallurgical studies suggest this operation[1].

MATHEMATICAL MODELLING

This method consists in a configuration of a mathematical model for the process alternatives. With this purpose we use semi-empiric models of principal unitary operations involved.

Grinding

This operation can be simulated mineral by mineral and size by size with a model called of p-order accumulative, by using concepts of Harris[2,3] and Horst, Bassarear[4]. The principal model characteristics are the following:

1.- The mill feed is characterized by a global granulometric distribution and the contents distribution for the constituent mineral species.

2.- After a mean residence time in the mill, the discharge also shows overall granulometric distributions and contents distribution for the constituent mineral species.

3.- The volumetric flow of the pulp is needed in the calculation of the mean residence time in the mill.

4.- The pulp transfer through the mill is defined by the following equation:

$$W_{i,j}(T) = W_{i,j}(0) \exp(-F_j K_{i,j} T^{p_{i,j}}) \qquad (1)$$

where T is the mean residence time in the mill, in min. $W_{i,j}(0)$ and $W_{i,j}(T)$ are the feed and discharge massic flow in the mill for the j component of the ore and the size coarser than the i opening mesh, respectively. $K_{i,j}$ is the accumulative specific grinding rate, and $p_{i,j}$ is the grinding kinetics order of the specie j of a size coarser than the mesh i. F_j is a scale factor applicable when the grinding is simulated at industrial scale, from the grinding kinetics derived of the tests made at laboratory scale.

Classification

The classification is the more important unitary operation, because in a closed grinding circuit of polymetallic ores, the classifier (hydrocyclone) operates according to the particles specific gravity as well as the particles size. In the case of the dense ores, this consideration is very important.

There are several mathematical models for the hydraulic classification by hydrocyclones. In this case, we need a hybrid model since it combines the virtues of three different models in one. These models are the following:

-Lynch and Rao classification model

This model found a wide success in the ores industry to predict satisfactorily the classification operation under any one process condition as the operation variables and the hydrocyclones design.

Two equations of the classification general model of Lynch and Rao[5] are applied:

* Percent of recirculated water, R_f :

$$R_f = K_3 \text{ spigot}/WF + K_4 / WF + K_5 \qquad (2)$$

where K_3 and K_4 are constant coefficients of the model and K_5 is a plant constant, spigot is the orifice diameter of coarse discharge of the hydrocyclone in cm. and WF is the water flow of dilution in the pulp feed to hydrocyclone in lt/min.

* Overall corrected cut size, d_{50c} :

$$\log d_{50c} = K_6 VF + K_7 \text{spigot} + K_8 \text{inlet} + K_9 FPS + K_{10} Q + K_{11} \qquad (3)$$

where K_6 to K_{10} are constant coefficients of the model, K_{11} is a plant constant, VF and inlet are the orifice diameters of fine discharge and feed of the hydrocyclone in cm., respectively. FPS and Q are the percent solids by weight and the volumetric flow of the pulp feed to the hydrocyclone in lt/min., respectively.

-Plitt classification model

The Plitt model[6] relates the classification index of mineral species to the respective corrected cut size, $d_{50c,j}$ in accordance to the following equation:

$$Y'_{i,j} = 1 - \exp(-(\ln 2)(d/d_{50c,j})^{n_j}) \qquad (4)$$

where $Y'_{i,j}$ is the classification index, d and $d_{50c,j}$ are the geometric middle size of particles of the specie j between i and i+1 meshes and corrected cut size of the specie j, respectively. n_j is the slope of the regression line.

-Finch model

This model makes possible the connection between the two above mentioned models, since Lynch and Rao model only predicts the overall classification efficiency of the classified material. The Finch model[7] equation is the following:

$$\ln(d_{50c,j}) = -K_1 \ln(P_j - P_1) + K_2 \qquad (5)$$

where K_1 is a constant coefficient of the model and K_2 is a plant constant. P_j and P_1 are the specific gravities of the specie j and the medium suspension, respectively.

-Simulation model

The models of the basic unitary operations described were coded and a digital simulator was developed. The digital simulator produced the following results:

1.- Pulp volumetric flows and water in each one of the grinding circuit lines.

2.- Mass flows of solids in each line.

3.- Pulp volumetric flow pumped to the hydrocyclones.

4.- Solids percent by weight and volume in each line.

5.- Overall granulometric distribution in each line.

6.- Granulometric distribution by species in each line.

7.- Mean residence time in the mill.

8.- Recirculating load by specie.

9.- Corrected overall cut size by specie.

METHODOLOGY

Experimental method

A representative sample of 50 kg. of "Jales de Santa Julia" was used. This sample was quartered to obtain 10 identical samples of 2.17 kg. The "jales" were subjected to a batch grinding for 2 and 8 minutes. After this operation, the material was wet and dry sieving to guarantee good particle size separation. The different fractions obtained were analyzed to determine the silver content by Atomic Absorption Spectrometry (AAS).

Grinding Kinetics Calculation

The grinding kinetics parameters were estimated by lineal regression of the experimental data applying the following model:

$$\ln\ln(C_{i,j}(0)/C_{i,j}(t)) = P_{i,j}\ln t + \ln K_{i,j} \qquad (6)$$

$$\begin{cases} 1 \leq i \leq m; \ 1 \leq j \leq 2 \\ m = 6; \ n = 2 \end{cases}$$

where $C_{i,j}(0)$ and $C_{i,j}(t)$ are the accumulative fractions coarser than the mesh i at time t = 0 and t = t (min) of the batch grinding of specie j.

Simulation

The applied methodology to the simulation with the developed digital simulator was carried out according to the following criteria:

(1) To obtain the maximum grinding degree no lower than 47.6% -400 mesh.

(2) To economize the dilution water in the classification step.

(3) To minimize the initial investment of the grinding plant.

(4) To operate in the permited capacity, mainly in the hydrocyclones.

RESULTS

Laboratory tests

Figure 1 shows the granulometric distributions of "Jales de Santa Julia" recently processed in the Ex-Hacienda de Loreto, Pachuca, México, and subjected to 0, 2 and 8 min. of grinding. Figure 2 includes the silver distribution size to size for 0, 2 and 8 min. of grinding. Figures 3 and 4 show the curves of accumulative percent coarser for all material and by specie (silver), respectively, for 0, 2 and 8 min. of grinding.

Simulation results

Preliminary simulations to define the circuit type to produce the desired grinding grade was carried out. The circuit in serie formed by a primary direct closed circuit and a secondary inverted closed circuit resulted from these simulation tests. The results are included in Figure 5. The nominal capacity of grinding was defined in 150 t/h of "jales". The appropiate dimensions of the mill are: 3 m. diameter and 3.5 m. length (outside dimensions). The primary and secondary mills are of the same size and of the overflow type.

The number of primary and secondary hydrocyclones are six. The dimensions are the same in all of them: (a) hydrocyclone diameter = 38.1 cm.; (b) inlet diameter = 11.43 cm.; (c) spigot diameter = 5.08 cm., and vortex diameter = 10.16 cm. The operation parameters to define are:

(a) Tonnage of the fresh feed, in t/h. (b) Granulometry of the feed ores. (c) Chemical composition of the valuable elements. (d) Grinding kinetics. (e) Densities of the mineral species of interest (Ag_2S and SiO_2). (f) Solid/water relation in the feed to the primary mill. (g) Density of the feed pulp to the primary hydrocyclones.

Figure 8 shows the percent curves corresponding to the size -400 meshes vs. hydrocyclone number only for the secondary circuit. Figure 7 shows the curves of water consumption in l/min. used in grinding of primary and secondary circuits vs. hydrocyclone number for constant densities of feed pulp to the hydrocyclone. Figures 8 and 9 illustrate the curves of corrected cut size, $d_{50c,j}$ vs. hydrocyclone number for constant densities of feed pulp to the hydrocyclones for silver and gangue respectively.

The mean residence time in the mill was 1.016 minutes and the scale factor, F_j used was 1.4 (the grinding in the industrial mill was 1.4 times more fast than in the laboratory mill). F_j was determined by the experience obtained with the grinding rate of this mineral type at industrial scale. According to Figure 5, the scale factor used, F_j in secondary grinding was of 1.2 because the material was approaching the grinding limit.

DISCUSSION

The selection of the grinding scheme consisting of a direct closed circuit (primary circuit) followed by a inverse closed circuit (secondary circuit) and six hydrocyclones of the dimensions above mentioned, was obtained according to the following considerations:

1.- Type and number of grinding circuits.

The primary circuit was selected of direct type considering that the "jales" require a conditioning of their surface previously to the flotation or any other separation method. This conditioning is referred to the removal of slimes as well as another products generated with a prolonged oxidation of the "jales". In this primary direct circuit all the material pass through the mill before the classification step. Otherwise, a fraction of the "jales" should be classified with the fines and the probability that they pass through the secondary mill should be very reduced. The grinding secondary circuit was selected of the inverse type to optimize the fines production in the global scheme of the grinding process. These two grinding circuits are sufficient to produce the grinding grade required for the silver optimum metallurgical recovery.

2.- Hydrocyclone number.

Simulations varying the hydrocyclone number in the grinding primary and secondary circuits were carried out. The criteria used to select the hydrocyclone optimum number were: (a) using the smaller spigot diameter, that produces the smaller corrected cut size, d_{50c}, and (b) having the pulp flow capacity that is recommended by the hydrocyclone manufacturer. Another restrictions of the hydrocyclones were taken into account: c) inlet flow capacity, and d) vortex flow capacity. However, the spigot showed higher sensitivity to the limit capacity of the process. Additionally, the water consumption in the grinding circuit was also considered. This way, although low densities of pulp produced corrected cut sizes of finer classification, the water consumption tends to be high as shown in Figure 8 at pulp densities lower than 1.50 t/m^3.

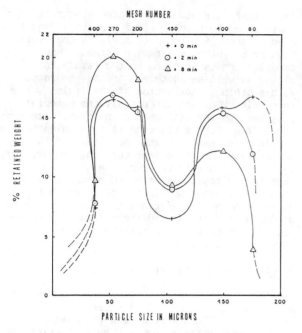

Fig. 1.- Granulometric distributions at 0, 2 and 8 grinding minutes.

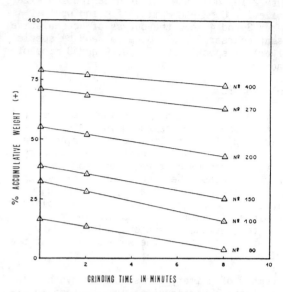

Fig. 3.- Percent in accumulative weight coarser vs. grinding time.

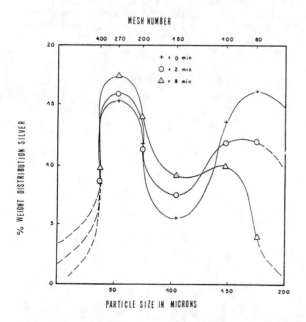

Fig. 2.- Granulometric distributions silver at 0, 2 and 8 grinding minutes.

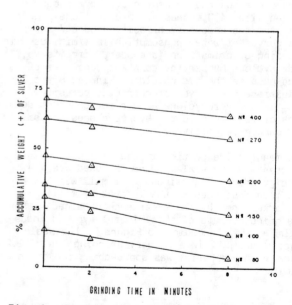

Fig. 4.- Percent accumulative coarser of silver vs. grinding time.

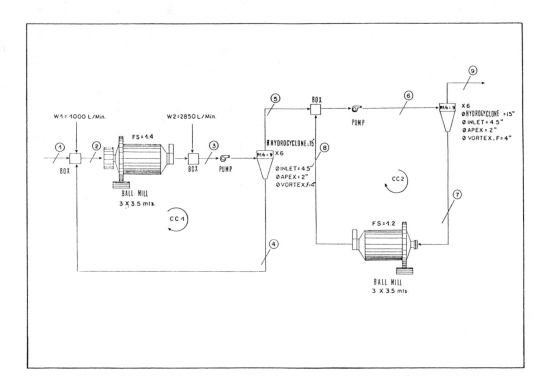

Fig. 5.- Process scheme of optimum grinding of Jales de Santa Julia

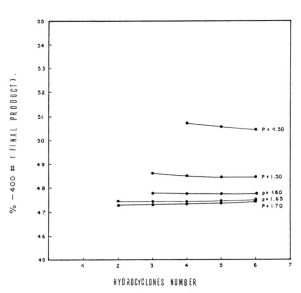

Fig. 6.- Optimum grinding degree % mesh -400
vs. hydrocyclones number at various
feed pulp densities, p.

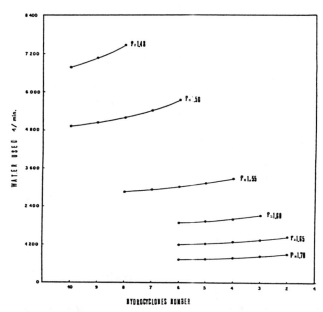

Fig. 7.- Water used in the grinding vs. hydrocyclones number at various feed pulp
densities, p.

24

References

1.- Espinosa De L. Estudio de ingeniería concep-
tual para la instalación de una planta de benefi
cio de los jales de Santa Julia. Unpublished report, Cía. Real del Monte y Pachuca, México,
1982.
2.- Harris C. The Alyavdin Weibull chart in
batch comminution kinetics. Trans.Inst.Min.Mat.
vol. 80, 1971. p. C42-4.
3.- Harris C. Chakravarti A. The effect of time
on batch grinding. Powder Technology. vol. 4,
1970, p. 57-60.
4.- Horst W. and Bassarear J. Use of ore grinda-
bility technique to evaluate plant performance.
Trans. AIME. vol. 260, 1976. p. 340-351.
5.- Modelling and scale-up of hydrocyclone classifiers. A. Lynch and T. Rao. In: 11th Inst.
Min. Process Congress. Cagliari, 1975.
6.- Plitt L. A mathematical model of the hydrocyclone classifiers. CIM Bulletin, vol. 69,
1976. p. 114-123.
7.- Finch J. and Matwijenko O. Individual mineral behaviour in a closed grinding circuit. CIM
Bulletin. vol. 70, 1977. p. 72-164.

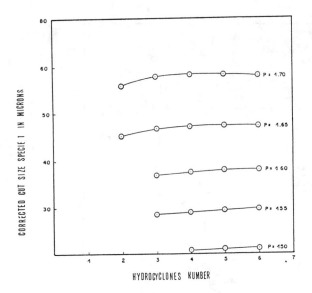

Fig. 8.- Corrected cut size specie 1 (silver) vs
hydrocyclones number at various feed
pulp densities, p.

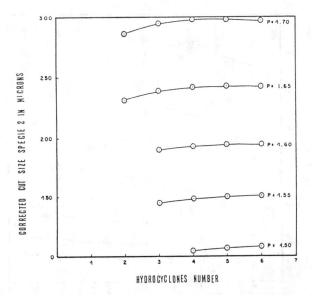

Fig. 9.- Corrected cut size specie 2 (gangue) vs
hydrocyclones number at various feed
pulp densities, p.

Lead and Tin

Lead blast-furnace evolution: a new approach

B. Madelin
Metaleurop Recherche, Trappes, France
S. Ferquel
Metaleurop SA, Fontenay-sous-Bois, France
J. L. Martin
Metaleurop, Noyelles-Godault, France

ABSTRACT

Metaleurop operates three primary lead blast furnaces adapted to the smelting of lead raw material. Their working ratios have been improved for the last ten years in particular for coke rate, lead bullion output, manpower and energy per ton. This, and the well known flexibility of this kind of furnace with regard to raw materials, allow them to increase its competitivity. In addition, this process has also to comply with the challenge of environmental regulations in particular for dust and SO_2 emissions.

The major target is the treatment of sintering offgases. It will be reached in the near future due to specific investments.

The most difficult target is the treatment of the lead blast furnace offgases. As the SO_2 level is very low, a specific process has to be developed.

Studies performed on this possible process pointed out a real chance of success if we can manage with the following points :
- feeding quality and repartition after the sealing of the blast furnace top,
- risk of accretions in the furnace as well as in the afterburning chamber,
- energy consumption with a nearly autogeneous post combustion,

The alternative is the use of completely new processes. Everybody knows that any really new industrial process needs a major research and development costs and that its optimization in order to reach a competitive position may last longer than expected. In addition, expected operating profits are insufficient to meet the very high investment required for these new technologies which, as far as we know, in the main, solve environmental problems.

Thanks to studies and tests, the traditional process will be able to meet the same environmental regulations with investment programmes far less costly and with increased flexibility of raw material treatment.

INTRODUCTION

Metaleurop operates three primary lead blast furnaces adapted to the smelting of lead raw material and one Imperial Smelting furnace delivering lead bullion as well as zinc bullion.

The three primary lead blast furnaces are located in Europe. One in Germany (Nordenham) another one in Spain (Cartagena) and the last one in France (Noyelles-Godault).

In order to improve the economic of these furnaces we have, for the last ten years, tended to improve technical and economical ratios such as coke rate, lead bullion output, lead and precious metals recovery, manpower and energy per ton of lead bullion. At the same time a real progress has been made in the working conditions such as lower dust content in the working atmosphere due to better gas cleaning and to the suppression of arduous tasks. For these purposes new investments have been made or new ways of working have been developed. For example : continuous slag tapping ; lowering or suppressing the number of furnace top cleanouts.

This positive evolution has to be carried on in order to satisfy increasingly stricter denvironmental standards. Therefore, lead producers will have to make investments. Nevertheless, they also know that the profitability of producing commercial lead has always been precarious, and will continue to remain so in the forthcoming years. As a result, investment capacity has to be limited. Lastly, the cost of new installations is always very high and a wrong decision is often fatal.

For all these reasons, Metaleurop is keeping

a close eye on the development of new technologies with regard both to their technical development and their economic feasibility :
- The new pyrometallurgical processes (QSL, Kivcet, Isasmelt, ...).
- The new hydrometallurgical chloride processes

The present status of these new processes and their heavy investment costs lead Metaleurop to retain its confidence in the traditional process and to develop the potential it still has.

THE EVOLUTION OF TRADITIONAL METALLURGICAL PROCESSES FACED WITH ECONOMIC PROBLEMS

Lead bullion production and coke ratio evolution
These two data seem to be a good example of lead blast furnace progresses in the last ten years. They clearly show us that a great deal of improvement can be achieved with a traditional process (data from one of our lead smelters).

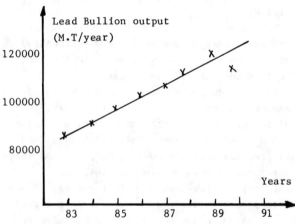

Fig. 1 Lead bullion output versus years

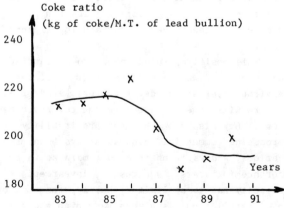

Fig. 2 Coke ratio versus years

This improvement is the consequence of numerous slight modifications in every field. In the following chapter we will try to focus on some of them.

Development and evolution in the last ten years
The capacity increase and the coke ratio decrease can be explained by a better operating ratio of the furnace and a better metallurgical control, as described hereunder :

Operating ratio
Ten years ago the four main problems affecting the operating ratio were :
1. The lack of furnace feed : the sinter plant was the bottle neck of the lead plant.
2. Difficulty of bottom furnacing in slag and matte tapping.
3. Frequent clean-outs of the top of the furnace to remove accretions.
4. Frequent failures, above all in the cooling system.
    Today the capacity of the sinter plant has been improved due to the investments carried out on feed preparation and return sinter treatment.
Concerning the bottom furnacing, the new continuous slag tapping allows a better working ratio, and has a very positive effect on working conditions.
    The frequency of clean-outs decreased markedly with the better design of the off-gas circuit, with the better physical and chemical feed control, and with the improvement of the feeding.
    A new design of the cooling water jackets has been adopted after simulation studies. Accordingly, their life time has increased strongly.

Metallurgical control
In a lead blast furnace, coke combustion is the main factor.
    The combustion of coke in front of the tuyeres supplies most of the heat consumed in the lead blast furnace. It sets some of the limiting conditions for the heat exchange in the shaft. On one hand, the heat delivered per mole of blast air is determined by the combustion ratio Oblast/C. On the other hand, the fraction of this heat supply transferred to the burden depends on the flame temperature of the combustion.
    Many fluctuations of the combustion parameters have been observed at industrial level. These variations were investigated in an experimental furnace, simulating, on a full scale

basis the region near one tuyere in the blast furnace. Cokes with different properties (size and reactivity) were compared under various blowing conditions (flow rate and oxygen content of the blast, tuyere diameter).

These trials, as a whole, pointed out the general features of the combustion. From the cooling box inwards, the coke stack exhibits two regions corresponding to the two successive steps of a "layered" combustion :
- exothermic combustion by the air oxygen ($C + O_2 \longrightarrow CO_2$). This region extends horizontally along the tuyere axis over 40 cm and vertically along the jacket over a few tens of centimeters. This height varies widely according to the permeability of the granular bed,
- endothermic solution loss reaction (Boudouard $C + CO_2 \longrightarrow 2CO$). This reaction occurs in the furnace center at the tuyere level, but at higher levels, in the front zone, near the jacket.

This means that, at any level above the tuyeres, the reducing power of the gases is higher in the central zone than in the peripheric zone. Moreover, the furnace center is colder than the periphery due to the predominance of endothermic reactions.

As it could be foreseen from earlier probings in industrial furnaces, our trials confirmed that some unburnt oxygen can be present in the ascending gases along the jacket, when the impulsion of the blast air at the tuyere becomes too weak. Then the oxygen content can remain higher than 5 percent over the whole height of the coke column. When this occurs in the blast furnace, the combustion provides less heat at the bottom of the shaft and may be completed high above the tuyere levels. This eventually leads to an early melting of the charge and to channelling.

The tests on the influence of the blowing parameters revealed the sensibility of the combustion behaviour of the coke. In particular, the close link with the blast flow through the tuyere suggests that the blast air distribution at the tuyeres is a fundamental parameter of the furnace.

The main results of the study relative to the influence of the operating parameters upon coke combustion were the following :

Table 1 Influence of the operating parameter upon the combustion ratio

	Coke size	Blast rate	Coke to burden ratio	Oxygen content of the blast
Variation of parameter	↗	↗	↗	↗
Influence upon Oblast/C	↗	∪	↘	↘

This study led to slight modifications in the industrial way of working, in particular :
- better utilization of air due to selected blast impulsion.
- better utilization of coke due to the quality control and to the optimisation of feeding.
- control of the lead charge quality and optimisation of feeding.

THE EVOLUTION OF TRADITIONAL METALLURGICAL PROCESSES FACED WITH ENVIRONMENTAL RESTRICTION

Having dealt with the technical aspect, we are going to present the environmental evolution concerning dust level and SO_2 emissions.

Dust in effluents and work place atmosphere
In all ore processing installations, dust, in the broadest meaning of this word, comes from two sources :
- Airborne dust released during handling and storage operations.
- Dust produced in situ in connection with the metallurgical operations.

Metaleurop lead plants have developed well-adapted filtering media and systems for each kind of process gas. Today the increase of air filtered volume has enabled us to comply with environmental restrictions. Of course, research for improving filtering media and systems will be carried on.

A comparison between a traditional plant and a new pyrometallurgical plant points out that both plants need a filtration capacity of the same magnitude.

Table 2

Operations	Traditional plant	New technology plant
Stocking, handing charge preparation	20 000	20 000
Sintering/ desulfurizing	20 000)	
	)	10 000
Reduction	10 000)	
Lead refining	30 000	30 000
Volume of air or gaz to be filtered (Nm^3/T market lead)	80 000	60 000

SO$_2$ emissions

Traditional metallurgical process consists of three main stages :
- roasting and sintering the lead bearing materials,
- reducing the sinter in a lead blast furnace,
- refining the lead bullion into soft lead.

In the first two stages, SO$_2$ bearing gases are mainly produced in the sintering stage (more than 90 % of sulfur is burned into SO$_2$), the remainder being produced at the reducing stage.

Sintering

The sintering machine, where the ore concentrates are continuously sintered, consists of two separate areas.

In the first one, known as the reaction zone, the gases emitted are "rich" (e.g. their SO$_2$ content reaches 4.5 % - 5.5 %). This allows the sulphur dioxide to be converted directly into sulphuric acid by the so-called "contact" process.

The second zone is where the sintered cake is cooled by a forced ventilation system. The gases collected are "poor" since their content is less than 2 %. They are often recycled in the circuit itself using a perfectly tried and tested technology. Yet, due to thermal effect, the productivity of the process is reduced by approximately 10 % because either the chain must be slowed down, or the sinter lead content must be reduced.

There are other solutions too, such as

mixing the poor gases with very rich gases from other installations on the same site, or the use of specific processes which produce sulphuric acid from poor gases.

Blast furnace

Lead sinter contains oxidized lead but also some remaining sulphur and sulphates. As a matter of fact due to thermodynamical aspects, it is not possible to fully complete the following reaction :

$$PbS + 3/2O_2 \longrightarrow PbO + SO_2$$

The sulphur remaining in the sinter is a little less than 10 % of the initial sulphur in lead concentrate.

During the carbon reduction, a small part of these 10 % is transformed into SO$_2$ and is found in the gases emitted at the top of the furnace. A greater part is found in the liquids tapped at the bottom of the furnace.

At present, the offgases of the lead blast furnace are diluted with air via the charging system, and the resulting gas is so diluted that it is not possible to process it from a technical point of view.

If we avoid this dilution, the SO$_2$ level can be higher than 1 % and then as explained hereunder, gases could be processed in a specific process or mixed with other richer gases.

Table 3 Comparison of the characteristics of the gas emitted with and without dilution

	With dilution	Without dilution
Gas delivery	80 000 Nm^3/h	15 000 Nm^3/h
Gas temperature	135 °C	650 °C
Speed of gas in the pipes	12.5 m/s	5.3 m/s
Dust concentration	16 g/Nm^3	83 g/Nm^3
Composition of gaz (in %)		
- CO in dry gas	0	10
- CO$_2$ "	7	25
- O$_2$ "	16	0
- N$_2$ "	76	63
- SO$_2$ "	0.3	1.5
- Content of H$_2$O	2.5	4

Apparently, a 1 % SO$_2$ bearing gas obtained without diluting could be treated in an acid plant using new processes, but the catalysts used in this process cannot deal with CO and dust.

Consequently, the process to be used to treat the blast furnace gases has to include the three following steps :
- sealing the top of the blast furnace,
- removing the CO by a post-combustion process,
- removing the dust.

These 3 steps cannot be taken very easily from a technical and economic point of view. The main aspects are the following :
- The sealing of the blast furnace top must not change the quality of the feeding of coke and sinter. The accretions we had in the past don't have to return with a new sealed top furnace. The channelling, with all its drawbacks, will also have to be avoided.
- The post-combustion of offgases could be done before or after dedusting. In the first case, post-combustion could be nearly autogenous, but we have to avoid any formation of new accretions and moreover the consequences on decreased operating ratio and unhealthy and arduous tasks due to the clean-out of dust accretions. In the second case, the cold gases have to be heated in order to avoid any additional energy consumption.

So, the real progress we obtained in the past by suppressing unhealthy and arduous tasks when removing accretions does not have to be cancelled out by the necessary SO$_2$ removal. This must be studied very carefully.

Therefore we have started off a study in order to find the best compromise to avoid these accretion difficulties which seem to be a problem for all the processes (including the new ones). We will give details about this study in the following part.

STUDY OF DESULFURIZATION OF THE PROCESS GASES
Currently, we can divide the SO$_2$ emission into three kinds of gases :
- The rich gases of the sinter plant. The SO$_2$ level (5 %) allows a treatment of these gases in a traditional acid plant.
- The poor gases of the sinter plant. The SO$_2$ level (1.5 %) is not high enough for a traditional acid plant. Mixing these gases with rich gases and/or recycling one part of them leads to a sufficient level in order to treat them right now in the improved new acid plant technology.
- The offgases of the blast furnace. As we said formerly, they have to be treated in a speci-

fic process that does not exist today as a whole but whose components do exist in other industries.

Sinter plant
In the sinter plant, the treatment of rich gases is easy. For the poor gases there are two alternative solutions :
- the first one is to recycle them in the circuit itself. This has the disadvantage that, due to thermal effect, the productivity of the process is reduced. In this case it is better to invest in a traditional, compact acid plant instead of a larger sinter machine.
- the second one is to mix the poor gases with the rich ones. In this case, you do not have to increase the sinter machine size, but you have to use a specific process which produces H$_2$SO$_4$ from poor gases.

Blast furnace
Various routes have been considered in the study of the afterburning and dedusting of the lead blast furnace offgases (preliminary treatment required before desulfurization in a H$_2$SO$_4$ plant). The two following routes are today considered as the most promising ones :

a)

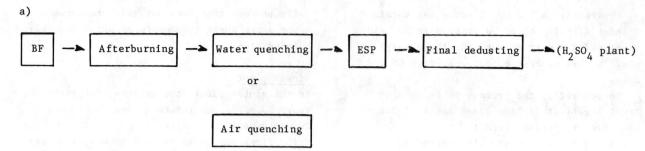

or

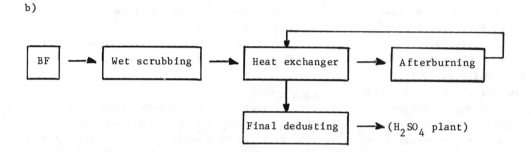

The advantage would be a quick combustion immediately after the blast furnace. The drawback is the risk of accretions by dust softening in the after-burning chamber.

b)

The advantage would be the very low dust quantities in the combustion chamber. The drawback is the recovery of sludges.

The first route has to be tested at pilot scale. The second one is rather similar to the one practised in the zinc Imperial Smelting Process.

In each case, theoretical mass and heat balances have been performed. We will give some details hereafter.

THE (AFTERBURNING + WATER QUENCHING) ROUTE

Principles of the process, and assumptions for mass and heat balances

The process is the following :

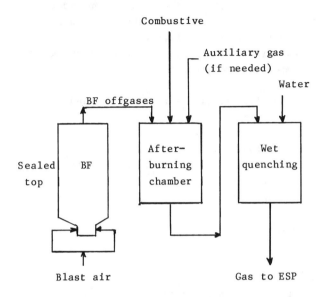

For mass and heat balances, the following assumptions have been made :

a) A temperature in the combustion chamber of 900 °C can be considered as a good compromise between the risks of dust soften (leading to accretions) and combustion kinetics. The combustion is assumed to be adiabatic. In reality, we can expect heat losses between 0,5 and 1 MW, which is equivalent to a T of gas of 50 - 100 °C.

b) The combustive in the after burning chamber will be air.

c) The rate of combustive injected into the combustion chamber is at least twice as much as the stoechiometric quantity (SQ) required to burn all the CO (this will ensure complete combustion). When the flame temperature at 2 SQ is above 900 °C, the rate of combustive is increased in order to decrease the temperature back to 900 °C by dilution. On the contrary, when the flame temperature at 2 SQ is below 900 °C, auxiliary combustible gas is injected into the chamber. We have assumed, for this auxiliary gas, a composition close to that of Groningue natural gas, The calorific value of this gas being 32 MJ/Nm^3.

d) The gases after combustion are quenched down to 400 °C, which is low enough to go through an Electrostatic Precipitator and to condense every condensable species into a solid form.

Results and discussion
Afterburning

The burden height of the three Metaleurop blast furnaces are slightly different. One can calculate that when operating the blast furnace at a low burden height, the top gas is sufficiently hot and rich in CO to reach spontaneously (with 2 QS) a flame temperature above 900 °C. Consequently, there is no consumption of auxiliary gas, but a dilution by combustive in excess.

On the contrary, when operating the blast furnace at an higher burden height, the top gas becomes colder and less rich in CO, so it becomes necessary to inject an increasing amount of auxiliary gas.

The average conditions of one of Metaleurop blast furnaces will be the following :
- flow rate of no diluted blast furnace off gases............. : 13 700 Nm^3/h
- % CO........................ : 7.6 %
- % SO$_2$........................ : 1.5 %

- Combustive air flow rate...... : 5 200 Nm^3/h
- Auxiliary gas flow rate....... : 70 Nm^3/h

- Flow rate of gas after combustion.................... : 18 400 Nm^3/h
- % SO$_2$........................ : 1.14 %

Water quenching

The amount of water to be injected and volatilized in order to quench the gases from 900 °C to 400 °C is approximately 0,245 l/Nm^3. The resulting water flow rates in our conditions, as well as the flow rates of quenched gases, would be :
- t water/h.... : 4,7
- Gas flow rate : 24 100 Nm^3/h
- % SO$_2$........ : 0,87

Of course, the quenching temperature can be changed by varying the water to gas ratio. By increasing it, we decrease not only the quenching temperature, but also the real gas flow rate treated in the Electrostatic Precipitator. Conversely, the H$_2$O content of this gas increases and prevent a good working of the filter.

THE (WET SCRUBBING + HEAT EXCHANGER + AFTERBURNING) ROUTE
Principles of the process and assumptions for the mass and heat balances

In this process, the blast furnace offgases are first dedusted by wet scrubbing before going to the combustion chamber, just as in the Imperial Smelting Process. The heat content of the burnt gases is partly recovered in an exchanger to preheat the gases to be burnt. The process is below sketched :

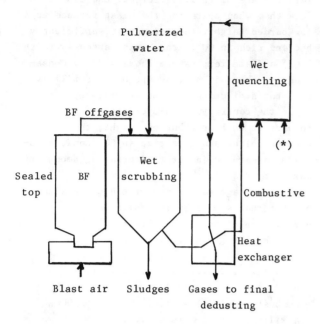

(*) Auxiliary gas (if needed)

The assumptions about the combustion chamber are those already described, except that, in each case, we have adjusted the temperature of preheating in order to reach a flame temperature of 900 °C with exactly 2 SQ and without any auxiliary gas. In consequence we have assessed the size of the required heat exchanger.

For wet scrubbing, the question concerns the water to gas ratio. This ratio can range from 0,5 to 10 $1/m^3$, according to the characteristics of the gases and of the scrubbing tower. In our calculations, we have adopted a value close to that practiced in our ISF plant of Noyelles-Godault, namely 2 $1/m^3$. The water temperature is taken equal to 20 °C. The wet scrubbing should allow to reach a residual dust content between 5 and 20 mg/m^3. As an acid plant requires less than 2 mg/m^3 of dust, a final dedusting of the burnt gases will have to be installed.

Results and discussion
Wet scrubbing
With a water to gas ratio fixed at 2 $1/m^3$ (as explained before) we calculate, in each case, the water flow rate and the temperature of the products (sludge + dedusted gas). The results are summarized hereafter for the average conditions of Metaleurop blast furnaces :

- Water consumption............. : 80.7 t/h
- Flow rate of dedusted gas.......: 14800 Nm^3/h
- Final temperature.............. : 45 °C

Afterburning
In the present case, thanks to the preheating of the process gases, there is no consumption of auxiliary gas in the afterburning tower :
- Flow rate of gases after wet
  scrubbing..................... : 14800 Nm^3/h
- Temperature of preheating...... : 615 °C

- Combustive air flow rate....... : 4600 Nm^3/h

- Flow rate of gases after post
  combustion.................... : 18900 Nm^3/h
- % SO_2........................ : 1.1 %

When comparing it with the conditions in the other route, the gas flow rate and SO_2 content after combustion appear to be of the same order of magnitude.

CONCLUSION
For the last ten years the traditional lead smelting process (sintering + blast furnace) has improved its technical ratios such as coke rate, lead bullion output, manpower and energy per ton. This, and its wellknown flexibility with regard to raw materials, allow it to increase its competitivity. In addition, it has also to comply with the challenge of environmental regulations, in particular for SO_2 emissions.

The first action to be undertaken is the treatment of sintering off gases. This implies specific investments and improvement of recycling loops.

The second target is to treat the lead blast furnace offgases. As the SO_2 level is very low, a specific process has to be developed. Today we think this process will involve three steps :
- sealing the top of the blast furnace,
- removal of CO by a post combustion process,
- removal of the dust.

Studies performed on this possible process pointed out a real chance of success if we can manage with the following points :
- keeping the feeding quality and repartition after sealing of the top blast furnace,
- avoiding the return of accretions we had in the past at the top of the furnace.
- trying not to increase the energy consumption with a nearly autogeneous post combustion,
- avoiding the formation of new accretions in the afterburning chamber and the consequences

on operating ratio and on the frequency of
unhealthy clean-out.

An alternative solution is the use of
completely new processes. Everybody knows that
any really new industrial process needs high
research and development costs and that its
optimization in order to reach a competitive
position may last longer than expected.

Expected operating profits are currently
insufficient to meet the often enormous investment needs required for these new technologies
which, as far as we know, mainly solve environmental problems.

Thanks to studies and tests, the traditional
process will be able to meet the same environmental regulations with investment programmes
far less costly and with increased flexibility
of raw material treatment.

Das QSL-Verfahren in Stolberg

L. Deininger
H. Neumann
'Berzelius'-Stolberg GmbH, Stolberg, Deutschland

SYNOPSIS

Als eine offensive Antwort auf die zunehmenden Forderungen für den Umweltschutz entschied die Metallgesellschaft AG im Sommer 1987, in ihrer Bleihütte in Stolberg die besonders emissionsträchtigen Teile Sinter- und Schachtofenanlage durch eine QSL-Anlage zu ersetzen.

Das QSL-Projekt beinhaltet insgesamt neben dem Reaktor mit Materialzufuhr, Mediendosierung, Abhitzekessel und Heißgasreinigung

1) Anlagen, in denen Arsen und Cadmium auf kürzestem Wege selektiv aus dem Prozeßgas als handelsfähige Produkte ausgeschleust werden können,

2) eine neue Schwefelsäuregewinnungsanlage mit Dampfüberhitzer am Kontaktkessel,

3) eine Energierückgewinnungsanlage, in der die Abwärme des QSL-Reaktors und der Schwefelsäureanlage gemeinsam für den Werkseinsatz verstromt wird und

4) eine von der Fa. Linde eigens dafür errichtete und betriebene Luftzerlegungsanlage, die aus ca. 100 m Entfernung von der Werksgrenze die Belieferung mit O_2 und N_2 über Rohrleitungen sicherstellt.

Der Anlagenkomplex wird von einer zentralen Meßwarte über ein modernes Prozeßleitsystem gesteuert und überwacht.

Mit dem QSL-Projekt in Stolberg wird beispielhaft demonstriert, wie durch spezifische Abstimmungen der einzelnen Nebenanlagen in ihrem Verbund zum QSL-Reaktor neben einer emissionsarmen Verfahrensweise ein Optimum an Energieausnutzung und eine abfallfreie Produktion erreicht wird.

Ende August 1990 begann der Probelauf mit Material, die offizielle Inbetriebnahme war am 29.11.1990.

Im Februar und März 1991 war ein geplanter Reparaturstillstand, in dem Mauerwerksschäden, resultierend aus dem stark schwankenden Inbetriebnahmebetrieb beseitigt wurden. Gleichzeitig wurden während des Stillstandes Erkenntnisse aus den ersten Betriebsmonaten konstruktiv umgesetzt.

Am 30.3. wurde die Anlage wieder in Betrieb gesetzt. Die Betriebsergebnisse entsprechen voll den Erwartungen. Es wurden Pb-Gehalte in der Endschlacke am Reaktorstich von bis zu 0,4 % erreicht. Damit wird das Verfahren auch für Betreiber interessant, die zur Verarmung von Schachtofenschlacken einen Elektroofen betreiben.

Im Mai 1991 wurden Versuche mit Erdgas als Reduktionsmittel erfolgreich abgeschlossen.

QSL-TECHNOLOGIE

Das Verfahren ist benannt nach den Professoren Queneau und Schuhmann sowie der Lurgi.

Es stellt einen kontinuierlichen Direkt-Bleischmelzprozeß dar, bei dem zwei grundsätzlich verschiedene pyrometallurgische Reaktionen in einem Reaktionsgefäß ablaufen, Abb. 1.

1. Das autogene Röst-Reaktions-Schmelzen schwefelhaltiger bleihaltiger Vorstoffe.

2. Die karbothermische Reduktion des in der Schlacke eingebundenen Bleioxides.

Bei dem QSL-Reaktor handelt es sich im Prinzip um ein zylindrisches Reaktionsgefäß, das um die

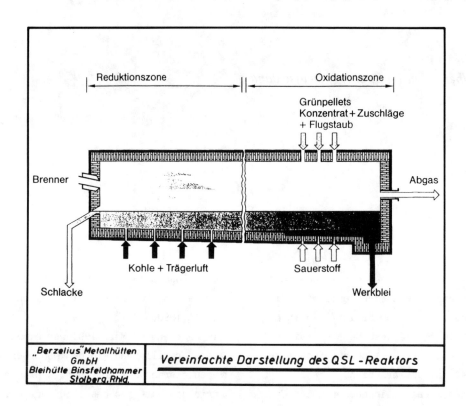

Abb. 1 Vereinfachte Darstellung des QSL-Reaktors

eigene Achse schwenkbar und zum Bleiabstich leicht geneigt ist. Im Oxidationsbereich wird technisch reiner Sauerstoff, in der Reduktions- zone Kohlenstaub in Verbindung mit Sauerstoff, ggf. mit Erdgas oder Stickstoff, über bodenbla- sende Düsen eingeblasen. Das metallische Blei, in dem sich die Edelmetalle gesammelt haben, fließt zum Abgasende des Reaktors und wird kon- tinuierlich abgestochen, während die Schlacke zum anderen Ende fließt und hier ebenfalls kontinuierlich abgezogen wird. Das hoch SO_2- haltige Abgas wird nach Kühlung in einem Ab- hitzesystem und Reinigung in einer Schwefel- säureanlage umgesetzt.

Prinzipiell läßt sich der Prozeß wie folgt be- schreiben:

In der Oxidationszone wird Bleisulfid oxidiert, im wesentlichen nach

$$PbS + 1\ 1/2\ O_2 = PbO + SO_2,$$
$$PbS + 2\ O_2 = PbSO_4,$$

ein Teil des vorlaufenden Bleis fällt nach

$$PbS + O_2 = Pb + SO_2$$
$$PbS + 2\ PbO = 3\ Pb + SO_2$$

$$PbS + PbSO_4 = 2\ Pb + 2\ SO_2$$

auch metallisch an.

In der Reduktionszone wird die hochbleioxidhal- tige Schlacke durch eingeblasenen Kohlenstaub reduziert nach

$$PbO + CO = Pb + CO_2.$$

In folgender Abb. 2 werden die Stoffströme zwischen konventioneller Verfahrensweise und dem QSL-Prozeß verglichen.

Man sieht sehr deutlich, mit welchen Mengen an emissionsträchtigen Kreisläufen der Vorstoff bei dem konventionellen Verfahren der Sinterröstung verdünnt werden muß, um die Sintertemperatur in Grenzen zu halten und ein schwefelarmes Pro- dukt zu erzeugen. In diesem Fall wird fast die dreifache Menge im Kreis geführt, während es beim QSL-Verfahren nur ca. 20 % sind. Daraus resultieren weniger Quellen zur Emission. Gleichzeitig ergeben sich geringere Abluft- und prozeßbedingt geringere Abgasmengen, die zu be- handeln sind. Damit werden also die Massenströme an emittierten Stoffen erheblich reduziert.

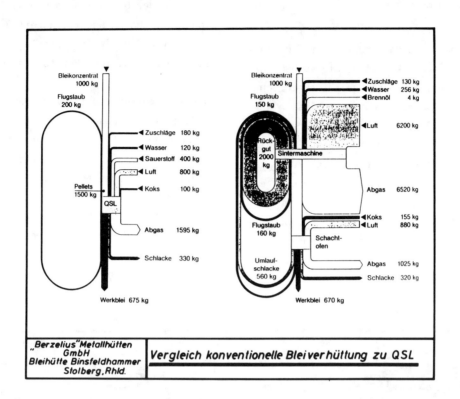

Abb. 2   Vergleich konventionelle Bleiverhüttung zu QSL

DAS QSL-PROJEKT

Geschichtliche Entwicklung

Mit einer 30.000 jato QSL-Demonstrationsanlage
bei Berzelius in Duisburg wurde von März 1981
bis Ende Januar 1986 die metallurgische und
technische Durchführbarkeit, die Umweltfreund-
lichkeit und die Wirtschaftlichkeit erfolgreich
nachgewiesen. Im Juli 1986 wurde für das QSL-
Projekt Stolberg der Antrag auf Genehmigung er-
stellt. Im August 1987 kam der Genehmigungsbe-
scheid. Vorstand und Aufsichtsrat der MG stimm-
ten im August der Realisierung des Projektes
zu. Umgehend wurden die Detailkonstruktionen be-
gonnen. Am 15.10.1988 erfolgte der erste Spaten-
stich. Der QSL-Reaktor, das Herzstück der Anlage, wurde am 8.8.1989 eingebaut. Ende August
1990 wurde der Probelauf mit Material gestar-
tet.

Die offizielle Inbetriebnahme war am 29.11.1990.
Im Februar und März 1991 war ein geplanter Reparaturstillstand, in dem Mauerwerksschäden,
resultierend aus dem stark schwankenden Inbetriebnahmebetrieb beseitigt wurden. Gleichzeitig wurden Erkenntnisse aus den ersten Betriebs-

monaten konstruktiv umgesetzt. Dabei wurde ein
zur Schlackenverarmung eingesetzter Elektroofen,
weil er im Verbund zu dem QSL-Reaktor nicht
wirtschaftlich betreibbar war, entfernt.

Übersicht

Mit dem QSL-Projekt

1) sind die besonders emissionsträchtigen Be-
   triebsteile Sinteranlage und Schachtofen
   durch den QSL-Reaktor ersetzt worden,

2) werden Arsen und Cadmium auf kürzestem Wege
   selektiv aus dem Prozeßgas als handelsfähige
   Produkte ausgeschleust,

3) wird zur Entfernung des SO_2 aus dem Prozeßgas eine neue Schwefelsäuregewinnungsanlage
   mit Dampfüberhitzer am Kontaktkessel betrieben,

4) wird die Abwärme des QSL-Reaktors und der
   Schwefelsäureanlage gemeinsam in einer Ener-
   gierückgewinnungsanlage gewonnen und für den
   Werkseinsatz verstromt,

5) werden beginnend mit mindestens 1/3 des Ge-
   samtvorstoffeinsatzes Reststoffe eingesetzt,

6) wird zur Belieferung mit O_2 und N_2 von der
   Fa. Linde eine ca. 100 m von der Werksgrenze

entfernt, eigens dafür errichtete Luftzerlegungsanlage betrieben.

QSL-Reaktor

Das Reaktorgebäude wurde parallel zur Sinteranlage errichtet, somit konnte an die vorhandene
Vormaterialzufuhr, Bunkeranlage, Mischeinrich-
tung und Entstaubungsanlagen angeschlossen werden. Der direkte Anschluß erfolgte an den vor-
handenen Aufgabebunker zum Sinterband. Von hier
aus gelangt die Vorstoffmischung nach Zumi-
schung von QSL-Flugstaub und Feinkohle über eine
Pelletiertrommel und zwei voneinander unab-
hängige Aufgabedosierungssysteme in den QSL-Reaktor. Die folgenden Abbildungen geben einen
Überblick über die Lage der einzelnen Anlage-
teile.

Der QSL-Reaktor hat insgesamt eine Länge von
33 m, in der Oxidationszone einen Durchmesser
von 3,5 m und im Reduktionsteil einen Durchmesser von 3 m. Er ist ausgelegt für einen Einsatz an bleihaltigen Vorstoffen von 150.000
jato.
Der QSL-Reaktor ist im Oxidationsbereich mit 6
bodenblasenden Düsen ausgestattet, die paarweise angeordnet sind und durch die technisch
reiner Sauerstoff in einem Stickstoffmantel als
Schildgas eingeblasen wird.
Die Reduktionszone ist mit insgesamt 8 paarweise
angeordneten, bodenblasenden Mehrstoffdüsen
ausgerüstet, durch die Kohlenstaub in Verbindung
mit Sauerstoff, ggf. mit Erdgas oder Stickstoff,
zur Reduktion des Bleioxides eingeblasen wird.

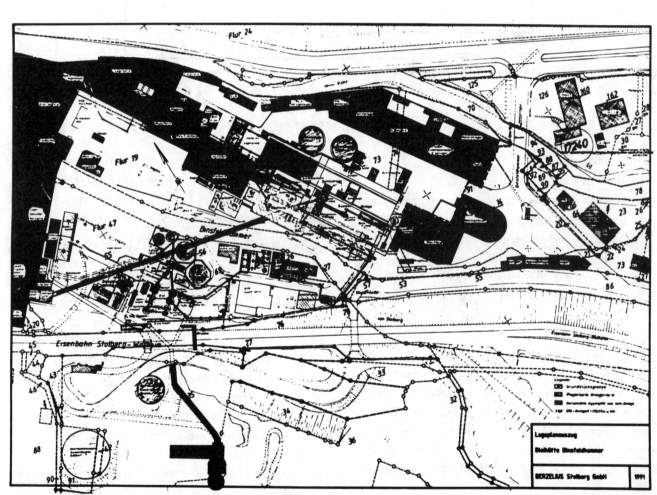

Abb. 3   Lageplanauszug

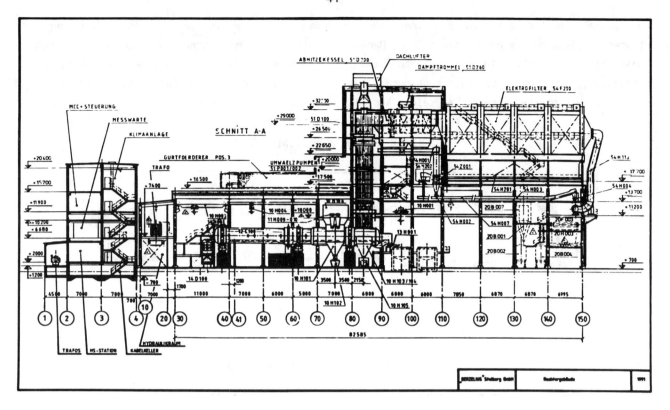

Abb. 4   Reaktorgebäude

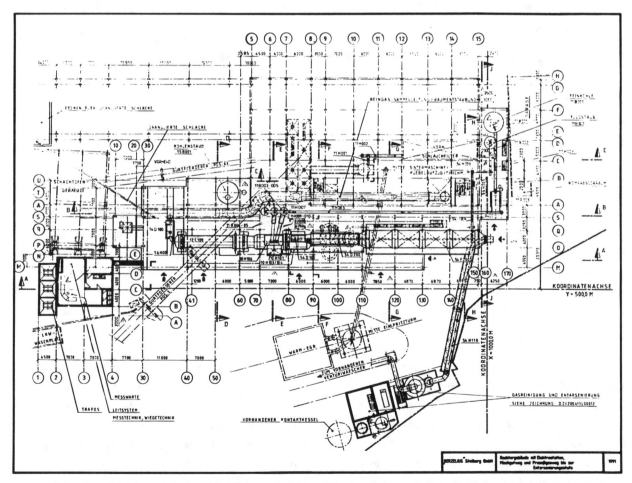

Abb. 5   Reaktorgebäude mit Elektrostation, Mischgutweg und Prozeßgasweg bis zur Entar-
senierungsstufe

Das gesamte Reaktionsgefäß ist mit einer feuerfesten Auskleidung aus hochwertigen Chrom-Mag-
nesia-Steinen versehen. Im Reaktionsgefäß sind
Einbauten, die im Schmelzbereich bestimmte
strömungsmechanische Wirkungen erzielen, Abb. 6.

einen beheizten Vorherd abgezogen. Im Vorherd
setzen sich mitgerissene Bleitröpfchen auf dem
Boden ab, sammeln sich zu einem Bleisumpf, aus
dem in gewissen Zeitabständen Blei abgestochen
wird. Die Schlacke gelangt dann in die vorhan-

QSL-REAKTOR VOR DEM UMBAU

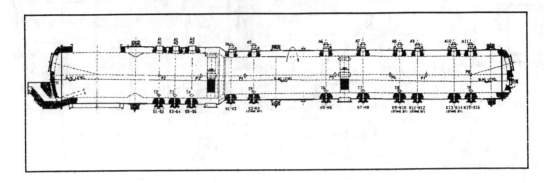

QSL-REAKTOR NACH DEM UMBAU

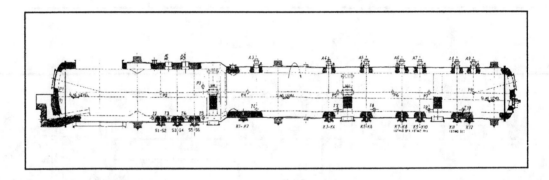

Abb. 6   Feuerfeste Auskleidung des Reaktors

Das reduzierte metallische Blei, in dem sich die
Edelmetalle gesammelt haben, fließt zum Abgasende des Reaktors und wird dort kontinuierlich
in Schmelzkessel abgezogen, in denen das Blei
anschließend direkt entkupfert wird. Die dabei
entstehenden festen kupferhaltigen Reaktions-
produkte werden von der flüssigen Bleioberfläche
pneumatisch abgezogen und in einem Zyklon mit
nachgeschaltetem Filter abgeschieden. Dieser
sogenannte Cu-Schlicker wird in verschließbare
Behälter gefüllt und gelangt in die Cu-Schlicker-
Laugung unseres Duisburger Werkes. Das vorent-
kupferte Blei geht im Flüssigtransport in die
Feinhütte.
Die Schlacke fließt zum anderen Ende des Reaktors und wird ebenfalls kontinuierlich über

dene Granulation.
Das gesamte Prozeßabgas - ca. 20.000 Nm^3/h mit
im Mittel 10 % SO_2 - wird vom bleistichseiti-
gen Ende des Reaktors abgesaugt und zunächst
in einem Abhitzekessel von ca. 1.200° C auf
ca. 400° C abgekühlt. Hierbei wird Sattdampf
mit 47 bar und 250° C erzeugt, der am Kontaktkessel der neuen Schwefelsäurekontaktanlage
auf 340° C überhitzt wird und dann in einem
Turbogeneratorsatz verstromt wird. Die max.
Leistung des Turbosatzes beträgt 4,5 MW.
Nach dem Abhitzekessel wird das Prozeßgas in
einem Heißgas-Elektrofilter (4-feldrig) auf
< 200 mg/Nm^3 vorgereinigt. Ein Teil des abge-
schiedenen Flugstaubes wird in eine Flugstaub-
laugung geführt.

Cd-Anlage

Es wird das bei "BERZELIUS" in Duisburg ange-
wandte Verfahren benutzt, das im wesentlichen
mit den folgenden Verfahrensschritten zu
charakterisieren ist, Abb. 7.

Das Verfahrensprinzip besteht darin, das im
heißen Reingas der EGR enthaltene As_2O_3 in
Schwefelsäure bzw. Waschsäure zu kondensieren,
in der Säure zu lösen und nach vorangegangener
Dekantierung der Feststoffe als technisch rei-

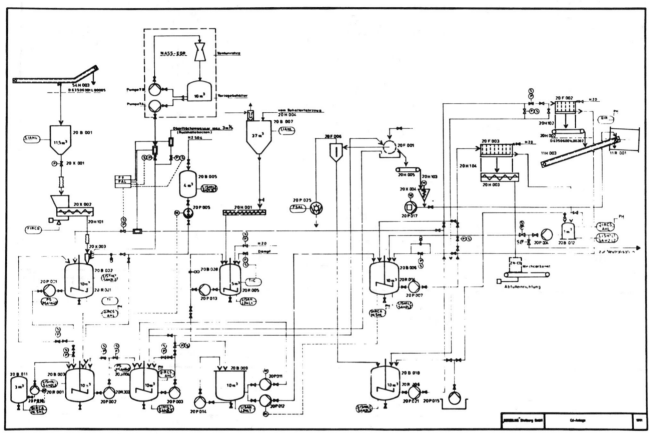

Abb. 7   Cd-Anlage

- Laugung in schwefelsaurer Lösung ggf. unter
  Zugabe von H_2O_2 bei pH 3 - 3,5,
- Abpuffern der Lauge auf pH 6,
- Filtration des bleihaltigen Laugerückstan-
  des und seine Rückführung in die Vorstoffmischung,
- Fällung des Cadmiums unter Zugabe von Soda-
  lösung bei pH 8 bis 9 in Form eines ver-
  kaufsfähigen Cd-Zn-Mischkarbonates.

Das Cd-Zn-Mischkarbonat mit ca. 40 % - 50 %
Cd wird über eine Filterpresse entwässert
und in einer automatischen Abpackvorrichtung
in Gebinden nach GGVS abgepackt.

As_2O_3-Fällung

Das vorgereinigte Prozeßgas wird nun in eine
Gasreinigungsanlage mit As_2O_3-Fällung geführt.

nes Produkt zu kristallisieren.
Die Arsenikkristalle werden über ein Bandfil-
ter entwässert und in einer automatischen Abpackvorrichtung in Gebinden nach GGVS abgepackt.

Energierückgewinnung

Die heißen Prozeßabgase des QSL-Reaktors werden in einen Strahlungskanal mit nachgeschaltetem Abhitzekessel von ca. 1.200° C auf ca.
400° C abgekühlt. Dabei wird Sattdampf mit
ca. 250° C und 47 bar erzeugt. Dieser Sattdampf wird in einen Überhitzer geführt, der
am Kontaktkessel der Schwefelsäureanlage steht.
Mit der Reaktionswärme aus der 1. Horde, die
bei der stark exothermen Reaktion

$$SO_2 + 1/2 \ O_2 = SO_3$$

entsteht, wird der Dampf auf ca. 340° C über-

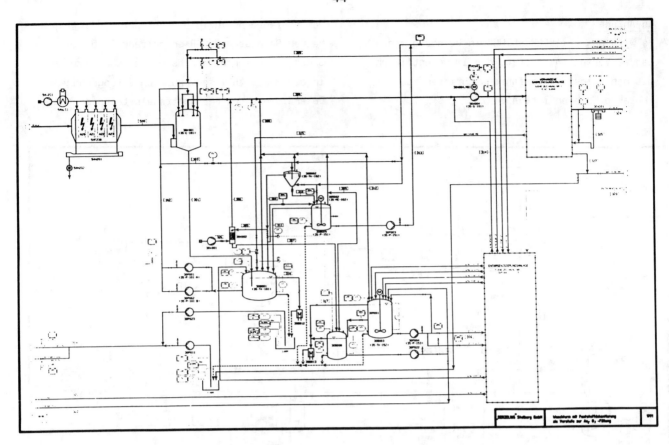

Abb. 8   Waschturm mit Feststoffdekantierung als Vorstufe zur As_2O_3-Fällung

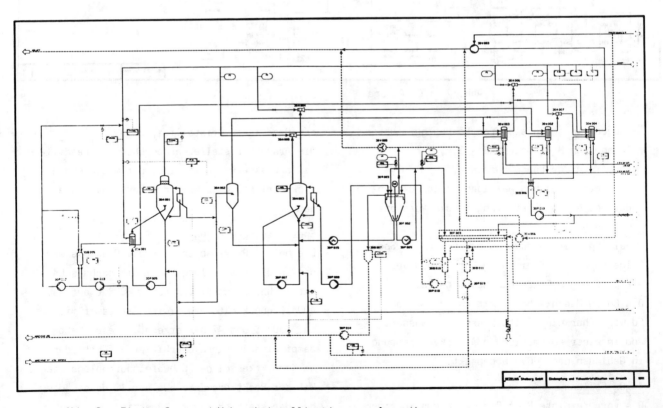

Abb. 9   Eindampfung und Vakuumkristallisation von Arsenik

hitzt. Ein primärenergiebefeuerter Hilfssatt-
dampfkessel mit separatem Hilfsüberhitzer über-
nimmt eine Regelfunktion, damit auch bei
schwankenden Betriebszuständen stets die volle
Dampfmenge bei gleicher Temperatur und glei-
chem Druck zum Betreiben des Turbo-Generator-
Satzes zur Verfügung steht. Um eine optimale
Betriebsweise sicherzustellen, wird der Hilfs-
sattdampfkessel ständig mit ca. 5 - 10 % seiner Leistung im unteren Regelbereich betrieben. Der Hilfsüberhitzer heizt den Dampf permanent auf ca. 370° C auf.
Die Leistung des Kondensations-Turbo-Genera-
torsatzes beträgt im Auslegungsfall 3,4 MW,
maximal können 4,65 MW erzeugt werden.

Damit wird die bei der Absorption des SO_3 in
Schwefelsäure abzuführende Lösungswärme voll
zur Energierückgewinnung genutzt.
Der rückgewinnbare Teil ist hier mit ca. 120
kW zu beziffern.
Das Gesamtenergiekonzept sieht vor, daß der
Turbo-Generator-Satz im Inselbetrieb gefahren
werden kann. Damit kann bei Netzausfall des
EVU und für diesen Fall im Werk vorgesehenen
differenzierten Lastabwurf von nicht lebens-
notwendigen Teilen des Altanlagenbereiches
der größte Teil der QSL-Anlage weiter betrieben werden.

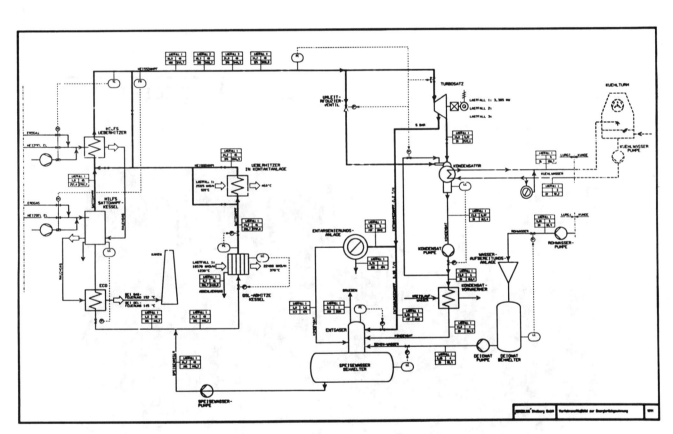

Abb. 10 Verfahrensfließbild zur Energierückgewinnung

Das im Kondensator der Turbine anfallende
Kondensat mit ca. 40° C wird am Säurekreislauf des Zwischenabsorbers über einen separa-
ten, geschlossenen Wasserkreislauf mit Platten-
kühlern auf ca. 80° C aufgewärmt, bevor es
über den Entgaser in den Speisewasserbehälter
gelangt.

Prozeßüberwachung

Der gesamte Anlagenkomplex wird von einem zentralen Leitstand aus, der als reine Bildschirm-
meßwarte ausgeführt ist, gesteuert und überwacht, Bild 12.

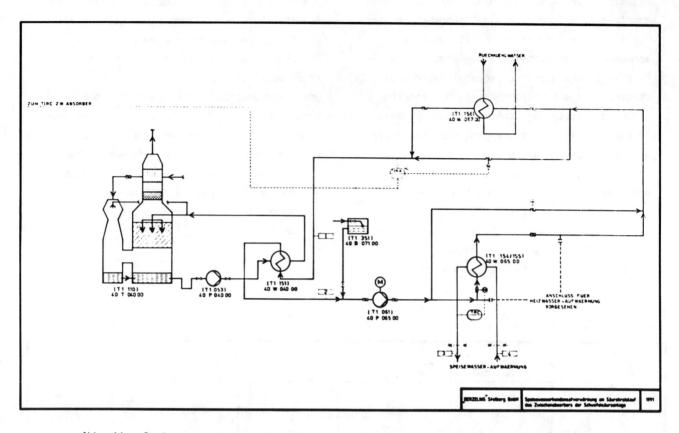

Abb. 11   Speisewasserkondensatvorwärmung am Säurekreislauf des Zwischenabsorbers der
Schwefelsäureanlage

Abb. 12   Meßwarte

Im folgenden Bild ist die Struktur des Prozeßleitsystems dargestellt, Abb. 13.

- Versorgung der Trendschreiber mit Meßdaten
  aus dem System

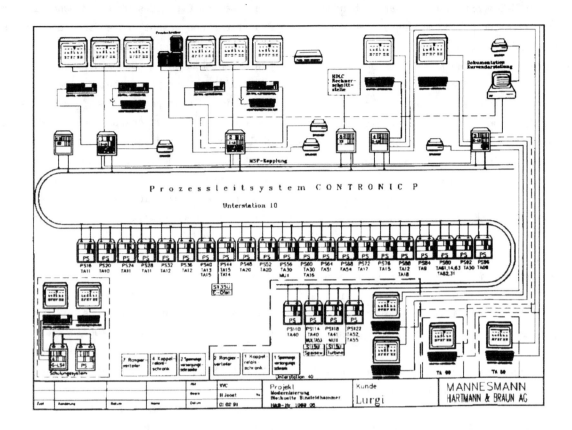

Abb. 13   Struktur des Prozeßleitsystems

Über einen redundant ausgeführten Systembus
sind die Prozeßstationen als Schnittstelle
zum eigentlichen Prozeß mit den Leitstationen
als Schnittstelle zum Menschen verbunden. In
den Prozeßstationen ist die Verarbeitung der
Funktionen Überwachung, Steuerung, Regelung
und Basic konfiguriert. Mit diesen Software-
funktionen werden die analogen und binären
Ein- und Ausgangskomponenten der eigenen Station bedient. Über Koordinator- und die Leitstationen können die Prozeßstationen unterein-
ander kommunizieren.
Die Leitstationen übernehmen die Aufgaben:
- Darstellung der Bilder mit ihren Prozeßdaten
- Bedienung des Prozesses über die Leittastatur
- Ausgabe der Protokolle auf Drucker und Bild-
  schirm
- Austausch von Meßwerten zwischen den Prozeßstationen
- Auslagerung der Kurven auf den Massenspeicher

An eine Leitstation können 3 Monitore und 2
Bedienteile angeschlossen werden.
Die Koordinatorstationen sind den Leitstationen ähnlich, verfügen aber über keine Monitore
und Bedienteile. In diesen Stationen laufen
Basic-Programme, die Daten aus mehr als einer
Prozeßstation benötigen und weitere System-
kommunikationen.
Mit den Leitstationen und einer Koordinatorstation direkt verbunden sind 2 Massenspeicher,
die aus Gründen der Redundanz über den absolut
gleichen Inhalt verfügen.
Die Massenspeicher, eigenständige Computer mit
eigenem Processor, sind für alle Archivierungen
zuständig. In ihnen werden alle Kurven und
Bilder in Dateiform sowie die Sicherungsdatei-
en der Stationen abgelegt und die Basic-Datei-
en verwaltet.
In der zentralen Meßwarte sind momentan 2 Be-
dienplätze mit jeweils 2 Monitoren und je 2

Bedienteilen, an die jeweils eine Konfigurier-
tastatur angeschlossen werden kann.
Über eine V24-Schnittstelle ist die Leitstation
1 mit einem IBM-kompertiblen Rechner verbunden.
Mit einem speziellen Programmierpaket, PC-Kopp-
lung und Contrend werden einmal u.a. wöchent-
lich die prozeßrelevanten Kurven auf dem PC
archiviert und dort auf eine optische Platte
ausgelagert, zum anderen können die Kurven in
beliebiger Zuordnung auf dem PC dargestellt
und auf einem Farbdrucker ausgegeben werden.
Eine weitere V24-Schnittstelle verbindet die
Leitstation 2 mit dem PC, von dem dann mittels
eines speziellen Programmpaketes CPUTI ein
Konfigurierdialog geführt werden kann.
Mit einem Dokumentationsprogramm, Condor C,
können alle Konfigurationsdaten aus dem Leit-
system analysiert und in Form von Fließbildern
dargestellt werden. Es wird somit die Möglich-
keit einer schnellen Überprüfung der Konfigu-
ration gegeben, was bei der Fehlersuche uner-
läßlich ist.
Gleichzeitig kann mit dieser Software die Do-
kumentation mit dem Iststand effektiv verglichen werden.
Zusätzlich ist über eine Rechnerschnittstelle
in der Koordinatorstation 8 und eine Modem-
strecke ein VAX-Rechner angebunden. In dieser
VAX ist eine bestimmte Software geladen, die in
Verbindung mit dem darauf abgestellten Interface KMV1A ausgewählte Prozeßdaten aus dem Leit-
system für das vorhandene Rechnerprogramm ver-
fügbar macht.
In den Teilanlagen Schwefelsäure und Energierückgewinnung bestehen noch vor Ort je ein
kleiner Leitstand, mit denen die Anlagen in
besonderen Betriebssituationen von vor Ort gefahren werden können.
Für Ausbildungszwecke existiert ein Schulungs-
system, bestehend aus einer Leitstation, einer
Prozeßstation und 2 Bedienplätzen mit je einem
Monitor, das auch an den Systembus angeschlossen ist, mit dem jedoch bezogen auf den Prozeß
nur beobachtet werden kann.
Leiteingriffe bei Schulungen werden über ein
Simulationsprogramm abgewickelt.

METALLURGISCHE RESULTATE

Die wichtigste, wenn auch überraschendste Erkenntnis aus dem Inbetriebnahmebetrieb war, daß
die Höhe des Bleibades über den Düsen in der
Reduktionszone wesentlich entscheidender für
ein gutes Reduktionsergebnis ist als die
Schlackenschichthöhe. Das widersprach der ur-
sprünglichen Konzeption, bei der das reduzierte
Blei durch die vorgegebene Neigung des Reaktors
von 0,5 % zur Oxidationszone und dem Bleistich
möglichst schnell zwischen den paarweise am
Reaktorumfang angeordneten Düsen aus der
Reduktionszone abfließen sollte. Mit der ur-
sprünglichen Konzeption waren entweder über-
haupt keine nennenswerten oder nur sehr
schlechte Reduktionsergebnisse erreichbar.
Während des Reparaturstillstandes im Februar/
März 1991 konnte zwar die Neigung des Reaktors
nicht mehr korrigiert werden, es wurde jedoch
vor dem Übergang der Reduktionszone zur Oxidationszone in dem Reaktormauerwerk ein 150 mm
hoher Wall installiert, der einen Mindestfüll-
stand an Blei in der Reduktionszone garantiert.
Aus der Abb. 14 ist deutlich der Einfluß der
Bleibadhöhe über den Düsen auf das Reduktionsergebnis zu entnehmen.
Die Ursachen hierfür sind in einem Wärme- und
Stofftransportproblem zu sehen. Vergleicht man
Blei und die entsprechende Schlacke, so fällt
die deutlich schlechtere Wärmeleitfähigkeit
der Schlacke auf. Aber nicht nur die Wärmeleitfähigkeit ist deutlich schlechter, sondern auch
die Wärmeeindringfähigkeit, definiert als die
Wurzel aus dem Produkt der beiden charakteristischen Größen Wärmeleitfähigkeit und Wärmekapazi-
tät je Volumen, ist bei der Schlacke deutlich
schlechter. Die Wärmeeindringfähigkeit charakte-
risiert das Vermögen, Wärme an die Kohle oder
das Gas abzugeben. Bei sonst gleichen Rand- und
Anfangsbedingungen ist also der Wärmeaustausch
zwischen Blei und Kohle bzw. Gas in jedem Falle
wesentlich stärker als zwischen Schlacke und
Kohle bzw. Gas. Hinzu kommt die niedrige Er-
starrungstemperatur des Bleis, damit ist der
Wärmeaustausch im Bleibad deutlich überlegen.

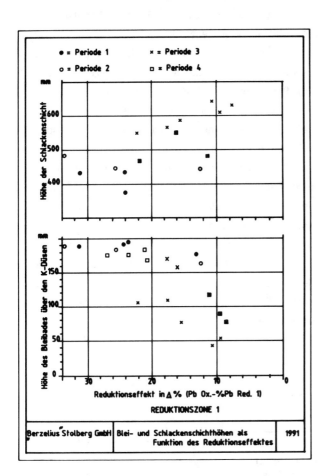

● = Periode 1   × = Periode 3
○ = Periode 2   □ = Periode 4

Höhe der Schlackenschicht

Höhe des Bleibades über den K.-Düsen

Reduktionseffekt in Δ % (Pb Ox.-%Pb Red. 1)
REDUKTIONSZONE 1

| Berzelius Stolberg GmbH | Blei- und Schlackenschichthöhen als Funktion des Reduktionseffektes | 1991 |

Abb. 14 Reduktionseffekt als Funktion des Blei
füllstandes

Gelangen Kohlepartikel mit den Fördermedien und einem gewissen Gehalt an Sauerstoff in das Bleibad, so werden sie entweder im direkten Kontakt oder in einer Gasblase von dem umgebenden Blei erwärmt. Die relativ guten Wärmeaustauschbedingungen im Bleibad gleichen einmal den lokalen Kühleffekt schnell aus und ermöglichen eine relativ schnelle Erwärmung.
Bei Temperaturen der Kohlepartikel von > 900° K setzt die Vergasung der Kohle ein. Spätestens hier bilden sich Gasblasen um die Kohlepartikel oder bereits vorhandene werden größer und steigen mit zunehmender Größe schneller auf.
Ist die Kohlevergasung beim Verlassen des Blei-

bades abgeschlossen, so gelangt ein hoch CO-haltiges Reduktionsgas bereits auf Schmelzbadtemperatur erwärmt in die Schlacke, womit hervorragende Reduktionsbedingungen geschaffen werden.
An dieser Stelle wird noch auf Untersuchungen zur Kohlevergasung im Bleibad hingewiesen, in denen die Verfasser von einer spontanen Vergasung der Kohle im Bleibad berichten, die gegenüber der Vergasung im Röhrenreaktor für den gleichen Umsetzungsgrad um 3 - 4 Zehnerpotenzen schneller abläuft. [1]
Gelangen nun Kohlepartikel mit den Fördermedien direkt in die Schlacke, so kann die lokale Kühlung, bestärkt durch die relativ schlechten Wärmeaustauschbedingungen in der Schlacke dazu führen, daß Kohlepartikel und auch Gasblasen mit einem festen Film aus erstarrter Schlacke überzogen werden. Die lokale Kühlung kann je nach Schlackentemperatur und Zusammensetzung so weit gehen, daß Kamine aus erstarrter Schlacke die Kohle direkt in den Gasraum leiten, wo sie für die Reduktion nicht mehr zu Verfügung steht. Im Falle eines erstarrten Schlackenfilms um Kohlepartikel und Gasblasen wird die Aufwärmung auf Vergasungstemperatur erheblich verzögert. Für die Reduktion werden Phasengrenzreaktion und Festkörperdiffusien geschwindigkeitsbestimmend.
Bei der Verwendung von Erdgas als Reduktionsmittel sind die Verhältnisse ähnlich. Bezüglich der lokalen Kühleffekte und ihre möglichen Folgen beim direkten Einblasen ins Schlackenbad laufen die Auswirkungen dramatischer ab, denn der thermische Zerfall nach

$$CH_4 = C + 2 H_2$$

hat einen hohen Wärmebedarf, der durch die partielle Oxidation nach

$$CH_4 + 0,5 \ O_2 = CO + 2 H_2$$

aufgrund der relativ schlechten Wärmeaustauschbedingungen im Schlackenbad lokal nicht ausgeglichen werden kann.
Betriebliche Beobachtungen bestätigen dies. So konnten bei dem Cominco-Reaktor in der Ruheposition des Reaktors noch Kamine aus erstarrter Schlacke auf den Düsen nachgewiesen werden, die dort als Elefantenrüssel bezeichnet wurden.

Eine weitere wichtige Erkenntnis war, daß bei gleichen Bedingungen mit einer von 8 auf 4 reduzierten Anzahl von Reduktionsdüsen die gleichen Reduktionsergebnisse zu erzielen sind. Mit der Unterteilung des Reaktorreduktionsbereiches in 3 Zonen, die durch die strömungsmechanische Wirkung ihrer Einbauten praktisch 3 Rührkessel darstellen, konnte auf die Nachbehandlung der Schlacke mit einem Elektroofen verzichtet werden. Es wurde ein gasbeheizter Vorherd installiert, der mit 4 qm Nutzfläche den gleichen Absetzeffekt bei mitlaufendem metallischen Blei erzielt wie vorher der Elektroofen. Er wurde in erster Linie deswegen installiert, um die unterschiedlichen Achsen Granulationsrohr und Reaktorstich nach dem Wegfall des Elektroofens zu verbinden.

Die folgende Abb. 15 zeigt den Verlauf des Bleiinhaltes der Schlacken in den ersten Betriebstagen nach der Wiederinbetriebnahme am 30.03.1991.

Bei scharf geführter Reduktion konnten Pb-Gehalte am Reaktorschlackenstich von bis zu 0,4 % erreicht werden. Diese Werte wurden bei einer Höhe des Bleibades in der Reduktion von 250 - 300 mm und einer Schlackenschichtdicke von 100 - 200 mm erzielt.

In Zusammenarbeit mit der Lurgi und unter Teilnahme der Cominco wurden im Mai 1991 Versuche zur Reduktion mit Erdgas durchgeführt.

Dabei konnte nochmals herausgestellt werden, welchen entscheidenden Einfluß ein genügend hohes Bleibad für die Reduktion und den Betrieb des Reaktors insgesamt hat. Bei einer Höhe des Bleibades in der Reduktion von 250 - 300 mm und einer Schlackenschichtdicke von 100 - 200 mm wurden Tagesmittelwerte an Blei in der Schlacke am Reaktorstich von < 2 % sicher erreicht.

Die folgende Abb. 16 gibt einen Überblick über den Umsetzungsgrad des Erdgases bei unterschiedlichen Bleibadhöhen und Sauerstoffverhältnissen[2]

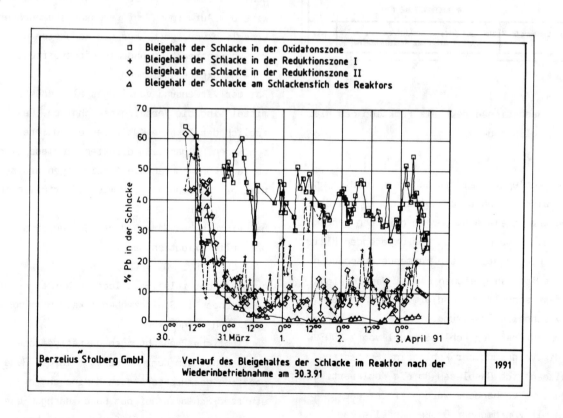

Abb. 15 Pb-Geh. der Schlacken an verschiedenen Tagen

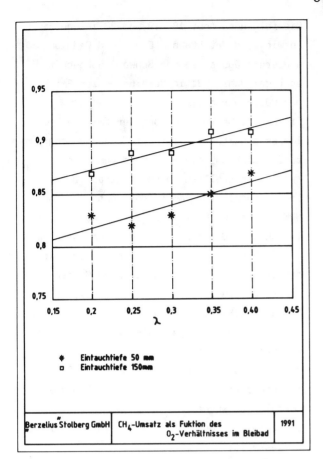

Abb. 16 Umsetzungsgrad des Erdgas bei unterschiedlichen Bleibadhöhen und Sauerstoffverhältnissen

In den folgenden Tabellen 1 + 2 sind die
wichtigen Prozeßdaten zusammengefaßt:

Tabelle 1

Vorstoff-

Einsatz: 150.000 jato
                  davon zunächst 2/3 Konzen-
                  trate und 1/3 Rückstände,
                  wie Pb/Ag-Rückstände, andere
                  Schlämme, Aschen, Gläser

Mittlere Zusammen-

setzung: 46   -   56   % Pb
                   2   -   5   % Zn
                  0,5   -   1   % Cu
                  0,2   -   0,6   % As
                  0,4   -   0,8   % Sb
                  0,02 -   0,06 % Cd

Werkbleiproduktion: 70.000 - 80.000 jato
                     mit im Mittel nur
                     0,04 % Sb
                     0,001 % Sn
                     0,004 % As

Tabelle 2

Reaktor

Oxidationszone

Badtemperatur: 1.050 - 1.100° C
Pb- Geh. d. Schlacke: 25 - 35 %
Flugstaubanfall: ca. 5 t/h bei im Mittel
                      25 t/h Aufgabe

Reduktionszone I

Badtemperatur: ca. 1.250° C
Pb- Geh. d. Schlacke   10 - 15 %

Reduktionszone II

Badtemperatur: ca. 1.250° C
Pb- Geh. d. Schlacke   5 - 10 %

Schlackenstich

Badtemperatur: ca. 1.250° C
Pb- Geh. d. Schlacke   < 2 %

Vorherd

Badtemperatur: ca. 1.250° C
Pb- Geh. d. Schlacke um ca. 20 % niedriger
nach Vorherd: als am Schlackenstich
                      Reaktor

Bei den Schlacken handelt es sich im wesentlichen um Eisensilikatschlacken, die bezogen
auf die Anteile der Hauptschlackenbildner FeO,
CaO, SiO_2 und Al_2O_3 im Bereich des Olivins
liegen. Das Verhältnis von CaO + MgO/SiO_2
liegt im Mittel bei 0,8 - 1,0 der Zn-Gehalt
der Schlacke bei max. 15 %.
Die Verbrauchszahlen sind als Vergleich gegenüber der Planung in Tabelle 3 aufgetragen.

Tabelle 3

Betrieb

Kohle:	115 kg/t	Vorstoff
Sauerstoff:	220 m3N/t	Vorstoff
Stickstoff:	65 m3N/t	Vorstoff
Erdgas:	10 m3N/t	Vorstoff
elek. Strom:	180 kWh/t	Vorstoff
Kalkstein:	60 kg/t	Vorstoff
Sand:	26 kg/t	Vorstoff

Planung

Kohle:	130 kg/t	Vorstoff
Sauerstoff:	240 m3N/t	Vorstoff
Stickstoff:	40 m3N/t	Vorstoff
Erdgas:	15 m3N/t	Vorstoff
elek. Strom:	150 kWh/t	Vorstoff

Es wird möglich, einen hohen Anteil an sekun-
dären Vorstoffen oder Reststoffen einzusetzen,
Abb. 17.

Der Energievergleich QSL-Anlage zur bisherigen
Verfahrensweise mit Sinteranlage und Schachtofen zeigt, daß der Energieeinsatz bei QSL
nur noch ca. 75 % beträgt. Mit der Optimierung
der Betriebsweise wird sich dieser Betrag
weiter senken lassen.

Relativ minderwertige schwefelhaltige Energie-
träger sind einsetzbar.

EMISSIONS- UND IMMISSIONSSITUATION

Seit Jahren wird mit einem Meßstellennetz um
das Hüttenwerk der Staub- und Schwermetall-
niederschlag gemessen. Im Nahbereich des
Hüttenwerkes mit einer Ausdehnung von insgesamt
4 km^2 sind es 16 Beurteilungsflächen von
0,25 km^2 Größe, dazu 48 Beurteilungsflächen
von 1 km^2 Größe über dem Stolberger Raum und
zusätzlich 13 weitere Beurteilungsflächen von
1 km^2 Größe, die zum Raum der Stadt Eschweiler
überleiten.

Die folgenden Abb. 18, 19 und 20 zeigen die
Entwicklung der Staub-, Blei- und Cadmium-
niederschläge sowie der Schwebstaubgehalte
seit dem Jahr 1981 in Verbindung zur Emission
des Hüttenwerkes.

Die deutliche Verbesserung der Emission- und
Immissionssituation ist auf Maßnahmenprogrammen
zurückzuführen, die auf Grund von regel-
mäßig aufgestellten Emissionskataster der
Landesanstalt für Immissionsschutz des Landes
NRW in unserem Werk durchgeführt wurden.

Im Jahre 1980 wurde das 1. Maßnahmenprogramm
abgeschlossen. Zur Überprüfung der Emissions-
situation nach dem durchgeführten Maßnahmenprogramm wurden von der Landesanstalt (LIS)
im Jahre 1981 erneut Messungen durchgeführt,
die die im Jahre 1974 erstellte Prognose im
wesentlichen bestätigte.

Auf Grund dieser Messungen wurde das 2. Maßnahmenprogramm durchgeführt, das 1986 abgeschlossen wurde.

Die Entwicklung der Emissionen ist in Abb. 21
dargestellt:

Ausgehend vom Stand 1974 ist eine drastische
Reduktion der Emissionen von Blei, Zink
und Cadmium zu verzeichnen. Die ohnehin schon
niedrige As-Emission - sie betrug 1974
278 kg/a- konnte zunächst geringfügig, im
Jahre 1986 jedoch stärker reduziert werden
und wird mit QSL nur noch ca. 64 kg/a be-
tragen.

Bei den Schwermetallen wurden die Forderungen
der TA-Luft bereits nach dem 2. Maßnahmenprogramm erreicht. Die drastische Senkung
der SO_2-Emission ist erst mit Einführung
des QSL-Verfahrens auf ca. 10 % des ur-
sprünglichen Wertes erreicht worden.

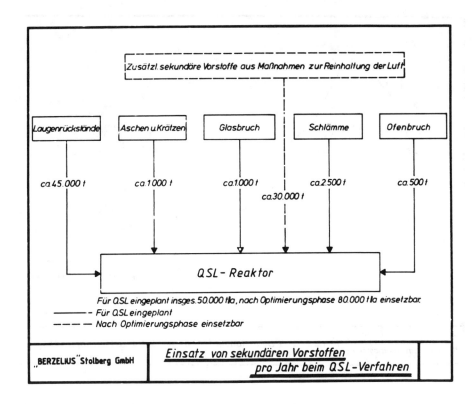

Abb. 17 Vorstoffeinsatz für QSL

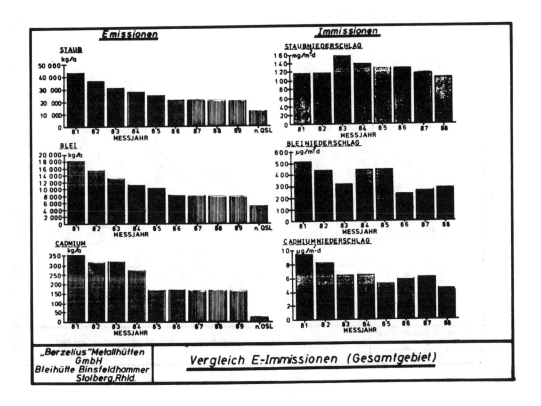

Abb. 18 Vergleich E-Immissionen

54

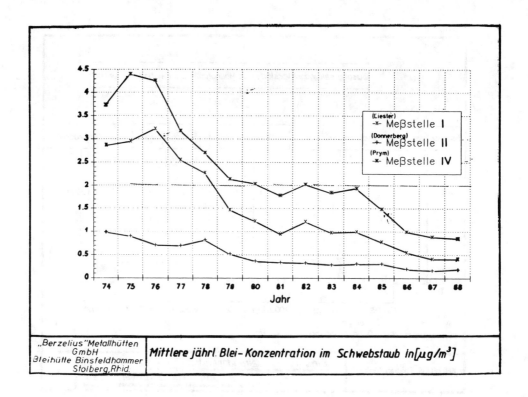

Abb. 19 Mittlere jährl. Bleikonzentration im Schwebstaub

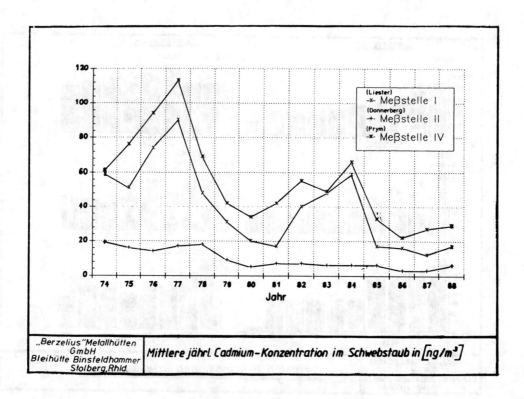

Abb. 20 Mittlere jährl. Cd-Konzentration im Schwebstaub

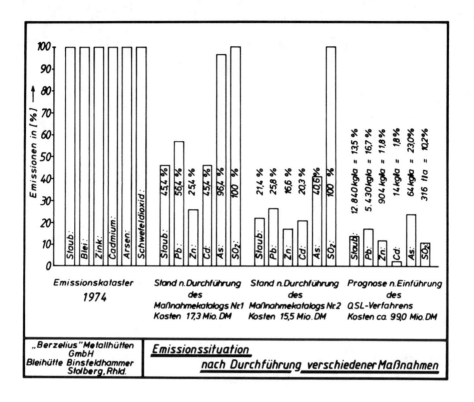

Abb. 21 Emissionssituationen nach Durchführung verschiedener
Maßnahmen

Für den Bereich Rohhütte ist die Emissions-
minderung gegenüber dem bisherigen Stand wie
folgt:

	Staub kg/a	Pb kg/a	Zn kg/a
Bisher	14.768	6.182	712
Mit QSL	3.381	1.452	395
Minderung in %	77	77	45
	Cd	As	SO2
Bisher	133	31	2.977
Mit QSL	7	4	190
Minderung in %	95	87	94

2. Untersuchungen zur Erdgasumsetzung und Kohlevergasung im Bleibad in Zusammenarbeit mit
dem Institut für Eisenhüttenkunde der
RWTH Aachen und der Lurgi.

LITERATURVERZEICHNIS

1. J. Schneider, Dissertation RWTH Aachen
   1974 Herstellung von Reduktionsgas für
   die Eisenhüttenindustrie durch Umsetzung
   feuchter Braunkohlen im kernenergiebe-
   heizten Bleibad.

Hydrogen reduction of cassiterite concentrates

G. Leuprecht
P. Paschen
Institute of Nonferrous Metallurgy, Montanuniversität Leoben, Austria

Synopsis

Hydrogen as a reducing agent shows, compared to carbon monoxide, some advantages both metallurgically and environmentally. A very precise adjustment of the reduction potential with hydrogen-steam mixtures allows reduction conditions under which it is possible to produce a low iron-containing tin with high tin yield, though the tin is not completely free of iron. The iron contents are between 0,4 and 1,6% at reduction temperatures between 700 and 900°C. With the use of B_2O_3 for stabilizing wustite and for conditions of liquid slag iron contents can be lowered to less than 0,05%.

The experiments prove that it is possible to reduce cassiterite to metal with a low iron content in a one-step process with a high reaction yield. The separation of metal from solid residue, however, is still an unsolved problem.

1. Introduction

The expensive reducing agent hydrogen has two decisive advantages compared to the cheaper carbon monoxide: the rate of diffusion is 4 times higher and thermal conductivity is 7 times higher. Considering the recycling possibilities of hydrogen from metallurgical off-gases by simple washing-out of steam in water

and the environmental acceptability of the reaction product H_2O, hydrogen metallurgy offers an excellent alternative to carbothermic reduction.

Only a few years ago did mankind become aware of the possible impacts of the so-called greenhouse effect on our climate. The well recorded increase of CO_2 content in our atmosphere causes – together with other more meteorological factors – a rise of the average temperature with possible consequences such as spreading of deserts, melting of polar ice and raising of sea water level.

A hydrogen energy carrier scenario is already propagated by numerous scientists and could counteract the greenhouse effect by decreasing CO_2 emissions. More use of hydrogen would increase technical scale tonnage production and, in consequence, reduce costs and prices.

2. Comparison of cassiterite reduction rate with hydrogen and carbon monoxide

It can be seen from the water-gas equilibrium

$$CO_2 + H_2 = CO + H_2O$$

that carbon monoxide ist the better reducing agent at temperatures lower than 824°C. Above this temperature hydrogen shows a higher affinity to oxygen. Thermodynamically, this means that hydrogen exhibits better reduction potential.

For the actual progress of a reaction or, in other words, for reaction rate, however, the kinetics are responsible. Our experiments were conducted with a Thai tin concentrate with 72% of tin. Although the thermodynamic conditions of hydrogen reduction are less favourable than for CO-reduction, we achieved much better reaction rates with hydrogen (Fig.1).

the following curves can be plotted from thermodynamic data[1] as a function of the ratio pH_2O/pH_2 and temperature (Fig.2).

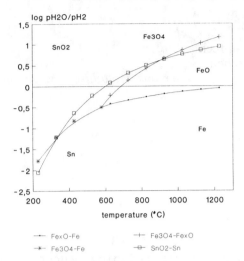

Fig.2-Stability areas for metals and oxides as a function of temperature and reduction potential

We used these theoretical data to establish our series of experiments to find out the actual rates of reduction as a function of temperature and reduction potential in a period of 60 minutes.

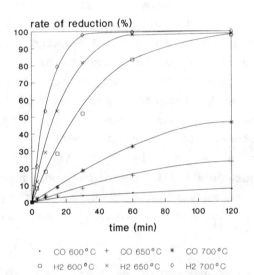

Fig.1-Rate of reduction for CO and H2 at different temperatures vs. time

These results show very clearly the superiority of hydrogen even at low temperatures, i.e. at unfavourable thermodynamic conditions.

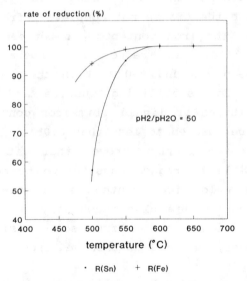

Fig.3- Rate of reduction for SnO2 and Fe2O3 at a reduction potential of 50

3. Tests for development of one-stage tin-reduction process

In the Richardson diagram the lines of ΔG^0 for SnO_2 and FeO or Fe_3O_4 are quite near to each other. In practice this means that with solid carbon as a reducing agent it is not possible to produce an iron-free tin and a tin-free slag in one step. For pure substances

Fig.3 demonstrates that at low temperatures and high reduction potentials hematite is reduced more quickly than cassiterite, whereas at reduction potentials pH_2/pH_2O of about 6 the lines are quite near to each other (Fig.4).

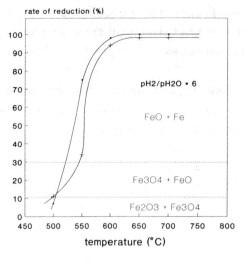

Fig.4-Rate of reduction for SnO2 and
Fe2O3 at a reduction potential of 6

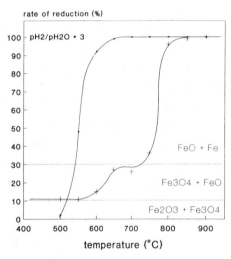

Fig.5-Rate of reduction for SnO2 and
Fe2O3 at a reduction potential of 3

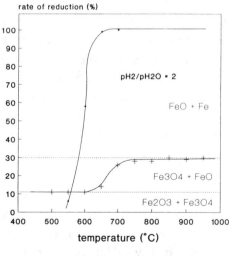

Fig.6-Rate of reduction for SnO2 and
Fe3O4 at a reduction potential of 2

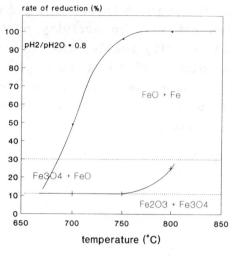

Fig.7-Rate of reduction for SnO2 and
Fe2O3 at a reduction potential of 0.8

Cassiterite reduction rate is considerable higher than that of hematite at all lower reduction potentials (Fig.5-7).

Compared to calculated data all Fe_3O_4-FeO and FeO-Fe transitions are shifted to higher temperatures, which is an indication of kinetic stoppage of reduction.

For pure substances (the activities of raw materials and final products are equal to 1) there is a relatively large area in which a selective tin reduction can take place from a mixture of SnO_2 and Fe_2O_3. But it must be stressed once again that this is true only for pure substances. In fact, there is a solubility of iron in liquid tin (Fig.8).

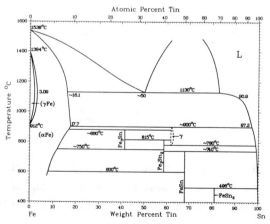

Fig.8-Phase diagram Fe-Sn [2]

It is for this reason that iron is produced not with an activity of 1 but with an activity smaller than 1 because it is dissolved in liquid tin. The amount of iron dissolved in tin depends on temperature and on reduction potential used.

$$FeO \quad + \quad H_2 \quad = \quad Fe \quad + \quad H_2O \quad (1)$$

$$\Delta G^0 = -R \cdot T \cdot \ln K$$

$$K = \frac{pH_2O \cdot a[Fe]_{Sn}}{pH_2 \cdot aFeO} \qquad aFeO = 1$$

$$a[Fe]_{Sn} = K \cdot \frac{pH_2}{pH_2O} = e^{-\frac{\Delta G^0}{R \cdot T}} \cdot \frac{pH_2}{pH_2O}$$

The same, of course, is valid at low temperatures for the transition of Fe_3O_4 to Fe:

$$\frac{1}{3} \quad Fe_3O_4 \quad + \quad \frac{4}{3} \quad H_2 \quad = \quad Fe \quad + \quad \frac{4}{3} \quad H_2O \quad (2)$$

$$a[Fe]_{Sn} = e^{-\frac{\Delta G^0}{R \cdot T}} \cdot \left(\frac{pH_2}{pH_2O}\right)^{\frac{4}{3}}$$

The maximum iron activities in liquid tin can be calculated from the two equations at different reduction potentials (Fig.9).

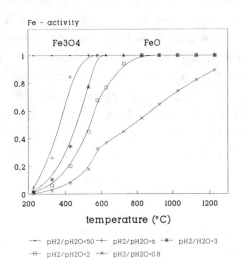

Fig.9-Iron activities as a function of temperature for different reduction potentials

Fig.9 shows that theoretical iron content in tin depends on the reduction potential used and on the reduction temperature.

4. Experiments for production of low iron containing tin from synthetic concentrates

All these experiments were conducted in a special tube furnace. The synthetic concentrate was produced by mixing SnO_2 (Merck, pure, Art.7818) and Fe_2O_3 (Merck, pro analysi, Art.3924). The final composition was:

67,5% Sn,

10,0% Fe,

22,5% O_2.

The amount of material was always 4g per test. After weighing the concentrate into a small Al_2O_3 boat, the boat was loaded and put into the heating-zone of the furnace under a nitrogen atmosphere. After reaching reduction temperature (the control was effected by a thermocouple in the interior of the tube furnace directly above the boat) the nitrogen stream was switched off and the reducing gas stream was switched on. The steam content of the reducing gas was adjusted by introducing hydrogen in a thermostatically controlled water bath. The steam saturated hydrogen is fed to the furnace tube through a heated pipe for the specific reduction time. A fully continuous gas analysis set sucks off part of the off-gas stream from the furnace. According to this, there is a continuous control of H_2, H_2O and O_2 content. After the test period hydrogen is switched off and nitrogen is fed to the furnace tube. The boat is pulled out of the heating zone and is cooled to a temperature of about 100°C by nitrogen. It is then taken out of the furnace. The evaluation of the reduction rate is done by measuring the loss of weight, which corresponds to the oxygen loss of material. Small metallic tin droplets are produced that are examined metallographically with respect to iron-tin phases and their iron content is analysed by atomic absorption spectrography.

4.1 Experiments in FeO area

In line with Fig.6, pure Fe_2O_3 is only reduced to wustite at 900°C and a pH_2/pH_2O of 2. As described above iron will be reduced following reactions (1) and (2). This iron can be found in metallic tin in the form of the intermetallic compounds $FeSn_2$, FeSn, etc. Fig.10 shows the reduction rate as a function of reduction time.

After 8 minutes 100% of SnO_2 has already been reduced.

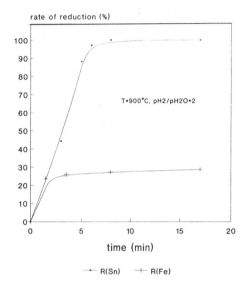

Fig.10-Rate of reduction vs. reduction time

The iron content of the metallic tin droplets amounts to 1,5% after a reduction time of 8 minutes. The maximum solubility of iron in liquid tin at 900°C is 2,8%[2].

4.2 Experiments in Fe_3O_4 area

It is possible to reduce Fe_2O_3 only to magnetite at high pH_2O/pH_2 ratios and at relatively low temperatures (see Fig.7; SnO_2 is reduced to 100%).

Fig.11 shows the reduction rate as a function of reduction time. Although the iron solubility in tin amounts to about 1% at a temperature of 750 °C , our own experiments indicate an iron content of only 0,4%.

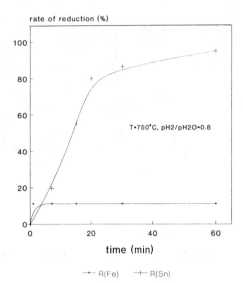

Fig.11-Rate of reduction vs. reduction time

Because of the low reduction potential (pH_2/pH_2O = 0,82) a complete SnO_2 reduction needs much longer, in this case 60 minutes.

4.3 Experiments in FeO area with simultaneous stabilizing of wustite

The idea for doing these tests is to lower the FeO activity to values smaller than 1 by complex formation of FeO in the slag. The problem here was that we conducted our tests at relatively low temperatures - far below normal slag melting points. A compound formation of wustite with SiO_2 in the form of fayalite $2FeO*SiO_2$ or hercynite $FeO*Al_2O_3$ is only possible by solid-state diffusion. This kind of diffusion needs much time, which was not available. The FeO formed from magnetite has to be complexed immediately to avoid further reduction by contact with liquid tin. In our tests 2 to 3 hours at 950°C were necessary to complex at least a large amount of FeO as fayalite or hercynite. The advantages stated by Bear and Caney[4] could not be confirmed by us. Different factors are the basis for the complex formation of wustite with B_2O_3.

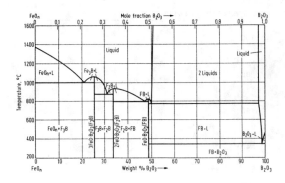

Fig.12-Phase diagram FeO-B2O3 [3]

The low melting point of B_2O_3 (450°C) renders tests with liquid slag possible and shows the following results:

reduction temperature: 900°C
reduction potential $pH_2/pH_2O = 2$
H_2 flowrate: 120 Nl/h
stoichiometric amount of B_2O_3 for formation of FeB_2O_4
reduction rate: 100% (for SnO_2)
reduction time: 30 minutes
iron content in tin: 0.05%

The low iron content in produced tin indicates a complete wustite complex formation in the form of FeB_2O_4. There is no reason to fear SnO_2 slagging together with FeO. The reduction rate is identical with that of tests with pure substances. For industrial practice, however, it is difficult to imagine the use of B_2O_3, because there is no material that resists this aggressive liquid slag phase.

5. Outlook

It is possible to reduce synthetic cassiterite concentrates at low temperatures with high tin recovery and with low iron contents in tin. The separation of the metal from the solid residue has still to be resolved and further research is required.

References

1. Barin I. Thermodynamical Data of Pure Substances. VCH Verlagsge-sellschaft, Weinheim, 1989

2. Binary Alloy Phase Diagrams, Massalski T.B. and others. American Society for Metals, Metals Park, Ohio, 1986

3. Schlackenatlas, The Verein Deutscher Eisenhüttenleute, Verlag Stahleisen m.b.H., Düsseldorf, 1981

4. Bear I.J., Caney R.J.T.,Selective reduction of a low-grade cassiterite concentrate, IMM 85 (1976) C139 - 146

Hydrometallury

Ammonium jarosite formation in ferric chloride leaching processes

J. E. Dutrizac
CANMET, Ottawa, Canada

SYNOPSIS

Ammonium jarosite is readily formed from $FeCl_3$-NH_4Cl leaching solutions containing an independent source of sulphate. The principal factors affecting NH_4-jarosite formation are pH, temperature and the concentrations of sulphate and $FeCl_3$. At 98°C in 4 M NH_4Cl media, no precipitate forms when the pH is <0.4; pH values >1.8 result in difficult-to-filter products. Only traces of jarosite are precipitated at 50°C, but the amount of product increases systematically with increasing temperatures above this value. A minimum sulphate concentration is required for NH_4-jarosite precipitation, and the critical sulphate concentration is a function of the ferric ion concentration. When excess sulphate is present, the amount of jarosite increases directly with the $FeCl_3$ concentration. The addition of jarosite seed accelerates the precipitation reaction. The jarosite precipitation reaction is relatively independent of the concentrations of NH_4Cl (2.0-5.0 M) and the $FeCl_2$ reaction product (0.0-1.0 M $FeCl_2$). Regardless of the $PbCl_2$ concentration of the solution, the jarosite products precipitated from 4 M NH_4Cl solutions have Pb contents <0.2%. The Pb content of the ammonium jarosite exceeds 1% only when the NH_4Cl concentration is <2 M. Regardless of the AgCl concentration of the solution, the silver content of the ammonium jarosite is always <0.01% Ag. The low levels of Pb and Ag incorporation in the ammonium jarosite are attributed to the strong anionic complexation of these elements in concentrated chloride media.

INTRODUCTION

Jarosite precipitation ($MFe_3(SO_4)_2(OH)_6$ where M is K^+, Na^+, NH_4^+, Ag^+, $\frac{1}{2}Pb^{2+}$, etc.) is widely used in the zinc industry to remove iron from acidic zinc sulphate processing solutions (1,2). The jarosite process is commonly carried out at about 95°C in the presence of jarosite seed to yield a precipitate which can be readily filtered and washed. Ammonium jarosite and sodium jarosite are the species most commonly precipitated although lead-silver jarosite is formed unavoidably at the higher temperatures (~155°C) associated with the O_2-H_2SO_4 zinc pressure leaching process (3). Although most of the developmental work on jarosite precipitation has focussed on sulphate media, it is known that jarosites can be readily precipitated from chloride solutions provided that an independent source of sulphate is available (4). In fact, the Duval CLEAR process (5) used potassium jarosite precipitation at 150°C in the second stage of a two-stage $CuCl_2$-$FeCl_3$-KCl-$NaCl$ leach to control sulphate as well as excess iron. The process ran for several years on a demonstration plant scale, and few problems were encountered with the jarosite precipitation part of the circuit. Furthermore, turboaeration is often used to maintain the iron balance in $FeCl_3$ leaching processes (6), and the iron precipitate made in the turboaerator consists principally of jarosite and FeO.OH. The presence of jarosite in the turboaerator product helps to control the sulphate build-up in the chloride leaching solution although complementary methods of sulphate control may be necessary to maintain a low sulphate concentration (7).

Because they produce elemental sulphur rather than SO_2 gas, $FeCl_3$ and/or $CuCl_2$ leaching technologies continue to be developed for a variety of sulphide ores and concentrates. The technique is especially suited for pyrite-rich feeds because the pyrite is not significantly attacked. In this regard, CENIM (Centro Nacional de Investigaciones Metalurgicas, Madrid, Spain) and LNETI (Laboratorio Nacional de Engenharia e Tecnologia Industrial, Lisbon, Portugal) have recently developed a two-stage ammonium chloride leach process for the treatment of pyritic Zn-Pb-Cu-Ag bulk concentrates that solubilizes the required metals while leaving the pyrite in the leach residue (8). The two-stage leaching process is carried out at 105°C in concentrated NH_4Cl media and under 1 atm O_2 pressure. The use of concentrated ammonium chloride solutions offers advantages with respect to pH control, the rate the leaching, and the subsequent recovery of Zn and Cu by solvent extraction methods. These advantages could lead to the commercial acceptance of the CENIM-LNETI process as well as the development of other NH_4Cl-based leaching technologies. The presence of ferric chloride and sulphate ion in the moderately acidic NH_4Cl solutions likely to be employed in an ammonium chloride leaching process could result in the precipitation of ammonium jarosite, $NH_4Fe_3(SO_4)_2(OH)_6$. Although some jarosite precipitation is desirable for the control of sulphate and impurities, the formation of excessive amounts of jarosite would result in the unacceptable loss of both ammonium ion and the ferric chloride lixiviant. Furthermore, the processing solutions will probably contain dissolved $PbCl_2$ and AgCl, and the incorporation of Pb or Ag in the structure of the ammonium jarosite, which is discarded in the leach residue, is an undesirable possibility.

The above discussion indicates the importance of understanding the jarosite precipitation reaction in concentrated ammonium chloride media and of defining the level of impurity incorporation in the ammonium jarosite precipitate. To this end, a systematic study of the parameters affecting the formation of ammonium jarosite in concentrated ammonium chloride media was carried out. The extent of Pb and Ag incorporation in the ammonium jarosite precipitate was also determined, and the results of these investigations are summarized in this report.

EXPERIMENTAL

Reagent-grade chemicals were used for all the experiments. Most of the experiments were carried out at 98°C in glass reaction vessels fitted with titanium baffles. In all experiments, appropriate concentrations of NH_4Cl, $FeCl_3$, $PbCl_2$, AgCl, etc. were heated for a suitable period in a stirred solution, the pH of which was adjusted with HCl or Li_2CO_3. A three-point calibration procedure, with buffer solutions at pH = 4, 2 and 1, was used to calibrate the pH meter. Lithium sulphate was added as the independent sulphate source, and in general, lithium compounds were chosen because lithium

does not form an end-member jarosite-type compound (9). One litre of solution was used for each test, and the concentrations employed bracketted those likely to be employed in an ammonium chloride leaching process. At the end of the experiments, the jarosites were filtered and washed with copious amounts of water to remove any entrained solution. The pH was not controlled during either the precipitation or washing operations. Finally, the jarosites were dried at 110°C prior to analysis. Because the synthesis solutions often contained dissolved $PbCl_2$, the possibility of $PbSO_4$ precipitation exists. Accordingly, the products were checked for possible $PbSO_4$ contamination by the use of a Guinier deWolff focussing X-ray camera and cobalt radiation. Previous studies have shown that about 0.25% $PbSO_4$ is readily detected as a "trace" constituent by this technique.

RESULTS AND DISCUSSION

Effect of solution pH

Figure 1 illustrates the effect of the initial pH of a 0.5 M $FeCl_3$ – 4.0 M NH_4Cl – 0.3 M Li_2SO_4 – 20 g/L $PbCl_2$ solution on the amount and composition of the products formed at 98°C after 24 h of reaction. Negligible amounts of product are made at pH values <0.4, and this situation prevails even for NH_4Cl concentrations as high as 6.0 M or as low as 0.5 M NH_4Cl. Previous studies on the precipitation of Na-jarosite from 3 M NaCl media (4) showed that no product was precipitated for free acid concentrations >0.1 M HCl, and this observation is in general agreement with the current results. The amount of product increases in a nearly linear manner with increasing pH values above 0.4. When the initial solution pH is >1.8, however, iron hydrolysis and precipitation occur at or near room temperature, and the precipitates are difficult to filter. The natural pH of a 0.5 M $FeCl_3$ – 4.0 M NH_4Cl – 0.3 M Li_2SO_4 solution is 0.8-0.9, and as a result, most of the subsequent experiments were done using an initial pH of 0.8.

Jarosite formation generates large amounts of acid, and as a consequence, high initial acidities reverse the precipitation reaction and result in reduced product yields.

$$NH_4^+ + 3Fe^{3+} + 2SO_4^{2-} + 6H_2O \rightarrow$$

$$NH_4Fe_3(SO_4)_2(OH)_6 + 6H^+ \qquad [1]$$

In the current experiments, initial pH values <0.4 are sufficient to prevent the precipitation of any product. If the pH is raised too high, however, other hydrolysis reactions occur, and poorly filterable ferric hydroxide is formed.

$$Fe^{3+} + 3H_2O \rightarrow Fe(OH)_3 + 3H^+ \qquad [2]$$

Although the amount of product increases significantly as the pH increases, the product composition is essentially independent of pH. Over the pH range 0.4 to 1.8, the products contained 3.7% NH_4^+, 34.5% Fe and 40.0% SO_4^{2-}; these values correspond approximately to those of $(NH_4)Fe_3(SO_4)_2(OH)_6$: NH_4^+ 3.75%, Fe 34.93%, SO_4^{2-} 40.05%. Although all the synthesis solutions contained 20 g/L $PbCl_2$, the products consistently contained <0.2% Pb. Guinier X-ray diffraction analysis of the products made over the pH range 0.4 to 1.8 confirmed that only ammonium jarosite was precipitated. Lead sulphate was not detected in any of the products, and analysis of the solutions at the completion of the experiments showed that virtually all of the lead remained dissolved in the concentrated chloride media.

The Influence of Precipitation Temperature

Figure 2 shows the effect of the synthesis temperature on the amount and composition of the products made by heating a 0.5 M $FeCl_3$ - 4.0 M NH_4Cl - 0.3 M Li_2SO_4 - 20 g/L $PbCl_2$ solution at pH = 0.8 for 24 h.

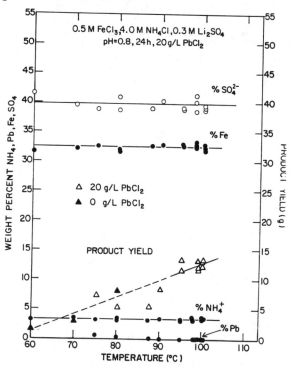

Fig. 2 – Variation of the amount and composition of the products with synthesis temperature in the absence of jarosite seed.

Although only a small quantity of material is formed at 60°C, higher temperatures result in the precipitation of significant amounts of jarosite. It is interesting to note that in the corresponding ferric sulphate-alkali sulphate systems (11), no product is formed in 24 h at temperatures <80°C. The implication is that jarosite formation in chloride media is in many respects "easier" than in the corresponding sulphate system despite the fact that pH control is more critical in chloride solutions. The products prepared over the temperature range 60-100°C have relatively constant ammonium, iron and sulphate contents which are indicative of ammonium jarosite. Although the Pb contents of the products made at temperatures >80°C are always <0.2% Pb, the products formed at 80 or 75°C contained greater amounts of lead. Guinier X-ray diffraction analysis of these two samples showed the presence of trace amounts of $PbSO_4$ and $PbCl_2$ which presumably crystallized from the concentrated chloride solutions at the lower temperatures employed. Because of the $PbSO_4/PbCl_2$

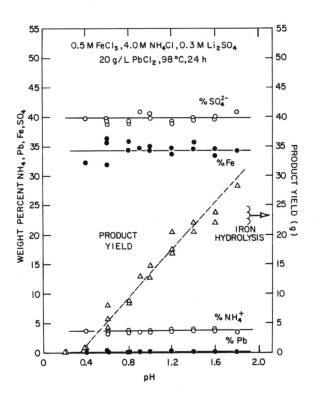

Fig. 1 – Effect of the initial pH of the synthesis solution on the product yield and product composition.

68

contamination, the lower temperature tests were repeated (solid triangles) using 0.5 M FeCl₃ - 4.0 M NH₄Cl - 0.3 M Li₂SO₄ solutions containing no PbCl₂. The results of these tests are consistent with those obtained in the presence of 20 g/L PbCl₂, except for the minor contamination caused by PbSO₄ or PbCl₂. The data of Figure 2 clearly show that high temperatures promote jarosite precipitation. Unfortunately, high temperatures are also required for the effective leaching of refractory sulphide minerals such as chalcopyrite or tetrahedrite. Consequently, some ammonium jarosite precipitation is likely to occur in any FeCl₃-NH₄Cl leaching process.

Effect of Sulphate and Ferric Ion Concentrations

The effect of the initial sulphate concentration of the FeCl₃ - 4.0 M NH₄Cl - 20 g/L PbCl₂ solution on the amount and composition of the products made at 98°C and pH = 0.8 was determined. Figure 3 presents the results obtained when the solutions contained 0.5 M FeCl₃, and Figure 4 shows the corresponding data generated from 0.2 M FeCl₃ media. In both systems, there is a minimum sulphate concentration required before any jarosite is precipitated. In 0.5 M FeCl₃ solutions the critical sulphate concentration is approximately 0.05 M whereas in 0.2 M FeCl₃ media the critical sulphate level is about 0.075 M. Thus, the critical sulphate concentration seems to increase as the FeCl₃ concentration decreases. For sulphate concentrations less than the critical value, only trace amounts of an amorphous reddish phase are produced. Above the critical sulphate concentration, progressively greater amounts of yellow ammonium jarosite are formed. In fact, the amount of product increases linearly with increasing sulphate concentration for solutions containing either 0.5 M FeCl₃ (Figure 3) or 0.2 M FeCl₃ (Figure 4). As would be expected, the slope of the product yield versus sulphate concentration curve is greater at the higher FeCl₃ concentration, but the ratio of the slopes (~1.3) is less than the ratio of the initial FeCl₃ concentrations (2.5). The products made from 4.0 M NH₄Cl - 20 g/L PbCl₂ solutions containing either 0.2 or 0.5 M FeCl₃ have virtually the same composition. Furthermore, the product compositions are essentially independent of the Li₂SO₄ concentration; the lead contents are consistently <0.2% Pb. All the compositions reflect ammonium jarosite which was the only phase detected by Guinier X-ray diffraction analysis.

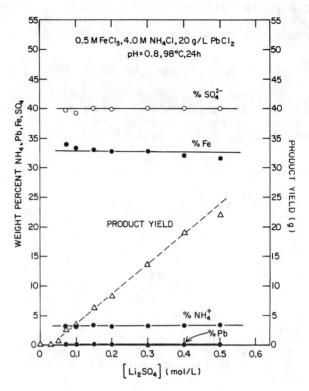

Fig. 3 - Effect of the Li₂SO₄ concentration on the amount and composition of the products made in 0.5 M FeCl₃ media.

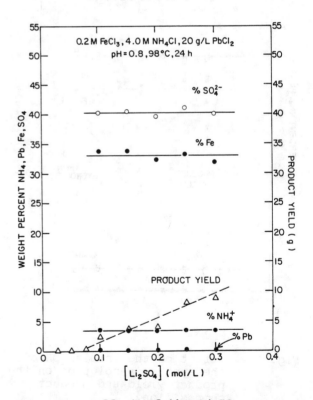

Fig. 4 - Effect of the Li₂SO₄ concentration on the amount and composition of the products made in 0.2 M FeCl₃ media.

Previous studies on the synthesis of sodium jarosite from 0.5 M $FeCl_3$ - 0.1 M $FeCl_2$ - 3 M NaCl solutions also indicated a critical sulphate concentration below which no jarosite precipitation occurred (4). In those studies, the critical sulphate concentration was found to be ~0.05 M, in general agreement with the results of the current investigation. A minimum sulphate concentration for jarosite formation in chloride media seems to be a general phenomenon, and the implication is that jarosite precipitation will not reduce the sulphate concentration of a chloride leaching solution to low levels. The practically attainable sulphate concentration is likely to be about 0.1 M or ~10 g/L SO_4^{2-} (7).

Figure 5 shows the effect of the initial $FeCl_3$ concentration on the amount and composition of the products made at 98°C from 4.0 M NH_4Cl - 0.3 M Li_2SO_4 - 20 g/L $PbCl_2$ media at pH = 0.8. The amount of product increases in a linear manner with increasing $FeCl_3$ concentration, and the product yield curve extrapolates through the origin. That is, there does not seem to be a critical ferric ion concentration for jarosite formation provided adequate sulphate is available in solution. Although the amount of precipitate increases with increasing $FeCl_3$ concentration, the composition of the product is relatively constant at ~3.2% NH_4^+, ~32.5% Fe and ~39.5% SO_4^{2-}. It is significant that there is no pronounced change in the iron content of the precipitate as the iron concentration of the solution increases from 0.1 to 0.7 M $FeCl_3$. The Pb contents of the products made from solutions containing >0.2 M $FeCl_3$ were consistently <0.1% Pb. By contrast, the Pb contents of the three products made at 0.2 M $FeCl_3$ were 0.13, 0.16 and 0.43% Pb; the corresponding analyses for the products made in 0.1 M $FeCl_3$ media were 0.27 and 1.30% Pb. Although Guinier X-ray diffraction analysis showed all the products to consist only of ammonium jarosite, the lead-rich products had slightly different unit-cells indicative of an enhanced level of ½Pb^{2+} substitution for NH_4^+ in the ammonium jarosite structure. It seems that low ferric ion concentrations, and their concomitant low product yields, favour an enhanced level of Pb incorporation in the ammonium jarosite although the extent of Pb incorporation is always relatively small. The amount of lead incorporation depends on the jarosite species formed as well as on the concentrations of the solution species. For example, when sodium jarosite was precipitated from 0.5 M $FeCl_3$ - 3 M NaCl - 0.3 M Li_2SO_4

solutions containing 20 g/L $PbCl_2$, the resulting precipitate contained ~0.8% Pb (4).

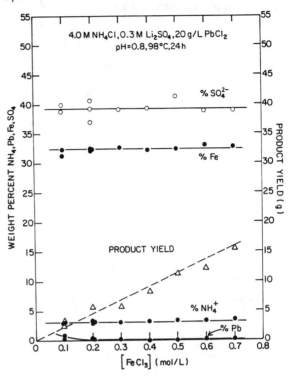

Fig. 5 - Variation of the product yield and product composition with the $FeCl_3$ concentration of the solution in the presence of excess Li_2SO_4.

Effect of Retention Time and Seeding

The jarosite precipitation experiments were done in a batch mode. The solutions were prepared at room temperature and were then heated to the 98°C reaction temperature. About 1 h was required to heat the charge, and some reaction inevitably occurred during the heating-up period. This point is reflected in Figure 6 which shows the amount and composition of the products as a function of the retention time at 98°C. A modest amount of precipitate is noted at "zero" time, and this reflects the material precipitated during the 1 h heating-up period. Prolonging the reaction at 98°C for 6-8 h results in the progressive formation of more jarosite. Thereafter, there is only a slight increase in the product yield for retention times as long as 48 h. Previous work on the precipitation of alkali jarosites in ferric sulphate media (11) showed that ~15 h was necessary to achieve nearly steady state conditions at 97°C in the sulphate system. A significantly shorter time is required to achieve steady state conditions in chloride media although the final amount of precipitate is less than that in the sulphate system.

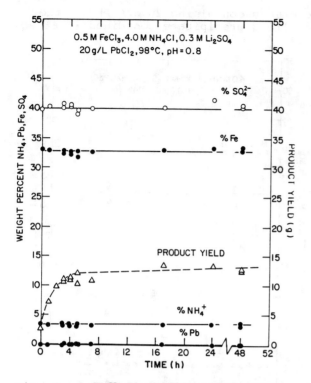

Fig. 6 - Influence of retention time at
98°C on the amount and
composition of the products
made in the absence of
jarosite seed.

The composition of the products
made in 0.5 M FeCl$_3$ - 4.0 M NH$_4$Cl -
0.3 M Li$_2$SO$_4$ - 20 g/L PbCl$_2$ media is
essentially independent of the retention
time for periods at least as long as
48 h. All the precipitates contain
~3.5% NH$_4^+$, 33% Fe and ~40% SO$_4^{2-}$; the
lead contents are consistently <0.2% Pb.
All the precipitates were shown by
Guinier X-ray diffraction analysis to
consist only of ammonium jarosite.

It is known that jarosite
precipitation in sulphate media is
accelerated by the presence of jarosite
seed; a minimum 100% seed addition is
commonly employed to maximize the
precipitation kinetics (12). The effect
of seeding in the current experiments
was studied using a 15 g/L addition of
ammonium jarosite seed. The reaction
in the FeCl$_3$-NH$_4$Cl-Li$_2$SO$_4$-PbCl$_2$ system
usually generates about 12 g of ammonium
jarosite, and hence, the 15 g/L of added
jarosite seed corresponds to a 125% seed
addition. Figure 7 shows the effect of
retention time on the amount and
composition of the products made at 98°C
in the presence of 15 g/L ammonium
jarosite seed. The product yield is the
weight of the final precipitate less the
weight of the seed initially added to
the reactor. Over 50% of the reaction
occurred at "zero" time; i.e., during

the one-hour heating-up period. The
amount of product increases with
prolonged retention time for about 2 h,
but thereafter the product yield
increases only gradually. Comparison
with the analogous experiments done
without seed additions (solid triangles
and Figure 6) shows that steady state
conditions are achieved more rapidly in
the presence of seed, but that the
product yield is ultimately the same
whether seeding is employed or not.

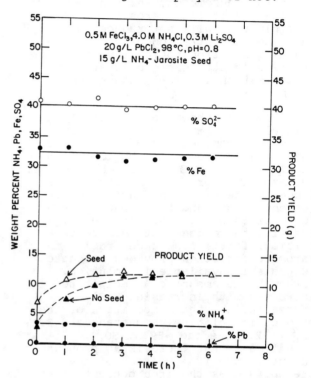

Fig. 7 - Influence of retention time at
98°C on the amount and
composition of the products
made in the presence of
15 g/L jarosite seed.

X-ray diffraction analysis of the
products made in the presence of seed
indicated only ammonium jarosite
regardless of the retention time. The
analytical data presented in Figure 7
indicate that the composition of the
ammonium jarosite is independent of the
retention time, and is virtually
identical to that determined in the
absence of seed, Figure 6.

Figure 8 illustrates the effect of
the synthesis temperature on the amount
and composition of the ammonium jarosite
precipitated in the presence of 15 g/L
of ammonium jarosite seed. In the
presence of seed, a significant amount
of precipitation occurs at 60°C; in
fact, a trace amount (~0.4 g) of product
was formed even at 40°C. Comparison
with the similar experiments carried out
in the absence of seed, Figure 2, shows
that the presence of seed partly

overcomes the slow jarosite precipitation kinetics at the lower temperatures. At the higher temperatures, 90-100°C, the amount of product made in the presence of seed is approximately the same as that made in the absence of an initial seed addition. This is a reflection of the fact that steady state conditions are achieved within 8 h at the higher temperatures whether or not seeding is used, Figure 7. Seeding is most beneficial under those conditions where the amount of precipitate is under kinetic control, and those regions include short retention times and low temperatures. When the precipitation system is at steady state, near equilibrium conditions, the amount of product is the same in the presence or absence of seed.

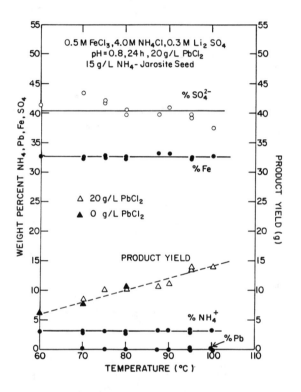

Fig. 8 – Variation of the amount and composition of the products with synthesis temperature in the presence of 15 g/L jarosite seed.

The products made in the presence of seed at temperatures above 70°C were shown by Guinier X-ray diffraction analysis to consist only of ammonium jarosite. At the lower temperatures, traces of $PbSO_4$ and/or $PbCl_2$ were detected, and this is a reflection of the reduced solubility of these species. Accordingly, the lower temperature experiments were repeated without any $PbCl_2$ in the synthesis solution. These experiments are indicated by the solid triangles which follow the general trend established by the products made in the

presence of 20 g/L $PbCl_2$. Only ammonium jarosite was detected by Guinier X-ray diffraction analysis of the products made from the lead-free solutions. Although there is some scatter in the analytical data for sulphate, the overall trend suggests that the product composition is independent of the synthesis temperature. The compositions are virtually identical to those realized in the absence of seed, Figure 2, and indicate ammonium jarosite.

Effect of the Concentrations of NH_4Cl and $FeCl_2$

A high ammonium chloride concentration is required in any $FeCl_3$-NH_4Cl leaching process to retain $PbCl_2$ and $AgCl$ in solution, to provide a source of "ammonia" in the subsequent metal recovery operations and to increase the boiling point of the lixiviant. The effect of the NH_4Cl concentration on the amount and composition of the products made at 98°C from 0.5 M $FeCl_3$ – 0.3 M Li_2SO_4 – 20 g/L $PbCl_2$ media at pH = 0.8 is illustrated in Figure 9. In the absence of NH_4Cl, a relatively large amount of product rich in lead but deficient in NH_4^+ is precipitated. As the NH_4Cl concentration increases to ~1 M, the amount of product declines sharply; the amount of product decreases more gradually for further increases in the NH_4Cl concentration. The lead content of the product decreases abruptly as the NH_4Cl concentration increases from 0.0 to 1.0 M, but thereafter decreases gradually. The Pb content of the precipitates made from solutions containing >2.0 M NH_4Cl was <0.2% Pb. The NH_4^+, Fe and SO_4^{2-} contents of the products increased as the NH_4Cl concentration of the solution increased from 0 M to ~1 M, but remained relatively constant at the higher ammonium chloride concentrations.

Guinier X-ray diffraction analysis of the products made from solutions containing <0.5 M NH_4Cl indicated major amounts of both lead jarosite and $PbSO_4$; the elevated lead contents reflect these compounds. The products made from solutions containing 0.5 to 2.0 M NH_4Cl consisted of ammonium jarosite with minor, and progressively decreasing, amounts of $PbSO_4$. The precipitates made from solutions containing >2.0 M NH_4Cl consisted only of ammonium jarosite, and this observation is consistent with the analytical data.

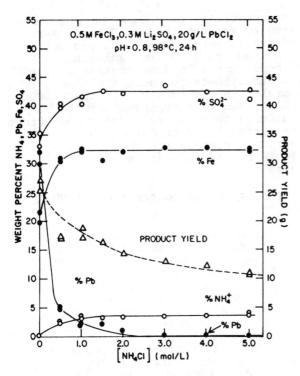

Fig. 9 - Effect of the NH₄Cl
concentration on the amount
and composition of the
precipitates made under
conditions where the total
chloride concentration varied
with the NH₄Cl concentration.

The presence of PbSO₄ in the
precipitates made from solutions having
NH₄Cl concentrations <2 M is a
reflection of the low total chloride
concentration of those solutions that
reduces the solubility of PbSO₄. To
investigate the influence of the NH₄Cl
concentration on the jarosite
precipitation reaction under conditions
of constant total chloride
concentration, a series of experiments
was carried out in solutions containing
different concentrations of NH₄Cl, but
always having [NH₄Cl] + [LiCl] = 4.0 M.
The results of these experiments are
summarized in Figure 10. The amount of
product increases slightly as the NH₄Cl
concentration increases, and this
behaviour is in marked contrast to the
abrupt reduction in product yield
observed in the corresponding tests done
at variable total chloride
concentration, Figure 9. In the absence
of NH₄Cl, that is in a solution
containing 4.0 M LiCl, a product
containing ~11% Pb and 0.0% NH₄⁺ is
formed. X-ray diffraction analysis
indicated only Pb-jarosite, and this
observation is consistent with synthesis
studies recently carried out on the
formation of Pb-jarosite in chloride
media (13). As the concentration of
NH₄Cl increases, the NH₄⁺ content of the
precipitate increases, and the Pb

content drops. This trend continues to
~0.3 M NH₄Cl, at which concentration a
product containing ~3.1% NH₄⁺ and 0.4%
Pb is produced. Further increases in
the NH₄Cl concentration result in only
modest changes in the ammonium and lead
contents. The iron and sulphate
contents of the precipitates increase
slightly as the NH₄Cl concentration
rises from 0.0 to 0.3 M. Lead sulphate
was not detected in any of the
precipitates by Guinier X-ray
diffraction analysis. Figure 9 and
Figure 10 indicate that only a modest
NH₄Cl concentration in the processing
solution is required to form ammonium
jarosite in preference to lead jarosite.

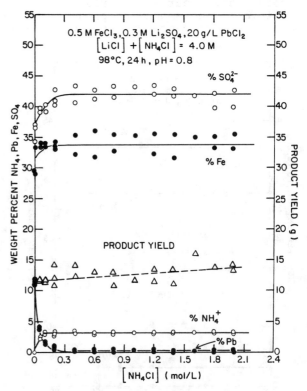

Fig. 10 - Effect of the NH₄Cl
concentration on the amount
and composition of the
precipitates made from
solutions having a constant
total chloride concentration.

In any iron chloride leaching
process, the FeCl₃ lixiviant is reduced,
at least in part, to FeCl₂ by reaction
with the sulphide minerals.

$$2FeCl_3 + PbS \rightarrow 2FeCl_2 + PbCl_2 + S° \quad [3]$$

$$4FeCl_3 + CuFeS_2 \rightarrow 5FeCl_2 + CuCl_2 + 2S° \quad [4]$$

Hence, the effect of ferrous chloride on
the jarosite precipitation reaction in
concentrated chloride media is relevant.
In this regard, Figure 11 illustrates
the effect of the FeCl₂ concentration on
the amount and composition of the
products made at 98°C from solutions

containing 0.5 M FeCl₃ - 4.0 M NH₄Cl -
0.3 M Li₂SO₄ and 20 g/L PbCl₂ at
pH = 0.8. The amount of product
increases steadily with increasing FeCl₂
concentration, but there is considerable
scatter in the product yields. Both
phenomena are due to the uncontrolled
oxidation of some of the FeCl₂ to FeCl₃
during the 24 h precipitation
experiments; increasing FeCl₃
concentrations precipitate more
jarosite, Figure 5. Although the amount
of product varies somewhat with the
ferrous chloride concentration, the
composition of the products is virtually
independent of FeCl₂ concentrations in
the 0-1 M FeCl₂ range. Furthermore, the
product composition is that of ammonium
jarosite with <0.2% Pb substitution for
the ammonium ion. Guinier X-ray
diffraction analysis of the various
precipitates indicated only ammonium
jarosite.

Incorporation of Lead and Silver in the Jarosite Precipitate

It is known that complete solid
solution series exist among the alkali
jarosites and lead jarosite (14) and
among the alkali jarosites and silver
jarosite (15,16). Because both PbCl₂
and AgCl are moderately soluble in
concentrated chloride media, the
structural incorporation of Pb and/or Ag
in the ammonium jarosite is a
possibility. The effect of the PbCl₂
concentration on the extent of Pb
incorporation in the ammonium jarosite
made at 98°C from 0.5 M FeCl₃ - 4.0 M
NH₄Cl - 0.3 M Li₂SO₄ solutions at
pH = 0.8 was investigated, and the
results are summarized in Figure 12.

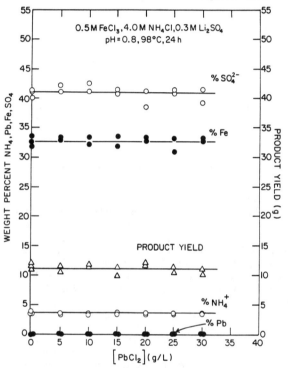

Fig. 12 - Influence of the PbCl₂
concentration on the amount
and composition of the
ammonium jarosite; the lead
analyses are given in Table 1.

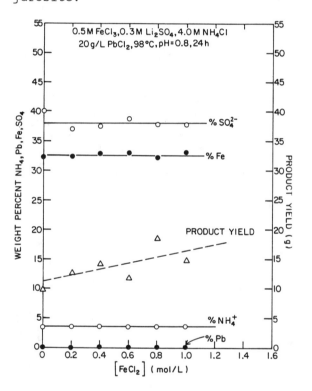

Fig. 11 - Variation of the product yield
and product composition with
the FeCl₂ concentration of the
solution. The tests were done
in air, and partial oxidation
of the FeCl₂ likely occurred.

The presence of ferrous chloride in
a FeCl₃-NH₄Cl leaching solution will
have no direct effect on the
precipitation of ammonium jarosite.
Some of the FeCl₂ may oxidize to FeCl₃,
however, and the ferric ion will
generate additional amounts of jarosite.

Clearly, there is no significant
variation in the amount of product as
the PbCl₂ concentration of the solution
increases from 0 to 30 g/L PbCl₂. The
composition is essentially that of
ammonium jarosite which was the only
phase detected by Guinier X-ray
diffraction analysis. Although the
ammonium, iron and sulphate contents are
essentially independent of the PbCl₂
concentration, Table 1 shows that the Pb
content of the ammonium jarosite
increases systematically as the PbCl₂
concentration of the solution increases.
The Pb content, however, reaches a value
of only 0.22% Pb even in a solution
nearly saturated in PbCl₂; i.e., in a

74

30 g/L PbCl$_2$ solution. The implication is that only modest losses of lead will occur in the ammonium jarosite precipitated from concentrated NH$_4$Cl media.

Table 1. Effect of the PbCl$_2$ Concentration of the Solution on the Pb Content of the Ammonium Jarosite

PbCl$_2$ Concentration (g/L)	Average Pb Content (%)
0	0.01
5	0.03
10	0.07
15	0.08
20	0.10
25	0.13
30	0.22

Silver chloride is moderately soluble in concentrated chloride media, and as a consequence, silver dissolves in most chloride leaching processes. Figure 13 shows the effect of dissolved AgCl on the amount and composition of the products made at 98°C from 0.5 M FeCl$_3$ - 4.0 M NH$_4$Cl - 0.3 M Li$_2$SO$_4$ solutions at pH = 0.8. None of the solutions contained dissolved PbCl$_2$. The data show that neither the amount of precipitate nor its composition are affected by AgCl concentrations as high as 1.2 g/L, a value near the saturation limit of AgCl in the solutions employed. The product composition is essentially that of ammonium jarosite which was the only phase detected by Guinier X-ray diffraction analysis. The Ag contents of the ammonium jarosite varied from 10 to 40 ppm Ag. Clearly, silver is not significantly concentrated in the ammonium jarosite product.

Figure 14 presents the analogous results obtained when jarosite was precipitated at 98°C from 0.5 M FeCl$_3$ - 4.0 M NH$_4$Cl - 0.3 M Li$_2$SO$_4$ - 20 g/L PbCl$_2$ solutions at pH = 0.8 that also contained 0.0 to 1.2 g/L AgCl. Increasing the AgCl concentration of the solution from 0.0 to 1.2 g/L AgCl has no significant effect on either the amount or composition of the precipitate. The analytical data suggest ammonium jarosite, which was the only phase detected by Guinier X-ray diffraction analysis.

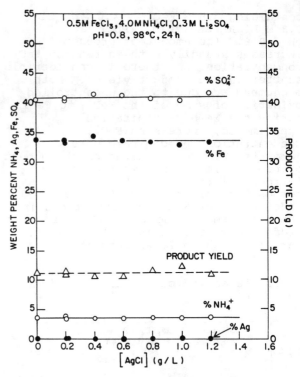

Fig. 13 - Effect of the AgCl concentration on the amount and composition of the products made in the absence of dissolved PbCl$_2$. The Ag contents are <40 ppm.

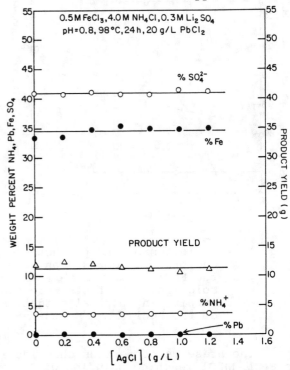

Fig. 14 - Effect of the AgCl concentration on the amount and composition of the products made in the presence of 20 g/L PbCl$_2$. The Pb and Ag contents are given in Table 2.

Table 2 lists the Pb and Ag contents of the various precipitates, and it is clear that neither Pb nor Ag are significantly concentrated in the ammonium jarosite. The lead contents vary from 0.05 to 0.12% Pb, and the Ag contents never exceed 40 ppm Ag.

Table 2. Effect of the AgCl Concentration of the Solution on the Ag and Pb Contents of the Ammonium Jarosite

AgCl Concentration (g/L)	Ag Content (ppm)	Pb Content (%)
0	<10	0.06
0.2	10	0.05
0.4	30	0.11
0.6	20	0.12
0.8	40	0.11
1.0	30	0.12
1.2	30	0.11

It is well established that both silver and lead form stable jarosite-type compounds (2). In chloride media, however, both Ag and Pb are strongly complexed as a variety of chloro-species (17).

$$Ag^+ + Cl^- \rightarrow AgCl^° \quad Log\ K_1 = 3.5 \quad [5]$$

$$Pb^{2+} + Cl^- \rightarrow PbCl^+ \quad Log\ K_1 = 1.5 \quad [6]$$

The high level of complexation of Ag and Pb in concentrated chloride media, coupled with the relatively low solubilities of AgCl and $PbCl_2$, results in very low concentrations of "free" Ag^+ and Pb^{2+} ions which seem to be required to form jarosite. The consequence is that lead, and especially silver, remain in solution and do not extensively precipitate as jarosite-type compounds. By contrast, free ammonium or alkali ions are abundant and result in the precipitation of ammonium or alkali jarosites.

CONCLUSIONS

Ammonium jarosite, $NH_4Fe_3(SO_4)(OH)_6$, is readily precipitated from concentrated $FeCl_3$-NH_4Cl media containing Li_2SO_4 as the independent sulphate source. The precipitation reaction is strongly dependent on the initial pH of the solution. At 98°C, no precipitate forms when the pH is <0.4. The amount of jarosite increases linearly with increasing pH above this value, but difficult to filter products are formed when the pH is >1.8. Although the amount of product strongly depends on the pH, the composition of the ammonium jarosite is essentially pH independent. The amount of jarosite decreases steadily as the temperature is reduced, and only traces of product are made at temperatures <60°C. The product composition, however, remains essentially independent of temperature over the 60-100°C temperature range. An independent source of sulphate is required to form ammonium jarosite, and there is a critical sulphate concentration below which no jarosite forms. The critical sulphate concentration is ~0.05 M when the $FeCl_3$ concentration is 0.5 M, but rises to 0.075 M when the $FeCl_3$ concentration is reduced to 0.2 M. When excess sulphate is present, the amount of jarosite increases linearly with increasing $FeCl_3$ concentration; also, the composition of the jarosite is relatively constant for $FeCl_3$ concentrations >0.1 M. The ammonium jarosite precipitation reaction proceeds rapidly in chloride media, and steady state conditions are realized within 6-8 h at 98°C. The presence of jarosite seed considerably reduces the time needed to achieve steady state conditions, and also allows jarosite precipitation to occur at temperatures as low as 40°C. If the processing solution contains dissolved $PbCl_2$, high total chloride concentrations (>2 M NH_4Cl) are required to prevent the contamination of the jarosite with $PbSO_4$ and/or $PbCl_2$. At high total chloride concentrations, the extent of Pb incorporation in the ammonium jarosite is <0.2% Pb regardless of the $PbCl_2$ concentration. Silver is not significantly incorporated (Ag <40 ppm) in any of the jarosites made from concentrated chloride media, and the presence of $PbCl_2$ in the solution does not increase the level of Ag incorporation. The low Ag and Pb contents of the ammonium jarosite are attributed to the strong complexation of these metals by chloride ion that results in the retention of both Pb and Ag in the processing solution.

ACKNOWLEDGEMENTS

The assistance of D.J. Hardy with the experimental program and J.L. Jambor and P. Carriere with the X-ray diffraction studies is gratefully acknowledged.

REFERENCES

1 - V. Arregui, A.R. Gordon and
    G. Steintveit, The jarosite process
    - past, present and future. In
    Lead-Zinc-Tin '80, J.M. Cigan,
    T.S. Mackey and T.J. O'Keefe, eds.,
    TMS, Warrendale, pp. 97-123, 1979.

2 - J.E. Dutrizac, The physical
    chemistry of iron precipitation in
    the zinc industry. In Lead-Zinc-
    Tin '80, J.M. Cigan, T.S. Mackey
    and T.J. O'Keefe, eds., TMS,
    Warrendale, pp. 532-564, 1979.

3 - J.E. Dutrizac and T.T. Chen, A
    mineralogical study of the jarosite
    phase formed during the autoclave
    leaching of zinc concentrate. Can.
    Metall. Q. 23, pp. 147-157, 1984.

4 - J.E. Dutrizac, Jarosite formation
    in chloride media. Proc.
    Australas. Inst. Min. Metall. 278,
    pp. 23-32, 1981.

5 - A.W. Fletcher, Production of an
    acid forming fertilizer from CLEAR
    plant iron residue. In Iron
    Control in Hydrometallurgy,
    J.E. Dutrizac and A.J. Monhemius,
    eds., Ellis Horwood, Chichester,
    pp. 689-694, 1986.

6 - R. Raudsepp and M.J.V. Beattie,
    Iron control in chloride systems.
    In Iron Control in Hydrometallurgy,
    J.E. Dutrizac and A.J. Monhemius,
    eds., Ellis Horwood, Chichester,
    pp. 163-182, 1986.

7 - J.E. Dutrizac, Sulphate control in
    chloride leaching processes.
    Hydrometallurgy 23, pp. 1-22, 1989.

8 - J.L. Limpo Gil, A.L. Martin, A.H.
    Fernandez and J.A. Priego,
    Hydrometallurgical recovery of
    metals from complex sulphides.
    Span. Patent ES 545,698, 16 January
    1986.

9 - J.E. Dutrizac and J.L. Jambor,
    Behaviour of cesium and lithium
    during the precipitation of
    jarosite-type compounds.
    Hydrometallurgy 17, pp. 251-265,
    1987.

10 - J.A. Ripmeester, C.I. Ratcliffe,
     J.E. Dutrizac and J.L. Jambor,
     Hydronium ion in the alunite-
     jarosite group. Can. Mineralogist
     21, pp. 435-448, 1986.

11 - J.E. Dutrizac, Factors affecting
     alkali jarosite formation. Metal.
     Trans. 14B, pp. 531-539 (1983).

12 - R.V. Pammenter and C.J. Haigh,
     Improved metal recovery with the
     low-contaminant jarosite process.
     In Extraction Metallurgy '81, Inst.
     Min. Metal., London, pp. 379-392,
     1981.

13 - J.E. Dutrizac, The precipitation of
     lead jarosite from chloride media.
     Hydrometallurgy, in press, 1991.

14 - J. Kubisz, A study on minerals of
     the alunite-jarosite group. Polska
     Akad. Nauk, Prace Geol. 22, pp.
     9-93, 1964.

15 - J.E. Dutrizac and J.L. Jambor,
     Behaviour of silver during jarosite
     precipitation. Trans. Inst. Min.
     Metall. 96, pp. C206-C218, 1987.

16 - J.E. Dutrizac, Jarosite-type
     compounds and their application in
     the metallurgical industry. In
     Hydrometallurgy Research
     Development and Plant Practice,
     K. Osseo-Asare and J.D. Miller,
     eds., TMS, Warrendale, pp. 531-551,
     1982.

17 - J. Bjerrum, G. Schwarzenbach and
     L.G. Sillen, Stability Constants,
     Part 2: Inorganic Solids. The
     Chemical Society, London, 1958.

Jarosite precipitation during acid pressure leaching of zinc leach residue—application of factorial design and characterization of leach residue

Sridhar Acharya
Shashi Anand
R. P. Das
Regional Research Laboratory, Bhubaneswar, Orissa, India

SYNOPSIS

The results obtained on iron rejection as jarosite at elevated temperatures have been reported. The effects of three leaching parameters namely, temperature, acid concentration and amount of ammonium sulphate were quantified using 2^3 full factorial design. The regression equations obtained for iron and zinc dissolution were tested for their adequacy to fit the experimental data by Fisher's test. The extrapolations beyond the studied levels did not yield satisfactory results. Additional experiments are reported using oxygen for oxidation of Fe(II) to Fe(III) and its rejection as jarosite. Ammonium sulphate, sulphuric acid and temperature were varied in the narrow range to get maximum zinc recovery with minimum iron dissolution. The best results were obtained under the following conditions : 210°C, 30 g/l $(NH_4)_2SO_4$, 120 g/l H_2SO_4, pO_2 412 kPa for 1.2 g/l iron with 94% zinc extraction. Sodium sulphate was found to be more effective compared to ammonium or potassium sulphate. Dissolution of minor impurities like Cu, Ni, Co and Cd were not affected during high temperature jarosite precipitation. The various iron phases identified were ammonium and plumbo jarosites with minor amounts of hematite. The advantages of this process have been mentioned.

Electrolytic zinc is commercially produced through roast-leach process where the zinc sulphide concentrate is roasted in a smelter followed by leaching the calcine (neutral leach) with 85-93% zinc recovery[1]. The residue containing zinc ferrites is further treated with spent electrolyte in the hot acid leaching step to dissolve zinc, but most of the iron also dissolves. The dissolved iron is usually removed by processes such as Goethite[2-11], Hematite[12-16] and Jarosite[17-26] for the final recovery of zinc from such leach liquors.

The major research and development work is continuing for improvement in Jarosite process[27-31] to reduce the process steps and the amount of neutralizing agent. The conversion process developed by Outokumpu is a modification of the Jarosite process and iron precipitation takes place simultaneously in the reactor[32,33]. Much details are not available on this process, but it is believed that long residence times in the conversion reactor are necessary[1]. On the other hand, in the Hematite process, efforts are continuing for a single step leaching with iron rejection at high temperatures (HTC)[34] in the presence of oxygen. The direct pressure leaching of zinc sulphide concentrate has been studied with iron rejection as jarosite[35].

As most of the zinc plants are based on roast-leach process, it was felt necessary to improve the residue leaching step with simultaneous iron rejection as jarosite at moderate temperatures (170-190°C)[36] with an objective to reduce (a) the amount of initial acid (b) the amount of neutralizing agent (zinc calcine) (c) iron level in the leach liquors and (d) number of solid-liquid separation stages. From this study it was found (i) high temperature favoured zinc dissolution and iron precipitation (ii) the order of monovalent cations for iron rejection as jarosite was $Na^+ > NH_4^+ > K^+$, (iii) increase in monovalent cation concentration increased iron precipitation and (iv) high acid concentration decreased iron precipitation. In the present work the significance of leaching parameters and their interactions were quantified using factorial design. Various iron phases in the leach residue were identified by X-ray diffraction technique. Additional experiments were conducted to minimize iron contamination.

PROCESS CHEMISTRY

The ponded leach residue accumulated over years at Hindustan Zinc Ltd., Udaipur, India, was found to contain zinc phases as zinc ferrite $ZnFe_2O_4$, zinc manganate $ZnMnO_3$, zincite ZnO, zinc oxysulphate $Zn_3O(SO_4)_2$ and sphalerite ZnS. Iron was distributed as maghemite, γ-Fe_2O_3, green rust $Fe(II)_{3.6}Fe(III)_{0.9}(O, OH, SO_4)_2$, plumbo-jarosite $Pb_{0.5}Fe_3(SO_4)_2(OH)_6$ and hematite α-Fe_2O_3[37].

During sulphuric acid leaching most of the zinc and iron phases would dissolve to different extents depending on the reaction conditions. α-Fe_2O_3 would remain unattacked. The various chemical reactions would be :

$$ZnFe_2O_4 + 4H_2O \longrightarrow ZnSO_4 + Fe_2(SO_4)_3 + 4H_2O \quad ...(1)$$

$$ZnMnO_3 + H_2SO_4 \longrightarrow ZnSO_4 + MnO_2 + H_2O \quad ...(2)$$

$$ZnO + H_2SO_4 \longrightarrow ZnSO_4 + H_2O \quad ...(3)$$

$$ZnS + 4Fe_2(SO_4)_3 \longrightarrow ZnSO_4 + 8FeSO_4 + 4H_2SO_4 \quad ...(4)$$

$$ZnS + 4 MnO_2 + 4H_2SO_4 \longrightarrow ZnSO_4 + 4MnSO_4 + 4H_2O \quad ...(5)$$

$$Fe(II)_{3.6}Fe(III)_{0.9}(O, OH, SO_4)_2 + 3H_2SO_4 \longrightarrow 3.6 FeSO_4 + 0.45 Fe_2(SO_4)_3 + 4 H_2O \quad ...(6)$$

$$2FeSO_4 + MnO_2 + 2H_2SO_4 \longrightarrow Fe_2(SO_4)_3 + MnSO_4 + 2H_2O \quad ...(7)$$

$$2Pb_{0.5}Fe_3(SO_4)_2(OH)_6 + 6H_2SO_4 \longrightarrow PbSO_4 + 3Fe_2(SO_4)_3 + 6H_2O \quad ...(8)$$

The dissolved Fe(III) would precipitate as hematite, basic iron sulphate $Fe(OH)SO_4$ and jarosite[38] as :

$$Fe_2(SO_4)_3 + 3H_2O \longrightarrow Fe_2O_3 + 3H_2O \quad ...(9)$$

$$Fe_2(SO_4)_3 + 2H_2O \longrightarrow 2Fe(OH)SO_4 + H_2SO_4 \quad ...(10)$$

$$3Fe_2(SO_4)_3 + 2MeSO_4 + 7H_2O \longrightarrow 2MeFe_3(SO_4)_2(OH)_6 + H_2SO_4 \quad ...(11)$$

Tozawa and Saski[38] have studied the effects of co-existing sulphates on the precipitation of Fe(III) at elevated temperatures (170-200°C) from ferric sulphate solutions. At lower acidity Fe_2O_3 and at higher acidity $Fe(OH)SO_4$ are favoured. The presence of monovalent cations lead to precipitation of jarosite at higher acidity and even at lower acidity, there exists a very narrow region of ferric oxide. The experimental conditions can be selected to precipitate iron preferentially as hematite, basic ferric sulphate or jarosite.

The leaching of residue with simultaneous precipitation of Fe(III) as jarosite would take place as :

$$3ZnO.Fe_2O_3 + 7H_2SO_4 + Me_2SO_4 \longrightarrow$$
$$3ZnSO_4 + 2MeFe_3(SO_4)_2(OH)_6 \quad ...(12)$$

where Me is Ag^+, $0.5\ Pb^{2+}$, H_3O^+, K^+, Na^+ or NH_4^+. Ag^+, Pb^{2+} and H_3O^+ already exist in the system and addition of K^+, Na^+, or NH_4^+ would promote jarosite formation.

EXPERIMENTAL

Leaching experiments were conducted in a two gallon capacity Parr reactor having provisions for gas inlet, outlet, sampling, agitation and temperature control. The required amount of water, zinc leach residue and ammonium, sodium or potassium sulphate were charged to the reactor and heated to the desired temperature at constant agitation of 600 rpm. Experiments were conducted in 150 g scale using 15% wt./vol. pulp density. The leaching time was counted on attaining the required temperature. Samples were drawn at regular intervals, filtered immediately and appropriately diluted for zinc, iron, copper, nickel, cadmium, manganese and cobalt estimation by Perkin-Elmer atomic absorption spectrophotometer.

Zinc leach residue samples were provided by M/s. Hindustan Zinc Ltd., Udaipur, India. Chemical composition of a typical zinc residue used in the present study is given in Table I. All reagents used were of AnalaR grade.

Table 1 Chemical Composition of typical ponded zinc leach residue

Element	Wt.%
Zn	17.5
Fe	28.8
Cu	0.220
Ni	0.027
Co	0.018
Mn	0.90
Cd	0.32
Ag (ppm)	464
Pb	3.8
Acid insolubles	6.8

X-ray diffraction patterns were obtained with automatic powder diffractometer using a copper target and a scanning speed of 2° (2θ)/min. in the range of 15 to 65°. The peaks were matched with powder diffraction file data to identify the various mineral phases.

DESIGN OF EXPERIMENTS

Leaching experiments were carried out according to a 2^3 full factorial design. The three leaching variables were : temperature, acid concentration and amount of ammonium sulphate. The upper and lower levels of temperature were fixed as 170 and 190°C respectively. The two levels of acid concentrations chosen were 140 and 160 g/l (close to the concentration of spent electrolyte). Ten and twenty g/l of ammonium sulphate were taken as the lower and upper levels of monovalent cation concentration (higher amounts of ammonium sulphate result in the precipitation of zinc ammonium sulphate[37]). The base level conditions were : temperature 180°C, acid concentration 150 g/l and ammonium sulphate 15 g/l.

RESULTS AND DISCUSSION

Regression equations for iron and zinc dissolution

Leaching results obtained for half-an-hour and two hours retention time using 2^3 full factorial design for iron dissolution are shown in Fig.1. Matrix used for evaluation of linear and interaction co-efficients is given in Table II. The full regression equation[39] can be written as :

$$Y_{Fe(0.5h)} = b_0 + b_1X_1 + b_2X_2 + b_3X_3 +$$
$$b_{12}X_1X_2 + b_{23}X_2X_3 + b_{31}X_3X_1 \quad ...(13)$$

where $Y_{Fe(0.5h)}$ is the percent iron dissolution at half-an-hour, b_1, b_2 and b_3 are the linear co-efficients, b_{12}, b_{23} and b_{31} are the interaction co-efficients. The third order interaction term is neglected. X_1, X_2 and X_3 are the dimensionless coded factors for temperature, acid concentration and ammonium sulphate concentration respectively. On evaluating the interaction and linear

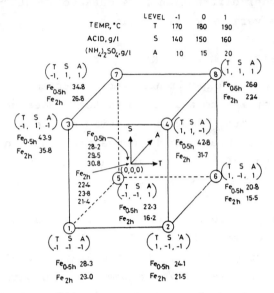

Fig. 1: 2^3 factorial presentation for %Fe dissolution at 0.5 and 2 hours.

co-efficients by standard procedure[39] and testing for significance by student 't' test, the regression equation (13) becomes :

$$Y_{Fe(0.5h)} = 29.33 + 5.37X_2 - 5.64X_3 + 3.1X_2X_3 \qquad ...(14)$$

Table II Matrix for 2^3 full factorial design used to evaluate linear and interaction co-efficients.

N	X_o	X_1	X_2	X_3	X_1X_2	X_2X_3	X_3X_1
1	+1	-1	-1	-1	+1	+1	+1
2	+1	+1	-1	-1	-1	+1	-1
3	+1	-1	+1	-1	-1	-1	+1
4	+1	+1	+1	-1	+1	-1	-1
5	+1	-1	-1	+1	+1	-1	-1
6	+1	+1	-1	+1	-1	-1	+1
7	+1	-1	+1	+1	-1	+1	-1
8	+1	+1	+1	+1	+1	+1	+1

X_1, X_2, X_3 are the dimensionless coded factors for temperature, acid concentration and $(NH_4)_2SO_4$ concentration. N is the number of observations.

It is seen from eqn. (14) that with increase of acid concentration, iron extraction increases, whereas increase in ammonium sulphate decreases iron in solution. The range of temperature selected does not significantly affect iron dissolution. The interaction between acid concentration and ammonium sulphate shows positive significance.

Similarly, the regression equation for iron dissolution for two hours retention time was estimated[39] to be :

$$Y_{Fe(2h)} = 24.11 - 1.34 X_1 + 5.11 X_2 - 3.84 X_3 \qquad ...(15)$$

Comparing the regression equations (14) and (15), it is observed that with increase in leaching time, temperature shows negative significance, and the interaction co-efficient for acid and ammonium sulphate no longer remains significant. Positive effect on iron dissolution for acid concentration and negative effect for ammonium sulphate remain significant with some variation in numerical value of these linear co-efficients.

Leaching results for half-an-hour and two hours retention times for 2^3 full factorial design for zinc dissolution are shown in Fig.2. The regression equations after evaluating the significant co-efficients by student '-t-' test will be given below :

$$Y_{Zn(0.5h)} = 78.80 + 3.58 X_1 + 4.92 X_2 - 2.17 X_2X_3 \qquad ...(16)$$

$$Y_{Zn(2h)} = 85.08 + 3.52 X_1 + 2.54 X_2 - 2.64X_2X_3 + 3.67X_3X_1 \qquad ...(17)$$

Regression equations (16) and (17) show, (i) positive dependence for zinc dissolution on temperature (ii) positive dependence on acid concentration, (iii) linear co-efficient for ammonium sulphate to be insignificant for half-an-hour and two hours zinc extractions. The interaction co-efficient for acid and ammonium sulphate concentration show negative co-relation with zinc dissolution. At 2 hour retention time, interaction co-efficient for temperature

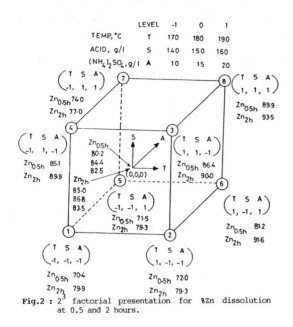

Fig.2 : 2^3 factorial presentation for %Zn dissolution at 0.5 and 2 hours.

and ammonium sulphate concentration also become significant.

Testing the adequacy of the regression equation (14)-(17)

The adequacy of eqn. (14) was tested by Fisher's test[39] to see how it fits the observations. The variance ratio[39] F_1 was calculated to be 3.34. The tabulated value of Fisher's F for α =0.05 (95% confidence level) and r_1 X r_2=4X2 degrees of freedom is 19.3 (r_1 is evaluated by substracting the number of significant co-efficients in the regression equation from total number of observations and r_2 is equal to the number of repeat tests for base level minus one). Since $F_1 < F_{1-\alpha} (r_1, r_2)$, the estimated regression equation (14) would fit the experimental data adequately. The adequacy for other equations (15), (16) and (17) was also evaluated similarly.Table III shows the comparison between the calculated variance ratio and tabulated values. In all cases, the variance ratio is much less than the tabulated values, hence, all the regression equations should fit experi-

mental data.

Table III Comparison of experimentally obtained variance ratio with the tabulated values of Fisher's F[39]

Regression eqn.no.	r_1	r_2	Experimental variance ratio		Tabulated values of Fisher's F
(15)Fe(2h)	4	2	F_2	1.89	19.3
(16)Zn(0.5h)	4	2	F_3	9.16	19.3
(17)Zn(2h)	3	2	F_4	4.80	19.2

Comparison of iron and zinc dissolution regression equations

From the comparison of Zn(2h) and Fe(2h) regression equations, the following observations are made

(i) With the increase in leaching temperature, zinc dissolution increases and iron contamination decreases.

(ii) Increase in acid concentration results in increase of both zinc and iron dissolution.

(iii)With the increase in ammonium sulphate concentration, the linear co-efficient for zinc dissolution is not significant but iron dissolution decreases with increase in $(NH_4)_2SO_4$.

(iv) Iron dissolution is independent of interactions of leaching parameters, whereas simultaneous increase in acid and ammonium sulphate concentration will decrease zinc extraction. Simultaneous increase in leaching temperature and ammonium sulphate concentration will result in higher zinc extraction. From these results, it is inferred that by carefully selecting the levels of temperature, acid concentration and ammonium sulphate, it should be possible to optimise zinc extraction with minimum iron dissolution. Acid concentration should be maintained at a minimum level for the desired zinc dissolution with temperature and ammonium sulphate to be maintained at a higher level.

Prediction from regression equations

Regression equations from (14) to (17) were used to predict extractions both within the cubical region shown in Fig.1 and 2 and also extrapolated outside these levels. The results are shown in Table IV. (The regression equations were first converted to the natural

contamination within the cube and for the extrapolations of leaching parameters, the regression equations are not adequate, additional experiments were conducted in the following directions :

(i)   Use of oxygen partial pressure for the oxidation of Fe(II) to Fe(III) to facilitate iron rejection.

(ii) Increase of leaching temperature to 210°C.

(iii) Decrease in acid conc. to 110-130 g/l.

Table IV Comparison of experimental results with those of predicted extractions from regression equations under various leaching conditions.

Leaching conditions			Fe(0.5h)		Fe(2h)		Zn($\frac{1}{2}$h)		Zn(2h)	
Temp. °C	Acid Conc.	Amm. sul.	Exp.	Pred.	Exp.	Pred.	Exp.	Pred.	Exp.	Pred.
190	140	30	17.7	16.1	12.1	5.9	80.4	77.5	104.5	84
190	140	40	12.2	11.1	10.0	-1.9	79.6	80.3	118.4	88
180	150	15	29.5	29.2	22.5	20.3	80.1	72.8	84.5	85
175	150	18	26.1	25.8	20.1	22.4	74.3	68.8	81.7	80.9
190	160	nil	47.4	60.8	39.8	48.4	84.8	83.9	87.5	93.4

scale and then used for predicting extractions). Extrapolation to higher levels of ammonium sulphate (> 20 g/l) show that actually obtained extractions for iron and zinc for half-an-hour match well with those obtained from equations. The values obtained for Fe(2h) and Zn (2h) did not match with the predicted values (-ve extractions for iron suggest that there should be no iron in solution even with < 40 g/l $(NH_4)_2SO_4$ and similarly all the zinc should have been extracted using 30 g/l $(NH_4)_2SO_4$). Predictions within the cubical regions match very well, except that Zn(0.5h) predict less extractions than actually obtained ones. Extrapolation to higher temperatures will lead to non-compliance of Fe($\frac{1}{2}$h) regression equation as it is independent of temperature between 170-190°C. In the absence of $(NH_4)_2SO_4$, actually obtained extractions are more than the predicted ones. Since there does not exist a region for low iron

(iv) Role of different cations at high temperature and low acid concentration.

(v)   Effect of jarosite seeding.

Effect of Oxygen

The leaching results obtained at 190°C using 40 g/l of $(NH_4)_2SO_4$ and 140 g/l of H_2SO_4 under oxygen partial pressure of 412 kPa are shown in Fig.3. Zinc extraction increased from 85 to 88% and

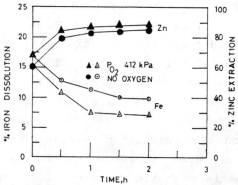

Fig.3 : Effect of pO$_2$ on Fe & Zn dissolution.Condn. 40g/l $(NH_4)_2SO_4$,140g/l H_2SO_4,190°C & pO$_2$ 412kPa.

iron dissolution decreased from 10 to 6.8%.

Combined effect of acid and oxygen

Acid concentration was decreased from 140 g to 120 g/l while keeping oxygen partial pressure at 412 kPa; $(NH_4)_2SO_4$ 40 g/l and leaching temperature at 190°C. With decrease in acid concentration, iron decreases from 6.8 to 3.2% with marginal decrease in zinc extraction from 89 to 86% (Fig.4).

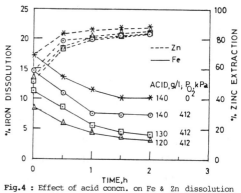

Fig.4 : Effect of acid concn. on Fe & Zn dissolution at 190°C, pO_2 412 kPa, 40g/l $(NH_4)_2SO_4$.

Effect of $(NH_4)_2SO_4$ concentration at 130 g/l H_2SO_4 and 190°C

The results shown in Fig.5 indicate that iron contamination would remain high with $(NH_4)_2SO_4$ concentration lower than 40 g/l. However, zinc extraction remains unaffected with the decrease of $(NH_4)_2SO_4$.

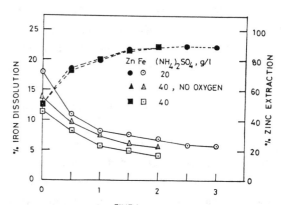

Fig.5 : Effect of $(NH_4)_2SO_4$ on Fe & Zn dissolution at 190°C using 130g/l acid and pO_2 412 kPa.

Effect of $(NH_4)_2SO_4$ concentration at 210°C and 120 g/l of H_2SO_4

Ammonium sulphate was varied between 0 to 30 g/l at 210°C using 412 kPa of pO_2. The results for iron and zinc extractions are shown in Fig.6. It is seen that about 19% iron remains in solution in absence of ammonium sulphate and it decreases to 8, 4.8 and 2.5% with increase in ammonium sulphate to 10, 20 and 30 g/l respectively. With increase in iron precipitation, zinc extraction also increases, perhaps due to release of acid as shown by eqn.(9), (10) or (11).

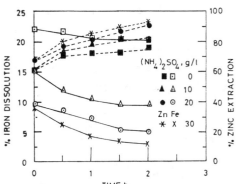

Fig.6 : Effect of $(NH_4)_2SO_4$ on Fe & Zn dissolution at 210°C using 120g/l acid pO_2 412 kPa.

Effect of acid concentration at 210°C and 30 g/l $(NH_4)_2SO_4$

By using 120 g/l sulphuric acid and 30 g/l $(NH_4)_2SO_4$, it was possible to decrease iron contamination to 2.8%, it

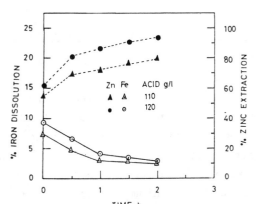

Fig.7 : Effect of acid concn.on Fe & Zn dissolution at 210°C using 30g/l $(NH_4)_2SO_4$ and pO_2 412 kPa.

was thought to further decrease the acid content to 110 g/1. The results obtained are shown in Fig.7. Iron dissolution maginally decrease from 2.8 to 2.5%, but zinc extraction decreases from 94 to 80% by decreasing initial acid concentration from 120 to 110 g/1. These results show that at 210°C minimum acid required with maximum zinc extraction is 120 g/1.

Effect of monovalent cation concentration

Effect of various monovalent cations for iron rejection is shown in Fig.8. By using 20 g/1 of sodium sulphate iron content could be brought down to 2.5% whereas to achieve almost similar iron levels 30 g/1 of ammonium or potassium sulphate were required. This confirms the earlier observation[38] that sodium sulphate is most effective to reject iron as jarosite.

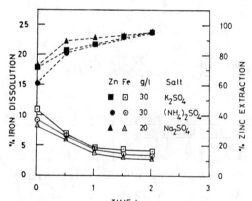

Fig.8 : Effect of monovalent cations on Zn & Fe dissolution at 210°C, 120g/1 H_2SO_4 & 412 kPa of pO_2.

Effect of seed during iron rejection

Usually jarosite precipitation is carried out in presence of seeds. In the present work, the leach residue obtained after acid pressure leaching (with simultaneous iron rejection as jarosite) of the ponded residue was used as the seed as it contains both plumbo and ammonium jarosites. Fig.9 shows initially more iron dissolves but with progress of leaching time, the rate of iron rejection becomes faster in

presence of seed. At the beginning a part of the jarosite seed might be dissolving which reprecipitates due to the

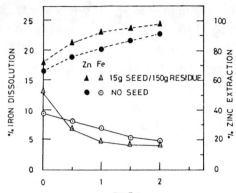

Fig.9:Effect of seed on Zn & Fe dissolution at 210°C 120g/1 H_2SO_4, 20g/1 $(NH_4)_2SO_4$, 10 wt% seed & pO_2 412 kPa.

depletion of acid with time. In addition to it no information is available on the use of jarosite seeds at high temperature so the quantification of seeding result is difficult at this stage. Zinc recovery improves from 90 to 97%. The residue used as seed also contains residual zinc which might have dissolved and resulted in apparent higher zinc extractions. These aspects need further investigations.

Behaviour of minor impurities during iron rejection as jarosite at high temperature

Solution composition with respect to Zn, Fe, Cu, Ni, Co and Cd are shown in Table V. During iron rejection as jarosite at high temperature (210°C) impurities like Cu, Ni, Co and Cd are not much affected. At leaching temperature of 210°C, using 120 g/1 acid, Cu, Ni and Co marginally increase.With the decrease in acid concentration to 110 g/1, copper decreased to 210 ppm.

In an earlier study[37], iron level of leach solution obtained under condition S.N(1) of Table V was reduced through a jarosite precipitation step at 95°C. The acid was neutralized using ZnO (in actual practice zinc calcine is used) to achieve a pH of 1.5. The

Table V Solution composition with some minor impurities under different leaching
conditions (15% p.d., 2h Time)

Sl. No.	Acid g/l	pO_2 kPa	Temp. °C	NH_4^+,K^+ or Na^+ g/l	Zn g/l	Fe g/l	Cu ppm	Ni ppm	Co ppm	Cd ppm
1	140	nil	170	nil	21.3	16.4	260	38	14.8	423
2	112	412	210	30(NH_4^+)	210	1.12	216	44	16.5	340
3	120	412	210	nil	20.0	8.8	-	-	-	-
4	120	412	210	30(NH_4^+)	24.6	1.21	280	42	20	370
5	120	412	210	20(Na^+)	24.5	1.27	310	40	18	380
6	120	412	210	30(K^+)	24.5	1.49	310	40	18	380

Ag^+ and Pb^{2+} were less than 2 ppm in solution

requirement of ZnO was 40 g/l. Further to provide nucleation for jarosite precipitation 50g of pressure leach residue/litre of solution was required with 25 g/l of ammonium sulphate. The contents were then heated at 95°C for 1 hour to bring down the iron level to about 1 g/l[37]. Other impurities like Cu, Ni, Co and Cd remained unaffected. The present work definitely provides advantage in terms of (i) acid requirement, (ii) amount of neutralizing agent and (iii) number of S/L separation stages. Further work need to be carried out for minor impurities removal to make it compatible with the neutral leach solution composition facilitating the treatment of these solutions in the existing zinc plant circuits.

IDENTIFICATION OF PHASES IN LEACH RESIDUE

X-ray diffraction pattern obtained for a typical leach residue (conditions S.N. (2) of Table V) is shown in Fig.10. The major iron phase is ammonium jarosite[40] with α-Fe_2O_3 as minor phase. Plumbo-jarosite lines are also very distinct in the pattern.

CONCLUSIONS

(i) 2^3 full factorial design for experiments was used to quantify the linear and interaction co-efficients for Fe(0.5h), Fe(2h), Zn(0.5h) and Zn(2h). The regression equations

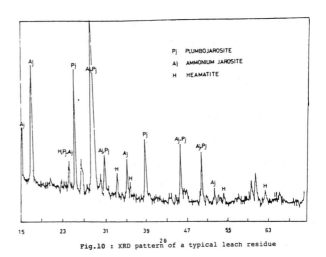

Fig.10 : XRD pattern of a typical leach residue

were found to be adequate to fit the experimental data by Fisher's test. However, the extrapolation to leaching conditions beyond the studied levels did not yield satisfactory results.

(ii) Iron dissolution decreased from 10 to 6.8% in the presence of 412 kPa of O_2 while keeping other leaching conditions similar.

(iii) By varying the amounts of ammonium sulphate and acid at 190 and 210°C in presence of O_2, conditions obtained for minimum iron dissolution and maximum zinc extraction were : 210°C, 30 g/l $(NH_4)_2SO_4$, 120 g/l acid with iron 1.2 g/l and 94% (24.6 g/l) zinc extraction.

(iv) Sodium sulphate was found to be most effective for jarosite precipitation as compared to potassium and ammonium sulphates.

(v) XRD pattern revealed the presence of ammonium and plumbo-jarosites with minor amounts of α -Fe_2O_3 in the leach residue with high temperature jarosite precipitation.

(vi) Minor impurities like Cu, Ni, Co and Cd were not affected during iron rejection as jarosite at 210°C.

(vii) High temperature jarosite precipitation would yield advantage with respect to acid requirement, neutralizing agent requirement and number of solid-liquid separation stages.

ACKNOWLEDGEMENTS

The authors are thankful to the Director, Regional Research Laboratory, Bhubaneswar, for his permission to present this paper. Authors also wish to thank Shri M.K. Ghosh for his help in XRD work.

References

1. Monhemius A.J., The electrolytic production of zinc, In Topics in Nonferrous Extractive Metallurgy, R.Burkin ed., Blackwell Scientific Publications, Oxford,1980,p.104-130.

2. Societe de la Vielle Montange, Belgian Patent 724 214, Application , November 20, 1968.

3. Van Den Neste E. Metallurgie Hoboken -Overpet's zinc electrowinning plant. Canadian Institute of Metallurgy,Bulletin vol.70, No.784,1977,p.173-185.

4. Boxall J.M. and James S.E., Experience with the goethite process at National zinc, In : Iron Control in Hydrometallurgy, J.E. Dutrizac and A.J.Monhemius eds., Ellis Horwood Ltd., Chichester, 1986, Part VI, p.676-688.

5. Bryson A.W. and te Riele W.A.M., Factors that affect the kinetics of nucleation and growth and the purity of goethite precipitates produced from sulfate solution, ibid., part IV, p.377-390.

6. Andre J.A. and Masson N.J.A., The goethite process in treating zinc leaching residues. Paper presented at American Institutionof Mining, Metallurgical and Petroleum Engineers 102nd Annual Meeting, Chicago, 1973.

7. Davery P.T. and Scott T.R., Removal of iron from leach liquors by goethite process, Hydrometallurgy, Vol.2, 1976, p.25-33.

8. Knobler R.R., Moore t.I., and Capps R.L., The new electrolytic zinc plant and residue treatment of National Zinc Company, paper presented at the American Institution of Mining, metallurgical and petroleum Engineers Annual Meeting, New Orlens, February, 1979.

9. Boxall J.M., Litz, L.M. and Wiese M.M., A new Oxygen reactor system for the goethite zinc leach residue process, Journal of Metals, vol.36, No.8, 1984, p.58-61.

10. Capps R.L. and Boxall M.J., Zinc leach residue treatment and solution purification at the National Zinc Company, Paper presented at the American Institution of Mining, Metallurgical and petroleum Engineers Annual Meeting, Los Angeles, February 1984, paper A 84-90.

11. Electrolytic Zinc Company of Australasia Ltd., Australian Patent 424095, application May 19, 1970.

12. Von Ropenak A. Hematite-the solution to disposal problem - an example from the zinc industry,Ref.4,Part VII, p.730-741.

13. Gordon A.R., Improved use of raw materials, human and energy resources in the extraction of zinc. In : Advances in Extractive Metallurgy, London : Institution of Mining and Metallurgy, 1977, p.153-160.

14. Tsunoda S., Maeshiro I., Emi F., and Sekine K., The construction and operation of the Lijima electrolytic zinc plant. Paper presented at the American Institution of Mining, Metallurgical and petroleum Engineers Annual Meeting, Chicago, 1973, The Minerals, Metals and

Materials Society, paper No.A 73-65.

15. Anon Environmental considerations and the modern electrolytic zinc refinery, Mining Engineering, November 1977, p.31-33.

16. Wuthrich and Von Ropenak, The electrolytic zinc plant of Ruhr - Zinc GmbH, in American Institution of Mining Metallurgical and Petroleum Engineers World Symposium on Mining and Metallurgy of Lead and zinc, TMS-AIME, PA, vol.2,1970, p.247-268.

17. Austriana de zinc S.A., Spanish Patent 304 601, Application October 12, 1964.

18. Det Norske Zinkkompani A/S, Norwegian Patent 108 047, Application April 30, 1965.

19. Dutrizac J.E. Jarosite type compounds and their application in the metallurgical industry, in : 3rd International Symposium on Hydrometallurgy, K. Osseo-Asare and J.D. Miller eds.,Atlanta Georgia, March 6-10, 1983, p.531-551.

20. Steintveit G., Electrolytic zinc plant and residue recovery Det Norske Zinkkompani A/S In : World Symposium on the Mining and Metallurgy of lead and zinc, New York : American Institution of Mining, Metallurgical and Petroleum Engineers, 1970, p.223-246.

21. Haigh C.J. and Pickering R.W., The treatment of zinc plant residues at the Risden works of the Electrolytic Zinc Company of Australia Ltd., ibid., p.423-447.

22. Steintveit G., Treatment of zinc leach plant residues by jarosite process In : Advances in Extractive Metallurgy and Refining, London : Institution of Mining and metallurgy, 1971.

23. Wood J. and Haigh C., Jarosite process boosts zinc recovery in electrolytic plants, World Minerals, vol.25, September 1972, p.34-38.

24. Gordon A.R. and Pickering R.W., Improved leaching technologies in the electrolytic zinc industry, Metallurgical Transactions, vol.6B, 1975, p.43-53.

25. Electrolytic Zinc Company of Australia Ltd., Australian Patent 401472, Application, March, 1965.

26. Matthew I.G., Haigh C.L. and Pammenter, R.V. Initial Pilot evaluation of Low-contaminant Jarosite Process in Hydrometallurgy - Research, Development and Plant Practice, K. Osseo-Asare and J.D. Miller, editors, TMS-AIME, New York, 1983, p.553-567.

27. Au Yeung S.C.F. and Boltman G.L., Iron control in process developed at Sherrit Gordon Mines, In : Ref. 4, Part I, p.131-151.

28. Pammenter R.V., Kershaw M.G., and Horsham T.R., The low contaminant jarosite process - further developments and the implementation of the process, in : Ref. 4, Part VI, p.603-617.

29. Scott J.D., Donyina D.K.A., and Mouland J.E., Iron - the good with the bad - Kidd creek zinc plant experience, In Ref. 4, Part VI, p.657-675.

30. Cenozaki A., Sato K., and Kuramochi S., Effect of some impurities on iron precipitation at the Iijima Zinc Refinery, In Ref. 4, Part VII, p.742-752.

31. Emmett R.C., Thickening and filtration techniques for dewatering iron-bearing minerals and precipitates. In : Ref. 4, Part VI, p.593-602.

32 Outokumpu Oy : British Patent 1464447, Application February 12,1973.

33. Huggare, T.L., Fugleberg, S. and Rastas J., How Outokumpu conversion process raises zinc industry, World Minerals. vol.27, February 1974, p.36-42.

34. Ismay A., Stanley R.W., Shink D., and Daoust D., The high temperature conversion process, in : Ref.4, Part VI, p.618-639.

35. Aranco H. and Doyle F.M., Hydrolysis and precipitation of iron during first stage pressure leaching of zinc sulphide concentrates, In : Ref. 4, Part III, p.409-430.

36. Acharya S., Anand S., and Das R.P., Iron rejection during acid pressure

88

leaching of zinc leach residue. Paper
presented at the <u>International conference on Chemical Metallurgy 91</u>, Bombay,
India. 9-11 January, 1991.

37. Acid pressure leaching of zinc
leach residue. Report submitted by
Regional Research Laboratory, Bhubaneswar to M/s. Hindustan Zinc Ltd.,Udaipur
India.

38. Tozawa K. and Sasaki K., Effect
of coexisting sulphates on precipitation
of ferric oxide from ferric sulphate
solutions at elevated temperatures,
In : Ref. 4, Part IV, p.454-476.

39. Experimental Optimisation in Chemistry and Chemical Engineering. S.
Akhnazarova and V. Kafarov. Translated
from Russian by V.M. Matskovsky and
A.P.Repyer, Mir. Publishers, Moscow,
1982, p.151-158, 300, 303.

40. Mineral Powder Diffraction, published by Joint Committee on Powder Diffraction Standards, International Centre
for Diffraction Data, Pennsylvania,
1980, File No.26-1014.

Selective recovery by precipitation of selenium and mercury from acid leaching solutions

D. D'Hondt
N. Gérain
A. Van Lierde
Unité des Procédés, Université Catholique de Louvain, Louvain-La-Neuve, Belgium

SYNOPSIS

This paper concerns the hydrometallurgical treatment of lead, selenium and mercury muds coming from the gas cleaning stage of zinc sulfide concentrates roasting. In a first step, the muds are leached at a pulp dilution of 250 g/l and a temperature of 90°C with a 150-200 g/l HCl and 100-150 g/l HNO_3 solution. Such conditions give outstanding results both in recovery (96.7 % for Se and 99.1 % for Hg) and selectivity towards lead : after cooling the solution to about 25°C, 97.5 % of this metal remains in a 40 % Pb residue.

Preliminary tests have shown that high purity chemical precipitates of selenium and mercury can be obtained by using a SO_2 solution. When sulfur dioxide is added to selenium in a ratio of 3.3/1 (E = 670 mV SHE), Se and Hg are coprecipitated as $Hg_3Se_2Cl_2$ with recoveries higher than 98 %.

A careful study of the crystals' stability shows that they can be decomposed to metallic selenium and soluble mercury chloride with a very high efficiency by keeping them under moderate oxidative conditions (900-910 mV SHE). 90 % of the selenium remains in a 88 % Se precipitate assaying less than 5 % Hg. The mercury can thereafter be recovered by precipitation as Hg_2Cl_2 using iron powder as the reducing agent : with a consumption of 850 g Fe/kg Hg, the mercury yield reaches 97 % and the precipitate assays 83 % Hg_2Cl_2.

INTRODUCTION

Most zinc sulphide ores contain traces of selenium and mercury. During roasting, the mercury compounds are decomposed and the mercury vapors go into the flue gas together with the SO_2 and the selenium dioxide produced by air oxidation of the selenides. Many processes have already been developed to recover both the mercury and selenium.[1-2] At Vieille Montagne's Balen zinc plant, they are extracted in the scrubbing towers of the gas cleaning circuit as muds with small amounts of lead, zinc and other minor elements.

These muds constitute a serious environmental problem[3] but are also an attractive source of both selenium and mercury. Many detailed studies have been published on the leaching of selenium or mercury from various origins, the purification of solutions by ion exchange or solvent extraction and the metal precipitation by electrolysis, SO_2 reduction or cementation.[4-18]

This paper deals with methods well suited to the treatment of such substances at a low tonnage but at an industrial scale :

- the direct atmospheric leaching of both selenium and mercury in high selectivity conditions ;

- the selective precipitation of selenium by SO_2 and mercury by cementation on iron.

MATERIALS

The mud sample was supplied by Vieille Montagne. Its chemical composition and the distribution of lead, mecury and selenium between the different particle size fractions are given in Table I.

TABLE I

Chemical analysis of muds and particle size fractions (wt.%)

Component	Feed	Size fractions (mesh Tyler)			
		+100	-100+200	-200+400	-400
Pb	21.75	9.50	14.57	19.00	24.65
Se	7.45	5.59	8.47	6.43	7.65
Hg	16.10	8.90	14.57	14.72	17.73
Zn	8.16	-	-	-	-
Fe	4.26	-	-	-	-
Cu	0.70	-	-	-	-
Cd	0.10	-	-	-	-
S	9.10	-	-	-	-
CaO	5.33	-	-	-	-
SiO_2	3.55	-	-	-	-
Al_2O_3	1.74	-	-	-	-
Weight (%)	100.00	12.10	6.30	6.10	75.50

The - 400 # fraction represents 75 % of the weight of the whole sample ; its Pb and Hg grades are increased by about 13 % in relation to the original feed but for selenium the enrichment is lower than 4 %.

METHODS AND PROCEDURE

No physical or chemical pretreatments of the muds were performed before leaching.

The leaching tests were always done with similar conditions of agitation in a 2 liter glass vessel. During hydrochloric leaching, this vessel was topped by a condenser so that HCl returns into the slurry. At the end of the experiments, the pulp was filtered on a Büchner and the cake washed with slightly acidified water with a liquid/cake ratio of one. The metal recoveries were calculated from balance sheets on the solution and dried residue.

The treatments of the leach liquor by SO_2 and the cementation were done in an one liter erlenmeyer equiped with a small magnetic agitation bar. SO_2 was always added as a saturated solution and the iron for cementation as a - 50 μm powder.

The reaction vessels had water jackets which were used for heating when necessary. The temperatures (regulated to within 1°C) were measured by a thermometer suspended in the solution and the redox potential by an Ag/AgCl electrode. All the potential values are converted to standard hydrogen tension (SHE).

RESULTS AND DISCUSSION

Leaching of selenium and mercury

The first leach experiments were carried out at 90°C for 3 hours with sulfuric, hydrochloric or nitric acids at a solution volume/solid ratio of 3. Table II gives the best performances obtained with each acid alone or with mixtures of H_2SO_4 and HNO_3. These leaching systems lead to especially bad results in spite of high acid consumptions.

TABLE II

Results of leaching with sulfuric, nitric or
hydrochloric acids (vol. solution/solid ratio : 3/1,
temperature : 90°C, time : 3 h)

Acid concentration (g/l)			Leach recovery (%)		
H_2SO_4	HNO_3	HCl	Pb	Se	Hg
200	-	-	0.5	0.6	0.1
-	200	-	1.4	44.9	0.6
-	-	300	5.1	1.9	10.2
200	200	-	0.1	41.5	3.0

On the other hand, excellent performances can be reached by using mixtures of various proportions of HCl and HNO_3 as shown in Table III.

TABLE III

Leaching results with mixtures of HCl and HNO_3
(vol. solution/solid ratio : 3/1,
temperature : 90°C, time : 3 h)

Acid concentration (g/l)		Leach recovery (%)		
HNO_3	HCl	Pb	Se	Hg
200	100	3.2	94.3	99.0
200	150	2.7	96.9	98.7
200	200	8.0	97.5	99.9
50	200	4.7	89.6	97.6
100	200	5.8	97.8	97.9
150	200	8.0	97.5	99.9

As soon as the acids' concentration reach 150-200 g/l for HCl and 100-150 g/l for HNO_3, selenium and mercury are dissolved with very high recoveries of around 98 % whereas lead leaching yields are lower than 9 %. Both zinc and iron are also leached with a high efficiency (95-98 %).

Bringing the HNO_3 concentration down to 50 g/l has no effect on the mercury dissolution but slightly decreases the selenium leach recovery from 98 to 89 %.

During the cooling of the slurry, at least 50 % of the soluble lead cristallizes as $PbCl_2$ [11] and goes into the residue whose weight corresponds to about 53 % of that of the original mud. For a lead recovery as high as 97.5 %, the Pb, Se and Hg grades correspond respectively to 40 %, 0.32 % and 0.3 %. The final liquor, resulting from the mixing of the leaching and washing solutions, has the following concentrations : 1.55 g/l Pb ; 20.6 g/l Se ; 45.1 g/l Hg ; 10.9 g/l Fe ; 22.15 g/l Zn and 2.5 to 3.1 g/l H^+ according to the amount of acid added during the leaching. About 50 % of the acids remain free at the end of the reaction and all attempt to really reduce the high acidity by a counter-current leaching unfortunately failed because of the precipitation of an important fraction of selenium.

Treatment of the solution

Divalent mercury ions stay soluble over a wide range of pH (0-8) whereas selenium already precipitates at pH around 5-6.[11-12] It should, therefore, be possible to separate these two elements, with a high efficiency, by adding lime or caustic soda until the pH rises to 6. Such an approach quickly appeared unsuitable, here, for two reasons :

* the lack of selectivity during the selenium precipitation : the precipitates assay only 16 to 23 % Se for 1.5 to 2.8 % Pb ; 0.5 to 1.5 % Hg ; 10 to 12 % Fe and 2 to 7 % Zn ;

* a too high consumption of neutralizing agents.

Therefore, selenium precipitation by a SO_2 reduction according to reaction (1) has been investigated.[8-14]

$$H_2SeO_3 + 2SO_2 + H_2O = Se° + 2SO_4^{--} + 4H^+ \quad (1)$$

Reduction of the metallic selenium by SO_2

Figure 1 shows the effect of SO_2 additions on the precipitation yield of selenium, mercury and lead for reactions performed at 50°C without any control of the redox potential. The SO_2 addition clearly results in a bulk precipitation of both the selenium and mercury whatever the amount of SO_2 added. However, the highest recoveries, 98 % for selenium and 99 % for mercury, are obtained when SO_2 is added in amounts corresponding to about 3.3 kg SO_2 for 1 kg selenium. The redox potential is then stabilized at about 680-700 mV SHE. Lead ions are also precipitated but with much lower yields. These bulk precipitates essentially contain Se, Hg, Cl and Pb as shown in Table IV.

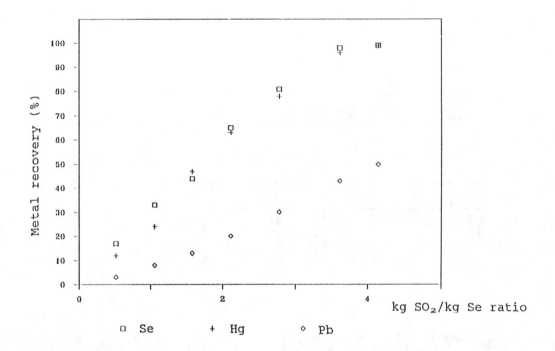

Fig. 1. Variation of the metal recovery as a function of kg SO_2/kg Se ratio.

TABLE IV

Chemical composition of bulk precipitates

kg SO_2/kg Se	Grades (%)				
	Pb	Se	Hg	Cl	Total
0.5	0.7	27.0	60.5	7.2	95.4
1.1	0.7	29.9	58.9	7.0	96.5
2.1	0.7	26.8	59.9	7.1	94.5
3.6	1.0	27.2	60.0	7.5	95.7

Mineralogical investigations (X-ray diffraction and microprobe) show that the selenium is present in a metallic form and as $Hg_3Se_2Cl_2$ crystals in which the oxidation states are respectively - 2 for selenium and + 2 for mercury. As the standard potentials (E°) of the electrochemical equilibria

$Se^{2-}/Se°$; Hg^{2+}/Hg^+ and Hg^{2+}/Hg correspond respectively to + 910, + 907 and + 852 mV, it would be possible to transform the mercury chloroselenides into insoluble metallic selenium and soluble Hg_2Cl_2, by rising the redox potential of the liquor to about + 900 mV. As shown in Figure 2, this could be easily done by holding in suspension the precipitates formed by the SO_2 treatment of the solution, for one hour under strong agitation. Indeed, as soon as the redox potential reaches 900-910 mV, all the mercury and lead remain in solution whatever the amount of SO_2 added during the reduction stage, while the selenium is precipitated with recoveries that increase with the SO_2 consumption. When the SO_2 addition corresponds to about 3.5 kg/kg Se, the precipitation of the selenium reaches about 93-94 % and the final precipitate contains 89-91 % Se ; 0.1 % Pb and 3 to 5 % Hg. The liquor, diluted owing to the addition of the SO_2 solution, presents the following composition : 0.93 g/l Pb ; 1 g/l Se ; 31 g/l Hg ; 7.63 g/l Fe ; 15.5 g/l Zn and 2 g/l H^+.

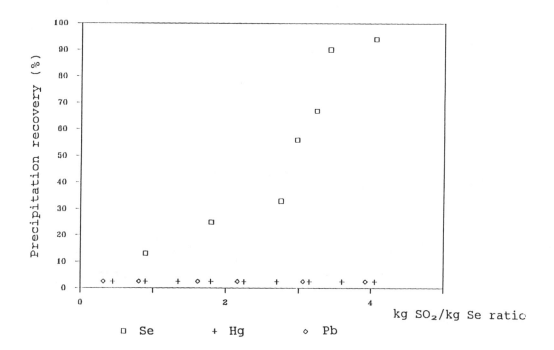

Fig. 2. Variation of the precipitation recoveries versus the ratio kg SO_2/kg Se when the precipitate is strongly agitated (with vortex) for one hour (final potential : 900-910 mV)

Cementation of mercury[11]

By looking at the position of mercury in the Nernst redox potential table, it seems clear that its cementation on various less noble metals should cause no problems. Iron powder has been choosen because it is cheaper and above all because the resulting reaction is much less exothermic.

All the experiments have been done, without any modification of the pH or the temperature, with the liquor obtained after the total precipitation of the selenium. The time of reaction was one hour after which the temperature had reached about 35°C. The results reported in Figure 3 clearly show that very high recoveries can be reached with moderate consumptions of iron powder, in spite of the high acidity of the liquor. As expected, almost all the selenium[15] and a fraction of the lead[16-17-18] are deposed together with the mercury.

94

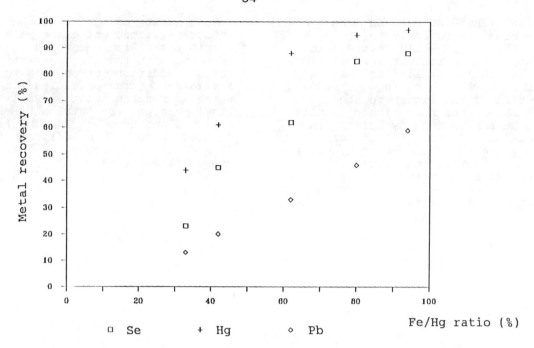

Fig. 3. Variation of mercury, selenium and lead recoveries versus the value of the ratio iron powder (g)/mercury (g) in solution

With 660 kg iron per ton of mercury, about 93 % of the mercury, 35 % of the lead and 70 % of the selenium are already coprecipitated by iron. By rising the iron powder consumption to 850 kg/t, the mercury recovery is increased to 97.5 % while the lead and selenium extractions yield respectively 50 and 87 %. As shown in Table V, the mercury deposits are mainly composed of Hg_2Cl_2 and to a lesser extent of selenium, lead and of course iron. For appropriate values of the iron excess, the calomel grade of the deposit easily reaches about 85 %.

TABLE V

Chemical composition of precipitates

Fe/Hg (wt.%)	Grades (%)					
	Hg	Se	Pb	Cl	Fe	Zn
33.5	69.0	2.1	0.7	11.8	0.8	0.3
46.3	73.7	2.4	1.3	12.5	1.3	0.1
62.0	75.4	2.65	0.8	13.4	1.8	0.1
80.0	72.0	3.13	1.5	12.6	4.0	0.15
92.7	67.4	2.93	2.2	11.0	9.7	0.2

CONCLUSIONS

The main conclusions are :

1) selenium and mercury can be leached, at 90°C, by using an acid solution of HCl (150 to 200 g/l) and HNO_3 (100 to 150 g/l). The recoveries are higher than 98 % and there is a high selectivity towards lead : after cooling the slurry, 97.5 % of the lead is recovered in a 40 % Pb residue ;

2) selective separation of selenium can be achieved by :

- treating the leach liquor with SO_2 until all the selenium and mercury are precipitated as $Hg_3Se_2Cl_2$ crystals ;

- keeping the crystals under moderate oxidative conditions (900-910 mV SHE) ; $Hg_3Se_2Cl_2$ is decomposed into selenium and soluble mercury chloride with a very high efficiency : 90 % of the selenium remains in a 88 % Se precipitate assaying less than 5 % Hg.

3) mercury can finally be recovered by precipitation as Hg_2Cl_2 owing to the addition of iron powder as a reducing agent. For an iron consumption of 850 kg/kg Hg, the mercury yield reaches around 97 % in a precipitate assaying 85 % Hg_2Cl_2.

Moreover, this treatment seems all the more interesting since it would allow to :

- recycle an important fraction of HCl and HNO_3 to the leaching stage after a treatment of the residual liquor with lime to precipitate most of the metallic cations (Fe, Zn ...) and by acidification with sulfuric acid ;

- suppress all stockage of polluting substances by recycling the hydroxide precipitates of low weight to the roasting stage.

References

1. A. Kuivala, J. Poijarvi. Sulfuric acid washing removes mercury from roaster gases. Engineering and Mining Journal, 10, 1978, p. 81-84.

2. G. Steinveit. The Boliden-Norzink mercury removal process for purification of roaster gases. Lead Zinc Tin'80, TMS AIME, 1980, p. 85-96.

3. T. Louderback. Selenium and the environment. Minerals Industries Bulletin, vol. 18, n° 3, 1975.

4. Amax Inc. Selenium rejection during acid leaching of matte. United States Patent, 3959097, 1976.

5. F. Habashi. Leaching of selenides and tellurides. Principles of extractive metallurgy, vol. 2, 1970, p 52-53 and p. 139-144.

6. Inco. Process for the recovery of selenium. European patent application, 80301993.4, 1980.

7. C. Nunez, F. Espiell, M. Cruells. Leaching of cinnabar with HCl-thiourea solutions as the basis of a process for mercury obtention. Metallurgical Transactions B, vol. 17B, 1986, p. 443-448.

8. K.N. Subramanian, N.C. Nissen, A. Illis, J.A. Thomas. Recovering selenium from copper anode slimes. Society of Mining Engineers, vol. 11, 1978, p. 1538-1542.

9. H.G. Vazarlis. Hydrochloric acid-hydrogen peroxide leaching and metal recovery from a Greek zinc-lead bulk sulphide concentrate. Hydrometallurgy, n° 19, 1987, p. 243-251.

10. G. Yildirim, F.Y. Bor. Hydrometallurgical treatment of a copper refining slime rich in both selenium and tellurium. Erzmetall, n° 4, 1985, p. 196-199.

11. A. Ballester, E. Otero, F. Gonzalez. Mercury extraction from cinnabar ores using hydrobromic acid. Hydrometallurgy, n° 21, 1988, p. 127-143.

12. R.G. Holdich, G.J. Lawson. The solubility of aqueous lead chloride solutions. Hydrometallurgy, n° 19, 1987, p. 199-208.

13. A.J. Monhemius. Precipitation diagrams for metal hydroxides, sulphides, arsenates and phosphates. Transactions of the Institution of Mining and Metallurgy, vol. 86, 1977, p. C 202-206.

14. J.E. Hoffmann. Selenium and tellurium : rare but ubiquitous. Journal of the Minerals, Metals and Materials Society, vol. 41, n° 7, 1989, p. 32-48.

15. M. Pourbaix. Atlas d'équilibres électrochimiques à 25°C. Gauthier Villars, 1963, p. 421-427 ; p. 554-559.

16. L.M. Kabanova, B.V. Teplyakov. Selenium and tellurium recovery from low grade solutions by cementation on copper. The Soviet Journal of Non Ferrous Metals, vol. 5, n° 8, 1964, p. 78-79.

17. W.N. Marchant, R.O. Dannenberg, R.T. Brooks. Selenium removal from acidic waste water using zinc reduction and lime neutralization. US Bureau of Mines 1978, Report of investigations 8312.

18. P.K. Sahoo, P.C. Rath. Recovery of lead from complex leach residue by cementation with iron. Hydrometallurgy, n° 20, 1988, p. 169-177.

Bioleaching of complex sulphides with different cultures of mesophilic microorganisms

E. Gómez
F. González
M. L. Blázquez
A. Ballester
Departamento de Ciencia de los Materiales e Ingeniería Metalúrgica, Facultad de Ciencias Químicas, Universidad Complutense, Madrid, Spain

SYNOPSIS

Complex sulphides are extensively studied to find the best process permitting total recovery of contained metals. The worldwide reserves are very important. Spain has 250 millions t. and this country and Portugal have approximately 70% of EEC reserves in copper, zinc and lead.

The work approachs Extractive Metallurgy of these raw materials using bioleaching with different microorganisms to put in solution contained metals. So a comparative study of leaching capacity of different bacteria cultures on different ores has been carried out.

The following bacteria cultures have been used:
1) Mixed natural culture isolated from sulphide mine waters.
2) Two pure cultures of Thiobacillus ferrooxidans.
   The following ores were attacked:
1) Three differential flotation concentrates of copper, zinc and lead, respectively.
2) Two bulk flotation concentrates. Mixed natural culture obtained the best leaching efficiencies in relation with solubilization of contained metals. These results are explained bearing in mind than natural culture produced a combined attack as a consequence of the presence of different kind of microorganisms in the same medium. In addition, mixed bacteria were naturaly better adapted to the medium than pure one.

INTRODUCTION

The last few decades have witnessed an increasing demand on the part of industry for metallic materials. At the same time the mining industry has had to face increasingly difficult problems associated with the rising costs of extraction, low or unstable prices of metals in international markets and a growing social awareness of the environmental pollution caused by treatment plants of sulphurized raw materials. This situation is compounded by the increasing lower metal content of ores and the rise in paying-off costs of industrial installations and in the price of energy.

The extractive metal industry, then, is taking steps to maintain profitability and at the same time comply with the stricter control of its atmospheric emissions.

Due to this situation the bioleaching of metallic sulphides can sometimes play an important role as an alternative to the classical extraction processes.

In the case of complex polymetallic sulphides (a submicroscopic and very complex association of chalcopyrite, sphalerite, galena and other less important components in a pyrite matrix), a study of new alternatives for their environmentally acceptable and financially viable extraction is of great interest due to the large world wide reserves. From the point of view of the Spanish economy, it is estimated that 70% of EEC reserves of copper, zinc and lead are to be found in the south west of the Iberian Peninsula in the form of complex sulphides (some 250 million tons).

From an extraction point of view, these minerals present three basic problems:
1) The presence of a great quantity of pyrites in the minerals; 2) a very complex and extremely fine association of the mineral phases and 3) the presence of minority compounds (such as silver and gold) which may have and important influence on the profitable treatment of these minerals.

To solve these problems there are three possible methods of concentration by flotation which might be used: 1) differential flotation to obtain concentrates rich in Cu, Pb and Zn; 2) semibulk flotation to obtain Cu-Pb or Zn-Pb concentrates and 3) bulk flotation to obtain Cu-Pb-Zn concentrates. Of these methods, semibulk flotation is only applicable in specific cases and for specific deposits, while differential and bulk flotation can be applied to every kind of deposit. However, although differential flotation is widespread in the industry, the commercialization of the concentrates is difficult. For this reason a possible alternative treatment of these complex sulphides would be by bulk flotation. However, there is the inconvenience that no technologically satisfactory procedures exist for this treatment, although its use would present great savings because it would eliminate the need for fine grinding which is necessary when differential concentrates are obtained.

To the above-mentioned problems, hydrometallurgy and, more specifically, bioleaching can offer possible solutions and interesting studies are currently being undertaken (2, 3, 4, 5).

However, despite the interest being shown in

bioleaching processes, there is still no comple
te and fundamental knowledge of the mechanisms
and in particular of the specific contribution
of the microorganisms involved.

Bearing this in mind, the present paper summa
rizes the work carried out with one of the bacteria existing in a natural culture from which
it was isolated in order to determine its leach
ing behaviour and compare it with several strains of collection (American Type Culture Collec
tion-ATCC), and with that of the natural mixture itself, from which the microorganism was iso
lated.

The said microorganisms, after isolation, was
characterized morphologically and physiologica-
lly by microscopic techniques and by growth in
selected media. In addition, a molecular genetic
study was carried out (6), although this aspect
is not included in the present paper.

This comparative study was carried out using,
in the first place, several sulphides concentra
tes obtained by differential flotation. This was
done in order to have available a reference when
evaluating the bioleaching efficiency of the
said microorganisms on a complex sulphide obtai
ned by bulk flotation.

The results obtained in this experimentation
with pure and natural cultures represent one
step more towards the full understanding of the
process, a necessary precondition for obtaining
maximum benefits.

MATERIALS AND METHODS

Primary mineral materials
The following mineral concentrates were used:
- Differential lead concentrate
- Differential zinc concentrate
- Differential copper concentrate
- Mixed concentrate (a mixture of differential
concentrates)
- Bulk concentrate (obtained by flotation)

These ores were provided by Andaluza de Piri
tas, S.A. (Aznalcollar, Spain). X ray diffraction was used to characterize these concentrates
and it was demonstrated that the following sulphides phases appeared in all of them: Galena,
sphalerite, pyrite, chalcopyrite and chalcocite,
although, logically, in different proportions.
A quantitative chemical analysis of the principal metallic elements was also carried out (Table I).

MINERAL CONCENTRATE	%Cu	%Zn	%Pb	%Fe
Diff. Copper Conc.	16.1	7.3	4.8	31.9
Diff. Zinc Conc.	0.5	40.2	0.9	15.0
Diff. Lead Conc.	1.4	6.3	45.9	13.7
Mixed Conc.	4.3	35.6	7.3	15.5
Bulk Conc.	2.2	35.0	8.3	19.0

Table I.- Quantitative chemical analysis of the
principal metallic elements present in the concentrates.

Some bioleaching experiments were carried out
using chemically pure ferrous sulphide (II), in
order to observe the behaviour of the microorganisms on an almost pure compound, thus eliminating any interactions which might be caused by
other metallic cations present in the concentrate.

Biological material
Microorganisms: The following bacterial cultures
were used in the bioleaching experiments:
- A natural mixed culture from the waters of the
Aznalcollar (Sevilla) mine, consisting of Thio-
bacillus s.p. ferrooxidans, Leptospirilum fe-
rrooxidans and Thiobacillus thiooxidans.
- Thiobacillus sp. isolated from the above mixed
culture.
- Thiobacillus ferrooxidans ATCC 23270

Culture Medium: The solution of salts reported
by Silverman and Lundgren, denominated 9K medium, modified without iron, was used (7).

The different concentrates described above
were the energy source for the microorganisms.
Pulp density was 3.3%. In previous experiments
related to the growth of isolated microorganisms
the culture medium was supplemented with a solu
tion of 4% $FeSO_4$ as nutrient.

Bioleaching experiments
Adaptation of cultures: In order to have available suitable inocula for the bioleaching experiments, each of the cultures used was previously adapted to the corresponding mineral substrate. 150 ml of nutrient medium and 5 g of
the mineral concentrate to be studied were intro
duced into a 250 ml capacity flask. After the
pH was adjusted to 2.5 and the inoculation was
then carried out with 5 ml of the corresponding
culture. Incubation was realized with orbital
stirring (150 min^{-1}) and at 37 ºC. Once a high
density microbial population was observed by
optic microscopy, the process was repeated by
inoculating a new culture from the previous one,
although always with sterile medium and ore.
When this process had been repeated three times,
the cultures were considered to be totally adapted to the environmental conditions of the experiment.

Experimental conditions: The experiments were
realized in similar conditions to those of the
culture adaptation, although in this case the
inoculum of each experiment was obtained from
the corresponding microorganism already adapted
to the mineral being studied. The experimental
conditions are summarised in Table II.

REACTOR	250 ml flask
Ore (g)	5
Medium (ml)	150 (9K medium)
Temperature (ºC)	37
Stirring (rpm)	150
Initial pH	2.5
Inoculum (ml)	5

Table II.- Experimental conditions in the bioleaching experiments.

Apart from the inoculated tests, other experiments were carried out in the same conditions
but without inoculation in order to obtain the
corresponding controls for the process.

The experiments were controlled in the manner
detailed below: periodically, stirring was stopped and the evaporated liquid replaced by steri
le distilled water. After the solid had sedimen

ted, a 1 ml sample of solution was taken. So, stirring was continued and the samples were then suitably diluted for analysis by atomic absorption spectrophotometry in order to determine quantitatively the metallic elements of interest.

When the experiment had finished, the pulp was filtered and the solid retained in the filter repeatedly washed in distilled water. This, when dry, was analyzed by X ray diffraction.

RESULTS AND DISCUSSION

Growth experiments of the isolated Thiobacillus sp.

Previous to the experiments detailed in this paper, one of the microorganisms present in the initial natural mixed culture was isolated. This, (called Thiobacillus sp.) was characterized morphologically and submitted to a series of physiological tests. This microbiological study was completed with other genotype studies which are not included in this paper.

Before the bioleaching experiments were carried out, growth test with Thiobacillus sp. isolate in pure state were realized at different temperatures and pH (its morphology is shown in Figure 1) in order to know the most suitable values of these variables for its optimum growth.

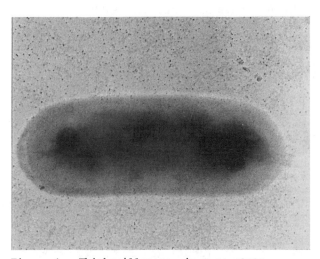

Figure 1.- Thiobacillus sp. in pure state.

As regards pH, 4 experiments were carried out in 100 ml reactors using each time 30 ml of 9K medium with ferrous ion and, as inoculum, 1 ml of a culture enriched in Thiobacillus sp. Incubation took place in a culture stove, without stirring, the temperature being maintained at 37 ºC. The initial pH values of the medium were 2, 2.5, 3 and 3.5. Bacterial growth was followed by optical microscopy. The best results were obtained with the culture at pH 2.5; this value also produced an earlier growth. (See Table III).

		DAYS					
	2	4	6	8	10	12	15
pH = 2	-	-	-	-	+/-	+	++
pH = 2.5	-	-	+/-	+/-	+	++	++
pH = 3	-	-	-	-	-	+/-	+
pH = 3.5	-	-	-	-	-	-	+/-

Table III.- Growth of the culture enriched with Thiobacillus sp. at different pH values.
+ :growth; - :no growth; +/- :few cells.

As regards temperature, two experiments were carried out in the same experimental conditions mentioned above. Initial pH in both cases was 2.5 and incubation carried out in a culture stove without stirring at 30 ºC and 37 ºC, respectively. Growth was also followed by optic microscopy and only weak growth observed in the initial 4-6 day period in both cases.

However, from the 8th day onwards, population density was much higher in the experiment at 37 ºC. (Table IV).

		DAYS						
		2	4	6	8	10	12	15
TEMPERATURE	30 ºC	-	-	+/-	+/-	+/-	+	+
TEMPERATURE	37 ºC	-	-	+/-	+	+	++	++

Table IV.- Growth of the culture enriched in Thiobacillus sp. at different temperatures.
+ :growth; - :no growth; +/- :few cells.

When these experiments were previously carried out, the corresponding culture adaptations were realized at different pH values and temperatures to prevent the conditions in which the original culture was being incubated from influencing the results.

As a result of these previous experiments, the optimum conditions were fixed at pH 2.5 and 37ºC.

Bioleaching experiments

These experiments served to ascertain the leaching capacity of the isolated microorganism, Thiobacillus sp., and to compare this capacity with that of another bacterium of the same specie from a collection of cultures and, therefore, not adapted to the bioleaching process. They also served to compare the behaviour of the isolated microorganism with that of the natural mixed culture from which it comes. This information might help to clarify the contribution of Thiobacillus sp. to the bioleaching process, when there is no in-

teraction with other species of bacteria normally found in the mixed culture.

Influence of stirring:
Two series of experiments were realized, using
isolated Thiobacillus sp.; one with and the other
without stirring. Four different substrates were
used: differential concentrates of Cu, Zn and Pb
and a chemically pure ferrous sulphide.

In all cases the concentration of copper, zinc
and iron in solution was followed except when pu_
re FeS used in which case iron alone was measured.

Results are shown in Figure 2.

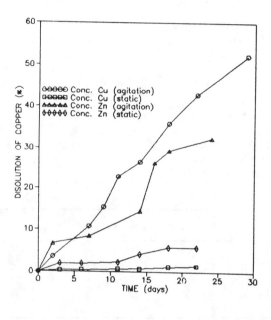

(a)

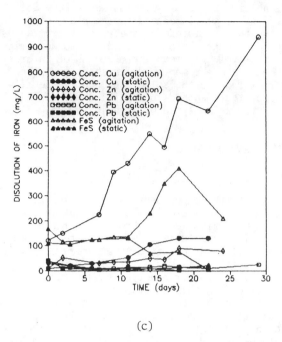

(c)

Figure 2.- Influence of stirring on bioleaching
of (a) cooper; (b) zinc; and (c) iron, from
different raw material inoculating with a Thiobacillus sp. culture.

These results show the clear influence of
stirring in the metal dissolution, especially
in the case of copper from the differential concentrates of copper and zinc. The influence is
also pronounced in the dissolution of zinc, although less markedly due to the preferential pro_
cess of chemical dissolution in addition to the
process of biological dissolution that this metal undergoes. As regards the dissolution of
iron, its concentration in solution is more erra_
tic, due to the fluctuations in pH, which can
cause the precipitation of some insoluble products in the form of both ferric compounds and
jarosites. This phenomenon explains the irregular tracing of the graphic representation.

A poor metallic dissolution was also observed
from the differential concentrate of lead, which
in the case of copper was so small that no ana-
lyzable concentrations were obtained. For this
reason it is not represented on the corresponding graph. This can be explained by the toxicity of lead towards the bacteria which can be pro_
ved by microscopic observation of the liquids,
because much lower bacterial population densities in tests on lead concentrate than in tests
on copper and zinc concentrates and on pure FeS
were observed. Furthermore, galena, the principal phase with lead in the concentrate, when
dissolved in a medium with sulphate ion precipitates $PbSO_4$, which is deposited on the surface
of the ore, hindering any subsequent bacteria
attack on the rest of solid thus preventing disso_
lution.

Influence of the type of microorganism:
To study the influence of the type of microorganism on the bioleaching process, a series of experiments was carried out with different microor_
ganism acting on differential concentrates of
copper, zinc, lead and on FeS. The bacteria cultures used were: isolated Thiobacillus sp. cultu-

re; T.ferrooxidans ATCC 2370 culture and the natural mixed culture from which the Thiobacillus sp. had been isolated.

In each experiment, an inoculum of 5 ml of the particular culture previously adapted to the corresponding mineral substrate was used and the bacterial population periodically observed by optical microscopy.

To evaluate the leaching activity of the bacteria, sterile tests were realized on each of ᷉ the mineral substrates used in the same experimental conditions as for the inoculated tests.

The solution samples taken in these experiments, were analyzed for copper, zinc, and iron. The results are shown in Figures 3 to 12.

In the first three figures (3, 4 and 5) the dissolution of copper from the three differential concentrates is shown. These results clearly illustrate the different leaching properties of each culture. In all these cases, it is the mixed culture which produces the greatest extraction of copper, reaching 90% in the case of the differential concentrate of copper.

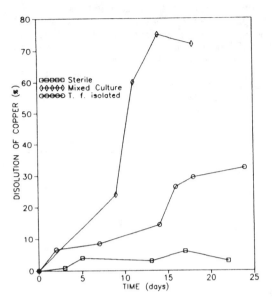

Figure 4.- Influence of the type of microorganisms on bioleaching of copper from a zinc differential concentrate.

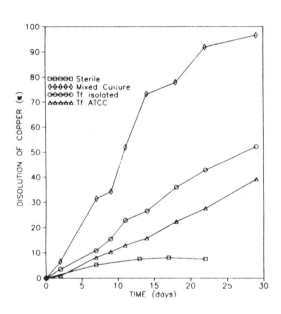

Figure 3.- Influence of the type of microorganisms on bioleaching of copper from a copper differential concentrate.

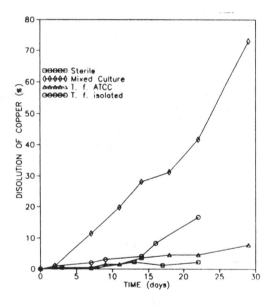

Figure 5.- Influence of the type of microorganisms on bioleaching of copper from a lead differential concentrate.

There is a manifest difference between the results obtained using the mixed culture and those with the two cultures of pure species. Isolated Thiobacillus sp. does not even reach 50% of the extraction rate of the mixed culture and the results for T.ferrooxidans ATCC 23270 are even lower. In the case of the differential concentrate of zinc, the results obtained with this latter microorganism are not represented as the values are of little significance.

The results for the lead concentrate show a similar metal extraction rate to that of the sterile test, except in the case of the mixed cultu

re. The commom factor of the lead concentrate ex
periments, is the decline in bacterial population density with respect to the same culture of
microorganism with other different substrates.

The dissolution of zinc is represented in figures 6,7 and 8. Once again it can be seen that
the extraction rate of this metal is highest
with the mixed culture, except in the case of
the copper concentrate, where the values for the
three cultures are similar.

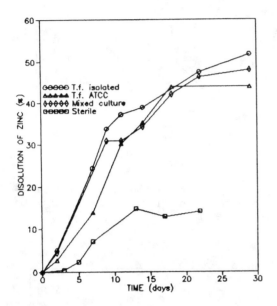

Figure 6.- Influence of the type of microorganisms on bioleaching of zinc from a copper diffe
rential concentrate.

With the other two concentrates, although the
extraction rate of zinc is higher for the mixed
culture, the differences in relation to the cultures of pure species are not so pronounced as
those recorded for copper.

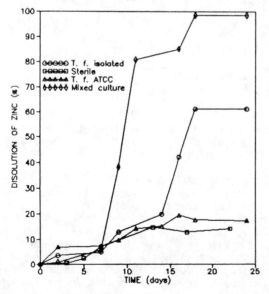

Figure 7.- Influence of the type of microorganisms on bioleaching of zinc from a zinc differential concentrate.

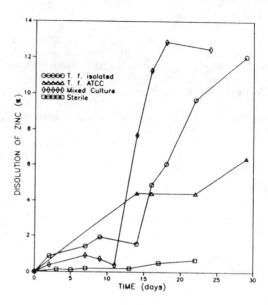

Figure 8.- Influence of the type of microorganisms on bioleaching of zinc from a lead differential concentrate.

The evolution of the iron concentration is re
presented in figures 9, 10, 11 and 12. Again,
there is, generally speaking, a greater extraction of iron by the microorganisms of the mixed
culture. However, there are no great differences
in the attack on the copper concentrate Figure
9), contrary to that which occurs on the concentrates of zinc and lead (Figures 10 and 11). In
these two cases a great difference can be observed between the extraction rates with the mixed
culture and those with pure species, which do
not even reach 100 mg/l. The extraction of iron
from pure FeS is shown in Figure 12. The irregu
larity of the curve is again due to the precipitation of jarosites produced by the fluctuation
in pH throughout the experiment.

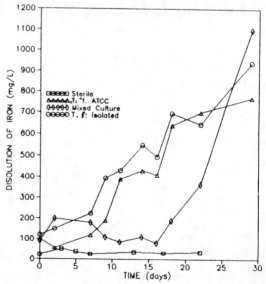

Figure 9.- Influence of the type of microorganisms on bioleaching of iron from a copper differential concentrate.

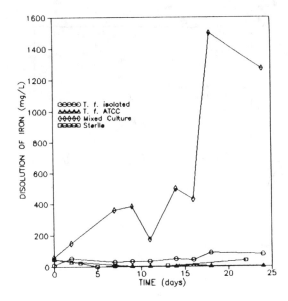

Figure 10.- Influence of the type of microorganisms on bioleaching of iron from a zinc differential concentrate.

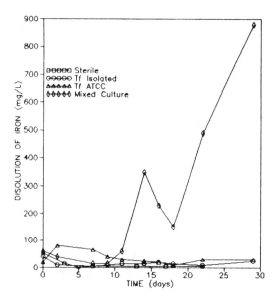

Figure 11.- Influence of the type of microorganisms on bioleaching of iron from a lead differential concentrate.

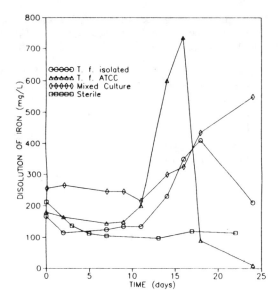

Figure 12.- Influence of the type of microorganisms on bioleaching of iron from pure FeS.

The conclusion, then, from these experiments is that the natural mixed culture has a greater bioleaching capacity, since it is made up of microorganisms perfectly adapted to the precess.

Furthermore, the several species of bacteria present in the culture, cooperate in the achievement of this highest bioleaching capacity.

It is also clear that, in spite of the process of adaptation to which they were submitedd before each experiment, the bioleaching capacity of the T.ferrooxidans ATCC 23270 bacteria is less than that of the isolated strain from mixed culture. This might be due to the loss of part of its capacity to obtain energy from mineral substrates after hundreds of generations growing in 9K medium with ferrous ion. This microorganism can be considered the type strain of the Thiobacillus ferrooxidans species, having been maintained in collections for many years.

The results also show the differences between the different cultures used as regards their resistance to high concentrations of heavy metals. In the lead concentrate, for example, the percentage of this metal is 45.92%, a very high figure, which is responsible for a steep decline in the population densities of the cultures, although this is less pronounced in the case of the mixed culture than in the other cultures of pure species. This, together with the formation of a layer of lead sulphate which is deposited on the ore, prevents the extraction rates reaching the levels of those obtained with the other concentrate.

Bioleaching of the mixed concentrate:
The above experiments were carried out using different differential concentrates of complex sulphides. However, interest in the bioleaching process lies also in the treatment of bulk concentrates. For this reason, as an approximation of a bulk concentrate, one was prepared by mixing the differential concentrates. This synthetic ore was denominated mixed concentrate.

In a series of experiments this mixed concentrate was attacked with the three cultures in question: natural mixed culture; isolated Thio-

bacillus sp. and T.ferrooxidans ATCC 23270.
Results of the periodical analysis of samples
are shown in figures 13, 14 and 15.

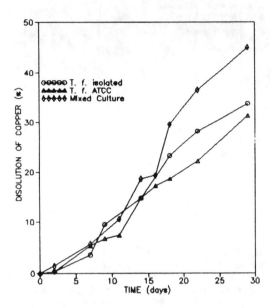

Figure 13.- Influence of the type of microorganisms on the bioleaching of copper from a mixed
concentrate.

Microscopic observation of the bacterial population during the experiments showed, unusually, a scant growth with the mixed culture. Despite this fact, an anlysis of copper in the solu
tion (Figure 13) showed greater indices of disso
lution in the case of the natural mixed culture
than when cultures of pure species were used, although extraction rates of this metal with all
three cultures were not very high. The similarity of the results obtained with both species of
Thiobacillus should be noted. All this may be due
to the fact that in the mixed concentrate, the
relative quantities of each metal are more similar than those of the differential concentrates,
which means that it is less toxic for the less
adapted microorganisms, such as T.ferrooxidans
23270. In fact, microscopic observation during
this experiment showed a greater rise in bacterial population density of this culture compared with
the above mentioned experiments with differential
concentrates.

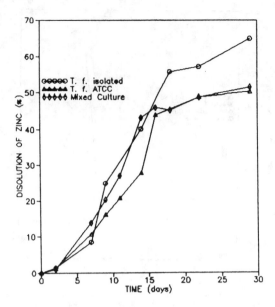

Figure 14.- Influence of the type of microorganisms on the bioleaching of zinc from a mixed
concentrate.

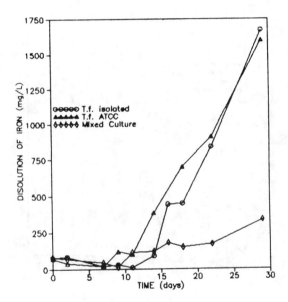

Figure 15.- Influence of the type of microorganism on the bioleaching of iron from a mixed con
centrate.

Figure 15 shows that with the mixed culture
there was practically no significant dissolution
of iron from the concentrate, while the other ex
periments with Thiobacillus produced a high quan
tity of iron in solution. This discrepancy can
not be explained by the metal being precipitated
in the form of jarosite since, when the diffraction X ray tests of the solid residues of experi
ments with mixed culture and with isolated Thiobacillus sp. are compared, it can be clearly observed ,in the former case, that iron is not pre
cipitated in this form. It is probable, therefore, that the low yield obtained in this experiment with the mixed culture in only due to the
poor development of the microorganisms.
This fact is further supported by the informa

tion presented in Figure 14, in relation to the dissolution of zinc. A poorer yield is obtained with the mixed culture than with isolated Thiobacillus sp. and is slightly higher than with T.ferrooxidans 23270.

Bioleaching of the bulk concentrate:
After studying the bioleaching process with the artificially prepared mixed concentrate of complex sulphides, another series of experiments was carried out using a true bulk concentrate of complex sulphides, in order to observe the behaviour of the microorganisms on this sort of extremely complex raw material, thus contributing to attempts at obtaining the best possible yield.

The experiments were carried out in the same experimental conditions as previously mentioned and inoculations were made with each one of the microorganisms. Microscopic observation also revealed high bacterial populations. Figures 16, 17 and 18 shows the extraction rates of copper, zinc and iron.

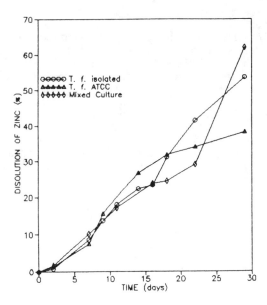

Figure 17.- Influence of the type of microorganisms on the bioleaching of zinc from a bulk concentrate.

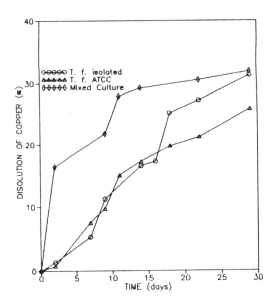

Figure 16.- Influence of the type of microorganisms on the bioleaching of copper from a bulk concentrate.

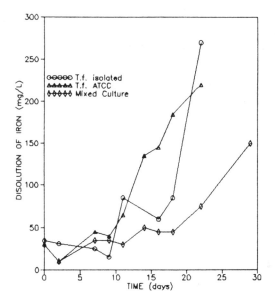

Figure 18.- Influence of the type of microorganisms on the bioleaching of iron from a bulk concentrate.

The evolution of the copper concentration in the solution is depicted in Figure 16 in which it can be seen that the mixed culture is much more active in the first part of the experiment, although the extraction rates with this culture and isolated Thiobacillus sp. tend to even out after day 18, reaching similar values at the end of the experiment (32.0% and 31.4% respectively). These results are similar to those obtained with the mixed concentrate, but considerably lower than those with the differential concentrates of copper and zinc.

For the dissolution of zinc (Figure 17), the three cultures show similar behaviour in the first phase and begin to differ after day 20, with the mixed culture producing the highest dissolution of zinc (62.1%).

Figure 18 shows the extraction of iron. The dissolution of this metal is poor for the three cultures even after 10 days. However, a second phase then begins in which the cultures of pure species dissolve the iron rapidly while the mixed culture does so more slowly. This is similar to what happens with the mixed concentrate, and is clearly shown by a X ray diffraction study of the solid residues of the experiments inoculated with the mixed culture and with isolated Thiobacillus sp. (Figures 19 and 20). In neither of these X ray recordings, were peaks corresponding to jarosites observed. This indicates that the greatest part of the pyrites continues to form part of the ore, which is logical considering its cathodic behaviour in relation with the other sul phide phases of the bulk concentrate (8). The va riations in the peaks corresponding to PbS should also be noted, their diminution being much great- er in the case of the residue of the concentrate attacked by the mixed culture. Furthermore in this same residue it can be observed that the peaks corresponding to lead sulphate (PbSO$_4$) are of greater intensity than those of the residue of the isolated Thiobacillus sp. culture. The sa me occurs with ZnS, and the X ray recording of the bulk concentrate residue attacked by the na- tural mixed culture presents a greater diminution in intensity relative to the principal peaks of the sphalerite (ZnS) than the residue of the bulk concentrate attacked by the pure species. This indicates a greater leaching capacity on the part of the mixed culture in relation to these compo- unds.

Influence of the type of concentrate:
In order to analyze the influence of the nature of the mineral substrate in relation to its attakc with the different cultures a comparative study was carried out of the preceding experi- ments. The conclusions are the following:
- The copper concentrate is the substrate from which the three culture types extract the best yields of copper, a good perfomance in zinc and iron dissolution also being obtained.

- Lead concentrate, as mentioned already, is the substrate worst assimilated by the different bacteria, very poor levels of extraction being obtained for the three metals studied (copper, zinc and iron).
- With the differential zinc concentrate high levels of metal dissolution are obtained, which are, however, inferior to those of the copper concentrate. The difficulty which T.ferrooxidans ATCC has in growing in this concentrate has to be noted.
- As regards the mixed and bulk concentrates, the latter has the worst results for the extrac tion of the different metals, apart from the al- ready menthioned case of the differential lead concentrate. With the mixed concentrate, the extraction values are similar to those obtained with the differential zinc concentrate and,using optical microscopy, the population densities of the different bacteria cultures are observed to be very high, except in the already mentioned case of the mixed culture.
- As regards pure FeS, given its nature, the pre sence of iron has only been analyzed in solution an extremely variable behaviour of this metal being found due to the changes in pH, which does not give us much information. However, of note is the good growth of the bacteria cultures ob- tained when this compound is used as a source of metabolic energy.

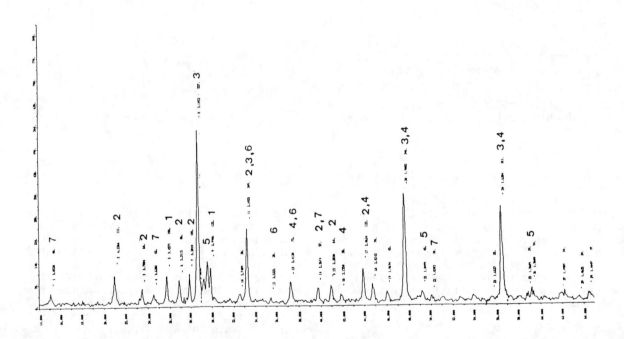

Figure 19.- X ray diffraction recording of the solid residues after attacking bulk concentrate in the presence of a natural mixed culture.
1: PbS; 2: PbSO$_4$; 3: ZnS; 4: FeS$_2$; 5: CuFeS$_2$; 6: CuS$_2$; 7: Jarosite

107

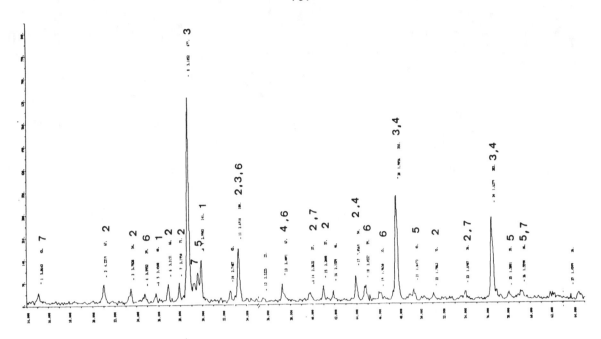

Figure 20.- X Ray diffraction recording of the solid residues after attacking bulk concentrate in the presence of a Thiobacillus sp. isolated. 1: PbS; 2: PbSO$_4$; 3: ZnS; 4: FeS$_2$; 5: CuFeS$_2$; 6: CuS$_2$; 7: Ja rosite.

References

1. Carranza F. Las pintas españolas: Un recurso natural poco explotado. Químicos del sur. Vol 4, 1985. p 13-16
2. Torna A.E.: Leaching of metals, Biotechnology Vol. 6th. Special Microbial Processes (Rehm H.J., Reed G. eds). Chapt. 12, 1988, p. 367-399.
3. Chaundhury G.R., Das R.P.: Bacterial Leaching complex sulphides of copper, lead and zinc. International J. mineral Processing. Vol. 21, 1987 p. 57-64.
4. Suzuki, T., Lizama H.H., and Key, J.: Bacterial leaching of a complex sulphide ore. 25th OIM Conference of Metallurgirst. Toronto-Ontario 1986 Canadá (Reprient obtained from the authors).
5. Krehs-Yurll B.A., Tsai C.B., Wu R. Hilligan, P.A. and Troncoso N.J.: Metal value recovery from metal sulphide containing ores. United States Patent No. 4,740,243. 1988.

6. Goméź E.: Biolixiviación de sulfuros complejos. Trabajo de Licenciatura. Universidad Complu tense de Madrid. 1990.
7. Silverman, M.P. and Lundgren, D.D.: Studies on the chemo-autotrophic iron bacterium ferrobacillus. I. An improved medium and a harvesting procedure for securing high cell yields. J. Bacterial Vol. 77, 1959, p. 642-647.
8. Jyothi, N., Sudnak, N., and Natarajan K.A.: Electrochemical aspects of selective bioleaching of sphalerite and chalcopyrite from mixed sulphides. International Journal of Mineral Processing Vol. 27, 1989, p. 189-203.

Recovery of copper by dump leaching with use of bacteria and cementation at the Vlaikov Vrah mine, Bulgaria

S. N. Groudev
Research and Training Centre of Mineral Biotechnology, University of Mining and Geology, Sofia, Bulgaria
V. I. Groudeva
Department of Microbiology, Faculty of Biology, University of Sofia, Sofia, Bulgaria

SYNOPSIS

The ore piles in the proximity of the Vlaikov Vrah mine, Bulgaria, contain about thirty million tons of waste low-grade copper sulphide and mixed ores. In 1972 industry began using these ores for recovering copper by means of microbial leaching. The present paper discusses the results obtained during the 18-year period of the industrial copper dump leaching operation. The flowsheet is presented and each separate component of the complex production process is described.

It was found that the copper leaching rate directly depended on the amount and activity of the chemolithotrophic bacteria which occured in the ore dumps and in the circulating solutions. On their part, this amount and activity depended on a complex of environmental factors. The attempts to enhance leaching by improving some of these environmental factors as well as by introducing laboratory-bred bacterial mutants into the dumps are described.

INTRODUCTION

At present, the microbial leaching of copper is applied on an industrial scale mainly as dump leaching of mining wastes. That practice is growing in importance and operations of this type are located in more than twenty countries in the five continents. Accurate estimates are not available to evaluate the relative contribution of the microbial leaching to the world total copper production. Torma (I) estimated that the copper produced by dump leaching amounts to about 25 % of the total copper production in the U.S.A. with largest operations being located in the western states.

Many studies have been carried out on both the theoretical and practical problems of dump leaching. Most of the achievements in this field are summarized in recent publications (2,3). Little is known, however, on the mixed microbial communities associated with the leach dumps as well as on the possibility to improve the leaching process by monitoring the microbial activity.

The industrial copper dump leaching operation at Vlaikov Vrah, Bulgaria, is one of the largest operations of this type in Europe. A lot of information on this operation has been published, especially on its microflora, which was a subject of intensive studies (4-6).

The present paper discuss the most essential data obtained during the twenty-year period of the industrial copper dump leaching operation. Conclusions about the possibilities to improve the biological leaching in such type of operations are presented.

DUMPS AT VLAIKOV VRAH COPPER MINE

The ore piles in the proximity of the Vlaikov Vrah mine contain about thirty million tons of waste low-grade copper sulphide and mixed ores from open cut mining.

There are several ore dumps at Vlaikov Vrah but the leaching operation is connected with the largest among these dumps, the so called dump No 3. This dump has been formed on moderately steep hill without any ground preparation. The formation of the dump started in 1960 and was finished in 1979. The material was hauled from the open-pit mine to the dump by trucks and was dumped from the top edge of the dump. As a result of this, the dump is built as a series of thin, sloping layers. It must be noted that the waste ores have not been piled in a strict chronological succession. Furthermore, these ores have been dumped with relatively little control on the size and shape of the dump as well as without any control over particle size in the dump.

The dump has the shape of a truncated cone and consists of three high lifts. Its surface area is about 350 000 m^2 and its maximum thickness is about 60 m. The material placed on dump was run-of-mine material and includes boulders as large as I - 2 m in diameter and weighing many tons. However, most of the material is less than 500 mm in diameter and includes many fine particle sizes. There was a tendency for boulders to roll down the slope. For that reason, the foot of the dump is composed mainly of boulders. However, all ore sizes are distributed through the dump.

The initial copper content of the dump was about 0.05 - 0.IO % but some parts of the dump had a higher copper content - up to O.I5 %. The main copper-bearing minerals were chalcopyrite, covellite and chalcocite. Pyrite was well represented, and quartz and feldspars were the basic minerals of the host rock.

The observations conducted in the dump area during I968 - I97I showed that after rainfall, acid drainage waters with a high content of copper and iron ions (over 0.5 and 2 g/l, respectively) and of bacteria related to the species Thiobacillus ferrooxidans and T. thiooxidans flow out of the dump. That provoked laboratory and field studies of the possibilities to leach copper from the dump by means of bacteria. Data about these studies have been published elsewhere (5,7).

LEACHING OPERATION

The leaching operation started in I972. In operation the solution containing chemolithotrophic bacteria, dissolved oxygen, sulphuric acid and iron ions is pumped to the top of the dump. The solution percolates through the dump and dissolves copper. The amount of sulphuric acid formed as a result of the oxidation of sulphide minerals is sufficient to maintain the pH in the ranges which are favourable for the growth and oxidative activity of the acidophilic chemolithotrophic bacteria. Dump effluents are sent to the precipitation plant where copper is removed by cementation with iron. The depleted solution is recycled to the dump. Data about those solutions are shown in Table I.

Depleted solutions together with make up water are introduced onto the dump by means of spraying and flooding. The spraying is carried out by means of metal sprinkler heads, while the flooding is carried out by ponds formed on the surface of the dump by bulldozer. Selected areas of the dump surface are saturated for a given period, followed by a rest period with no solution addition. In practice, a dump section is irrigated with leach solution until such time as the copper content of the pregnant solution obtained from that section falls below a predetermined level. Then the solution application on that section is stopped and is commenced on another. The duration of the irrigation periods initially was in the range of I - 3 months. The duration steadily decreased in the course of time reaching IO - 20 days for most dump sections at the end of leaching. The optimum irrigation rate for dump sections consisting of mixed ores was in the range of 20 - 40 l/ton.day, while for dump sections consisting of sulphide ores was in the range of IO - 20 l/ton. day. The ratio between the duration of the rest periods to that of the irrigation periods steadily increased in the course of time and was in the range of 2 : I to 6 : I. The increased ratio was due to the necessity to improve the aeration inside the dump. This necessity is connected with the progressive exhaustion of the oxide minerals, but mainly with the increasing encasement of the ore. The advantage of the long rest periods was that solution was only recirculated for a small percentage of the time and higher copper concentrations in solution were obtained. In most dump sections rest periods as long as 2 - 3 months were not detrimental to the bacterial activity since a film of solution remained around the ore particles. Unfortunately, shortage of water often necessitated rest periods of many months as well as irrigation rates much lower than the optimum ones. The total surface irrigated at a given moment ranged from 3 - 5 000 m^2 in I972 to about 60 000 m^2 in I98I - I986. The maximum relative portion of the ore being leached at a given moment was approximately I5 % from the total amount of the ore in the completely built dump.

It was found that spraying gave better leaching presumably because it provided more uniform distribution of the solutions as well as more dissolved oxygen. However, the evaporation rates were greater than those observed at flooding and together with the seepage losses amounted to IO - 20 %. Spraying also showed some advantage in rates of

Table I Data about leach solutions at the Vlaikov Vrah
         copper dump leaching operation

Year	pH	Cu, g/l	Fe^{2+}, g/l	Fe^{3+}, g/l
		Solution to dump		
1968 - 1971		rainfall		
1972 - 1975	3.0-4.3	0.05-0.24	2.4-3.9	0.05-0.2
1976 - 1985	3.1-4.5	0.04-0.12	2.6-4.2	0.07-0.3
1986 - 1990	3.4-4.6	0.01-0.04	2.3-4.1	0.10-0.3
		Solution from dump		
1968 - 1971	2.1-4.4	0.3-0.9	0.03-0.2	1.6-2.3
1972 - 1975	1.9-3.0	1.0-3.2	0.01-0.2	1.7-3.0
1976 - 1985	2.0-3.4	0.5-1.5	0.01-0.3	1.8-3.2
1986 - 1990	2.6-3.5	0.2-0.4	0.01-0.3	1.4-2.8

percolation. The time taken for effluent to appear after application of solution to the top of the dump depended mainly on the height of the dump, the horizontal distance, the character of the material being leached and of the solution being applied. The percolation rates differed in the different dump sections but usually were in the range of 0.5 - 3 m/h. The dump permeability steadily decreased in the course of time. Thus, in 1972 effluents appeared from a 23 m high dump section II hours after the solution application, while in 1985 they appeared from the same dump section 18 hours after the solution application. The decreased permeability was due to the precipitation of iron salts on the surface of the dump and within the dump as well as to the formation of clay layers within the dump. These iron precipitates shielded the mineral surface from bacterial attack, while the clay layers prevented efficient leaching of the underlying minerals.

Most of the iron in solution applied to the dump was in the ferrous state because most of the solution is derived from precipitation plant tailing. These tailing solutions are usually recirculated to the dump without purification. However, some oxidation of iron occurs during storage of solutions and thus precipitates some of the contained iron. Sealing of dump surface by precipitated iron salts is overcome by ripping the top 0.5 - 1 m of the dump. Furthermore, these 0.5 - 1 m top layers periodically were removed from some dump sections. Iron salt precipitation within the dump was impossible to overcome. On passing through the dump, almost all the ferrous iron in solution turns into ferric iron as a result of oxidation by bacteria. Some of the ferric iron

hydrolyses and there is a net loss of iron in solution in passing through the dump. It must be noted that some of the iron in the pregnant solution is derived from the dissolution of the iron-bearing minerals within the dump. In principle, it is possible to decrease the iron precipitation by an increase in irrigation acidity. It was found that the release of iron from some dump sections was improved by such treatment. This was, however, not always connected with an increase in copper release. Application of acid even helped to improve permeability of some dump sections but it seemed that this was caused by altering the character of slime and not by redissolving iron salts.

The rates and the final degrees of copper extraction were different in the different dump sections. In dump sections consisting of mixed ores the initial annual copper extraction was of the order of 8 - 12 % of the total amount of copper in the relevant sections. The rates usually declined sharply after 40 - 45 % extraction had been obtained. The final copper extractions from such dump sections were in the range of 50 - 60 %. Usually 10 - 12 years of leaching were needed to achieve such extraction levels. The leaching can continue further but at extremely low rates which are economically not attractive.

In dump sections consisting of chalcopyritic ore the annual extraction of copper never exceeded 3 %. The cumulative extractions after 10 - 12 years of leaching were always less than 20 %. It was evident that leaching in such dump sections was not profitable. However, in some cases, it was possible to combine the leaching of dump sections consisting of mixed ores with that of chalcopyritic dump sections. This was carried out by passing effluents from the mixed ore sections through the chalcopyritic sections.

The progressive decrease in the leaching rates resulted in more prolonged rest periods in order to achieve a predetermined copper grade. The copper grade depended also on the temperature of leach solutions as the temperature in winter can be considerably lower than in summer. Heat is generated by the oxidation reactions in the dump and the temperature in some pyrite-rich-dump sections exceeded sometimes 80°C. However, the temperature of the pregnant solutions from the dump is relatively constant (I2 to I6°C) and decreases below that level only after torrential rain. For that reason, the leaching was efficient even during the coldest winter months when the air temperature on the dump site sometimes was as low as -5°C.

A test dump section containing IOO OOO tons of mixed ore with an initial copper content of 0.I5 % was leached over the course of IO years. Then this dump section was taken down with a bulldozer. Samples of the leached material were obtained for every 3 m of depth. There was no evidence of solution penetration into the larger rocks in the dump section. It was found that leaching of copper had been taken place to about I2 m below the surface, to the clay layer formed within that dump section. Secondary enrichment had taken place on the underlying ore.

The removal of the IO - I5 m top layers, consisting of leached ore, as well as of the almost impervious horizontal clay layers from different dump sections improved or even initiated the leaching of the underlying ores.

Attempts to increase the copper content of the pregnant solutions by recycling or by irrigating the dump with solutions, where most of the iron was in the ferric state, were unsuccessful in most cases.

COPPER RECOVERY FROM PREGNANT SOLUTIONS

Copper was recovered from pregnant solutions by means of precipitation with metallic iron. Scrap iron from different sources is used for that purpose. The precipitation is carried out in concrete cementation launders with pneumatic agitation. Individual cementation cells may be isolated from the flow so that the precipitated

copper may be washed via the flushing holes to the drainage basin.

Characterization of the cementation process is shown in Table II.

The retention time of treated solutions changed during the years. The retention time for obtaining a desired level of copper recovery (usually in the range of 90 - 95 %) depended mainly on the concentrations of copper and iron in the pregnant solutions as well as on the pH and temperature. The amount of iron consumed per kg of copper precipitated was always higher than the theoretical consumption and ranged from 2.0 kg for solutions containing more than I g/l of copper to 3.5 kg for solutions containing below 0.5 g/l of copper. Consequently, the copper grade of the cement copper decreased from about 82 % to 59 - 66 %. It was found that the addition of chloride ions in 50 mg/l concentration to the copper-bearing solutions improved the cementation process.

The cement copper was allowed to settle in basins. It was removed mechanically from those basins and dried before shipment to the smelter. Drying is carried out on concrete drying pads constructed adjacent to the settling basins and by heating with infrared lamps or hot plates. The final product contains from I5 to 20 % moisture.

MICROFLORA OF THE DUMPS AND RECIRCULATING SOLUTIONS

Regardless of the selective environmental conditions, the microflora of the industrial operation is characterized by considerable variety of its species composition (Table III). Acidophilic chemolithotrophic bacteria were the prevalent microorganisms in the dump and recirculating solutions. T. ferrooxidans was the most widely distributed and the most numerous microbial species. It is known that from a taxonomic point of view the species T. ferrooxidans can be regarded as a rather heterogeneous population characterized by a marked physiological elasticity (3). T. ferrooxidans strains isolated from the industrial operation differed markedly from each other with respect to their ability to oxidize sulphide minerals, ferrous iron and elemental sulphur as well as to their growth rate and yield and in the way they reacted to some environmental factors. Some of the strains were very efficient oxidizers of the above-mentioned inorganic substrates. A more detailed information about the T. ferrooxidans strains

Table II   Data about the cementation process at the Vlaikov Vrah
           copper dump leaching operation

Variable	1972 - 1973	1974 - 1980	1981 - 1985	1986 - 1990
Recovery of copper from pregnant solutions, %	85-95	91-97	88-95	86-93
Retention time, min	240-360	45-60	60-80	70-90
Iron consumed per kg of copper precipitated, kg	2.0-3.0	2.0-3.2	2.8-3.5	3.0-3.5
Copper content of the cement copper, %	71-82	68-82	60-70	59-66
Copper content of the tailing solutions, mg/l	50-240	40-160	10-50	10-40

In 1972 - 1973 the cementation plant worked without pneumatic
agitation.

in Vlaikov Vrah has been published earlier (4).

T. thiooxidans and Leptospirillum ferrooxidans were invariably found but almost always in lower numbers than T. ferrooxidans. The growth of T. thiooxidans in such systems probably depends on the elemental sulphur produced by chemical and electrochemical processes. Mineral leaching by T. ferrooxidans could be enhanced by T. thiooxidans. The facultatively chemolithotrophic bacterium T. acidophilus was also typically present, usually in lower numbers than T. thiooxidans. Mixed cultures of T. thiooxidans (or of T. acidophilus) with L. ferrooxidans leach efficiently some sulphide minerals, especially pyrite.

Moderately thermophilic chemolithotrophic bacteria related to Sulfobacillus thermosulfidooxidans and the genus Thiobacillus were found in some rich-in-pyrite dump regions, in which the temperature exceeded 40°C. In these regions the moderate thermophilic chemolithotrophs were the prevalent microorganisms. These bacteria sometimes occured in mixed populations with the mesophilic chemolithotrophs in temperature zones at 15 - 40°C, i.e. far below their temperature optimum. In these mixed populations the number of the moderate thermophiles was usually lower than that of the mesophilic chemolithotrophs, especially T. ferrooxidans. The isolated moderate thermophiles were able to oxidize ferrous iron, elemental sulphur and various sulphide minerals at rates similar to those achieved by the most active strains of T. ferrooxidans isolated from this industrial operation.

Extremely thermophilic chemolithotrophic bacteria were isolated from three pyritic ore samples in 1978. The samples were taken from zones in which the temperature had exceeded 60°C. The isolates were related to the genus Sulfolobus (Acidianus). They were able to oxidize ferrous iron, elemental sulphur and sulphide minerals also at rates similar to those achieved by the most active local strains of T. ferrooxidans at lower temperatures. The numerous attempts to isolate later such extreme thermophiles from different dump regions were unsuccessful. It must be noted, however, that there is little evidence to suggest that extreme thermophiles would be typical inhabitants in dump leaching operations.

Some Thiobacillus species, capable of oxidizing S°, thiosulphate, and polythionates at neutral pH were sporadically isolated from samples of leach liquors and ores. These bacteria were more numerous in the initial phases of dump leaching when the prevailing pH of bulk solution was close to neutral. Their low numbers and inability to oxidize sulphide minerals suggest that they do not play an important role in the industrial operation.

Many heterotrophic microorganisms as well as some phototrophic sulphur bacteria and algae were found but their role is more or less incidental. Some heterotrophs, mainly of the genus Acidiphilium, are close associates of T. ferrooxidans and utilize organic substances formed by the chemolithotrophic bacteria. These heterotrophs are typical in microbial consortia found in dump leaching operations.

Table III   Microflora of the copper dump leaching operation at
Vlaikov Vrah mine

Microorganisms	Solution to dump	Dump ore	Solution from dump
	Cells/ml	Cells/g	Cells/ml
T. ferrooxidans	10^2-10^7	$10-10^8$	10^2-10^8
T. thiooxidans	$1-10^4$	$1-10^6$	$10-10^5$
T. acidophilus	$0-10^2$	$0-10^4$	$1-10^3$
L. ferrooxidans	$1-10^4$	$1-10^6$	$1-10^6$
Bacteria capable of oxidizing S° at neutral pH (T. thioparus, T. neapolitanus, T. denitrificans)	$0-10^2$	$0-10^3$	$0-10^3$
Moderately thermophilic chemolithotrophic bacteria (Thiobacillus spp., S. thermosulfidooxidans)	$0-10$	$0-10^5$	$0-10^3$
Extremely thermophilic chemolithotrophic bacteria (Sulfolobus, Acidianus)	$0-1$	$0-10^3$	$0-10^2$
Saprophytic bacteria (Acidiphilium, Pseudomonas, Bacillus, Aerobacter, Caulobacter, etc)	$10-10^3$	$1-10^4$	$1-10^3$
Fungi and yeasts (Cladosporium, Penicillium, Trichosporon, Rhodotorula, etc)	$0-10^2$	$0-10^3$	$1-10^3$
Algae	$0-10$	$0-10$	$0-10^2$
Protozoa	$0-10$	$0-10^2$	$0-10^2$

Their number was always lower than that of T. thiooxidans. Other heterotrophs were found sporadically and always in lower numbers than the chemolithotrophic bacteria.

Sulphate-reducing bacteria were found in some deep layers of the leach dump, characterized by low redox potential.

The basic "reservoir" of T. ferrooxidans in the industrial operation is the ore in the dump. The bacterial distribution is mostly confined to the upper layers (the top 0.5 - 1 m) of the dump with densities as high as in excess of 10^8 bacteria/g of ore. The number of T. ferrooxidans decreases with increasing depth and in deeper layers (8 - 10 m) is negligible. In some cases these bacteria were found in zones as deep as 30 m, mainly in association with voids in which were formed as a result of the channeling of the leach solutions. The presence of impervious clay layers, formed

within the dump, considerably influenced the distribution of bacteria. The compacted layers tend to retain bacterial cells as leach solutions infiltrate through these zones. This effect causes stratification and zonation of bacteria, with a total oxidizing activity which may be much higher than that achieved under simulated laboratory conditions (8).

There was no clear relationship between the numbers of bacteria in dump effluents and those within the dump. Usually, T. ferrooxidans counts in the effluents were lower than those in the upper layers of the dump. This is presumed to have resulted from the infiltration effect during the percolation of the leach solutions through the ore material. Dumps, with chemolithotrophic bacteria attached to ore particles, may be regarded as continuous-flow systems which, once in a steady state, display little change in population levels over long periods. An equilibrium exists between the cells attached to the ore particles and those in solution. Although this equilibrium is subject to transitory changes upon rainfall, bacterial counts in dump effluents reflected the progress of the leaching process. High numbers of T. ferrooxidans in dump effluents usually denoted active bacterial leaching. On the other hand, the number of these bacteria in dump effluents sometimes displayed fluctuations which have no predictable pattern and the cause of which was unknown.

Changes in pH and flow rate affected the number of T. ferrooxidans in dump effluents. After heavy rainfall, or after a sharp increase in the flow rate, bacterial numbers in effluents decreased and the total amount of bacteria washed out from the dump increased. Usually, bacterial numbers were elevated after rest periods in dump irrigation. The addition of ammonium and phosphate ions as nutrients as well as of surface active agents to the leach solution usually had no effect on the number of T. ferrooxidans and on the leaching rates.

The number of iron-oxidizing bacteria sharply decreased upon cementation of copper with scrap iron, due to coprecipitation of bacteria. The barren solution after cementation passed to the holding ponds. The residence time in these ponds was too short for efficient iron oxidation. As a result, and combined with subsequent dilution of leach liquor with make up water, bacterial numbers in barren, leach solution, were lower compared with pregnant effluents.

The microflora of the dump and solutions passing through the ore material changed in the course of time. The number of acidophilic chemolithotrophic bacteria increased together with the progressive increase in leaching. Simultaneously, the number of heterotrophs and of bacteria capable of oxidizing sulphur at neutral pH steadily decreased. After reaching a maximum, both the number of acidophilic chemolithotrophs and the rate of leaching were maintained relatively stable for long periods of time. Then the environmental conditions started to become less favourable for the acidophilic chemolithotrophs, mainly due to the exhaustion of the exposed sulphide minerals and to the encasement processes. As a result of this, the number of these bacteria started to decrease. On the other hand, the number of heterotrophs and of alkalophilic chemolithotrophs started to increase.

A laboratory-bred mutant strain of

T. ferrooxidans possessing a high leaching activity versus the mixed and sulphide ores from Vlaikov Vrah was introduced into a IOO OOO ton section of the dump in I974 - I975 (4). The introduction itself was successful but the leaching rate did not increase. This was due to the fact that the main rate-limiting factors inside the dump were related to mass transfer of oxygen and fluxes of reactants and products within the ore material being leached and not to the genetically determined ability of bacteria to oxidize sulphide minerals.

Oxygen availability can be considerably increased by proper design and construction of dumps with optimum shape and size. Furthermore, the method of leach solution application to the surface of the dump, the amount of fines present in the ore, and the degree of compaction greatly influence the amount of oxygen that can penetrate into the dump. For these reasons, the future progress in this area is likely to involve improvements in dump design to permit improved control of biological leaching rather than the utilization of genetically improved bacteria (3).

More important than the determination of the cell number was the determination of the level of bacterial activities in situ in ores and natural solutions and at natural temperatures. This was due to the finding that in some cases there was no relationship between the number and activity of bacteria. The measurements of in-situ rates of bacterial iron oxidation and carbon dioxide fixation were the most suitable ways to evaluate bacterial activity in acid leaching environments (Table IV). It is known that in such systems the bacterial activity depends not only upon one definite factor but upon a complex of factors (9). Factors such as temperature, pH, concentrations of copper ions and of some essential nutrients, etc markedly affected the bacterial activity. The magnitudes of some of these factors in situ differed from the optimum physiological magnitudes. Such findings are of a great importance since the bacterial activity can be enhanced by an artificial improvements of some of the rate-limiting environmental factors, especially pH and flow rate of the leach solution. Nowadays, this is the only applicable way to increase dump bioleaching under industrial conditions.

Table IV Iron oxidation and $^{14}CO_2$ fixation in situ by means of
chemolithotrophic bacteria in the copper dump leaching
operation at Vlaikov Vrah mine

Sample tested	Iron oxidation, Fe^{2+} oxidized, g/l	$^{14}CO_2$ fixation, Radioactivity of bacteria, counts/min.ml solution (or g of ore)
Dump effluents + Fe^{2+} (9 g/l) at 14°C	0.82 - 3.12	2 800 - 12 500
Barren solutions from the cementation plant at 14°C	0.51 - 1.40	1 500 - 5 300
Leach solutions to dump at 6 - 8°C	0.21 - 0.50	300 - 800
Ore suspensions in 9K nutrient medium (with 9 g/l Fe^{2+}) at 14°C	0.77 - 4.40	3 200 - 16 100
Ore suspensions in 9K nutrient medium (with 15 g/l Fe^{2+}) at 30°C	2.08 - 11.15	8 200 - 39 200

The iron oxidation was tested in static Erlenmeyer flasks.
The $^{14}CO_2$ fixation was measured by the technique described
by Karavaiko and Moshniakova (10) by using $Na_2^{14}CO_3$ added
to the microbial culture being tested in 16 ml glass bottles
capped with rubber serum stoppers. The incubation time was
5 days. Details are presented elsewhere (6).

References

1. Torma A.E. Impact of biotechnology
on metal extractions. Mineral Processing and Extractive Metallurgy Review,
vol. 2, 1987, p. 289-330.
2. Dump and underground biological
leaching of ores. G.I. Karavaiko, G.
Rossi, A.D. Agate, S.N. Groudev and Z.
A. Avakyan, eds. Moscow: Vneshtorgizdat, 1990.
3. Mixed cultures in biological leaching processes and mineral biotechnology. O.H. Tuovinen, B.C. Kelley and
S.N. Groudev. In: Mixed cultures in
biotechnology. New York: McGraw-Hill
Co., 1991, in press.
4. Observations on the microflora in
an industrial copper dump leaching
operation. S.N. Groudev, F.N. Genchev
and S.S. Gaidarjiev. In: Metallurgical
applications of bacterial leaching and
related microbiological phenomena.
New York: Academic Press, 1978, p.
253-274.
5. Large-scale application of biological copper dump leaching at Vlaikov
Vrah, Bulgaria. S.N. Groudev, F.N.

Genchev, S.S. Gaidarjiev and V.I.
Groudeva. In: Proceedings of the
XIVth International Mineral Processing
Congress, Toronto, 1982, p. IX.8.1 -
IX.8.13.
6. Microflora of two industrial
copper dump leaching operations. V.I.
Groudeva, S.N. Groudev and D.V. Vassilev. In: Dump and underground
biological leaching of ores. Moscow:
Vneshtorgizdat, 1990.
7. Industrial copper dump leaching
operation at Vlaikov Vrah mine,
Bulgaria. D.V. Vassilev and S.N.
Groudev. In: Dump and underground
biological leaching of ores. Moscow:
Vneshtorgizdat, 1990.
8. Beck J.V. The role of bacteria
in copper mining operations. Biotechnology and Bioengineering. vol. 9,
1967, p. 487-497.
9. Karavaiko G.I., Abakumov V.V.,
Krasheninnikova S.A., Mikhailova T.L.,
Piskunov W.P. and Khalezov B.D. Ecology and activity of microorganisms
upon metal dump leaching. Prikladnaja

<u>Biohimiya i Mikrobiologiya</u>. vol. I7, I98I, p. 73-80 (in Russian).

I0.   Karavaiko G.I. and Moshniakova S. A.   A study on chemosynthesis and rate of bacterial and chemical oxidative processes under conditions of copper - nickel ore deposits of Kolsky penin- sula.   <u>Mikrobiologiya</u>. vol. 40, I97I, p. 55I-557 (in Russian).

Ferrite behaviour in the processing of complex copper–zinc sulphide concentrates

B. S. Boyanov
R. I. Dimitrov
Department of Chemistry, University of Plovdiv, Plovdiv, Bulgaria

SYNOPSIS

With the aim of developing of technology of processing of Bulgarian copper-zinc sulphide concentrates a number of laboratory investigations have been carried out:

1. Kinetic studies on the process of ferrite formation involving zinc, copper and cadmium oxides;
2. Study of the solubility of the ferrites obtained in sulphuric acid solutions;
3. Sulphatization of the ferrites by means of $(NH_4)_2SO_4$ and $FeSO_4$.

The results obtained are interpreted in terms of hydrometallurgical processing of zinc and copper-zinc calcines and semi-finished products.

INTRODUCTION

The processes of ferrite formation in the systems $CuO-Fe_2O_3$, $ZnO-Fe_2O_3$ and $CdO-Fe_2O_3$ decrease the effectiveness of leaching of useful components from zinc, copper and copper-zinc concentrates[1,2]. The study of the kinetics and mechanism of solid state reactions occurring between the oxides of copper, zinc, cadmium and iron is of importance for the search for new technological solutions for preventing the occurrence of undesirable processes during the processing of various raw materials in non-ferrous metallurgy, as well as for the development of new materials having specific magnetic, electric, mechanical and optical properties[3,4].

The two indicated practical areas of significance of ferrite formation substantially differing from one another, define one of the purposes of this study: to synthesize individual and mixed copper, zinc and cadmium ferrites and to examine their solubility in dilute sulphuric acid solutions under conditions corresponding to the industrial process of zinc and copper-zinc calcine leaching.

The sulphatization process is an effective method for processing of zinc and lead cakes, pyritic calcines and powders from non-ferrous metallurgy. Different sulphates, H_2SO_4, spent electrolytes, gases containing SO_2, SO_3 and O_2, etc., are used as sulphating agents[5,6]. The study of the solid-phase interaction between the sulphating agents and the main components of the processed raw material is of importance for the efficient usage of the agents and for achieving a high degree of sulphatization. For this purpose the reactions of H_2SO_4, $(NH_4)_2SO_4$ and $FeSO_4$ with the main and admixture metals of the treated by-products, copper, zinc and cadmium ferrites and their mixed ferrites have been studied.

RESULTS AND DISCUSSION

The ferrites under study have been synthesized by standard ceramic technology, the conditions of thermal treatment being chosen on the basis of data in the literature and our previous investigations[7]. The X-ray phase analysis performed shows that single-phase samples of copper, zinc and cadmium ferrites have been obtained. For example, the crystal lattice parameter (a) of ferrites $Zn_xCd_{1-x}Fe_2O_4$ changes linearly depending on the composition (Fig.1), which is evidence that mixed zinc and cadmium ferrites are being formed.

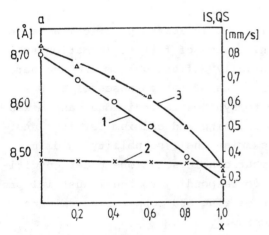

Fig.1. Changes in the crystal lattice parameter (1), isomer shift (2) and quadrupole splitting (3) in relation to ferrite composition of $Zn_xCd_{1-x}Fe_2O_4$.

The Mössbauer spectra of the ferrites were processed by means of a personal computer programme and the values of the following hyperfine interactions were determined: isomer shift (IS), quadrupole splitting (QS) and intrinsic magnetic field (H_{eff}). They are presented in Table I.

At room temperature, the Mössbauer spectrum of the cubic $CuFe_2O_4$ is a sextet, and those of $ZnFe_2O_4$ and $CdFe_2O_4$ are doublets. All ferrites were obtained by quenching into water down from 1000 °C.

Table I. Hyperfine interactions in ferrites

Ferrite	IS, mm/s ($\pm 0,01$)	QS, mm/s ($\pm 0,01$)	H_{eff}, kOe ($\pm 1,0$)
$ZnFe_2O_4$	0,34	0,32	0
$CuFe_2O_4$ (cubic)	0,35	0,11	471
$CdFe_2O_4$	0,36	0,79	0

The isomer shift data in table I are presented according to metal iron.

The kinetic data on interactions in the systems $ZnO-Fe_2O_3$[7], $CuO-Fe_2O_3$[8] and $CdO-Fe_2O_3$ were processed by means of kinetic equations based on certain models of interactions in the solid state[10]. In this series of experiments, the isothermal curves were processed by using the kinetic equations of: Jander (F_1); Ginstling-Brounshtein (F_2); Dunwald-Wagner (F_3); Tamman (F_4); Kröger-Ziegler (F_5); Ginstling-Brounshtein-Halbert (F_6); Kolmogorov-Eroffeev (F_7). The processing procedure has been described in detail[7]. The calculated kinetic parameters of ferrite formation in the three studied systems are summarized in Table II.

Notable is the great difference in the activation energy values, E_a, for $CdFe_2O_4$ as compared to those for $ZnFe_2O_4$ and $CuFe_2O_4$. Comparison of kinetic data of the three systems shows that more complete interaction between CdO and Fe_2O_3 occurs at lower temperatures, which leads to lower E_a values.

The zinc content in zinc sulphide concentrates is 50-60 %, and that of Cd 0,2-0,3 %. The substitution of zinc in the crystal lattice of ß-ZnS by cadmium is predominantly isomorphous. Sometimes cadmium forms separate minerals as well. The following heterogeneous and solid

Table II. Values of k_o and E_a obtained by different equations[7-9]

Interaction	Temperature range, ^{o}C	Equation	k_o	E_a, kJ/mol
ZnO + Fe$_2$O$_3$ ⟶ ZnFe$_2$O$_4$	600-750	F_1	$1,47 \times 10^3$	284
	600-750	F_2	$6,45 \times 10^2$	278
	800-1000	F_6	$3,16 \times 10^3$	112
CuO + Fe$_2$O$_3$ ⟶ CuFe$_2$O$_4$	850-1000	F_4	$6,40 \times 10^3$	117
CdO + Fe$_2$O$_3$ ⟶ CdFe$_2$O$_4$	750-950	F_4	$1,41 \times 10^5$	55
	750-950	F_6	$9,48 \times 10^3$	68

state reactions can possibly take place during the oxidation of ß-ZnS with admixtures of Cd, Cu, Fe, etc.:

$MeS + 3/2 \, O_2 \longrightarrow MeO + SO_2$, Me = Zn, Cd, Cu, Co, Ni, etc.

$2FeS + 7/2 \, O_2 \longrightarrow Fe_2O_3 + 2SO_2$

$MeO + Fe_2O_3 \longrightarrow MeFe_2O_4$

Natural ß-ZnS always contains iron as an admixture. When iron exceeds 10 % the minerals are called marmatites. They have a varying composition[11] and are expressed by the general formula nZnS.mFeS. During their oxidation the conditions for ferrite formation are most favourable. The presence of other admixtures leads to formation of ferrites of a complex composition. This is confirmed by the composition of the magnetic fraction separated from the insoluble residue from hydrometallurgical leaching of zinc calcine. It has the following composition (in mass. %): Zn - 17,12; Fe - 44,35; Cu - 3,25; Cd - 1,37; Mg - 1,80, etc. The cadmium and copper contents are about 5 times higher than those of the initial raw material.

Data from X-ray phase analysis show that the prevailing phase in the magnetic fraction is MeFe$_2$O$_4$, where Me = Zn, Cu, Cd, Mg, Fe. There is a certain amount of Fe$_2$O$_3$ and non-oxidized ß-ZnS. The Mössbauer spectra of the magnetic fraction (Fig.2) reveal the presence of a doublet, with IS = 0,35±0,01 mm/s and QS = 0,42±0,01 mm/s, as well as of a

sextet with data corresponding to α-Fe$_2$O$_3$. No changes in IS and QS values are observed after thermal treatment the sample up to 1000 ^{o}C.

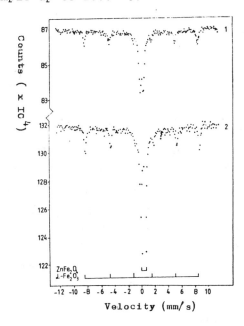

Fig.2. Mössbauer spectra of the magnetic fraction at room temperature: initial (1) and thermally treated up to 1000 ^{o}C (2).

The major phase in the magnetic fraction is ZnFe$_2$O$_4$, with admixture ions of Cu^{2+}, Cd^{2+}, Mg^{2+}, etc. On account of this the values of isomer shift and quadrupole splitting differ, though to a very slight degree, from those of the pure zinc ferrite.

The results concerning magnetic fraction composition support the conclusions made by R. Stoitsova and V. Karoleva for

the decrease of cadmium solubility due
to the formation of polymetallic sparing-
ly soluble ferrites[12].

As regards the solid state interactions in the $ZnO-CuO-Fe_2O_3$ system it is established that $ZnFe_2O_4$ formation occurs
ahead of $CuFe_2O_4$ formation. In this respect, it was interesting to examine the
interaction between $CuFe_2O_4$ and ZnO. The
chemical and X-ray phase analyses of the
final products (1000 ^{O}C, 600 min) show
that an exchange reaction has occured
between ferrite and oxide. In general,
this can be written in the following way:

$CuFe_2O_4$ + ZnO $\longrightarrow$ $Cu_xZn_{1-x}Fe_2O_4$ + xZnO +
+ (1-x)CuO

On the basis of results obtained and
the processed Mössbauer spectra, the
conclusion can be made that at temperatures of 900-1000 ^{O}C the values of (x)
is 0,2-0,3.

Experimental data indicate that the
presence of ZnO in the calcine from fluid bed roasting of zinc concentrates
(920-960 ^{O}C) will cause a decrease in the
degree of $CdFe_2O_4$ and $CuFe_2O_4$ formation.

A greater amount of $CdFe_2O_4$ should
be present in the oxidized particles
that have stayed for a shorter time in
the furnace, and also in those containing cadmium as an independent sulphide
mineral. The exchange reaction occuring
between $CdFe_2O_4$ and ZnO will be limited
by the zinc ion diffusion towards $CdFe_2O_4$
through the product layer containing
$ZnFe_2O_4$ and CdO[9].

With solid state reactions involving
zinc, copper and iron oxides, the increase of temperature leads to an increase
in the amount of $ZnFe_2O_4$ obtained as compared to $CuFe_2O_4$. In the studied temperature range (800-1000 ^{O}C), which is typical for the roasting of zinc and copper-
zinc sulphide concentrates the ferritic
ZnO is 3-4 times more than the ferritic
CuO.

The problem of the effect of ferrite

composition on their solubility in sulphuric acid solutions is left open.
This provided grounds for a detailed
study on the solubility of mixed copper-
zinc and zinc-cadmium ferrites. The effect of ferrite composition, temperature (t), time (τ) and acidity ($C_{H_2SO_4}$)
on the degree of leaching of zinc
(η_{Zn}), copper (η_{Cu}), cadmium (η_{Cd})
and iron (η_{Fe}) had been examined.

The obtained dependences have the
form presented in Fig.3.

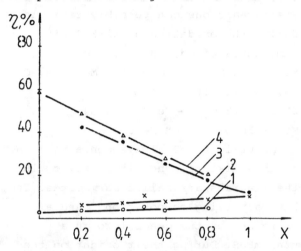

Fig.3. Degree of leaching of Zn and Cu
from $Cu_xZn_{1-x}Fe_2O_4$ depending on composition and temperature at $C_{H_2SO_4}$ = 14 % and
τ = 60 min:1,2- η_{Zn}, η_{Cu} (50 ^{O}C);
3,4- η_{Cu}, η_{Zn} (90 ^{O}C).

With regard to solubility of mixed
copper-zinc ferrites of the kind
$Cu_xZn_{1-x}Fe_2O_4$ following general conclusions could be drawn. The influence of
temperature on the degree of leacning of
copper and zinc is most important. By
increasing the temperature of leaching
from 50 ^{O}C up to 90 ^{O}C the degree of
leaching of the elements Cu and Zn raises 2 to 5 times.

The effect of concentration of sulphuric acid is stronger at 90 ^{O}C. At
50 ^{O}C with an increase of $C_{H_2SO_4}$ from
7 % up to 14 %, η_{Zn} and η_{Cu} increase
by 2-4 %, while at 90 ^{O}C this increase
is about 10 %.

The change of ferrite composition has following effect on η_{Zn} and η_{Cu}. Compared to mixed ferrites, pure $ZnFe_2O_4$ (x = 0) has the highest solubility as regard zinc at 90 $^{\circ}$C under otherwise equal conditions. Pure copper ferrite $CuFe_2O_4$ (x = 1) at 90 $^{\circ}$C has the lowest solubility regarding copper compared to all other ferrites. From these results the following assumption about the expected solubility of copper and zinc from zinc and copper-zinc calcines could be made. During processing of zinc calcines their copper content is under 1 % and $ZnFe_2O_4$ is obtained almost pure. A certain amount of admixture ions (Cu, Cd, Mg, Fe) are observed in it. From the data obtained it is to be expected that this ferrite will have the highest solubility as regards zinc and copper.

If copper in the concentrates is in the form of separate minerals (mainly $CuFeS_2$) a ferrite with a certain amount of admixture is obtained, but its behavior will be similar to this of pure $CuFe_2O_4$. In that case a low solubility of copper from ferrite may be expected.

Copper-zinc concentrates differ from zinc and copper concentrates in the very fine mixing of the elements in initial ore, which makes their separation by flotation practically impossible. When processing concentrates with 5-8 % copper content, the copper and zinc solubility from mixed ferrites in high concentrations of sulphuric acid will be relatively high. Lower leaching efficiencies will be achieved when leaching calcines where copper is a main component (15-20 %) and the zinc content is lower (10-15 %).

When increasing leaching time the values of η_{Zn} and η_{Cu} increase. For example, when leaching ferrites of the kind $Cu_xZn_{1-x}Fe_2O_4$ at 50 $^{\circ}$C by means of 7 % H_2SO_4 solution and at a duration of 420 min., η_{Zn} and η_{Cu} values are about 3 to 4 times higher than these at 60 min. duration.

On the basis of results obtained for zinc and cadmium leaching from mixed $Zn_xCd_{1-x}Fe_2O_4$ ferrites, the effect of temperature (t), time (τ) and acidity (C) factors has been studied by means of a 2^3 factorial experiment. The mixed ferrite used for the purpose had the following composition $Zn_{0,8}Cd_{0,2}Fe_2O_4$. The basic levels and interval of factor variation are the following: t=70$\pm$20 $^{\circ}$C, τ =120$\pm$60 min, and C=10,5$\pm$3,5 %.

By using Student's criterion, it was established that all coefficients in the regression equations are significant, on account of which the model obtained is entirely adequate.

The obtained equations for the degree of leaching (η) for the different metals are of the following type:

For zinc: $\eta_{Zn}= -4.27+0.13t-0.60\tau -5.87C+ +0.012t.\tau +0.115t.C+0.046 \tau.C-0.0008t.\tau.C$

For cadmium: $\eta_{Cd}= 11.24-0.14t-0.73\tau - -7.67C+0.014t.\tau +0.147t.C+0.059\tau.C-0.001. t.\tau.C$

For iron: $\eta_{Fe}= -18.05+0.31t-0.46\tau -4.91C+ +0.01t.\tau +0.104t.C+0.041\tau.C-0.0007t.\tau.C$

In spite of the heterogeneous evaluation of regression coefficients, the conclusion can be drawn that in the interval of factor variation the effect of temperature on the degree of metal leaching is the greatest.

The interactions between H_2SO_4, $(NH_4)_2SO_4$, $FeSO_4$ and MeO (Me = Zn, Cu, Cd, Ca, Mg, Pb), $Mé_2O_3$ (Mé = Fe, Al,), $Me"Fe_2O_4$ (Me" = Cu, Zn, Cd) and mixed ferrites of Cu, Zn, and Cd ($Cu_{0,5}Zn_{0,5}Fe_2O_4$, $Cu_{0,5}Cd_{0,5}Fe_2O_4$, $Zn_{0,5}Cd_{0,5}Fe_2O_4$) were examined by using different methods: DTA and TGA, quasi isobaric isothermal and X-ray phase analyses, Mössbauer spectroscopy.

When using sulphuric acid as the sulphating agent, the behavior of individual ferrites of zinc, copper and cadmium is of greater interest. Some experimental results are shown in Table III.

as a sulphatizing agent.

Table III. Sulphatization of ferrites by means of sulphuric acid

System	Molar ratio	Temperature, ^{o}C	Degree of sulphatization, %
$ZnFe_2O_4 + H_2SO_4$	1 : 0,3	700-720	77
	1 : 1	765-785	89
$CuFe_2O_4 + H_2SO_4$	1 : 0,3	715-735	48
	1 : 1	715-735	59
$CdFe_2O_4 + H_2SO_4$	1 : 0,3	765-780	73
	1 : 1	765-780	92

It is evident from the results that the stability of the studied ferrites as regards the effect of sulphuric acid increases in the following order:
$$CdFe_2O_4 \longrightarrow ZnFe_2O_4 \longrightarrow CuFe_2O_4.$$

The sulphatization of mixed ferrites shows that the presence of 3 sulphitizing elements complicates the process strongly. Up to 740 ^{o}C mixed ferrites of copper and zinc and of copper and cadmium sulphatize in approximately the same way. This can be explained by the similar character of the formation and dissociation processes of $Fe_2(SO_4)_3$ and $CuSO_4$.

During sulphatization of the mixed ferrite $Zn_{0,5}Cd_{0,5}Fe_2O_4$ "a transfer" of SO_3 from $Fe_2(SO_4)_3$ to ZnO and then from formed $ZnSO_4$ towards CdO is observed.

The sulphatization process with $(NH_4)_2SO_4$ occurs at lower temperature compared to $FeSO_4$. Almost all sulphates are obtained in considerable degree at a temperature of up to 650 ^{o}C whereas with $FeSO_4$ that is possible at its complete dissociation - above 700 ^{o}C (Fig.4 and 5).

Zinc ferrite is sulphatized more completely by $(NH_4)_2SO_4$ and extensive dissociation of $ZnSO_4$ occurs at a temperature above 750 ^{o}C. About 45 % of zinc is in sulphate form under the action of $FeSO_4$

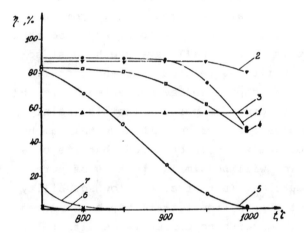

Fig.4. Obtained $MeSO_4$ in the $(NH_4)_2SO_4 +$ +MeO system depending on the temperature: 1-$CdSO_4$; 2-$PbSO_4$; 3-$CaSO_4$; 4-$MgSO_4$; 5-$ZnSO_4$; 6-$Al_2(SO_4)_3$; 7-$CuSO_4$.

On the grounds of the results obtained it may be assumed that the mechanism of the sulphatization is as follows:
$$(NH_4)_2SO_4 \longrightarrow 2NH_3 + H_2O + SO_3$$
$$ZnFe_2O_4 + 4SO_3 \longrightarrow ZnSO_4 + Fe_2(SO_4)_3$$
$$3ZnFe_2O_4 + Fe_2(SO_4)_3 \longrightarrow 3ZnSO_4 + 4Fe_2O_3.$$

Copper and cadmium ferrites have aproximately the same behavior during sulphatization both with $(NH_4)_2SO_4$ and $FeSO_4$. A difference in the degree of sulphate dissociation is observed at 500-600 ^{o}C. It is about 20 % lower when $(NH_4)_2SO_4$ reacts with the ferrites. The

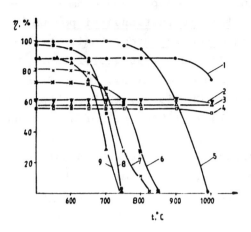

Fig.5. Obtained $MeSO_4$ in the $FeSO_4$+MeO system depending on the temperature: 1-$CdSO_4$; 2-$PbSO_4$; 3-$CaSO_4$; 4-$MgSO_4$; 5-$ZnSO_4$; 6-$Al_2(SO_4)_3$; 7-$CuSO_4$.

reason for this is probably the simultaneus sulphatization of MeO and Fe_2O_3 on account of the lower temperature of $(NH_4)_2SO_4$ dissociation. Upon increase of temperature, an additional amount of MeO is sulphatized by the liberated SO_3 as a result of iron sulphate dissociation.

The optimum temperature range for sulphatization of the studied ferrites is 600-650 ^{O}C. The degreesof sulphatization of $ZnFe_2O_4$, $CdFe_2O_4$ and $Cu Fe_2O_4$ are about 90, 80 and 70 % respectively.

The copper-zinc cake sulphatization process[13] was investigated for the purposes of copper-zinc concentrates processing following the scheme roasting-leaching-sulphatization. X-ray phase analysis and the leaching parameters of Cu, Zn and Fe from sulphate product in slightly acid solutions provide a basis for a suggested sulphatization scheme (Fig. 6).

The results obtained indicate that by selecting thermal processing conditions granules of cake and sulphatizing agent can be formed, in which 75-80 % of zinc and 80-85 % of copper are presented in water soluble forms at low solubility of iron. Copper in the solu-

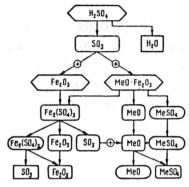

Fig.6. Fe_2O_3 and $MeO.Fe_2O_3$ (Me = Cu, Zn, Cd) sulphatization scheme.

tion can be separated from zinc by means of zinc powder cementation or fluid bed electrolysis. Using the latter method a copper concentration of $\sim 0,3$ kg/m³ prior to cadmium electrolysis can be achieved.

Following the scheme roasting - leaching - sylphatization copper and zinc leaching from copper-zinc sulphide concentrates reaches 94-96 %. Lead cake (containing $\sim$12 % Pb and $\sim$30 % Fe) obtained after leaching of granules is further processed in a lead smelter for utilization of lead and precious metals.

The proposed technology for processing copper-zinc concentrates[13,14] can be carried out using standard equipment.

REFERENCES
1. Naboichenko S.S. Smirnov V.I. Hydrometallurgy of copper, Moskow: Metallurgy, 1974, 271 pp.
2. Sergeev G.I. Khudyakov I.F. Lusakov A.A. Zak M.S. Gorbashov V.V. Influence of high temperature roasting of zinc concentrates in fluid bed on the zinc concentrates in fluid bed on the zinc ferrite formation. Non-ferrous metallurgy, no.5, 1984, p.67-71.
3. Ferrites. Proceedings of the ICF3. Center for academic publication Japan. 1981, 990 pp.
4. Tretyakov Yu.D. Principles of creating of new solid state materials. Inorganic materials. vol. 21, no.5, 1985,

p.693-701.

5. Snurnikov A.P. Ogienko A.C. Yurenko V.M. Mikhailova E.M. Method of sulphatization of zinc cakes. Patent USSR, No 200764, 40a, 19/30, 1966.

6. Lenchev A.S., Karavasteva M.S. Sulphatization roasting of zinc cake. Non-ferrous metallurgy, no.1, 1982, p.36-40.

7. Dimitrov R. Boyanov B. Investigation of solid state interactions in the systems ZnO-α-Fe_2O_3, $ZnFe_2O_4$-CuO and $ZnFe_2O_4$-CaO. Rudarsko-metalurski zbornik, vol.31, no.1, 1984, p.67-80.

8. Boyanov B. Study of solid state interactions in the system Cu-Zn-Fe-O. Chemistry journal of Plovdiv university, vol.24, no.1, 1986, p.305-316.

9. Boyanov B. Dimitrov R. Kinetic study on the process of ferrite formation involving zinc, copper and cadmium oxides. In: 1st international simposium Interprogress-metallurgy, Košice-Bratislava, 26-29 June, 1990, p.182-185.

10. Solid state Reactions, Yu.D. Tretyakov. Moskow: Chemistry, 1978, 359 pp.

11. Mineralogy. I.Kostov. Sofia: Science and art, 1973, 674 pp.

12. Stoitsova R. Karoleva V. Electron microscope study of cadmium containing products from zinc production. Metallurgy, vol.31, no.4, 1976, p.19-22.

13. Boyanov B. Dimitrov R. Dobrev N. Processing of copper-zinc cakes by sulphatization. Non-ferrous metals, no.5, 1986, p.34-36.

14. Boyanov B. Dimitrov R. Balabanov Ya. Hydrometallurgical processing of copper-zinc calcine. Non-ferrous metals, no.3, 1985, p.29-31.

Limitation of lead jarosite formation in the leaching of calcine from high-iron Zn–Pb–Cu concentrates

M. Pedlík
J. Jandová
Department of Chemical Metallurgy and Corrosion Engineering, Prague Institute of Chemical Technology, Prague, Czechoslovakia

SYNOPSIS

Complex sulphide ores, which because of their fine grain size and complexly intergrown mineralogy are not possible to upgrade into selective concentrates, come for the metallurgical treatment as bulk or mixed concentrates. Zinc-lead-copper concentrates with higher iron content need special modifications of the processing because of interactions of components during oxidative roasting and leaching by sulphuric acid solutions.

To decompose zinc ferrite, which is originated in the roasting, it is necessary to use high acid leaching during which not only zinc but all the iron go into solution. In leach solution the iron sulphate reacts with lead sulphate and lead-jarosite is precipitated. This compound lowers the lead content in the leach residue and reduces its value for the following lead production.

Most of the present studies about the formation of lead-jarosite connected its origin to temperatures above 100°C during autoclave oxidative leaching of ZnPb sulphide concentrates.

This paper describes the conditions of lead-jarosite formation during the leaching of ZnPbCu bulk calcine and the necessary modifications of the leaching process to hinder its formation.

Very fine grain size and complexly intergrown mineralogy gives only limited possibilities for a number of sulphide ores to recover selective concentrates by conventional benefication. Even modified flotation methods can prepare selective concentrates but with small yields, often accompanied by the loss of some important components, such as precious metals. This brings about a category of complex bulk concentrates, the basic problem of which – the separation of metals therein – is transferred to metallurgical processes.

In the last decade a number of papers and articles by prominent metallurgists about the treatment of complex sulphide ores show a growing interest in such materials for the prospective expansion of the raw materials base. As regards zinc, lead and copper, the treatment of bulk concentrates has so far been just a marginal matter due to the sufficiency of good quality selective raw materials, and is therefore rather a task for those who try to solve difficult metallurgical technologies, and for future measures.

However, there are some countries, where these complex materials actually represent the only source of raw materials for metal production and their

Czechoslovakia can count itself among these countries, as the mining of ZnPbCu ores treatable to selective concentrates of zinc and lead is practically at its end and the only applicable raw materials with zinc and lead contents are complex ZnPbCu concentrates.

Our effort to make the best of the mining and ore tratment capacities as well as to cover some of the domestic zinc consumption by our own production lead to a decision in the middle of the eighties to construct a plant for their treatment.

The mining capacities owners, cooperated with well-known metallurgical companies, such as Lurgi and Outkumpu in preparing the technology. Their cooperation, however, concerned only a part of the operations - oxidation roasting and separation of iron from the leach solution - not the entire technology.

In order to verify the technology which was not completely worked out on a laboratory scale, a pilot-plant was constructed with a daily 12 t concentrate capacity. The main device was a fluid-bed roaster for oxidation roasting, two-step leaching and the removal of iron.

In spite of a very unfavourable composition of the concentrates good results were achieved by oxidation roasting and the removal of iron from solution. Though several variants of the technology were checked up in the pilot-plant, unexpected difficulties arose during leaching as - besides the zinc solution with a high concentration of iron - the leach residue contained only half the amount of lead predicted by technological calculations.

The low quality of the leach residue hindering its further treatment to lead, was the reason for a detailed study of the chemistry and kinetics of leaching the bulk calcine. The study of reaction mechanism and occuring interactions

rendered interesting results, somewhat different from the so far published works. The research work was accomplished, even though mining and production of concentrates have been brought to a standstill under the new economic conditions in Czechoslovakia.

OXIDATION ROASTING OF POLYMETALLIC BULK CONCENTRATES

ZnPbCu bulk concentrates produced by collective flotation in several benefication plants contain the main traditional components - sphalerite, galena, chalcopyrite, accompanying minerals pyrite, silica, hematite and a small amount of other contaminating components. The very fine grain size of the minerals, intimately intergrown with gangue can produce - without any significant decrease in yield - a concentrate greatly contaminated, primarily by pyrite, its composition is given in Tab. 1.

Table 1. Chemical analysis of the bulk concentrates

Zn	19 -30%	Fe	16 - 22%
Pb	6 - 15%	SiO$_2$	3 - 5%
Cu	2 - 3%	S	30 - 34%

The presence of the above mentioned minerals is usual in zinc concentrates but their proportions in our concentrates are atypical. The low contents of zinc, high contents of lead, iron and silica are the sources of secondary reactions which in their composition significantly affect the course of oxidation roasting as well as the quality of the calcine.

The high lead and silica contents are very unfavourable for oxidation roasting as they form low melting PbO.PbSO$_4$- type eutectics with melting points around 800°C, and silicate 4-1 PbO.SiO$_2$ eutectics with melting points between 720 and 770°C, causing grain agglomerations and formation of lumps during roasting in the fluidized bed.

The oxidation of sphalerite produces active zinc oxide which not only combines with SiO_2 forming a silicate – willemite Zn_2SiO_4, but also forms a spinel bond with ferric oxide, creating a ferrite – $ZnFe_2O_4$. The unusually high iron contents accounts for the fact that almost 50% of zinc is bonded in ferrite, considerably decreasing the neutralization ability of the calcine and requiring a modification of the conventional leaching method.

Oxide interactions, being greater than usual due to the composition of the concentrate, cause the optimal temperature of the oxidation roasting to be a compromise between dead roasting, ferrite formation limits and liquid phase formation together with grain agglomeration.

Temperature dependence of the phase composition was observed by X-ray diffraction analysis which helped to define the temperature of oxidation roasting as regards the leachability of the calcine. Changes in the phase composition are given in Fig. 1.

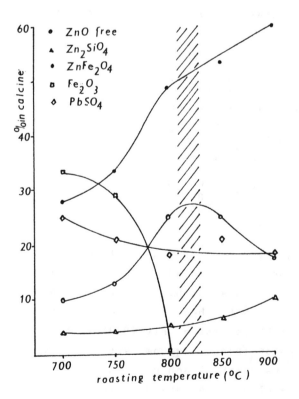

Fig. 1. Dependence of the calcine phase composition on the oxidation roasting temperature

Due to limiting factors the optimal temperature of roasting is 810-830°C, the chemical composition of the calcine can be as given in Tab. 2.

Table 2. Chemical analysis of the calcine sample produced in pilot plant

Zn	25.02%	Fe	20.08%
Pb	18.64%	SiO_2	5.76%
Cu	3.70%	S_t	4.58%

X-ray diffraction analysis enabled to dertermine the phase composition ot this calcine, containing 19% free ZnO – zincite, 50% ferrite $ZnFe_2O_4$ – franklinite, 7% silicate Zn_2SiO_4 – willemite, 24% sulphate $PbSO_4$ – anglesite. By means of the chemical phase analysis the composition was supplemented by an addition of 3% zinc sulphate, other components were in agreement with X-ray diffraction analysis.

The product of oxidation roasting contains, besides insoluble lead sulphate and a smaller amount of insoluble inert compnents, zinc in four different compounds – oxide, sulphate, silicate and ferrite. The individual compounds vary in dissolution in the H_2SO_4 solutions, or in spent zinc electrolyte, circulating in the usual technological flow sheet back to the multi-step leaching of the calcine.

The aim of leaching, even with the unusual composition, was to get the maximum of zinc and iron into solution, to separate iron in the form of jarosite and to recover the leaching residue with maximum lead content. The application of the conventional technique used for the treatment of classical zinc concentrates had to be considerably modified for the given concentrate.

LEACHING OF CALCINE

The different leaching kinetics of each zinc compound in the calcine require multi-step leaching with different concentrations of sulphuric acid and leaching temperatures. In hydrometallurgy of zinc, the conventional two- to three-step countercurrent leaching system by spent electrolyte containing 120-200 g l^{-1} free sulphuric acid had to by modified due to high iron contents in solution after strong acid decomposition of the ferrite bond. In removing iron from solution in the form of jarosite, half of the calcine had to be used for the neutralization of the acid liberated during the precipitation of jarosite according to equation

$$3Fe_2(SO_4)_3 + Na_2SO_4 + 12H_2O \rightleftharpoons$$
$$2Na[Fe_3(SO_4)_2(OH)_6] + 6H_2SO_4$$

After several modifications of the leaching system the flow sheet given in Fig. 2 was used.

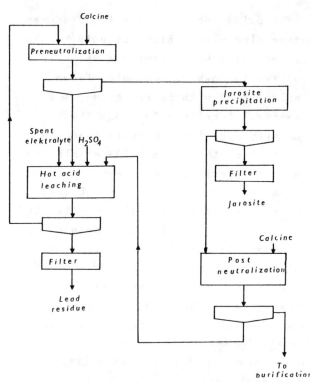

Fig. 2. Flow sheet of the applied two-step leaching

Due to low efficiency and slow kinetics of dissolution of high ferrite contents at 90°C, the temperature was increased to 105-110°C and low-pressure leaching was applied. The final parameters in pilot-plant tests are given in Tab. 3.

Table 3. Basic parametres of preneutralization and high acid leaching

Parameter	Preneutralization PN		High-acide leaching	
	input	output	input	output
Temperature °C	60 - 70°		105 - 110°	
Reaction time h	2 - 4		4 - 6	
free H_2SO_4 g.l^{-1}	60-70	10-15	120-200	60-70
concentration Zn^{2+}	80-90	120	50	80-90
concentration Fe^{3+}	35-40	35-40	0	35-40

During pilot-plant tests considerable differences were found as regards the expected leaching process, the main differences being low lead and high iron contents in the leach residue which threatened its further treatment to lead. Contrary to theoretically calculated lead contents above 50%, the residue contained only 20-30% of lead.

X-ray diffraction analysis disclose the presence of lead-jarosite in the leach residue which, contrary to lead sulphate with theoretical contents of 68.3% Pb, contains only 18.3% Pb at 29.6% of iron, this composition devaluates the leach product.

Metallurgists have devoted their attention to the formation and properties of this compound since the seventies when it was found that during direct pressure oxidation leaching of ZnPb sulphide concentrates most of lead in the leach residue is bonded in the form of jarosite. It is formed through the reaction of lead sulphate with ferric sulphate in solution according to equation

$$PbSO_4 + 3Fe_2(SO_4)_3 + 12H_2O \rightleftharpoons$$
$$Pb_{0.5}Fe_3(SO_4)_2(OH)_6 + 6H_2SO_4$$

Basic factors affecting its formation are acid concentration, ion strength of solution, temperature and time of contact $PbSO_4$ with solution copntaining a higher concentration of ferric sulphate. In a wider range of conditions its formation can be supported by seeding.

Most of the hitherto works about its formation connected its origin to temperature above 100°C during the autoclave processes. In our case, its formation already occured at the first technological variant when temperature did not surpass 90°C in any stage of the leaching process.

An investigation of the reaction conditions suitable for the formation of jarosite was linked with a complex study of leaching each zinc compound from the calcine and possible interactions in the individual leaching steps – in preneutralization and high acid leaching. On a laboratory scale we studied the effect of the variable parameters – concentration of free sulphuric acid, temperature and time – on the leaching kinetics and the formation of Pb-jarosite in both leaching steps.

Chemical phase analysis and X-ray diffraction analysis rendered it possible to make the process of leaching more accurate and to evaluate parameters for the formation of lead-jarosite.

According to works of J. E. Dutrizac

et al. (1, 2, 3) the basic factors effecting the formation of Pb-jarosite during leaching are - longer contacts of the overstoichiometric amount of lead sulphate with the ferric sulphate solution at temperature above 90°C and the concentration of free sulphuric acid below 0,3M H_2SO_4. The acid concentration as well as the total ion strength of the solution influences the solubility of lead sulphate and thereto the mechanism of jarosite formation which, according to (4) as well as our study of leach residues, consists of discrete crystals outside the surface of lead sulphate particles. The presence of other components in the leach solution, primarily zinc sulphate, further increase the low $PbSO_4$ solubility. The concentration of lead in solutions gained by leaching bulk calcine fluctuated up to 30 mg.l^{-1}.

Concentration of the individual conponents in our case create very good conditions for the formation of lead-jarosite, so that the decisive factor for its presence in leach residues will be temperature and concentration of free sulphuric acid. As regards the temperature range for its formation - above 90°C - hot acid leaching would be most beneficial, when the temperature range is from 90 to 110°C. Usually, however, the H_2SO_4 concentration does not drop below 60 g.l^{-1}. i.e. 0.6 M, this being twice the limit given in the mentioned articles.

From the standpoint of H_2SO_4 concentration preneutralization would in turn better correspond with the formation of jarosite, as the contents of free acid goes down below 10 g.l^{-1}, i.e. below 0.1 M. The maximum working temperature 60-70°C, however, is under the given limit for Pb-jarosite formation.

The differences between theoretical conditions for its formation and experimental setting during leaching of

roasted complex ZnPbCu concentrates, as
well as high contents of Pb-jarosite in
leach residues, were reasons for
studying the reaction mechanisms of both
the leaching steps of the complex
calcine.

Time dependence of the changes in
free H_2SO_4 concentration, corresponding
to the transfer of zinc and iron from
each compound present in the calcine
into solution is given in Fig. 3.

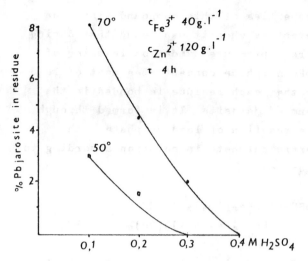

Fig. 4. Effect of final solution acidity
on the formation of Pb-jarosite

corresponding to the leaching of roasted
concentrates of the given composition,
a limited amount of Pb-jarosite is
formed at concentrations up to 0.3M
H_2SO_4.

Temperature has a dominating effect
on the formation of Pb-jarosite, this
being evident fom the dependences shown
in Fig. 5.

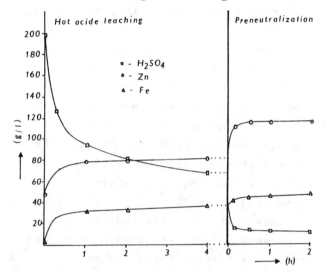

Fig. 3. Changes in the concentration of
H_2SO_4, Zn^{2+}, Fe^{3+} in solution
during hot acid leaching and
preneutralization

The decrease of free sulphuric acid
contents during leaching in both the
steps comports almost teoretically with
the increase of zinc and iron in
solution.

The effect of free H_2SO_4
concentration in the final step of
preneutralization, i.e. final acidity of
the leach, on the formation of
Pb-jarosite is given in Fig. 4.

Experiments have confirmed that in
the solutions with compositions

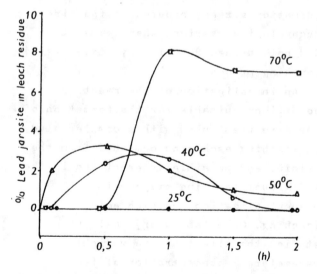

Fig. 5. Effect of temperature on the
formation of jarosite at final
solution acidity of 0.1M H_2SO_4

Observing the contents of jarosite in leach residue by X-ray diffraction analysis, the detection limit being 0.5 to 1%, some fluctuations of values were evident, influenced probably by redissolution of crystallic jarosite nuclei due to a local increase of the sulphuric acid concentration by the reaction product.

Due to the fact that the kinetics of leaching active free zinc oxide did not reveal any deterioration at lower temperatures, we continued to perform preneutralization at temperatures below 40°C.

Though a stable content of free sulphuric acid in solution to the second step - hot acid leaching - would correspond to a certain technological acid circulation and to the balance between leaching and electrolysis of the zinc solution, we studied the effect of free H_2SO_4 contents in the range from 150 to 250 g.l^{-1} H_2SO_4. With a constant addition of ferritic leach residue from

preneutralization (without Pb-jarosite) the above mentioned concentrations of H_2SO_4 corresponded to the final concentration of 20 to 100 g.l^{-1}.

Seeing that with the decreasing acidity of solution for leaching the kinetics of zinc ferrite decomposition also changes, the leaching tests were extended to 6 hours. Leaching was not performed with contents as in done in continuous acid leaching, but in the form of batch tests with decreasing acidity in the given range.

All tests were performed in an autoclave at a temperature of 105°C. The reaction time of 240 minutes was counted from the time the reaction temperature was reached, when the mixer was put into operation releasing a sample of leach residue from preneutralization, sealed in a plastic bag.

The final composition of the solution and leach residue from high acid leaching is given in Tab.4.

Table 4. Composition of solutions and leach residues at various input H_2SO_4 concentrations. Time of leaching 240 min, temperature 105°C, $ZnSO_4$ 48.0 g.l^{-1}

H_2SO_4[g.l^{-1}]		Solution composition			Leach residue	
input	output	g.l^{-1}Zn	g.l^{-1}Fe	mg.l^{-1}Pb	% ferrite	% Pb-jarosite
150	5	75.5	16.1	0,9	21	40.5
170	20	79.4	20.9	17.0	16.5	45.7
200	50	78.6	28.35	10.7	6.5	39.5
250	100	78.3	58.40	15.8	-	-

The kinetic curves for Zn ferrite decomposition and lead-jarosite formation are illustrated in Fig. 6.

The tests have confirmed that with the output concentration of H_2SO_4 below 50 g l^{-1} the Pb-jarosite forms at the given composition of solutions and leach residues even under the conditions of hot acid leaching. Supplementary tests verified that to prevent its formation the output concentration of free

sulphuric acid must not drop below 70 g l^{-1}.

As can be seen from Fig. 6 traces of Pb-jarosite in the leach residue appear soon after the first 5-10 minutes of leaching and its contents increase to 10% in the leach residue in the first hour. Afterwards a rapid increase of 40-60% was evidenced at the lower concentrations of free sulphuric acide.

Similar situation arises in hot acid leaching in case that a leach residue

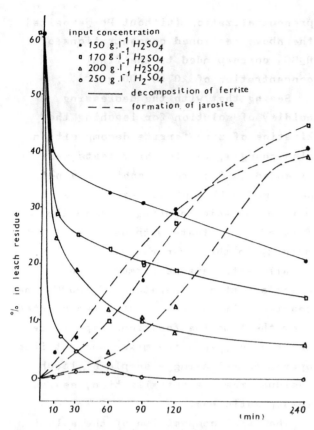

Fig. 6. The kinetic curves for hot acid
leaching of the ferritic
fraction of calcine at various
H_2SO_4 concentrations

from preneutralization already contains
seeds of Pb-jarosite. With 2-5% seeds
contents and concentration of H_2SO_4 in
high acid leaching 200/50 g.1^{-1}
(input/output) the amount of
Pb-jarosite increases, at 250/100 g 1^{-1}
a certain balance is reached without any
changes in contents, and at 270/120
g 1^{-1} even the original amount
dissolves and no further jarosite is
formed.

On the basis of the performed experiments
it was possible to determine the main
technological parameters of preneutralization and hot acid leaching
of complex high-iron Zn-Pb-Cu concentrates,
where no undesirable Pb jarosite is
formed.

Differences in the due compositions of
leach residues formed under technological
conditions of the initial and modified
parameters are given in Tab. 6.

Table 5.

Parameters	Preneutralization	Hot acid leaching
Temperature°C	40	105 - 110
H_2SO_4 input g.1^{-1}	70	= 220
output g.1^{-1}	30	= 70
Reaction time h	2	4 - 6

Table 6. Contents of the main elements in leach residues under
intitial and modified condition of leaching

Contents %	Main parameters Initial (Tab.3)	Modified
Pb	17 - 27	50 - 52
Fe	20 - 11	0.1 - 1.5
Zn	2 - 1	0.7 - 1.0
Cu	2 - 3	0.3 - 0.4
S	8 - 4	7.5 - 8.2

135

CONCLUSION

The chemical composition of complex high-iron ZnPbCu concentrates causes a number of undesirable interactions in oxidation roasting and leaching of the calcine. In oxidation roasting the negative effect of the composition can be restrained by choosing a temperature range which could be a compromise between the optimal roasting process and the formation of low-melting eutectics. In applying multi-step leaching of the roasted material, usually used for better quality ZnPb concentrates, we cannot restrain the formation of Pb-jarosite that considerably decreases the Pb contents and thus the total quality of the leach residue. We studied the kinetics of leaching of the roasted concentrates in the so-called preneutralization and hot acid leaching and observed factors influencing the formation of Pb-jarosite. At the given composition of the solutions - contents of Zn^{2+}, Fe^{3+} and the total ion strength the ratio $PbSO_4/Fe$ - temperature and contents of free sulphuric acid have dominating effect on the formation of Pb-jarosite. Modifying the conditions of both the leaching steps we can restrict the formation of Pb-jarosite and provide a Pb-leach residue with lead contents above 50%, yielding about 98% of zinc

into solution, these are the usual indeces in the treatment of good quality concentrates.

The performed experiments have confirmed that Pb-jarosite, the formation of which has up to now been only connected with direct preassure oxidation leaching of sulphide concentrates, was also formed in both the leaching steps of roasted complex high-iron concentrates.

Reference

1. Dutrizac J.E., Dinardo O. and Kaiman S. Factors affecting lead jarosite formation, Hydrometalurgy, 5, 1980 p. 305-324.
2. Dutrizac J.E. and Chen T.T. Canadian metallurgical quarterly vol. 23, no 2, 1984, p. 147-157.
3. Dutrizac J.E. and Dinardo O. The precipitation of copper and zinc with lead jarosite. Hydrometallurgy, 11, 1983, p. 61-78.
4. Dutrizac J.E. and Jambor J.L. Formation and characterization of argentojarosite and plumbojarosite and their relevance to metallurgical processing. Applied Mineralogy, Proceedings of the second international Congress on Applied Mineralogy in the Minerals Industry, February 22-25, 1984.

Nickel

Intensification of the reductive-roast ammonia leaching process for nickel lateritic ores

J. Jandová
M. Pedlík
Department of Chemical Metallurgy and Corrosion Engineering, Prague Institute of Chemical Technology, Prague, Czechoslovakia

SYNOPSIS

The dissolution behavior of Fe, Ni and the Fe-Ni system containing 10-50 pct Ni in ammoniacal ammonium carbonate solutions has been investigated with the aid of potentiodynamic polarization experiments and leaching tests. The investigated Fe-Ni system consisted of various amounts of pure metallic Fe, Ni, α-FeNi alloy containing maximum 5 pct Ni and γ-FeNi alloy containing cca 50 pct Ni.

The polarization measurements indicate that as pure metallic Fe, Ni so FeNi alloys show active and passive behaviour however Fe passivates more easily than Ni. The active dissolution region of FeNi alloy is placed between those of Ni and Fe and according with the increasing contents of Fe, resp. Ni approachs to the behaviour of pure Fe, resp. Ni.

The result of leaching tests correspond with those of polarization measurements and confirm that the dissolution of Ni from FeNi alloys in the range of high redox-potentials is influenced by the electrochemical effect of metallic iron. The presence of oxygen in leaching solution is favourable for nickel dissolution but exessive aeration must be avoid in practical ammonia leaching systems to prevent overstepping the critical redox-potential and thereby the passivation of the FeNi system. The ability of FeNi phase for the passivation increases with the rising amount of α-FeNi alloy and the falling nickel content in γ-FeNi alloy.

The reductive-roast ammonia leaching process has at present become one of the important methods to obtain nickel and cobalt from low-grade iron-nickel lateritic ores. This process, developed by Caron[1], consists of two major steps. The first is a roasting operation in which valuable metal oxides (i.e. Ni, Co) are selectively reduced to the more leachable metallic phases. The next step involves ammonia leaching during which the reduced metals disolve as various ammine complexes in an aerated ammoniacal ammonium carbonate solution.

The relatively low recovery of nickel, and expecially of cobalt which is the main drawback of this process, has been studied by several authors. This may be attributed besides other factors to the nickel and cobalt cooprecipitation with iron(III)-hydroxide as well as to its adsorption on the precipitated iron (III)-hydroxide and oxide leach residues. There are also indications [2-4] that an additional contributing factor is the electrochemical interaction between the metallic phases formed during the roasting process, i.e. iron and ferronickel or ferronickelcobalt alloys. Due to this interaction the alloy phases

139

may be passivated under certain conditions which results in a considerable decrease of their solubility in the leaching solution.

The objective of our work was to study the electrochemical effect of metallic iron on the nickel extraction during the iron-nickel alloy dissolution in ammoniacal ammonium carbonate solutions in dependence on the redox-potential of the leached system and the phase composition of metallic iron-nickel samples. For our study two experimental methods were used: anodic polarization measurements and leaching tests. Since the ammoniacal leaching of metallic phases is an electrochemical process akin to metal corrosion anodic polarization measurements which are frequently utilized in corrosion research may provide valuable information on the mechanism controlling the nickel dissolution in ammoniacal hydrometalurgical systems.

The objective of leaching tests was to confirm the validity of the obtained knowledge about the anodic dissolution of the studied iron-nickel samples under the condition of open circuit potential leaching. The further and most important aim of the tests was to define the redox-potential ranges of the leaching system in which the maximum nickel yield to the solution is reached without the danger of its passivation.

EXPERIMENTAL

For the polarization measurements and some of the leaching tests samples were prepared which simulated the metallic iron-nickel phases in the roast products common in the industry, and which at the same chemical composition differed in their phase composition. The metallic iron-nickel phase in reduced laterities is known[5-8] to consist of γ-ferronickel alloy containing cca 50 pct Ni and α-ferronickel alloy containing cca 3-5 pct Ni, whose amount de-

pends on the conditions of reduction. Part of prepared samples corresponded to the real reduced metallic phase but another part of those samples contained besides a small amount of iron-nickel alloy also pure metallic iron and nickel. This composition does not correspond to the real reduced phase but makes it possible to judge the electrochemical effect of metallic iron on the nickel dissolution on the bases of the phase composition of the measured samples. This was the reason why pure iron and nickel specimens were used for the anodic polarization measurements.

Sintered specimens were prepared from powdered iron and nickel (both min. 99,5 pct), of grain size under 50 um. The iron-nickel mixture were obtained by mixing iron and nickel powders at a predetermined weight ratio. The metal powders were pelletized under the pressure of 350 MPa. The pellets, which had the shape of tablets, were sintered at the temperature of 700 or 900°C and for the period of 2 or 5 hours, respectively, in a flow of hydrogen. Using different conditions of sintering we obtained sintered samples containing various amounts of ferronickel alloys. Sintered pellets were used for anodic polarization measurements. Closely before the experiments the samples were grinded with emery paper and polished with alumina. Part of the sintered pellets containing 30 and 50 pct Ni was crushed in a vibrating mill and used for the leaching tests. The grain size of the crushed pellets ranged from 300 to 400 um.

The structure of the sintered specimens was analyzed by means of electron microprobe and X-ray diffractometry. It was found that the structure of the specimens sintered at 700°C for 2 hours consists predominantly of α-iron and γ-nickel grains, and of a very small amount of γ-ferronickel alloy containing cca 40 pct Ni. The γ-ferronickel alloy was formed by way of iron dissolu-

141

tion in nickel at the grain boundaries.

The matrix of the specimens sintered at 900°C for 5 hours was formed by γ-ferronickel alloy grains with cca 50 pct Ni and by α-ferronickel alloy grains with cca 5 pct Ni. The ratio of the contents of γ-ferronickel alloy and -ferronickel alloy changed proportionately the initial specimen composition. The chemical and phase composition of specimens containing 30 and 50 pct Ni sintered at 100°C for 5 hours corresponds aproximately to the composition of metallic iron-nickel phase in reduced laterites. In contrast to the fine-grained structure of the real reduced metallic phase the structure of the sintered specimens was coarse-grained.

A nickel-iron specimen containing 54 pct Ni was prepared by reducing synthetic nickel(II)-iron(III)-hydrated oxide containing 26 pct Ni and 21 pct Fe in flowing hydrogen at 800°C for 3 hours. The mixed hydrated oxide was prepared in a common way for the preparation of pure goethit. This preparation is based on the precipitation of iron(II)-hydroxide from an iron(II)-sulphate solution by means of ammonia and slow oxidation of iron(II)-hydroxide to iron(III)-hydroxide. In our experiment was prepared the mixed hydrated oxide substituting the corresponding amount of iron(II)-sulphate by nickel(II)-sulphate. The structure of nickel-iron specimens prepared in this way (the structure was determined in the same way as in the case of sintered specimens) was fine - grained, consists of only γ-ferronical alloy with 54 pct Ni and correspondig to that of iron-nickel metallic phase in reduced laterites.

The characteristics of the prepared specimens are listed in table 1.

The anodic polarization measurements were performed in a three-electrode cell, which was controlled by an ECO Instruments Electrochemoscope II measurement system.

The reference electrode was a saturated calomel electrode (SCE), and the potentials reported in our paper were converted to the standard hydrogen electrode (SHE) scale. The working electrode was the studied specimen the exposed area of which was 1 cm². The auxiliary electrode was the platinum electrode. Each anodic polarization measurement (scan rate 1mVs⁻¹) was performed starting at the approximate equilibrium potential. All measurements were conducted at a room temperature.

The leaching solution contained 1,9kmolm⁻³ NH₃ and 1,4 kmolm⁻³ (NH₄)₂CO₃, had a pH of 10,4, and was continuously saturated with argon of air. The composion of the leaching solution corresponded to that used in the nickel plant in Sered, Czechoslovakia, and was the same for both in the anodic polarization measurements and the leaching tests.

The leaching tests took three hours. The leaching process was performed in a closed mixed leaching cell at 40°C. The redox-potential of leaching reactions was measured between the platinum electrode and the saturated calomel electrode and recorded. Prior to start of the leaching tests, the leaching solution was saturated with argon for 0.5 hours.

During the leaching tests the solution was saturated with air, oxygen or their mixture. The course of dissolution and the final value of the redox-potential was controlled by the amount of the bubbled gas. According to the composition of the leached sample and to the amount of aeration gas the final value of the redox-potential was established in the course of 0.5 - 1 hour and varied in the range ±20mV.

The ratio of solid and liquid phases during the leaching tests was 200-500, depending on the nickel content in the leached sample. By this way the influence of adsorption and cooprecipitation

Table 1. Characteristics of the iron-nickel samples

sample	Ni wt pct	the method of preparation	structure
Ni	100	sintration,700°C,2h	γ-Ni
Fe	0	sintration,700°C,2h	α-Fe
Fe-Ni10, 700°C,2h	10	sintration,700°C,2h	α-FeNi⁺,Ni,very limited amount of γ-Fe-Ni*
Fe-Ni30, 700°C,2h	30	– " –	– " –
Fe-Ni50, 700°C,2h	50	– " –	– " –
Fe-Ni10, 900°C,5h	10	sintration,900°C,5h	α-FeNi°, γ-FeNi▲, ratio : =8:1
Fe-Ni30, 900°C,5h	30	– " –	α-FeNi , γ-FeNi▲, ratio : =5:3
Fe-Ni50, 900°C,5h	50	– " –	α-FeNi , γ-FeNi▲, ratio : =1:9
Fe-Ni54,	54	reduction,800°C,3h	γ-FeNi containing 56 pct Ni

α-FeNi⁺ containing max. 3 pct Ni
α-FeNi° containing 3-6 pct Ni
γ-FeNi* containing cca 35 pct Ni
γ-FeNi▲ containing cca 50 pct Ni

upon the nickel extraction was practically eliminated.

RESULTS AND DISCUSSIONS

Polarization behaviour in an aerated solution for pure iron, nickel and iron-nickel mixtures containing 10, 30 and 50 pct Ni and sintered at 700°C for 2 hours is presented in Fig. 1.

Fig. 2 shows polarization behaviour of iron-nickel mixture containing 10, 30 and 50 pct Ni, and sintered at 900°C for 5 hours. The potentiodynamic polarization curves of all the measured sintered samples in an deoxygenated solution exhibit the same course as those in an aerated solution. They differ only from the polarization curves in an aerated solution in that their active regions are more widely spread and their beginning is shifted in the noble direction, i.e. to move possitive potentials.

The results of leaching tests for

sintered specimens are schematically presented in Fig. 3 - 6. Fig. 7 shows the dependence of the final nickel yield after three hours of leaching tests on the steady value of the redox-potential for iron-nickel specimens containing 50 pct Ni and sintered at 900°C for 5 hours.

The results of the leaching tests for iron-nickel specimens containing 54 pct Ni prepared by the reduction of mixed hydrated oxide containing 26 pct Ni and 21 pct Fe are presented in Fig. 8 - 10. The course of leaching curves at various leaching redox -potentials is shown in Fig. 8 - 9. The dependence of the final nickel yield after three hours of leaching on the steady value of the redox-potential is presented in Figure 10.

The possibility of passivating the metallic iron-nickel phase of reduced laterites during their leaching in ammo-

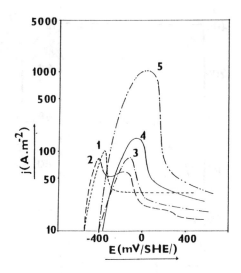

Fig. 1 Potentiondynamic polarization curves of sintered Fe-Ni mictures (700°C,2h) in aerated solution containing 1,9 kmolm^{-3} NH$_3$ and 1,4 kmolm^{-3} (NH$_4$)CO$_3$. Scan rate: 1mVs^{-1}.
1 - Fe, 2 - Fe-Ni10, 3 - Fe-Ni30, 4 - Fe-Ni50, 5 - Ni.

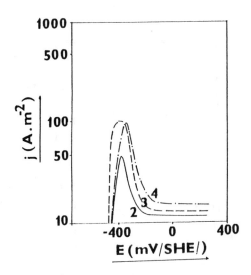

Fig. 2 Potentiondynamic polarization curves of sintered Fe-Ni mictures (900°C,5h) in aerated solution containing 1,9 kmolm^{-3} NH$_3$ and 1,4 kmolm^{-3} (NH$_4$)CO$_3$. Scan rate: 1mVs^{-1}.
2 - Fe-Ni10, 3 - Fe-Ni30, 4 - Fe-Ni50.

niacal ammonium carbonate solutions was confirmed both by the anodic polarization measurements and by their leaching

tests. The measured potentiondynamic curves for pure iron, nickel and all the studied iron-nickel specimens exhibit both active and passive regions.

The trend in the anodic polarization behaviour of nickel is similar to that of iron. Compared with iron, however, nickel shows a relatively wide active region shifted to more possitive potentials. The value of the maximum anodic current of a nickel electrode is 5-times higher than that of an iron electrode. On the other hand, the passive region of nickel is less developed than that of iron.

Polarization measurements of iron-nickel sintered mixtures indicate that the anodic dissolution is affected by the electrochemical behaviour of iron. The presence of iron in an alloy (or mixtures) leads to a decrease of active regions and their moving to more negative potentials as compared with pure nickel. The presence of iron also lowers the maximum anodic current density. These results suggest that at high anodic overpotentials the passivating of the dissolving iron-nickel specimens is very probable. The electrochemical effect of the metallic iron on the nickel dissolution is the more evident the higher content of the nickel present is bounded in the ferronickel alloy. The last mentioned influence concerns only the sintered samples. In roast reduced laterites, as known from the industrial production, the total nickel is bounded only in the ferronickel alloy.

The results of our leaching tests correspond with those of the potentiodynamic polarization measurements. The leaching curves obtained for the sintered specimens confirm the possible passivation of the metallic iron-nickel phase during the laboratory currentless leaching which corresponds to the condition of industrial leaching due to high redox-overpotentials The passivation of the leached metallic phase was observed

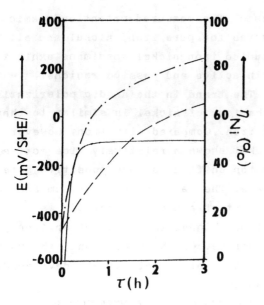

Fig. 3 The leaching behaviour of sintered Fe-Ni mictures (700°C,2h) and variation of redox-potential with time of the leaching system in aerated solutions: —— redox-potential, — — — Fe-Ni30, —·—·—Fe-Ni50.

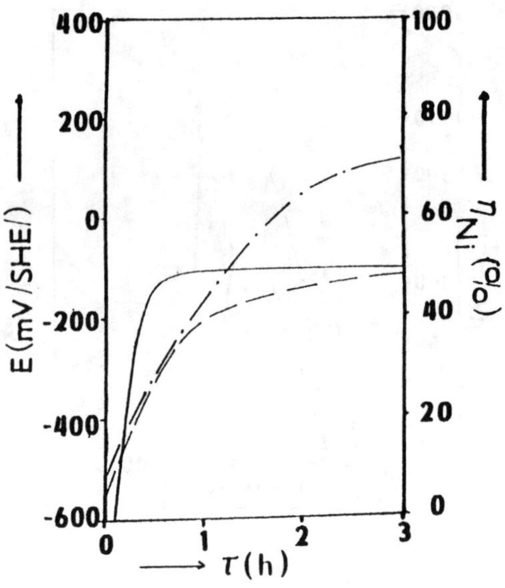

Fig. 5 The leaching behaviour of sintered Fe-Ni mictures (900°C,5h) and variation of redox-potential with time of the leaching system in aerated solutions: —— redox-potential, — — — Fe-Ni30,—·—·— Fe-Ni50.

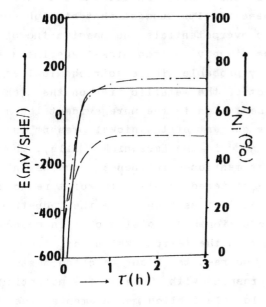

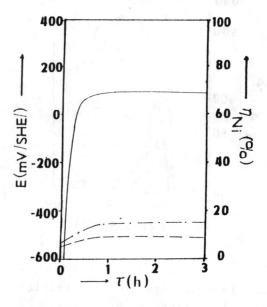

Fig. 4 The leaching behaviour of sintered Fe-Ni mictures (700°C,2h) and variation of redox-potential with time of the leaching system in aerated solutions: —— redox-potential, — — — Fe-Ni30, —·—·— Fe-Ni50.

Fig. 6 The leaching behaviour of sintered Fe-Ni mictures (900°C,5h) and variation of redox-potential with time of the leaching system in aerated solutions: —— redox-potential, — — — Fe-Ni30,—·—·— Fe-Ni50.

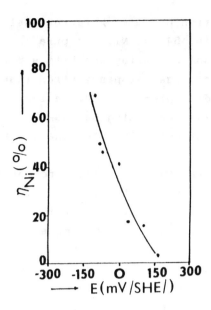

Fig. 7 The dependence of the nickel yield at the end of 3h leaching tests on the steady-state values of redox-potential of the leaching system for the sintered mictures Fe-Ni50 (900°C,5h).

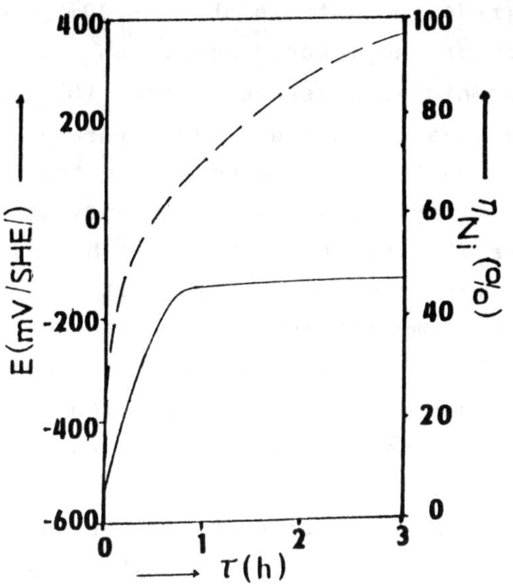

Fig. 9 The leaching behaviour of alloy Fe-Ni54 and variation of redox-potential with time of the leaching system in aerated solutions: ——— redox-potential, — — — Fe-Ni54.

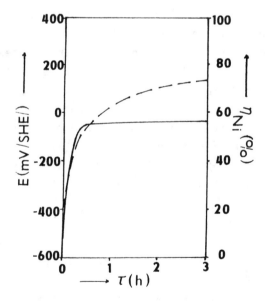

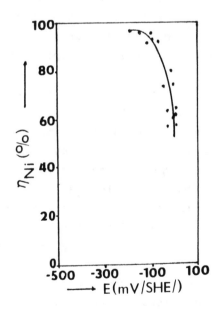

Fig. 8 The leaching behaviour of alloy Fe-Ni54 and variation of redox-potential with time of the leaching system in aerated solutions: ——— redox-potential,— — — Fe-Ni54.

Fig. 10 The dependence of the nickel yield at the end of 3h leaching tests on the steady-state values of redox-potential of the leaching system for the alloy mictures Fe-Ni54.

at redox-potentials above -100mV [SHE].
Further the leached metallic
iron-nickel phase passivates the easier
the greater part of nickel form its total quantity is bonded in the ferronickel alloy (this concerning only the samples sintered at 700°C for 2 hours), and
the more iron the specimen at the same
phase composition contains. The depende-
ces found for iron-nickel specimens containing 30 and 50 pct Ni, which according to their chemical and phase composition correspond to the reduced metallic phase in laterites, are of the greatest importance for industrial leaching.

Due to the slow nickel dissolution
from the sintered specimens the nickel
yield at the end of three hours of
leaching tests of unpassivated specimens
was low. However, the reults of those
tests are sufficient for our research
whose objective was to study the possible passivating of dissolved specimens,
which appears immediately at the beginning of the leaching, and not to study
the kinetics of nickel dissolution.

Special attention was paid to the
study of the dissolution of the
nickel-iron specimens prepared by reduction in the dependence on the leaching
redox potentials. This alloy specimens
containing 54 pct Ni have the same chemical and phase composition as the
γ-ferronickel alloy which besides
α-iron forms the major component of the
iron-nickel metallic phase of reduced
laterites and which practically contains
the total reduced nickel. The measured
leaching dependences, Fig. 8 - 9 indicate that the dissolution of ferronickel
alloy is strongly controlled by the
redox-potential of the leaching system.
An excessive aeration i.e. the potential
range above -50mV [SHE] must be avoided
because of the strong limitation of the
nickel dissolution due to the passivation of the ferronickel alloy. It was
found that the optimum redox-potential
ranges from -250 to -50mV [SHE] for the

dissolution of pure γ-ferronickel alloy
containing 54 pct Ni. The pure
γ-ferronickel alloy exhibits stronger
resistance against passivation than the
sintered samples (FeNi30, FeNi50,
900°C, 2h) containing besides
γ-ferronickel alloy also the α-iron
phase.

The results of our leaching measurements indicate that the conclusions concerning the possible passivation of sintered course-grained specimens during
the leaching process can be applied to
the reduced fine-grained iron-nickel
phase. The differences in the nickel
dissolution are caused by the different
ways of the preparation of sintered and
reduced specimens and their different
grain size. Both these mentioned factors
influence only the dissolution kinetics.

For industrial practice, i.e. for the
intensification of the reductive-roast
ammonia process of nickel lateritic
ores, the result of our research presented in this paper confirm and explain
the unfavourable effect of excessive
aeration of a leaching solution and the
necessity to mantain the
redox-potential of the leaching system
below -100mV [SHE]. Further it is useful
to control the reductive roasting of lateritic ores in order to minimize the
amount of α-iron in the reduced metallic phase and to ensure that the
γ-ferronickel alloy contain the maximum
possible amount of nickel.

References

1. Caron M.H. Ammonia leaching of nickel
   and cobalt ores. Journal of Metals,
   vol. 188, 1950, p. 67-90.
2. Osseo-Assare K., Lee W.J., Kim H.S.,
   Pickering H.W. Cobalt extraction in
   ammoniacal solution: Electrochemical
   effect of metallic iron. Metallurgical Trans.B, vol 14B, 1983, p.
   571-76.

3. Gon Quian, Cheng Weiwei, Fe Yuzkem,
   Lu Zhengya Chen Chiayung. Systemati-
   cal examination on behavior and interaction of iron and cobalt in the
   reduction roasting-ammoniacal
   leaching process for nickel oxide
   ores. Non-ferrous Metals (Quarterly,
   China) vol. 7, no. 1, March 1986, p.
   70-81.

4. Chia-Yung Che. Hydrometallurgy in
   China. In: Hydrometallurgy, Research,
   Development and Plant Practise: 3rd
   international Symposium of Hydrometallurgy, Atlanta, USA, 1986, p.
   65-85.

5. Ioffe P.A., Klementev V.V. Behavior
   of the iron compounds during the reduction roasting of nickel oxidic
   ores. Metally, no 2, p. 12-15.

6. Chandra D., Siemens R.E., Rund C.O.
   Electron-optical characterization of
   laterites treated with a reduction
   roast-ammoniacal leach system. In:
   TMS paper selection, paper no.
   A 78-23, 1976, p. 1-24. The Metallurgical Society of AIME, New York, USA.

7. De Graft J.E. The treatment of lateritic ores - a further study of the
   Caron process and the other possible
   improvements. Part I. Reduction of
   ores. Hydrometallurgy, vol.5, 1979,
   p. 47-65.

8. Jandová J., Pedlík M. Studies of metallic phase formed under the condition of lateritic FeNi-ores reduction. Kov. mater., 1991, in press.

Ecological pyro-hydrometallurgical technology of nickel pyrrhotite concentrate treatment with non-traditional and reagentless recovery of sulphur dioxide from low concentrated gases

A. G. Kitay
V. V. Mechev
Gintsvetmet Institute, Moscow, USSR
V. I. Volkov
Norilsk Mining and Metallurgical Integrated Works, Norilsk, USSR

SYNOPSIS

A non-traditional method of the poor sulphur dioxide collection by the pulp of a sulphur sulphide concentrate, the middling of an autoclave-oxidation technology of pyrrhotite concentrates has been considered. The technology developed joins organically the process cycle of Nadezhda metallurgical plant (NMP) of Norilsk mining-and-metallurgical integrated works and allows to increase considerably technical-and-economic indices thereof solving simultaneously ecological problems. The results of the first step of semicommercial tests of the flowsheet are represented. On the basis of the results obtained as well as the data of other authors main stages of the process of sulphur dioxide recovery and absorbent dissolution on the basis of iron oxides are offered.

The atmosphere pollution by harmful substances has become at present a main problem of the environment protection. One of the problems are acid rains caused by sulphur dioxide emissions in the atmosphere.

Norilsk mining-and-metallurgical integrated works ejects large amount of sulphur dioxide 60% of which is ejected with low-concentrated gases. The use of traditional methods of sulphur recovery because of geographical isolation of the integrated works leads to considerable capital and operating costs and consequently to rise in the cost of finished products. The results of a technical-and-economic comparison of sulphur recovery from poor gases containing 2-3,5% SO_2 by three versions, namely

- production of marketable sulphuric acid;
- production of technical sulphur with preliminary upgrading of gases by an ammonia-cycling method;
- neutralization by a lime suspension followed by gypsum production have shown that the most effective version

is the second one. The specific costs given amount to approximately 400 roubles per 1t of sulphur in finished products. The introduction of this method will lead to considerable increase of the manufacturing cost of the works finished products as a selling price of 1t of copper amounts to 65 roubles.

A unique location of the works requires the use of an unordinary solution of the problem indicated which is considered in the present paper conformably to the technology of Nadezhda metallurgical plant.

EXISTING AND OFFERED FLOWSHEET FOR THE TREATMENT OF NICKEL-PYRRHOTITE CONCENTRATES AT NMP

The non-ferrous metal production is realized by pyro- and hydrometallurgical technologies. Pyrrhotite concentrates are treated by an autoclave-oxidation technology (AOT). The AOT purpose is maximum separation of iron from a concentrate in a dump product as oxides, sulphur in a marketable product as elemental one and non-ferrous metals in a sulphide concentrate suitable for further pyrometallurgical treatment (fig 1).

A nickel and sulphide concentrate is treated by the technology of Outokumpu flash smelting followed by the elemental sulphur production from the gases by a methane method.

The main drawbacks of the flowsheets realized are
- high sulphur dioxide content (2,5%) in waste gases (off-gases of a methane process, converter gases etc.);
- formation of secondary iron sulphides owing to the interaction of iron oxides and sodium sulphide deteriorates the quality of a sulphide concentrate and increases the yield thereof. The flowsheet logic consisting in subsequent decrease of the iron sulphide content in the technology middlings is dis-

150

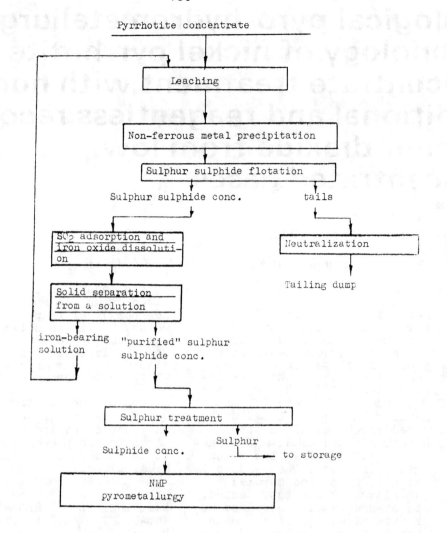

Fig 1. A flowsheet of pyrrhotite concentrate treatment and modernization thereof on the basis of the dissolution of iron oxides and sulphur dioxide absorption (new operation names are underlined)

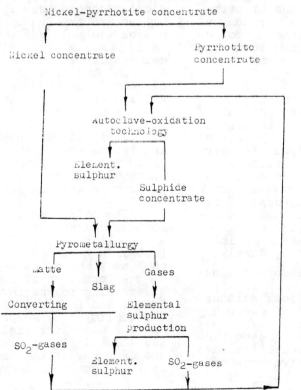

Fig 2. Ecological technology of nickel-pyrrhotite concentrate treatment

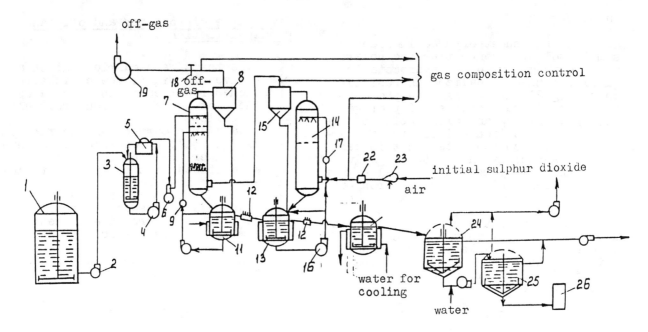

Fig 3. A plant of iron oxide leaching and recovery of waste gas sulphur dioxide:
1 - initial pulp receiver; 2 - pump; 3 - reactor; 4 - pump; 5 - breaker feeder; 6 - pump;
7,14 - absorber; 8 - drop collector; 9,17 - pulp flowmeter; 10,16 - pump; 11,13,21 - receiver-
-reactor; 12 - pH-Eh measurement; 18 - gate; 19 - exhauster; 20 - ball packing; 22 - gas flowmeter; 23 - gas mixer; 24,25 - thickener; 26 - filter

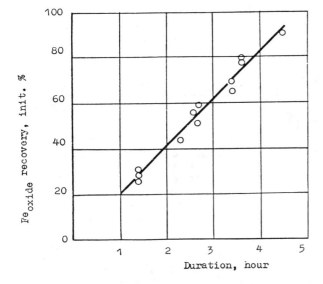

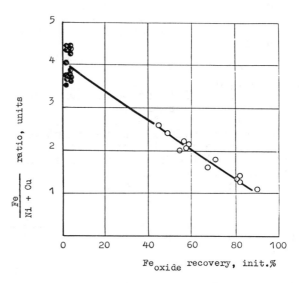

Fig 4. Dependence of iron oxide recovery on leaching duration at SO_2 content in an initial gas equal to 2,4-3,2 vol.%

Fig 5. Dependence of sulphide concentrate quality on effectiveness of iron oxide leaching from a sulphur sulphide concentrate

● initial sulphur sulphide concentrate
○ purified sulphur sulphide concentrate

turbed.

The flowsheet offered includes a simple solution converting the negative aspects of pyro- and hydrometallurgical flowsheets realized in positive ones (fig 2). It consists in the combination of the processes of poor gas desulphurization and dissolution of the absorbent, the iron oxides from a sulphur sulphide concentrate. In this case mutual "annihilation" of two negative components of hydrometallurgical and pyrometallurgical production, iron oxides and sulphur dioxide, takes place.

Main advantages of the flowsheet developed consist in the perfection of a metallurgical cycle of non-ferrous metal production and a method of sulphur dioxide recovery which at the same time is one of the main operations of hydrometallurgical production:
a) in a recovery method:
- absence of any expenditures for production, transport and storage of an absorbing agent for sulphur dioxide collection because the absorbing agent is a sulphur sulphide concentrate, one of the middlings of an existing technology;
- solid and liquid phases of the pulp of a sulphur sulphide concentrate after the treatment by gases do not require any expenditures for the utilization thereof and allow to obtain additional technical-and-economic advantages owing to the use thereof in an existing technology;
- high flexibility and stability of the process; the process can be realized at considerable fluctuations of the composition and volume of the gas treated and pulp at small fluctuations in indices;
b) in a metallurgical cycle:
- reduction of the yield of a sulphide concentrate, a main final product of autoclave-oxidizing technology and increase of the non-ferrous metal content in it by 1,5 times as compared to existing production; the treatment of such a concentrate in pyrometallurgical production will allow to obtain economic and ecological effect;
- increase of valuable component recovery and complexity of raw material use;
- reduction of reagent consumption and a manufacturing cost of market products.

APPARATUS AND RESEARCH PROCEDURE

The technology indicated has been tested on a laboratory scale and on a semicommercial plant of Norilsk integrated works.

Iron oxide dissolution and off-gas desulphurization

The tests were carried out on a plant given in Fig 3. The engineering approaches realized in the apparatus flowsheet of the absorption treatment are as follows:
- a countercurrent of the flows of pulp and sulphur dioxide; absorbers operate under the conditions of the phase countercurrent movement with multiple absorbent circulation;
- an apparatus with a movable packing was selected as an absorber and circulation receivers were hermetic reactors with turbine mixers.

The process parameters were established in a wide range:
Capacity:
- by pulp   300-1000 dm^3/hour
- by SO_2-gas   250-1000 nm^3/hour
- density of absorber reflux 30-130 m^3/m^2hour
- SO_2 content in initial gas   2,5-5,0 vol.%
- oxygen content   18,0-20,0 vol.%
- pulp temperature in circulating reactors- -veceivers   60-80°C

The following was used in semicommercial tests:
- the pulp of a pyrrhotite and sulphur sulphide concentrate of operating production; the latter contained, %: nickel - 6,0; copper - 1,5; iron - 25,0; total sulphur - 53,0; elemental sulphur - 43;
- a low-concentrated dried metallurgical gas (about 5,0 vol.% of sulphur dioxide) of sulphuric acid production.

Autoclave oxidation leaching of pyrrhotite concentrates, non-ferrous metal precipitation and sulphur sulphide flotation

The problem of the utilization of rich iron-bearing solutions produced in leaching of iron oxides from a sulphur sulphide concentrate by sulphur dioxide is of great importance in the flowsheet developed. The effectiveness of the solution thereof will determine in many respects the technical-and economic indices of the whole flowsheet. Iron oxidation (II) in iron (III) and hydrolysis of iron sulphate (III) at 400°K in the process of autoclave-oxidation leaching of pyrrhotite concentrates were selected as a primary direction of the research.

In order to determine the process indices under the conditions of combined treatment of iron-bearing solutions and a pyrrhotite concentrate all three operations indicated were tested.

Oxidation autoclave leaching of a pyrrhotite concentrate was carried out under continuous conditions in a four-section horizontal autoclave of 1,74 m^3 volume at 398-403°K and total pressure equal to 1515 kPa.

The precipitation process was realized in series of five apparatus of 0,63 m^3 capacity each. The pulp temperature in first four reactors was 363 ± 278°K and 333°K in the last one; a precipitant was an iron powder.

Sulphur sulphide flotation was carried out under continuous conditions in 36-cell flotation machine of the cell unit volume equal to 24 dm^3. The flotation was carried out according to the flowsheet including rougher, scavenger and four or two cleaner operations.

Lisintegration, sulphur flotation and sulphur smeltingsulphur treatment

Sulphur separation from a sulphur sulphide concentrate in operating production envisages subsequent realization of the following operations: desintegration and flotation of a treated pulp followed by production of sulphide and sulphur concentrates and sulphur smelting from the latter.

Lisintegration was realized under continuous conditions in a battery of three apparatus of 0,63 m^3 volume each. Sulphur flotation was carried out under continuous conditions in 36-cell flotation machine including rougher and scavenger flotation and 2 cleaner flotations of a sulphur concentrate. Sulphur smelting was carried out in autoclave-sulphur smelter of 0,18 m^3 volume.

RESEARCH RESULTS

Autoclave oxidation leaching of a pyrrhotite concentrate, non-ferrous metal precipitation and sulphur sulphide flotation

The comparison of the indices of combined treatment of an iron-bearing solution and pyrrhotite concentrate and a pyrrhotite concentrate only (the flowsheet realized) has shown that
- solid phase yield in autoclave leaching increases by 5-7%;
- iron content increases by 30-50 relat. %, including the increase of 3 - valency iron in the oxidized pulp solution;
- transfer of the sulphur of a pyrrhotite concentrate in elemental sulphur increases by 1-5 relat. %;
- degree of pyrrhotite decomposition, recovery of non-ferrous metals in a solution and specific oxygen consumption remain practically constant;

- specific consumption of a metallized precipitant will increase by 5-20% owing to increased iron (III) content in a liquid phase of an oxidized pulp;
- quality of dump tails and sulphur sulphide concentrate produced in sulphur sulphide flotation is somewhat better than the indices of the existing flowsheet.

Iron oxide leaching from a sulphur sulphide concentrate and sulphur dioxide absorption

Here are the results of the first step of the flowsheet development where a main object of the research was determination of the parameters ensuring maximum dissolution of iron oxides.

Iron is leached (90±5%) in 3,5 hours at the partial pressure of sulphur dioxide in an initial gas equal to 0,001-0,005 MPa (fig 4).

According to the data of a mineralogical analysis one of the factors preventing from complete leaching of iron oxides is "the capture" of the hematite newly formed and aggregations thereof by elemental sulphur globules with initial pyrrhotite. Elemental sulphur shields the hematite surface from a dissolved sulphur dioxide. "The capture" takes place, probably, in the process of autoclave-oxidation leaching of pyrrhotite concentrates at the moment of iron oxide and elemental sulphur formation.

The degree of iron oxide leaching determines the quality of a sulphide concentrate (fig 5) and the yield thereof. The improvement of these indices, as mentioned above, increases essentially the technical-and-economic indices of the treatment.

The optimal temperature of the leaching process allowing to leach iron oxides selectively is 345-355°K. Up to 95% of dissolved nickel and cobalt and 100% of copper is precipitated at these parameters and even at a high initial concentration of nickel in a solution (0,5-1,0 g/dm^3) in course of the treatment.

The elemental sulphur losses in leaching by sulphur dioxide determined by the content thereof in solid varied in wide ranges - from 0,5 up to 7,5%. They were the largest at high degree of iron oxide leaching and decreased temperature of the pulp treatment.

The study of an anionic composition of the solution indicates that the main forms of sulphur present in the solution are sulphate, thiosulphate and polythionate ones.

The effectiveness of leaching depends essentially on the design of an absorber acting as an aerating mixing device.

The efficiency of gas purification (65-85%) was obtained at two and three steps of absorption at the reflux density equal to 100 m^3/m^2 hour and height of a static packing layer equal to 290 mm.

As semicommercial tests indicated, the gas purification efficiency is influenced by the parameters determing the effectiveness and duration of the contact of gas and pulp in an absorber-reflux density, height of a static packing layer, free section of a support grate etc.

Main chemical reactions proceeding in leaching of iron oxides with the use of sulphur dioxide of dilute gases

The results obtained are not enough for construction of a reliable model of the chemistry and mechanism of iron oxide leaching in the process studied. Taking the data of literature and research results into account it is possible to represent a model explaining in a general form some results obtained.

We suppose that the first process stage according to the data of spectro-photometric research and literature (1) is sulphur dioxide dissolution in water:

$$SO_2 + x.H_2O \rightleftarrows SO_2 . xH_2O$$

$$SO_2 . xH_2O \rightleftarrows HSO_3^- + H_3O^+ + (x-2) H_2O$$

$$HSO_3^- \rightleftarrows SO_3^{2-} + H^+$$

50-100 g/dm^3 of sulphide dioxide was sent in a solution in carrying out the tests. The pH value was higher by 0,15-0,2 units than that of the initial one and decreased at the degree of oxide leaching over 85% only. It indicates that a reaction of protonation of a hydroxylated surface of the oxides takes place (2):

$$|O-Fe-OH+H_3O^+ \rightarrow | O-Fe^+ + 2 H_2O$$
$$S \qquad\qquad S$$

The SO_2 presence in the solution allows to assume the proceeding of the following reaction:

$$| O-Fe^+ + SO_2 \text{ ag} \rightarrow | O - FeSO_2^+$$
$$S \qquad\qquad\qquad S$$

The elemental sulphur loss observed in the tests is possible owing to the reaction:

$$S + SO_3^{2-} \rightarrow S_2O_3^{2-}$$

The proceeding of the reactions supposed above with the participation of

iron oxides and elemental sulphur shifts the equilibrium in SO_2-H_2O system. The desorption stage proceeds by the following directions:

$$| O-Fe^+ \rightarrow Fe - O^+ \text{ aq}$$
$$S$$

$$| - FeSO_2^+ \rightarrow O - FeSO_2^+ \text{ aq}$$
$$S$$

Non-ferrous metal, including copper, are precipitated from the solution:

$$2 Cu^{2+} + 2 S_2O_3^{2-} + H_2O \rightarrow Cu_2S + S^o + 4H^+ + 2SO_4^{2-}$$

Sulphur dioxide in the presence of oxygen and iron in the solution can be oxidized:

$$SO_2 + H_2O + 1/2 O_2 \xrightarrow{Fe^{3+}} SO_4^{2-} + 2H^+$$

$$SO_2 + 2Fe^{3+} + 2H_2O \rightarrow SO_4^{2-} + 4H^+ + 2Fe^{2+}$$

As a result of the increase of the pulp temperature up to 388^oK the concentration of under-oxidized sulphur compounds in the solution decreases

$$S_2O_2^{2-} + H_2SO_4 \rightarrow SO_4^{2-} + SO_2 + S^o + H_2O$$

and after-leaching of iron oxides separated takes place. The latter is confirmed by the results of our research.

Taking the experiment results obtained into account a total reaction describing the leaching process is as follows:

$$3 Fe_2O_3 + 8SO_2 + 2S^o \rightarrow 4FeSO_4 + FeS_2O_3 + FeS_4O_6$$

A stoichiometric coefficient of sulphur dioxide consumption amounts to 1,52 kg of SO_2/kg of oxidized Fe and of elemental sulphur - 0,19 kg of S^o/kg oxidized Fe.

Results of the process research-disintegration, sulphur flotation and sulphur smelting

It has been shown that in desintegration of a "purified sulphur sulphide concentrate containing 3,25% of oxidized iron the specific consumption of sodium sulphide for a reaction of sulphidizing of solid phase oxides is lower by 3,3 times than that of the existing flowsheet. In this case the elemental sulphur losses connected with side chemical interactions are reduced by 1,6 times approximately. In treating the concentrate containing 1% of oxidized iron (according to the research results the production of such a concentrate is practicable) the indices mentioned above can be improved.

The results of the flotation of disintegrated concentrates show that the quality of the concentrates produced is higher than in existing technology: the nickel content in a sulphide concentrate is 14,0% and 10,0% respectively.

The technology developed is planned for intoduction at Nadezhda metallurgical plant which allows
- to reduce sulphur dioxide emissions in the atmosphere by hundreds thous. t/year;
- to increase the concentrate treatment by 200 thous.t/year at pyrometallurgical capacities at the expense of the increase of the quality of the sulphide concentrate of autoclave-oxidation technology;
- to reduce the sodium sulphide consumption in autoclave-oxidation technology by 3% approximately.

The technical and economic indices have shown that the technology offered is the most economic as compared to alternative versions.

REFERENCES

1. Goldberk R.N., Parker V.B. Thermodynamics of solution of SO_2 (g) in water and aquous sulphur dioxide solution. Journal of research of the National Bureau of Standards, v. 90, N 5, 1985, p. 341.

2. Warren J.M., Hay M.G. Leaching of iron oxides with aquous solutions of sulphur dioxide. Transactions of the Institution of Mining & Metallurgy (Section C: Mineral processing. Extractive metallurgy), vol. 84, N 820, 1975, pp. 49-53.

3. Lung HA CHO. Removal of SO_2 with oxygen in the presence of Fe (III). Metallurgical Transactions B., vol. 17B, December, 1986, pp. 745-753.

Treatment of nickeliferous pyrrhotite

Liu Hanfei
Jinchuan Non-ferrous Metals Complex, Gansu, China

SYNOPSIS

A concept to leach nickeliferous pyrrhotite in a weak oxidation medium was proposed. In this selective leach process, Fe^{2+} is liberated into solution, elemental sulphur is dissociated from FeS, but other components in nickeliferous pyrrhotite— heavy metal sulphides and precious metal minerals, remain in undissolved state. Marvellously these undissolved minerals are agglutinated and captured by dissociated sulphur while the latter agglomerates in a form of sulphur—pearls. By means of this way, nickel, copper, cobalt, precious metals, even all components in nickeliferous pyrrhotite are expected to be recovered or made use of.

INTRODUCTION

In nickel sulphide ore deposit, pyrrhotite is frequently intergrown with pentlandite, chalcopyrite, and so on. They are floated together in the froth when dressing. The existing of pyrrhotite makes the nickel grade in concentrate decrase, increases the energy consumption in Smelter, also the load of smelting as well. Therefore INCO separated pyrrhotite out from nickel concentrate. But the separated pyrrhotite which is called as nickeliferous pyrrhotite in this paper, contains 0.5—1.0% of nickel, and some other nonferrous metals. To recover those nonferrous metals from nickeliferous pyrrhotite, many nickel producer with sulphide ore devoted themself to study on this material one after another.

There were various methods to treat nickeliferous pyrrhotite. In 1956, INCO roasted nickeliferous pyrrhotite in a reducing atmorsphere then leached the calcine with ammonia. In 1960, Falconbridge Nickel Mine Ltd employed sulphatized—roast and water reach to treat this kind of material[1]. USSR has also done some experiment in this field. China began studying on this merely recent years.

The key to the settlement of nickeliferous pyrrhotite treatment is that whether it is tenable in terms of economy.

The purpose of this study lies in overall recovering every component of nickeliferous pyrrhotite so that makes up the production expense with final products and by—products.

For this reason, selective leach was taken as a radical guideline in this study.

The experiment was in bench scale.

PRINCIPLE

Minerals

The main component of the studied material is pyrrhotite, the formula of which is

158

usually written as $Fe_{x-1}S_x$, or Fe_7S_8 or FeS.[1,2,3] For the convenience of thermodynamic caculation, FeS is typically expressed. Other components in the minority are pentlandite $(Fe,Ni)_9S_8$, chalcopyrite $CuFeS_2$, Millerite NiS, covellite CuS. Platinum exists in a form of PtAs. Palladium, gold and silver exist as metalic inter-compound.

Choose agent

The author has ever observed the effect of several agent on nickeliferous pyrrhotite. Sodium hydroxide does not react with pyrrhoyite, but some inorganic acids do, and the reaction can be expressed as followings

$$FeS + 2H^+ = Fe^{2+} + H_2S \qquad (1)$$

Nitric acid is so strong oxidative that can dissolve almost all component, never in the way of selective leaching. As far as the production scale is concerned, nitric acid is expensive, corrosive and hard on environment protection. Hydrochloric acid has the similar shortcommings as nitric acid does.

Fortunately sulphuric acid is cheap, has low volatility and low corrossoin. And, Jinchuan Nonferrous Metals Complex itself can produce sulphuric acid, therefore the latter was chosen as dissolvent.

Electrochemistry

The electrochemical reactions of components are as followings [2,5,6,7]

$$FeS = Fe^{2+} + S + 2e \qquad (2)$$
$$\varepsilon^0 = 0.065v$$

$$CuFeS_2 = Cu^{2+} + Fe^{2+} + 2S + 4e \qquad (3)$$
$$\varepsilon^0 = 0.0661v$$

$$CuFeS_2 + 8H_2O = Cu^{2+} + Fe^{2+} + 16H^+ + 2SO_4 + 16e \qquad (4)$$
$$\varepsilon^0 = 0.433v$$

$$CuS = Cu^{2+} + S + 2e \qquad (5)$$
$$\varepsilon^0 = 0.590v$$

$$Ni_3S_2 = 2NiS + Ni^{2+} + 2e \qquad (6)$$
$$\varepsilon^0 = -1.451v$$

$$Ni_3S_2 + 6H^+ = 3NiS + 6Ni^{2+} + 3H_2S + 6e$$
$$\varepsilon^0 = -0.714v \qquad (7)$$

$$NiS = Ni^{2+} + S + 2e \qquad (8)$$
$$^0 = 0.908V$$

It is obvious that the standard potential of pyrrhotite is different from that of others. That means the selective leach proccess can be realized. In order to carry out selective leaching, namely make reaction (2) occur in the preferential way, a suitable oxidant must be chosen for this purpose.

Choose oxidant

There are some option of oxidant, for instance, chlorine, oxygen and ferric ions. Their electrochemical reactions are as following [8]

$$Cl_2 + 2e = 2Cl^- \qquad (9)$$
$$\varepsilon^0 = 1.358$$

$$0.5O_2 + 2H^+ + 2e = H_2O \qquad (10)$$
$$\varepsilon^0 = 1.229$$

$$Fe^{3+} + e = Fe^{2+} \qquad (11)$$
$$\varepsilon^0 = 0.771$$

Comparing above all reactions reveals that chlorine, oxygen and ferric ions are not desirable oxidant because they have so positive potential that can not meet the requisite condition of selective leach.[9] So an attempt to find a weak oxidant was made. At last sulphur dioxide was tried. Its reaction is

$$SO_2 + 4H^+ + 4e = S + 2H_2O \qquad (12)$$
$$\varepsilon^0 = 0.451$$

It is sulphur dioxide that is used as the weak oxidant, which can dissolve FeS selectively, but can not attack $CuFeS_2$, CuS, NiS, and precious metal minerals. All these minerals will remain in residual.

Reaction (2) can be coupled with reaction (12) then a total reaction (13) occurs,

$$2FeS = 2Fe^{2+} + 2S + 4e \qquad (2)$$

$$+) \; SO_2 + 4H^+ + 4e = S + 2H_2O \qquad (12)$$

$$\overline{2FeS + SO_2 + 4H^+ = 2Fe^{2+} + 3S + 2H_2O \quad (13)}$$

Agglutinating action of element sulphur

A very interesting phenomenon was found during choosing oxidants. If oxidant used is oxygen, the dissociated elemental sulphur will natively be yellow and clean, will not mix with residual and silica, they separate from each other completely. If oxidant used is SO_2, the dissociated elemental sulphur will certainly be black, its surface is "dirty", especially this sulphur agglutinates together with undissolved heavy metal sulphide and the minerals of precious metals except silica.

In the point of view of the chemical reaction, when using oxygen as oxidant, there occurs reaction (14)

$$FeS + o.5O_2 + 2H^+ = Fe^{2+} + S + H_2O \qquad (14)$$

When using SO_2 as oxidant instead, there happens another reaction which can take place in two steps. The first step is

$$FeS + 2H^+ = Fe^{2+} + H_2S \qquad (15)$$

The second step is [10]

$$SO_2 + 2H_2S = 3/8S_8 + 2H_2O \qquad (16)$$

or $\quad SO_2 + 2H_2S = 3S + 2H_2O$

In other words, the resultant S obtained from SO_2 has different structure and different characteristics. Only does this sulphur can agglutinate undissolved component in a form of sulphur- pearl.

To control condition suitable for pearling of sulphur is significant operation in this study.

EXPERIMENT

Material and agent

Nickeliferous pyrrhotite was produced by the second dressing plant, Jinchuan Nonferrous Metals Complex. Its chemical analysis and microscope determination was listed in Table I, Table II and Table III:

Table I. The chemical analysis of nickeliferous pyrrhotite, %

Ni	Cu	Co	Fe	Pb	Zn	MgO	SiO$_2$	CaO	S
1.78	0.60	0.099	47.01	0.0032	0.014	6.60	7.88	1.41	25.00

Table II. The chemical analysis of nickeliferous pyrrhotite, gpt

Pt	Pd	Au	Ag	Os	Ru
0.43	0.27	0.28	5.18	0.092	0.15

160

Table III.Microscope determination, %

FeS	CuFeS	FeS	FeO	Nickel minerals	Gangue
67.35	5.47	11.98	5.21	2.52	0.54

Employed industrial sulphur acid and sulphur dioxide were produced in Jinchuan
Nonferrous Metals Complex.

Flowsheet and Procedure

The flowsheet is shown as Fig. 1.

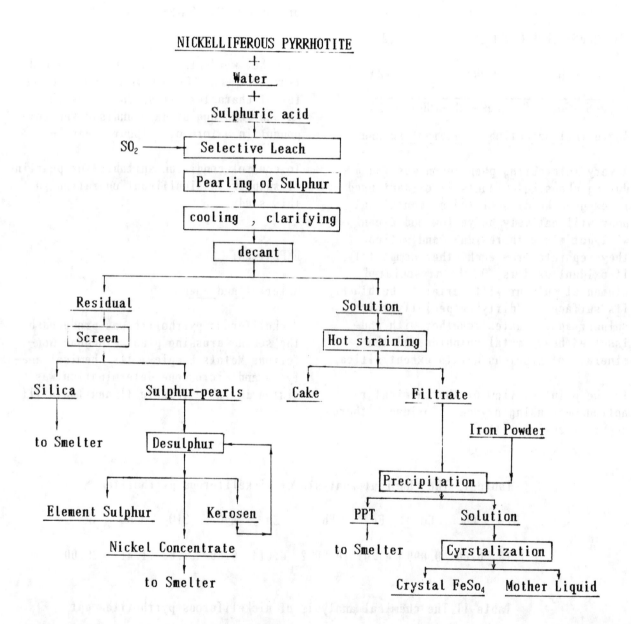

Fig.1. Tentative flowsheet for treatment of nickeliferous pyrrhotite

There is an example to explain how the procedure is.

Example

- Weigh 200 g of nickeliferous pyrrhotite, then put this sample into titanium barrel which will be moved in a minute into the autoclave.
- Pour 150 gpl of H_2SO_4 prepareed one day earlier into barrel, have the slurry well-distributed with a glass bar. Then cover the autoclave and make it tight by screws, start stirring.

- Lead SO_2 to the sealed system untill the pressure reaches 0.6 MPa. Then turn off the SO_2, switch on the electric circuit to heat.
- Hold the temperature at 90--110°C for 4 hours. This step is called as weak oxidation leach stage.
- The following is sulphur pearling stage. It was divided into two steps. Firstly, raise temperature to 130°C to make the resultant sulphur melt. Secondly decrease the temperature in a unhurried way down through the melting point of sulphur. In this interval liquid sulphur becomes sulphur pearls with diameter in about 2 millimeter. During this stage the sulphur pearls agglutinate together with undissolved minerals.
- Stop stirring when temperature is down to 110°C. Let the slurry clarify. When temperature reaches 80°C, relrase the valve untill the pressure become zero.
- Open the cover of autoclave, take out the barrel, decant the over-clarified solution, leave the sulphur pearls and silica in the bottom of autoclave.
- Strain the over-clarified solution in hot condition, then cool the solution to get ferrous sulphate crystal.
- Dump the sulpur-pearls and silica onto a screen. Sulphur-pearls leave on the screen but silica which is powder goes through the screen.

- Put the dried sulphur-pearls into kerosen, heat his mixture untill sulphur dissolves. The agglutinated minerals will precipitate when sulphur dissolves. Then pour out the kerosen to reclaim residual which is nickel concentrate containing heavy metals and precious metals minerals. Cool the hot Kerosen to seperate sulphur.

Equipment

A stainless steel autoclave heated by electricity, cooled with water and equiped with mechanical stirrer was employed. Its capacity is 2 litre. A movable titanium barrel was housed in to hold the tested slurry.

RESULT

· Solution after leaching

The ananlysis of solution in gpl was

Ni^{2+}	Cu^{2+}	Fe^{2+}	Fe^{3+}
1.08	0.0003	89.34	0.16

Correspondingly the leach rate in percentage was

Ni	Cu	Fe
29.34	0.28	90.50

Nearly one third of nickel was leached out because not all nickel exists as NiS. For instance, here are some Ni_3S_2 dissolved via reaction.[6,7] In this case Ni^{2+} must be precipitated with iron or magnesium powder. In the bench scale experiment, the precipitating of Ni^{2+} has not yet been carried out.

· Nickel concentrate

Resulted nickel concentrate contained 8.5% of nickel, the recovery of it was 80%. The precious metals grades in nickel concentrate in gpt were

Pt	Pd	Au	Ag
3.4	1.90	5.7	8.34

The enrichment rate of Pt and Pd was sevenfold.

• Crystals of ferrous sulphate

Crystals of ferrous sulphate produced in this experiment contained 17.65% of iron which was less than the theory value-- 20.14%, because purifying of leach solution has not been conducted.

• Silica residual

Silica residual contained 25.28% of sulphur because the finest sulphur-pearls went through the meshes and mixed with silica.

• Elemental sulphur

Elemental sulphur separated from Kerosen had yellow apperance yet was not analyzed.

Recovering of iron and magnecium

In this expriment, both crystal of ferrous sulphate and magnecium sulphate were obtained separately. A plan has been made to recover iron and magnecium--

. Roasting ferrous sulphate to produce sulphur dioxide and ferric oxide. The former should be returned to leach process, and the latter is the material of steel-making.
. Roasting magnecium sulphate to produce sulphur dioxide and magnesia. The former should be returned to leach process also, and the latter is to be chlorated to produce metallic magnecium.

But this plan has not been performed yet, cause there is not enough time before EMC'91.

CONCLUSSION

1. It is feasible in terms of theory and practice to liberate iron selectively and dissociate sulphur from nickeliferous pyrrhotite in the weak oxidation medium.

2. In the presence of SO_2, the resultant sulphur can agglutinate sulphide minerals of copper and nickel, and as well as precious metals minerals, but reject silica. Therfore the sulphur-pearling action can be used of to capture undissolved minerals.

3. This study provided one approach to totally recover every component in nickeliferous pyrrhotite.

REFERENCE

1. Joseph R. Boldt, Jr., The winning of nickel, pp 193-336; Longmans Canada Limited, Toronto, 1967.
2. K. A. Natarajan and Iwasaki, "Environmental leaching behaviour of copper-nickel bearing Duluth gabbro and flotation tailings," Hydrometallurgy, 10(1983), pp 339-342.
3. Mahesh C. Jha and Marcy J. Kramer "Recovery of gold from arsenical ores," Precious Metals: Mining, Extraction and Processing, eds. V. Kudryk, D. A. Corrigan and W. W. Liang, pp 337-365; The Metallurgical Society of AIME, Los Angeles, California, Feb. 27-29, 1984.
4. Michael J. Nicol, "The Non-oxidative leaching of oxides and sulphides: An Electrochemical Approach" Hydrometallurgy: Research, Development and Plant Practice, eds. K. Osseo-Asare and J. D. Miller, pp 243-260; The Metallurgical Society of AIME, Atlanta, Georgia, USA, March 6-10, 1983.
5. Hanfei Liu and Milton E. Wadsworth, "The electrochemical behavior of millerite in sulphuric acid electrolytes "Proceedings of the first symposium on physical chemistry of metallurgy, ed. Jian Han Ying, pp 360-363; The Society of nonferrous metals of China, Changsha, China, May, 1986.
6. M. H. Mao and E. Peters, "Acid pressure leaching of Chalcocite" Hydrometallurgy: Research, Development and plant practice, eds. K. Osseo-Asare and J. D. Miller, pp243-360; The Metallurgical society of AIME, Atlanta, Georgia, USA, March 6-10, 1983.
7. G. W. Warren, M. E. Wadsworth, S. M. EI. Raghy, "Anodic behaviour of chalcopy-

rrhite in sulphric acid" Hydrometallurgy: Research, Development and plant
practice, eds. K. Dsseo-Asare and J.D.
Miller, pp 261-276;The metallurgical
society of AIME, Atlanta, Georgia, USA,
March, 6-10, 1983.

8. Jian Han Ying, Metallurgical Electrochemistry, pp 242-243, Metallurgical
press, Beijing, 1983.

9. Hanfei Liu, et al., "Removal of base
metals from copper anode slime by a
two stage selective leach process";
Extractive Metallurgy and Material
Science, eds. Li Songren, Jin zhanpeng,
zhang Yonglian, pp II-63;Central South
University of Technology and Technical
University of Clausthals, Changsha,
China, Sept. 21-24, 1986.

10. Writting group, Inorganic chemistry,
Vol. 2, pp 107. People's education
press, Beijing, March, 1978.

Development of nickel smelting at Jinchuan Nickel Smelter, China

Liu Qingde
Jinchuan Nickel Smelter, Jinchang, Gansu, China

Abstract

Since the commissioning of Jinchuan Nickel Smelter(JNS) in 1968, numerous changes have been made to the process and operations resulted from the desire to increase throughput and recovery of valuble metals.

This paper gives some idea of process and metallurgical development at JNS and reviews the principal innovations to the roasting and smelting of concentrate, converting of L.G. matte and cleaning of converter slag.

The Jinchuan secondary extended project is under construction with the adoption of flash smelting technique. The future process is briefly introduced.

INTRODUCTION

Jinchuan Nickel Smelter(JNS) is part of a integrated nickel production enterprise called Jinchuan Non-ferrous Metals Complex, is located in the Jinchang city, Gansu province, north-west of china. Nickel and other by-products are extracted in the form of cast high grade matte from the concentrates produced by the concentrating mill, which handles the nickel-copper sulfide ore from three mines near the city.

Since the first electric furnace was commissioned in 1968, numerous changes have taken place due to expansion of production. The initial process(SHOWN IN FIGURE 1) consists of 3.6x52m rotary roasters, electric furnaces 3.6x7.7m Pieth-Smith converters. The concentrate feed to the smelter average 3.5-5% Ni, 1.8-2.8% Cu, 20-32% Fe, 12-23% S, 10-19% MgO, 13-21% SiO_2. In 1971, the new furnace was put into production with 0.5m wider and 0.2m higher in disign than the original furnaces, and lenthened 0.5m at the slag end in 1979.

For purpose of improving environment, reducing energy consumption and increasing production capacity, two fluid bed roasters were commissioned in 1985 and expanded with some process changes in 1989. The slag cleaning furnaces were commissioned in 1985 in order to clean the converter slag and increase mainly cobalt recovery. The current flowsheet of the Jinchuan Nickel Smelter is shown in figure 2.

Shortage of nickel required in China's basic and mechinery industries resulted in the decision of the secondary extended project construction started in 1987, adopting the flash smelting process originaly developed by OUTOKUMPU OY with the integrated furnace style practised in Karlgoolie Nickel Smelter of Western Mining Corporation in Australia. When it goes into production in 1992, the Smelter nickel throughput will be doubled.

It is the intention of this paper to give some ideas about the smelter process development and metallurgical innovations.

ELECTRIC FURNACE SMELTING

The parameters of the electric furnaces are shown in Table 1. As the new furnace is greater in hearth area than the original furnaces, it subsequently indicates the following advantages:

. Increased furnace life because of the end and side walls being further away from the electrodes and thus reducing refractories failure caused by its fury of melt turbulence.

. Reduced power consumption for per tonne of calcine fed to the furnace resulting from larger capacity.

. Decreased valuable metals losses to slag as a result of longer settling time of matte prills.

Many efforts have been made to minimise valuable metals losses to waste slag. Under normal operating conditions, nickel content in slag is considered as a function of the melt compositions (matte and slag) and the atmosphere or oxygen potential in the furnace. Taking portion of the operational data to analyse regressively, nickel percent in slag varies with assays of matte and slag accord- ing to the following equation:

$(Ni)_S = 0.1288 + 3.102 \times 10^{-3} (Ni)_m - 1.237 \times 10^{-3} (Cu)_m -4.624 \times 10^{-3} (Fe) + 6.881 \times 10^{-3} (S)_S + 6.107 \times 10^{-3} (Fe)_S -6.323 \times 10^{-3} (CaO)_S - 6.816 \times 10^{-3} (MgO)_S - 2.095 \times 10^{-3} (SiO_2)_S$

In respect to the composition of concentrates delivered to the smelter, silica was used as flux to slag with iron oxide and magnesia contained in the materials charged to the furnaces and formed by selective oxidation of iron sulphide in the furnaces. The investigation results of operation data have shown that controlling $SiO_2:(FeO+MgO)$ ratio, called "silica saturability" to 0.7 to 1.0 or $SiO_2:Fe$ ratio to above 1.4 can decrease slag nickel loss to a acceptable extent.

Study on molten slag and matte at normal conditions with sampling at different depth of the bath shown that there exist a layer of build-up mainly consisting of Fe_3O_4 at the slag-matte interface, which has the property of high viscosity and high melting point, and thus settlement and separation of matte from slag is restricted, with increasing slag losses. Therefore, it is neccessary to use reductants making a slightly reductive atmosphere or decreasing oxygen potential in the furnace to reduce the magnetite carried by returned slag or dampen its formation. Coarse coke at the size of less than 30mm was the option of the reductant in JNS, which was usually charged to near the sidewalls of a furnace at a rate of 1-3% of calcine blended with other feeds such as calcine, flux, reverts, by two drag chain conveyors. With the addition of coke, the oxidising rate of the electrodes and thus the electrode paste and shell consumption was reduced simultaneously.

The effect of slag composition on slag losses is reflected by its physical chemistry parameters such as viscosity(V), specific conductivity(k). The variations of V and k with molten slag compositions were examined at 1400 centigrade degrees and regressed as the following equations, respectively.

$$V = 6.9823 (\%SiO_2)/(\%FeO) -6.8221 \quad (poise)$$

$$k = 0.1694 (\%FeO)/((\%SiO_2)+(\%MgO))$$

$$+0.1614 \quad (ohm^{-1} cm^{-1})$$

Slag density (D) can be readily calculated with the densities of the solid metal oxides making up the slag.

$$Dest. = D_i (\%MeO)_i$$

As the slag density decresses with increasing temperature, the actual molten slag density must be less than that estimated , By regressing examined molten slag density with estimated one, its relation between molten slag density (Dmt) and the slag composition was:

$$Dmt = 3.87-1.42(\%SiO_2) -0.89(\%MgO)$$

PRACTICE OF FLUID BED ROASTING

In the second half of 1985, two 5m fluid bed roasters were commissioned in order to resolve the problems of low production capacity. High energy consump- tion and low desulfurisation rate and thus serious air pollution by the off gas containing low concentration of SO related to the original rotary kiln roasting process(Shown in Table 2). The major parameters both designed and examined during commissioning period of the initial fluid bed roasting process(see figure 3) are shown in table3. The grain size of calcine was found not satisfactory due to the cracking of the concentrate pellets immediately after they were was charged to the resters. It was later worked out that the pellet moisture is a main effect on the cracking as shown in the following table.

pellet moisture (%)	1.80	3.86	6.00
average pellet size (mm)	1.47	1.97	2.59
average calcine size (mm)	1.25	1.10	0.78
calcine size/pellet size	0.85	0.56	0.30

If the pellets were dewatered to less than 2% of moisture, the corresponding calcine grain size would be acceptable. Actual operation, however, was not achieved due to the restriction of the rotary dryer and worse working conditions rerulted from high dust produed by drying. At later time, addition of binding agent to the concentrate before pelletizing was tested but not much improvement was achieved.

In 1988, the pelletizing section was given up and replaced by feeding concentrate directly to the roaster after predrying to about 8% of moisture, and the roasters were rebult (see table 4) for the purpose of production expansion.

With fluid bed roasting, and its attributed high sulfur elimination, the sulfur dioxide content in the flue gas is increased(normally more than 5% after the electric precipitator) enabling it to be utilized to produce sulfuric acid, thus reducing atmospheric pollution, and permitting the furnace matte grade to be increased from 11% to 15% Ni approximately, with increase of furnace capacity.

CONVERTER OPERATIONAL INNOVATIONS

The furnace matte converting operation mainly performs the function of sepearating nickel and other valuable metals from iron by selective oxidation and subsequently enriching the valuble metals to the high grade matte. The conventional operation cycle in the JNS started with about 32-48 tonnes of furnace matte and blew for 50-60 minutes with adding some silica flux required for slagging and skimming a ladle of slag(about 15 ton) during this period. Then charging and skimming were done per 30-40 minutes. With 112-128 tons of furnace matte and 25-27 tons of cold scrap charged to the converter, blowing for some 330 min in total and approximately 150 tons of slag removed, the cycle was finished and the converter finally contained 30-35 tons of high grade matte. containing some 71% nickel plus copper. The results of investigations conducted in 1987 into the initial converting operation showed that the Ni. Cu. Co. concentrations in the corresponding matte and slag phases vary with the iron content in the matte phase according to relationships illustrated in figures 4 and 5 and the following regressive equations.

$(\%Ni)m = 48.87 - 0.74(\%Fe)m$

$(\%Cu)m = 25.97 - 0.467(\%Fe)m$

$(\%Co)m = 0.574 + 0.032(\%Fe)m - 7.62 \times 10^{-4}(\%Fe)m^2$

$(\%Ni)s = 3.25(\%Fe)m^{-0.531}$

$(\%Cu)s = 2.34(\%Fe)m^{-0.531}$

$(\%Co)s = 1.4(\%Fe)m^{-0.531}$

According to above rerults, it is essential for the high recovery of valuble metals into high grade matte that iron content of the matte is maintained at 16-23%. When the matte iron content is decreased to less than 10%, slag metal contents increase sharply.

The converting operation is, therefore, developed to control the iron content in matte phase at the level of more than 16 percent for each blow excepting "finish" blows by means of

a. Controlling each blow time

b. Adding furnace matte and blowing 5-10 minutes prior to skimming so as to clean the slag.

c. Adding appropriate amount of cold scrap to control the temperature.

With these measures, the nickel recovery is incraased by about 7 percent.

CONVERTER SLAG CLEANING PROCESS

The average compositions of converter slag in JNS are as follows.

	Ni	Cu	Fe	Co	S	CaO	MgO	SiO
"Green" blows	0.4	0.4	48	0.17	3.0	0.76	1.16	25
"Finished" blows	1.5	0.8	42	0.5	1.0	0.98	1.70	32

The distribution of nickel, copper, iron and cobalt among some minerals is listed below.

		Ni	Cu	Fe	Co
Metal and sulfide	"Green" blows	0.299	0.324	3.63	0.036
	"Finish" blows	0.849	0.579	1.53	0.034
Silicate and oxide	"Green" blows	0.110	0.050	32.04	0.125
	"Finish" blows	0.470	0.011	24.29	0.350
ferromagnetic	"Green" blows	-----	-----	6.55	-----
	"Finish" blows	-----	-----	7.58	-----
ferrite	"Green" blows	0.054	0.057	6.20	0.010
	"Finish" blows	0.240	0.134	6.78	0.101

The converter slag was previously returned to electric furnace for cleaning with low recovery of valuble metals, especially cobalt. In september of 1985, two 5000KVA rectangular slag cleaning furnaces were commissioned mainly handling 200 tons of final slag per day each furnace. Coarse coke 5-20mm is used as reductant at a rate of 4-6 percent of the charged slag to reduce metal oxides, calcine containing 16-19% of sulphur is charged at a rate of 15-20 percent of the slag to sulphidize the reduced metals into cobalt matte phase which is tapped, casted and delivered to the refinery for recovering cobalt, and silica flux is added to slag the iron oxide, finally skimmed and discarded. The main parameters of the furnace is listed in table 5. The furnace operation is conducted every 8 hours periodicaly with some 60 tons of converter slag charged and approximately 15 percent of cobalt matte produced containing 1.0-1.2 percent of cobalt, 12-13 percent of nickel. The furnace slag contains 0.1-0.2 percent of cobalt, 0.09-0.11 percent of nickel, 20-22 percent of sulphur. The cobalt enriching coefficient (ratio of matte cobalt content to charged converter slag cobalt content) is 3-3.5.

IMPLICATION OF FLASH SMELTING TECHNIQUE

In 1984, the flash smelting technique was adopted as the smelting process of the Jinchuan second extended project. Construction of the flash smelting engineering started in 1987 and will be finished in 1992.

On account of the high magnesia content in concentrate, the integrated furnace style developed by KNS of WMC has been introduced into the process, which mainly consisting of a reaction shaft, a settler, a uptake and an appendage. The process flowsheet and some parameters of the flash furnace are shown in figure 6 and table 6, respectively.

The process is designed to use oxygen enriched (about 42% O_2) and preheated ($200°c$) air and pulverized coal instead of oil to replenish reaction shaft heat when required. There are six consumable electrodes,(three in a phase)installed in the appendage, supplyed by two monophase transformers with 8000KVA of power. The smelting process will be controlled with the MOD-300 distributed control system manufactured by Taylor Company.

Table 1 Electric Furnace Details

		Original	Redesigned
Dimension	(m)	21.5×5.5×4.0	22×6×4.2
Hearth Unit Power	(KVA/m^2)	139.5	126
Tap Hole Hight From Bottom	(cm)	45	45
Skim Hole Hight From Bottom	(cm)	155	155
Bath Depth	(cm)	220	220
Matte Phase Depth	(cm)	70	70
Electrode Number		6	6
Electrode Diameter	(cm)	100	100
Power Input Per Ton of Clcine	(KWH)	650	590
Ni Loss in Slag	(%)	0.2	0.18
Matte Temperature	(°c)	1100-1250	1100-1250
Slag Temperature	(°c)	1300-1350	1300-1350

Table 2 Rotary Roaster Parameters

Dimention	(m)	3.6 52
Concentrate moisture	(%)	8-10
Concentrate Residual Time in Roaster	(min)	59
Oil Consumption	(Kg/hr)	1200
Calcine Ni Content	(%)	5.6
Calcine Temperature	(°c)	500-550
Sulpur Elimination	(%)	30-50
SO$_2$ Concentration in off Gas	(%)	< 3

Table 3 Original Fluid Bed Roasting Parameters

		Designed	Actual
Concentrate Pellet Moisture	(%)	6	5-6
Pellet Size	(mm)	-6	-6
Bed Temperature	(°c)	600	600-700
Flow Rate Through Bed	(m/sec)	2.65	---
Air To Feed Ratio	(Nm^3/Kg)	0.42	0.5
Calcine Size	(mm)	---	0.7
Sulphur Elimination	(%)	45	45-60
SO Concentration In Off Gas	(%)	11	8
Off Gas Temperature	(°c)	450-500	500

Table 4 New Fluid Bed Roaster Parameters

		Designed	Actual
Bed Height	(m)	1.4-1.7	1.55
Bed Temprature	(°c)	650±20	710±30
Flow Rate Through Bed	(m/sec)	2.56-2.76	3.0-3.1
Air Flow	(Nm^3/h)	20500-22000	24000
Air Pressure	(Pa)	15×10^3	(10-11.5)×10^3
Feed Size	(mm)	2	3.2-3.65
Feed Moisture	(%)	6-8	5-7
Off Gas Rate	(Nm^3/h)	24000-27000	29000
Off Gas Temperature	(°c)	650	680-710

Table 5 Slag Cleaning Furnace Details

Dimention	(m)	10.3×5.0×3.2
Electrode Number		3
Electrode Diameter	(mm)	900
Transformer Power	(KVA)	5000
Matte Hole Hight	(mm)	396
Slag Hole Hight	(mm)	926
Bath Depth	(cm)	130-150
Matte Phase Depth	(cm)	55-65
Slag Temperature	(°c)	1300-1350
Matte Temperature	(°c)	1200-1250

Table 6 Designed Flash Furnace Details

R/S Dimention	(m)	Diam 6 Hgt 6.4
Process Air Rate	(Nm^3/h)	32600
Cons. Burner Number		4
Settler Hearth Length	(m)	15.35
Appendage Hearth Length	(m)	16.35
Hearth Width	(m)	6.414
Bath Depth	(cm)	130
Slag Phase Depth	(cm)	80
Off Gas Flow Rate	(Nm^3/h)	62000
Off Gas Temperature	(°c)	1380
Cons. Composition	(Wt %)	Ni 7, Cu 4, Fe 41,
		S 27, MgO 6.5, SiO$_2$6.5

REFERRENCES

1. Jin Zhaimiao, etc, Relation betwen the mineralogy and process technology as per minerals of Jinchuan Mines, 1987.

2. Li Baoping Zhang Yichuan, Lu Xiaoping,

Non-ferrous metals extraction, 8 (1988), p. 27.

3. Liu Qingde, Gansu Non-ferrous metals, 3(1987), p. 7.

4. Liu Qingde, Wang Shuqing, Non-ferrous metals extraction, 4(1989), p. 14.

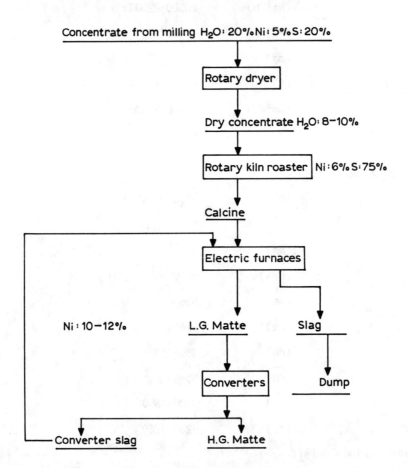

Fig. 1 INITIAL FLOWSHEET OF JINCHUAN SMELTER

173

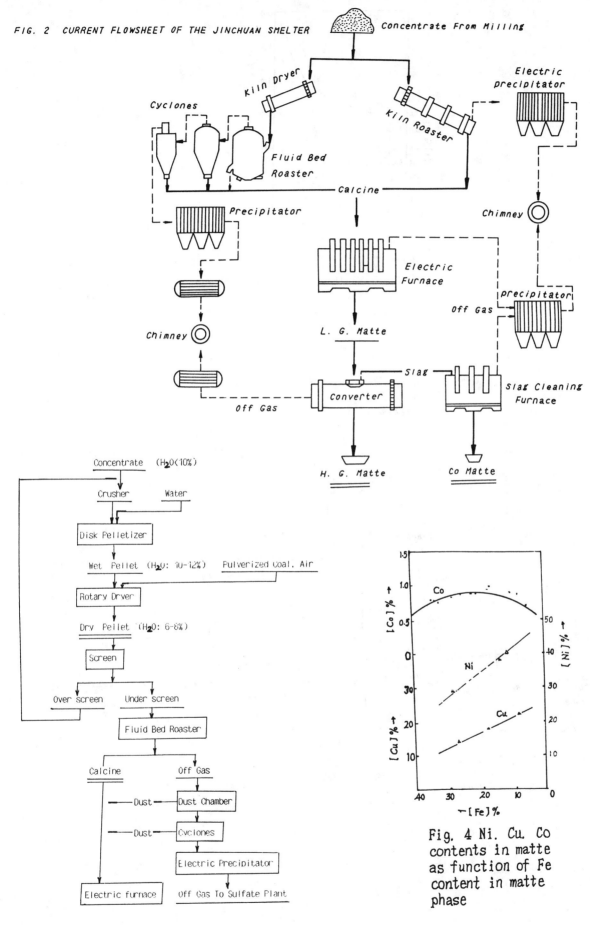

FIG. 2 CURRENT FLOWSHEET OF THE JINCHUAN SMELTER

Concentrate From Milling

Kiln Dryer

Cyclones

Kiln Roaster

Electric precipitator

Fluid Bed Roaster

Calcine

Chimney

Precipitator

Electric Furnace

Precipitator

Off Gas

Chimney

L. G. Matte

Slag

Off Gas

Converter

Slag Cleaning Furnace

H. G. Matte

Co Matte

Concentrate (H$_2$O<10%)

Crusher   Water

Disk Pelletizer

Wet Pellet (H$_2$O: 10-12%)   Pulverized Coal, Air

Rotary Dryer

Dry Pellet (H$_2$O: 6-8%)

Screen

Over screen   Under screen

Fluid Bed Roaster

Calcine   Off Gas

—Dust— Dust Chamber

—Dust— Cyclones

Electric Precipitator

Electric furnace   Off Gas To Sulfate Plant

FIG. 3 5M^2 FLUID BED ROASTER FLOWSHEET

Fig. 4 Ni. Cu. Co contents in matte as function of Fe content in matte phase

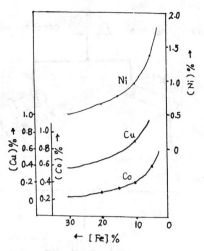

Fig. 5 Ni. Cu. Co
contents in conv. slag
as function of Fe
contents in matte
phase

FIG. 6 FLASH SMELTING FLOWSHEET OF JNS

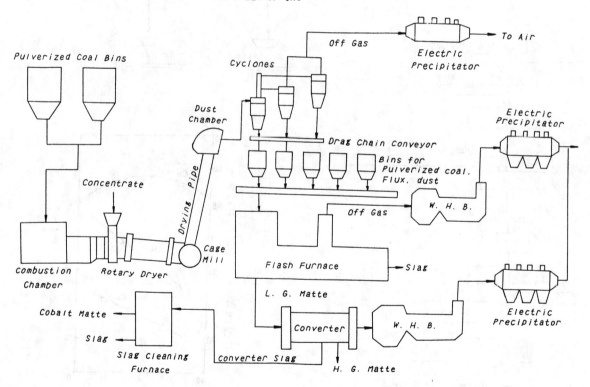

Extraction of Co(II) and Ni(II) with Cyanex 272

D. Maljković
Z. Lenhard
M. Balen
Faculty of Metallurgy, University of Zagreb, Croatia, Yugoslavia

SYNOPSIS

Extraction of cobalt(II) and nickel(II) with the commercial extractant Cyanex 272 (bis/2,4,4-trimethylpentyl/ phosphinic acid) diluted in kerosene or 2,2,4-trimethylpentane from sulphate and sulphate-chloride solutions was studied. The purpose of this investigation was the applicability of Cyanex 272 as an extractant in some extraction stages in the process of recovering cobalt and nickel from laterite ore (the Goleš locality, Yugoslavia).

Due to some merits of the ore leaching with sulphuric acid and possible use of different solvents besides Cyanes 272 (e.g. ethers, ketones) in different stages of the process, investigations were made in systems containing sulphuric acid and a mixture of sulphuric and hydrochloric acids. Prepared and real (leach liquor) samples were extracted with 10 % Cyanex 272 at 25 °C using initial phase volume ratio (organic/aqueous), r_i = 0.5 and 1. A mutual influence of cobalt and nickel on their distrubution under certain conditions was observed. In the case of the mixture of cobalt and nickel in sulphate (0.5 M) - chloride (4 M) solutions best separation was obtained with r_i = 0.5. An effect of sodium on the distribution ratio of cobalt was also found. In such system at certain conditions three stable liquid phases were observed. In the heavy organic phase (the third phase) cobalt concentration was higher than in the light organic phase.

INTRODUCTION

The phosphinic extractant bis (2,4,4-trimethylpentyl) phosphinic acid was developed by American Cyanamid Company under the trade name Cyanex 272.[1,2] Like that of corresponding phosphoric and phosphonic acids its use in solving problems of recovery and separation of cobalt and nickel from aqueous solutions was received with great attention. The interest was supported by results showing that the separation of cobalt and nickel improved with the change of extractant from phosphoric to phosphonic to phosphinic acid based derivatives.[3] Previous studies of Cyanex 272 were concerned with extraction stoichiometry, the structure of the formed compounds, equilibrium constants, the effect of pH, organic phase loading, phase modifier effect, temperature effect, kinetics of extraction and microemulsion formation. They also confirmed the high selectivity for cobalt over nickel as its major advantage.[4-13] The objects of investigation were mostly prepared samples containing cobalt(II) or nickel(II) or their mixture. Most data in literature refer to sulphate solutions but the application of Cyanex 272 in chloride [2,7,10] and nitrate [6] solutions has also been demonstrated. To describe the behaviour of Cyanex 272 in systems containing other metals several studies have been undertaken.[14-18] Unfortunately, not much information has been gained on possible effect of matrix composition on extraction of cobalt and nickel from real samples. Recently some results of Cyanex 272 use in extraction of cobalt and nickel from leach liquor have become available.[10,12]

The purpose of this investigation was to test the applicability of Cyanex 272 as extractant in some extraction stages in the process of recovering cobalt and nickel from Yugoslav laterite ore (the Goleš locality). Although Cynex 272 gives better results of separation from sulphate solutions, experiments were carried out in mixed chloride-sulphate media too. Namely, preliminary examinations showed that iron(III) had to be removed before the extraction of cobalt and nickel. The most convenient is extraction with some carbon-bonded oxygen-donor extractants (e.g. ether) from solutions of hydrochloric acid. In the presence of the elements that have to be removed (e.g. aluminium, chromium) it is possible by the precipitation-dissolution procedure, without introducing an additional operation, to return to sulphate medium wich is more convenient for extraction with Cyanex 272.

In this paper, results are presented of the extraction of cobalt and nickel with Cyanex 272 from sulphate and mixed chloride-sulphate solutions, and from pretreated leach liquors.

EXPERIMENTAL

Apparatus

A Radiometer pH-meter 28 was used for pH measurements. A Varian UV, visible spectrophotometer type M635 and a Perkin Elmer atomic absorption spectrometer type 503 were used for analytical measurements.

Reagents

Bis (2,4,4-trimethylpentyl) phosphinic acid, Cyanex 272, kindly supplied by American Cyanamid Company was used without further purification. Purity was found to be 89 %. Kerosene was obtained from INA-Oil Refinery, Sisak. All other chemicals were analytical reagent grade.

Samples and sample preparation

The samples extracted were solutions prepared from pure chemicals and pretreated leach liquors. The content of metals in prepared samples was analogous to that in real samples (leach liquors). Leach liquor samples were obtained by leaching ground ore (laterite ore the Goleš locality, Yugoslavia) of grain size < 0.125 mm in a device of own construction at room temperature with a mixture of sulphuric (1 M) and hydrochloric (8 M) acids. An average content of metals determined in leach liquor was: 0.25 g/dm³ for cobalt and 2.9 g/dm³ for nickel. Table 1 shows that besides cobalt and nickel leach liquor contains also other elements, which may disturb the course of extraction. Under appropiate conditions, at equilibrium pH values below 6, iron (III), aluminium(III), chromium(VI) and magnesium (II) with Cyanex 272 can be extracted, if they are not present in high concentration.[4,15] Therefore iron(III), which was present in leach liquor in high concentration (46.0 g/dm³), was extracted by a mixture of ethyl ether - n-butanol.[10] After its removal as result of the nature of the mixed solvent (high co-extraction of water in the organic phase followed by reduction of aqueous phase volume) cobalt and nickel concentrations in the sample increased. (Table 1)

In further procedure aluminium and chromium were removed from the solution by precipitation, with the addition of ammonium hydroxide until pH 8. The precipitate of hydroxides was then centrifuged and dissolved with 1 M sulphuric acid to free the coprecipitated cobalt. The precipitation-dissolution procedure was repeated several times depending on the quantity of free cobalt. By choice of acid for dissolution and by repetition

Table 1   Content of metals in leach liquor after pretreatment. Stages: A - removal of iron(III) by extraction, B - removal of aluminium(III) and chromium(VI) by precipitation

	Content of metals (g/dm³)	
	after stage A	after stages A and B
Co	0.35	0.06
Ni	4.22	1.01
Fe	0.005	traces
Al	2.44	0.002
Mn	1.78	0.42
Cr	0.35	0.001
Mg	6.50	2.22

of this procedure it was possible to obtain sulphate, chloride, chloride-sulphate or other appropriate medium for extraction with Cyanex 272 as the next step. By operations described the content of metals was sufficiently reduced. (Table 1). The centrifugates obtained were collected and extracted with Cyanex 272.

Methods of analysis

Concentrations of cobalt (using nitroso-R salt) and nickel (using dimethylglioxime and $K_2S_2O_8$) were determined by apsorbance measurements. Iron was determined by complexometric titration. All other metals were determined by atomic absorption spectrometry.

Extraction equilibrium procedure

The experimental conditions of extraction were chosen on the bases of previous examinations of various factors which influence the extraction of cobalt and nickel such as temperature, initial phase volume ratio, phase contact time, initial pH and diluent type.[10] Extraction was carried out in graduated cuvettes (15 cm³). Samples prepared by dissolution of metal salts or samples of pre-treated leach liquor were adjusted to appropriate pH and mixed in chosen ratio with a 10 % solution of Cyanex 272 in diluent (kerosene or 2,2,4-tri-methylpentane). The pH values were roughly adjusted with solid sodium hydroxide and finely with ammonium hydroxide. Systems were thermostated for 15 minutes at 25 °C and after that vigorously shaken four times for 30 seconds with 2-minute intervals. The pH values of the aqueous phase before mixing of phases and after their separation were taken as the initial and the equilibrium pH values respectively. Whenever possible metal content was determined in both (or in all of three) phases. If this was not possible, the concentration was calculated from the difference. On the basis of concentration values distribution ratio (D), recovery factor (R) and separation factor (α) were calculated.

RESULTS AND DISCUSSION

Extraction of nickel from sulphate solution

In extension of previous investigation[10] extraction of nickel from samples containing 1 g/dm³ of nickel was examined in the range of initial pH 7 - 10, at initial phase volume ratios (organic to aqueous), r_i, 0.5 and 1. Maximum extraction was obtained at pH 9 for both initial phase volume ratios. Recovery factor was much higher for r_i = 1. (Fig. 1).

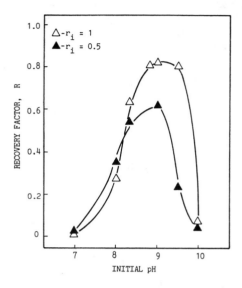

Fig. 1 The effect of initial pH value on extraction of nickel with 10 % Cyanex 272 in 2,2,4-trimethylpentane from sulphate solution. c_{Ni}^i = 1.0 g/dm³.

Extraction of cobalt and nickel from sulphate solution

Extraction of cobalt and nickel from a mixture prepared with Co/Ni ratio 10 was carried out in the range of initial pH values 6.5 - 9 (Fig. 2). A mutual action of cobalt and nickel was observed. (Fig. 3 and 4). The influence of nickel on cobalt was much stronger; it manifested itself as shifting, at lower pH values, of the curve that plots recovery factor for cobalt as a function of the pH (Fig. 3). The effect for the chosen pH values for both metals, expressed numerically, is given in Table 2.

Extraction of cobalt and nickel from chloride -sulphate solution

From samples containing mixture of cobalt and nickel in increased concentrations corresponding to their higher contents in leach liquor after removal of iron by ether extraction was carried out by Cyanex 272 dissolved in kerosene. This is more easily available diluent which exibits almost the same behaviour as 2,2,4-trimethyl-pentane.[10] Results are shown in Fig. 5 for the more favourable initial phase volume ratio 0.5.

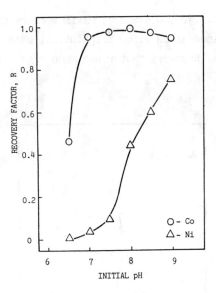

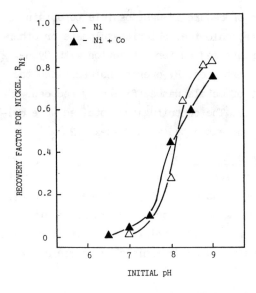

Fig. 2 The effect of initial pH value on extraction of nickel and cobalt from their mixture with 10 % Cyanex 272 in 2,2,4-trimethylpentane from sulphate solution. Experimental conditions:

$$c_{Co}^i = 0.09 \text{ g/dm}^3, \quad c_{Ni}^i = 1.0 \text{ g/dm}^3, \quad r_i = 1.$$

Fig. 4 Comparison of recovery factors for nickel obtained by single metal extraction and extraction in the presence of cobalt. Experimental conditions: 10 % Cyanex 272 in 2,2,4-trimethylpentane, sulphate solution $c_{Co}^i = 0.09 \text{ g/dm}^3, \quad c_{Ni}^i = 1.0 \text{ g/dm}^3, \quad r_i = 1$

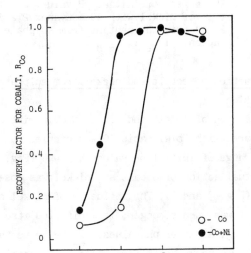

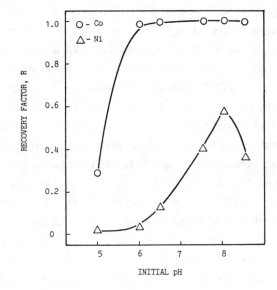

Fig. 3 Comparison of recovery factors for cobalt obtained by single metal extraction and extraction in the presence of nickel. Experimental conditions: 10 % Cyanex 272 in 2,2,4-trimethylpentane, sulphate solution, $c_{Co}^i = 0.09 \text{ g/dm}^3, \quad c_{Ni}^i = 1.0 \text{ g/dm}^3, \quad r_i = 1.$

Fig. 5 The effect of initial pH value on extraction of cobalt and nickel from their mixture with 10 % Cyanex 272 in kerosene from chloride-sulphate solution. Experimental conditions: $c_{Co}^i = 0.35 \text{ g/dm}^3, \quad c_{Ni}^i = 3.5 \text{ g/dm}^3, \quad r_i = 0.5$

Best results of extraction were obtained at pH 6.5 when distribution ratios were 569 for cobalt and 0.27 for nickel, giving a separation factor, α_{Ni}^{Co}, 2100.

Effect of sodium on distribution ratio

During extraction from chloride-sulphate solutions the effect of sodium content in the extraction system on the extraction of cobalt and nickel was also observed. Systems containing dif-

Table 2 Recovery factors (R) and distribution ratios (D) for cobalt and nickel obtained by extraction of single metals and their mixture with Cyanex 272 in 2,2,4-trimethylpentane. Initial volume ratio $r_i = 1$, initial concentrations of metals $c^i_{Co} = 0.08$ g/dm³ and $c^i_{Ni} = 1.0$ g/dm³

	initial pH	single		mixture	
		R	D	R	D
Co	7	0.146	0.17	0.955	21.14
	8	> 0.973	> 35.62	0.994	170.00
Ni	7	0.012	0.01	0.039	0.04
	8	0.274	0.38	0.443	0.79

Table 3 The effect of sodium content in extraction system on recovery factors of cobalt and nickel. Extraction with 10 % Cyanex 272 in kerosene from chloride (4 M) - sulphate (0.5 M) solution containing the mixture of cobalt and nickel. Experimental conditions: $r_i = 1$, initial pH = 8, $c^i_{Co} = 0.385 \pm 0.02$ g/dm³, $c^i_{Ni} = 3.060 \pm 0.06$ g/dm³

Na (g/dm³)	48	66	80	91
R_{Co}	0.823	0.930	0.964	0.978
R_{Ni}	0.316	0.484	0.592	0.652

Table 4 Results of extraction of cobalt and nickel with Cyanex 272 under conditions of third liquid phase formation. Experimental conditions: solvent - 10 % Cyanex 272 in kerosene, aqueous - chloride (4 M) - sulphate (0.5 M) solution, initial metal concentration - $c^i_{Co} = 0.35$ g/dm³, $c^i_{Ni} = 2.5$ g/dm³, initial pH = 9.0, $r_i = 0.5$.

	Metal concentration (g/dm³)			Partial distr. ratios	
	Aqueous	Light Org.	Heavy Org.	D^H_L	D^H_A
Co	0.0336	0.367	1.500	4.1	44.6
Ni	–	0.148	0.380	2.6	–

ferent contents of sodium (at constant pH value 8) were examined and results are given in Table 3. It is evident that with increase in sodium content the recovery factor of both elements increased.

The third phase appearance

Under certain conditions of extraction of cobalt and nickel with Cyanex 272 from chloride-sulphate solutions formation of three stable liquid phases was observed. The third phase i.e. the heavy organic phase, which was completely clear and well defined, contained higher concentrations of both metals than the other organic phase. This was much more obvious in the case of cobalt where the partial distribution ratio of the heavy organic phase to light organic phase was 4.1. (Table 4).

Extraction from real samples

After pretreatment of leach liquor samples, as described in the experimental part of the paper, extraction of Co, Ni, Mn and Mg with 10 % Cyanex 272 dossolved in kerosene at initial phase volume

Table 5 Results of extraction of cobalt, nickel, manganese and magnesium
from pretreated leach liquors with Cyanex 272

Initial pH	Initial phase volume ratio r_i	Initial concentration of metals (g/dm³)				Equil. pH	Distribution ratios			
		Co	Ni	Mg	Mn		D_{Co}	D_{Ni}	D_{Mg}	D_{Mn}
7.8	1	0.069	1.160	2.379	0.460	5.6	10.2	0.05	0.52	**
8.3	1	0.056	0.866	2.060	0.384	6.4	13.0	*	1.01	191
8.3	2	0.056	0.866	2.060	0.384	7.2	15.3	*	2.02	254
9.0	2	0.066	1.058	2.379	0.460	8.7	0.08	*	2.58	**

* too low concentration of nickel

** too high concentration of manganese

ratios 0.5 and 1 at pH values 7.8, 8.3 and 9.0 was examined. (Table 5).

The results show high separation of cobalt over nickel followed by total co-extraction of manganese. Co-extraction of magnesium was also present and its distribution ratio increased with increase in pH over the investigated range. Magnesium concentration was lowest at pH 7.8, with the separation factor of cobalt over magnesium, α_{Mg}^{Co}, 19.6.

CONCLUSIONS

Cobalt and nickel were extracted from sulphate solutions, from chloride-sulphate solutions and from pretreated leach liquor samples of laterite ore using Cyanex 272.

Mutual influence of extracted metals, the effect of sodium content on extraction and the formation of the third liquid phase were observed.

Satisfactory extraction and separation of cobalt over nickel were also obtained in mixed chloride-sulphate solutions.

Extraction from real samples was followed by some interferences which were caused by the presence of other components.

A study of procedures and optimal conditions of separation of manganese and magnesium from cobalt is in progress.

Acknowledgment

This work was performed with support of Self--Managment Council for Scientific Research of R. Croatia.

REFERENCES

1. W.A.Rickelton, A.J.Robertson and D.R.Burley, Selective Extraction of Cobalt from Aqueous Solutions also Containing Nickel, Using Phosphinic Acid Compounds, US Patent 4,348,367 (1982).

2. Cobalt - Nickel Separation Using CYANEX* 272 Extractant, Preliminary Technical Brochure, American Cyanamid Company (July 1982).

3. J.S.Preston, Solvent Extraction of Cobalt and Nickel by Organophosphorous Acids. Comparison of Phosphoric, Phosphonic and Phosphinic Acid Systems, Hydrometallurgy, 9, 115 (1982).

4. W.A.Rickelton, D.S.Flett and D.W.West, Cobalt - Nickel Separation by Solvent Extraction with Bis (2,4,4-trimethylpentyl) Phosphinic Acid, Solvent Extraction and Ion Exchange, 2, 815 (1984).

5. P.R.Danesi, L.Reichley-Yinger, C.Cianetti and P.G.Richert, Separation of Cobalt and Nickel by Liquid-Liquid Extraction and Supported Liquid Membranes with Bis (2,4,4-trimethyl-pentyl) Phosphinic Acid /Cyanex 272/, Solvent Extraction and Ion Exchange, 2, 781 (1984).

6. P.R.Danesi, L.Reichley-Yinger, G.Mason, L.Kaplan, E.P.Horwitz and H.Diamond, Selecti-vity-Structure Trends in the Extraction of Co(II) and Ni(II) by Dialkyl Phosphoric, Alkyl Alkylphosphonic and Dialkylphosphinic Acids, Solvent Extraction and Ion Exchange, 3, 435 (1985).

7. I.Szilassy, Gy.Miketa, K.Vadasdi, Some Experiences in the Separation of Co-Ni with PC 88A and Cyanex 272 in Mixed Sulphate - Chloride Media, International Solvent Extraction

Conference, ISEC'86, Preprints Vol. II, München, 1986, p. 519.

8. Xun Fu and J.A.Golding, Solvent Extraction of Cobalt and Nickel in Bis (2,4,4-trimethylpentyl) Phosphinic Acid /Cyanex 272/, Solvent Extraction and Ion Exchange, 5, 205 (1987).

9. Xun Fu, J.A.Golding, Equilibrium and Mass Transfer for the Extraction of Cobalt and Nickel from Sulfate Solutions into Bis (2,4,4-trimethylpentyl) Phosphinic Acid - - Cyanex 272, Solvent Extraction and Ion Exchange, 6, 889 (1988).

10. Da.Maljković, Z.Lenhard and M.Balen, On the Extraction of Cobalt(II), Nickel(II) and Iron(III) from Acidic Leach Liquors, International Solvent Extraction Conference, ISEC'88 Conference Papers, Vol IV, Moscow, 1988, p. 269.

11. E.Paatero, P.Ernola, J.Sjöblom and L.Hummelstedt, Formation of Microemulsion in Solvent Extraction Systems Containing Cyanex 272, International Solvent Extraction Conference, ISEC'88, Conference papers, Vol II, Moscow, 1988, p. 124.

12. Liu Daxing, Separation of Cobalt and Nickel from High Nickel Matte Leach Liquor by Solvent Extraction with Cyanex 272, International Solvent Extraction Conference, ISEC'90, Abstracts, Kyoto, 1990, p. 208.

13. Nai-Fu Zhou, Jinguang Wu, P.K.Sarathy, Fuan Liu and R.Neuman, A Comparison of Aggregates Formed by Organophosphorus Extractants in Solvent Extraction Systems, International Solvent Extraction Conference, ISEC'90, Abstracts, Kyoto, 1990, p. 38.

14. Ke-an Li and H.Freiser, Extraction of Lanthanide Metals with Bis (2,4,4-trimethylpentyl) Phosphinic Acid, Solvent Extraction and Ion Exchange, 4, 739 (1986).

15. V.M.Rao and S.K.Prasad, Some Investigations of Solvent Extraction of Chromium(VI) by Bis (2,4,4-trimethylpentyl) Phosphinic Acid (Cyanex 272), International Solvent Extraction Conference, ISEC'88, Conference papers, Vol I Moscow, 1988, p. 338.

16. E.Figuerola, N.Miralles, A.Sastre and M.Aguilar, Extraction of Zinc(II), Cadmium(II), Copper(II) and Iron(III) by Cyanex 272, International Solvent Extraction Conference,

ISEC'88, Conference papers, Vol I, Moscow, 1988, p. 318.

17. Xun Fu, Zhengshui Hu, Yide Liu and J.A.Golding, Extraction of Sodium in Bis (2,4,4-trimethylpentyl) Phosphinic Acid - Cyanex 272: Basic Constants and Extraction Equilibria, Solvent Extraction and Ion Exchange, 8, 573 (1990).

18. A.M.Sastre, N.Miralles and E.Figuerola, Extraction of Divalent Metals with Bis (2,4,4-trimethylpentyl) Phosphinic Acid, Solvent Extraction and Ion Exchange, 8, 597 (1990).

Copper

Technological and technical problems of copper production from chalcosine concentrates in a flash furnace

J. Czernecki
S. Sobierajski
Z. Smieszek
Institute of Non-ferrous Metals, Gliwice, Poland

INTRODUCTION

Polish copper concentrates, like the majority of concentrates produced in the world, contain 18 to 27% Cu, whereas the sulphur and iron contents are several times less, being 7 to 11% and 2 to 5%, respectively.

Chalcosine is the main copper-bearing mineral; bornite and chalcopyrite occur rarely. The chemical composition of the gangue, consisting mainly of aluminium silicates, quartz and dolomite, does not require the application of fluxes during smelting. The presence of the organic carbon (5-7.5%) and lead (1.5-3%) is characteristic for the Polish concentrates. The chemical composition of the concentrate has been given in Table 1.

Table 1 Chemical composition of copper concentrates (wt.%)

Cu	18 - 27
Fe	2 - 5
S	7 - 11
Pb	1.5- 3.0
As	0.06-0.30
Zn	0.4- 0.6
SiO_2	17 - 22
CaO	7 - 11
MgO	4.5-5.5
Al_2O_3	5.5-6.5
K_2O	1.2-1.8
C_{org}	5 - 7.5

In 1978, the process of direct copper smelting from these concentrates in a flash furnace started at Glogow 2 copper smelter. The flowsheet of the process is shown in Fig.1.

It includes:

A. Concentrate drying to achieve H_2O content of below 0.3 wt.%

B. Smelting the concentrate in a flash furnace using the oxygen-enriched air blow of 60-75 vol.% O_2, as a result of which the following are obtained:

   i   blister copper of Pb content of below 0.3 wt.%

   ii slag of Cu content of approx. 14 wt.%

   iii dust recycled to the flash furnace

   iv gases of high SO_2 content, used for sulphuric acid production

C. Slag cleaning in an electric furnace: the following are obtained:

   i   copper alloy of 10-18 wt.% Pb and 3-8 wt.% Fe

   ii waste slag, containing below 0.6 wt.% Cu

   iii dusts, containing 35 wt.% Pb and 12 wt.% Zn

D. Fire refining of the blister copper from flash smelting and of the converter copper.

The production of copper of good quality, i.e. containing less than 0.3 wt.% of lead requires a high oxidation of the concentrate components. The necessity of oxidizing the carbon, iron and sulphur as well as of some copper in the reaction shaft is the reason, for the oxygen utilization being an important process para-

185

186

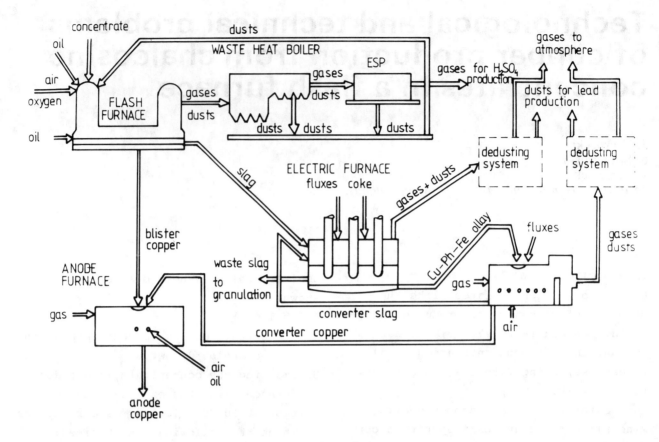

Fig.1 Anode copper production at Glogow 2

meter. The degree of oxygen utilization
affects not only the energy factors of
the process, but also the degree of the
transformation of SO_2 to SO_3 in the gas
cleaning (waste heat boiler, ESP) system
and its durability.

The high content of the oxidized copper
in slag (12-18 wt.%) affects unfavourably
the life of the refractory material in
the reaction shaft and settler of the
flash furnace. This effect was the reason
for introducing several changes in the
flash furnace design.

TECHNOLOGICAL PROBLEMS OF COPPER PRODUC-
TION BY FLASH SMELTING
Effect of lead in the concentrate on the
flash process parameters
Lead is the concentrate component, which
determines the parameters of the flash
smelting process. Its concentration in

the concentrate fixes the recessary oxygen potential of the system, which
affects the copper content in slag and
copper recovery in the whole process.
The slag, containing high contents of
cuprous oxide, has a strong corrosive
and erosive effect on the refractory.
The lead content in the charge to the
flash furnace ranges between 1.5 and
3 wt.%. The lead concentration in copper
is determined by the thermodynamic equilibrium of the reaction:

$$[Pb] + (Cu_2O) \rightleftarrows 2[Cu] + (PbO \qquad (1)$$

$$K_{p(1)} = \frac{a_{(PbO)} \, a^2_{[Cu]}}{a_{[Pb]} \, a_{(Cu_2O)}} \qquad (2)$$

whose free energy dependence on temperature is determined by reaction:

$$\Delta G^o_T = -10.240 + 2.7 \, T \qquad (3)$$

The lead content in copper, determined by eq.(2), is expressed by

$$X_{Pb} = \frac{X_{(PbO)} \cdot \gamma_{(PbO)}}{K_{p(1)} \cdot \gamma_{[Pb]} \cdot \gamma_{(Cu_2O)}} \cdot \frac{1}{X_{(Cu_2O)}} \quad (4)$$

where: X: molar fraction; γ: activity coefficient; ( ), [ ]: component contained in slag and metals, respectively.

The relationship between lead content in copper and cuprous oxide in slag is presented in Fig.2.

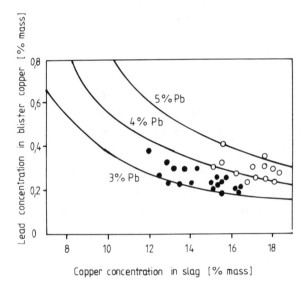

Fig.2 Relation of lead concentration in blister copper and copper concentration in a flash smelting slag.

The continuous lines represent the results of calculation relative to eq.(4) for slags containing 3.0; 4.0; 5.0 wt.% Pb, which refer to the following lead contents in the concentrate (wt.%): 1.6; 2.1; 2.7. The black points refer to lead contents in the blister copper, produced by smelting concentrates containing less than 2.1 wt.% Pb (below 4 wt.% Pb in slag); the circles refer to lead contents in concentrates containing more than 2.1 wt.% (above 4 wt.% Pb in slag). When the lead content in the concentrate does not exceed 2.1 wt.%, an oxidation potential, resulting in the presence of 13-15 wt.% Cu in slag is sufficient; in this case the blister copper produced

contains below 0.3 wt.% Pb.
The increased lead content in the concentrates is compensated by the increased copper oxidation degree. For higher lead contents (more than 2.1 wt.%), the required copper concentration in slag ranges between 16 and 18 wt.%.

Effect of organic carbon on the flash process parameters

The concentrate is the main energy carrier in the flash smelting. Over 80% of the heat introduced to the process is the thermal effect of the carbon and sulphide minerals oxidation reaction. The remaining energy is given by oil. Approx. 200 to 400 l of oil/hour are introduced to the settler, and some 100 to 1200 l/hour to the reaction shaft. The amount of heat burned in the shaft depends mainly on the process efficiency and on the concentrate chemical composition, i.e. mainly the carbon content. The thermal contribution of the concentrate oxidation exceeds the heat amount taken out by the process products. Therefore, as the smelting throughput increases, the amount of additional fuel introduced to the proces becomes reduced. (Fig.3). The amount of heat produced at

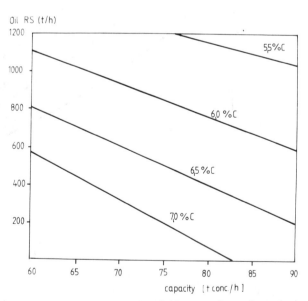

Fig.3 Dependence of oil consumption in a reaction shaft on capacity and C_{org} content in a concentrate (at oxygen concentration in a process air of 65 vol %)

burning the organic carbon is 2-3 times greater than the amount of heat produced during the reactions of burning the sulphide minerals in the concentrate.

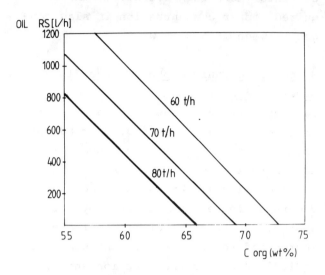

Fig.4 Oil consumption in a reaction shaft in dependence on a C_{org} content in a concentrate (oxygen concentration in a process air of 65 vol.%)

The varying concentrations of this component affect most significantly the energy balance in the flash process as well as the oil consumption in the reaction shaft. The change in the carbon content of 0.5 wt.% results in the change of oil amount in the process by 400 l/hour (Fig.4).

This results in the necesity of a frequent (every 2-4 hours) measuring the concentrate chemical composition and the slag temperature. The limited capasity the cooling elements and the refractory material determine, under certain conditions, the maximum process troughput (Fig.5).

For example, if the concentrate contains more than 7 wt.% C_{org}, the process output must be lower than 72 t/hour, even when using the process air containing 60 vol.% O_2.

Effect of the concentrate grain size

The degree of the concentrate oxidation in the reaction shaft depends on grain

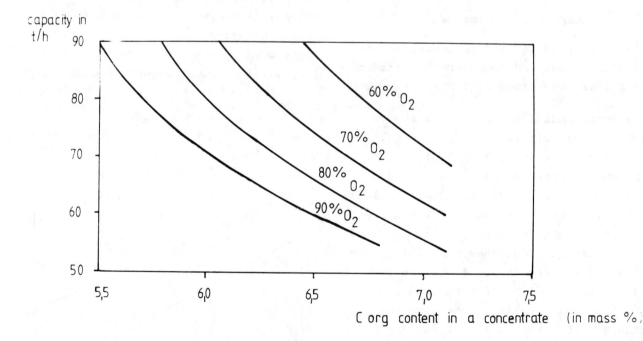

Fig.5 Maximum capacity of a flash furnace v C_{org} content in a concentrate and on oxygen concentration in process air.

size. The cupriferous minerals of finest grains become oxidized to cuprous oxide, the average grains to copper and cuprous oxide, whereas the largest grains become only surface oxidized. A high proportion of the largest grains results in many unfavourable phenomena in the process, mainly in a turbulent reaction course in the settler between the oxides and sulphides, which is the reason for the so-called "slag frothing".

In such a case, to achieve a determined lead content in copper it follows that a greater amount of finer grains is oxidised, which again leads to increased oxygen consumption in the process. Reduction of the proportion of the larger grains (over 0.75 mm) in the concentrate has been obtained from 10-14_ to 1-5%, by using sieves of smaller mesh at the final drying stages. The decreased share of the largest grains resulted in a decreased oxygen consumption to a value closer to the stoichiometric requirement. The oxygen utilization coefficient has increased from 85-90% to above 97%, and the oxygen concentration in gases behind ESP

has been reduced from 10-16 vol.% to 3-6 vol.% (Fig.6). Thus a reduction of the oil consumption by 200-300 l/h has been achieved. Also, a significal improvement of the operational conditions of the gas cleaning system (mainly the waste heat boiler) has been noted.

DESCRIPTION OF DESING CHANGES INTRODUCED IN THE FLASH FURNACE

The production of copper by direct flash smelting at Glogow 2 has for 6 years exceeded 100.000 tons/year (Fig.7).

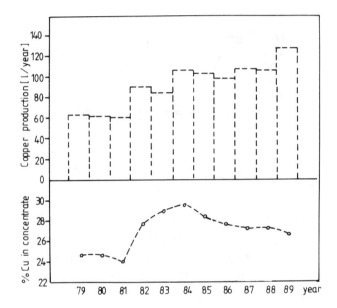

Fig.7 Copper production at Glogow 2

The smelter campaigns were approx.10 months, follwed by 2 months of repair and maintenance. The repair consisted in exchanging part of the refractory in the reaction shaft, walls and rool of the settler as well as the side walls of the hearth. The direct process differs from the classical technology of copper matte production by the fact, that the magnetite layer does not form on the walls and hearth of the furnace, which normally protects the refractory. On the contrary, due to the high cuprous oxide concentration, the slag is strongly corrosive and erosive. Therefore a large relatively fast wear of the refractory is observed,

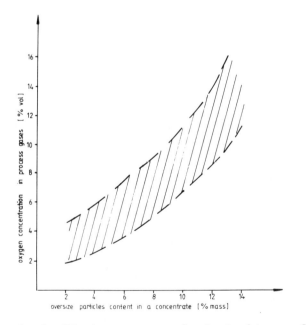

Fig.6 Effect coarse-grained strukture of a concentrate on oxygen concentration in process gases.

190

mainly in the reaction shaft and in the settler near it (Fig.8).

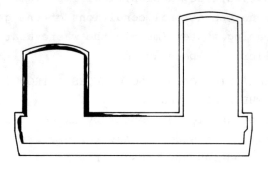

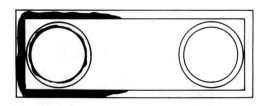

Fig.8 The areas of most serious attack of the flash furnace refractory.

The penetration of the liquid copper through the first refractory layer and consequently its pushing out is another serious problem. The life of the refractory in the areas of its strongest domage has been prolonged by applying a larger number of newly designed ceramic--metallic water jackets in the settler and the lower part of the reaction shaft. Moreover, some areas in these regions have been made from the fused-cast chrome-magnesite material. The new hearth is characterized by the larger amount of cooling designed into the furnace. This affect has been achieved by thinning the hearth, application of refractories of higher conductivity and introduction of forced cooling.
These steps allowed for maintaining the isotherm respective to the copper solidification temperature in the first, upper layer of the refractory, thus preventing the copper penetration through it. The life of the modernized hearth is 7 years. The design changes introduced

in the flash furnace allowed for a significant increase of its life and a compaign prolongation from 10 to 15-18 months.

SUMMARY

The direct copper production process in the flash furnace has been disigned to assure a continuous production of a good-quality blister copper; it is characterized by high efficiency and low fuel consumption. The process is still being developed to improve the life of the facilities. The new design solutions introduced so far prolonged the campaign at Glogow 2 from 10 to 15-18 months.

Flash technology for converting

Pekka Hanniala, Tuula Mäkinen, Markku Kytö
Outokumpu Engineering Contractors Oy, Outokumpu Research Oy, Espoo, Finland

1
INTRODUCTION

Most copper production processes starting from sulphidic raw materials comprise two separate steps: matte smelting and converting. Copper mattes produced in a so-called primary smelting furnace are converted to blister copper in the Peirce-Smith converter which has been the standard method in the copper industry for over 75 years. Although smelting and converting are carried out in separate reactors, the processes are combined resulting in costly investments in molten matte with transfer equipment (ladles, cranes etc.) and each other converter aisles.

In the transfer of large amounts of molten sulphidic materials to and from the converters a lot of revert (ladle shells and spills) and fugitive gases are generated. Cleanup of these solid materials and recycling them back to the converters for remelting requires additional manpower, ties up cranes and brings about additional fugitive emissions.

Until the end of the 1960s the greater part of the work on the way from concentrate to blister copper was performed in the converter. Practically, the purpose of the smelting unit was to produce molten material from the concentrate for converting. For example, reverberatory smelting has been characterized by production of low grade matte and dilute SO_2 gas. Therefore the heart of the smelter has been the converting process.

The Outokumpu Flash Smelting Process has been able to produce suitable gases for sulphur recovery and therefore sulphur fixation has been a natural part of the flash smelter. Only two of the 35 flash smelters licensed by Outokumpu exhaust process gases without sulphur recovery. Nowadays it is hardly possible to build a new smelter without the sulphur fixation facilities.

One alternative to fulfil today´s stringent environmental and hygienic requirements is to replace reverberatory furnaces by flash smelting units.

Magma Copper´s new flash smelter (Arizona, USA) can be mentioned as a typical example of reduction of operating costs and sulphur emissions. After the retrofit of old reverberatory smelter more than 96 % of the sulphur contained in the feed is captured.

It is not often possible to achieve significant improvements in the environment simultaneously with the improvements of the overall economy of the smelter. However, in the retrofit mentioned above, the net energy savings are 40-45 %.

In the future sulphur emission control is becoming tighter allover the world. Concerning the copper smelting Outokumpu is prepared to meet the new challenges. Upgrading from Peirce-Smith converting to the Kennecott-Outokumpu Flash Converting means similar technical and economical advantages as those achieved by replacement of a reverberatory furnace by a flash smelting furnace. Flash converting, like flash smelting, offers an unusual combination of benefits, better sulphur recovery and reduced capital and operating costs.

2
THE KENNECOTT-OUTOKUMPU FLASH CONVERTING PROCESS

Process description

The flow sheet of the Flash Converting process is diagramatically shown in Fig. 1. The concentrate is first smelted in the primary smelting furnace to produce high grade matte. The molten matte is tapped and granulated by means of high pressure water jets. After dewatering the matte granules are ground and dried. A grind of less than 80 % minus 100 mesh is sufficient for complete combustion. Slag from the concentrate smelting furnace can be subjected to separate slag cleaning operation for copper recovery. The fine-grained solid matte is oxidized and smelted in the flash converting furnace to blister copper and slag using pure oxygen or high oxygen enrichment. With the appropriate matte grade the flash converting process can operate autogenously without additional fuel. The sulphur content of blister copper can be controlled by the oxygen to concentrate ratio in the feed. Either silicate or lime-based slags can be used. A small slag amount, formed in the treatment of mattes with 65-78 % copper, is recycled back to the concentrate smelting furnace. Gas rich in SO_2, flows continuously through cooling and cleaning steps to the sulphur recovery plant.

All the unit operations in the process are based on proven technology. Matte for flash converting can be supplied from any type of furnace, reverberatory, electric or flash smelting furnace. The most recommendable concept is high grade matte production which is normal practice in the Outokumpu Flash Smelting process. Flash converting itself, which is carried out in a small flash furnace, utilizes the well proven Outokumpu Flash Smelting technology. Also in matte pretreatment demonstrated technology is used. Drying can be performed in simultaneous drying/grinding mills or standard rotary, flash, steam coil or fluid bed driers.

The benefits of flash converting are similar to those of flash smelting, but their significance becomes more pronounced in comparison with Pierce-Smith converting process which is costly, labour intensive, and a frequent source of sulphur emissions.

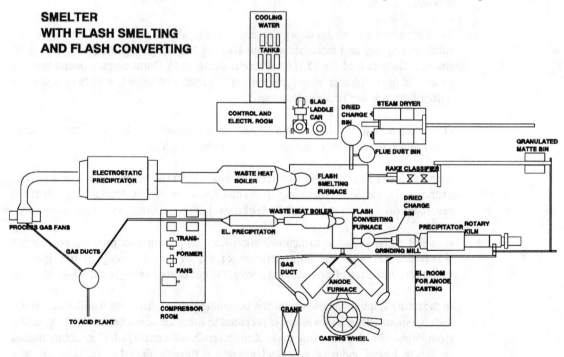

Figure 1. Kennecott-Outokumpu Flash Converting

Flash converting requires extra facilities for matte granulating, drying, and grinding, but these are compensated by savings due to the elimination of converters, converter aisles with cranes, ladles and other equipment, less gas cooling and cleaning, and especially the acid plant equipment size reduction. In addition, energy consumption in flash converting is less than that of Peirce-Smith plant, even though the matte is solidified and its latent heat is lost.

When considering the potential advantages of flash converting, it is important to take the whole smelter complex into consideration - smelting, converting and acid production. The different sections must be well matched from capacity point of view. This is important because about 35 % of the capital cost of a new smelter is for gas cleaning and sulphur fixation. Only 30 % of the total sulphur in the concentrate feed is eliminated in the converters, but because of the converter batch process, which produces fluctuating and high-volume gas flows and much fugitive gas, most of the gas which goes to the acid plant originates from the Peirce-Smith converters.

Development of the Flash Converting

In the development of the flash converting process the test results as well as practical experience of operating direct blister flash smelting process were utilized. As a matter of fact, the flash converting process is almost identical with a direct blister process smelting high-grade copper concentrates with low iron contents.

For the first time blister copper was produced directly in the pilot flash smelting furnace at the Metallurgical Research Centre of Outokumpu (currently Outokumpu Research Oy) in 1969 in connection with flash smelting tests with high-grade mattes. In 1973 extensive pilot test programme aiming at direct production of copper was accomplished using Polish and Zairean copper concentrates. These test campaigns which continued in 1976 with Polish concentrate, indicated clearly, that the direct blister process was technically and economically viable when processing low iron chalcocite concentrates. The first commercial scale flash furnace producing blister copper started operation at Glogow, Poland in 1978. Construction of Gecamines´ flash smelter in Zaire was started in 1977, but the work was suspended in 1979 because of the local civil war.

In 1973 mixtures of chalcocite and chalcopyrite concentrates were smelted in the pilot flash smelting furnace to blister copper. The testwork continued in 1978 using pure chalcopyrite concentrates. The tests succeeded technically very well. It was noticed that the role of the concentrate burner is of vital importance as to the oxygen efficiency, oxidation degree of the slag phase and flue dust generation. In spite of the good test results, large copper circulation through the slag cleaning back to the smelting step worsens the profitability of the direct copper production in a single flash smelting furnace in case of chalcopyrite concentrates.

In 1984 a pilot test campaign was carried out for the Australian bornite-digenite concentrate originating from the Olympic Dam ore body. Both high-grade matte and direct blister production were tested. The latter one was found to be the most recommendable alternative. Based on the know-how and the data acquired from the pilot test campaigns, the process and detailed engineering design of the flash furnace and specification of the key equipment were supplied by Outokumpu under the licence and engineering service agreements. Construction of the smelter commenced in 1987 and direct blister flash smelting started in 1988 at Olympic Dam in Australia.

The furnace design at Olympic Dam is very close to that of a flash converting unit. The mineralogical composition of the Olympic Dam concentrate is similar to that of high-grade matte - high copper content, iron in hematite, and sulphides. In fact, proposed strategy for future operations at Olympic Dam originally included installation of a new, larger furnace to produce high grade matte from the lower-grade chalcopyrite concentrate, which will be mined later. This solidified matte will be converted to blister copper in the existing, small flash furnace now being used to produce blister from high-grade bornite concentrate.

The flash smelter can be used to produce blister copper from concentrate in one unit, but as explained above this is only feasible in the case of some certain concentrates. An alternative process which has much wider application potential is flash converting jointly developed by Outokumpu and Kennecott Corp.

Following the small scale tests by Kennecott, it was decided to confirm the flash converting process and develop commercial plant design parameters on a pilot scale. The pilot tests were conducted by Outokumpu soon after the Olympic Dam test campaign in 1984. Over 600 tons of matte with the copper content of 55 and 70 % was smelted in the Outokumpu pilot plant at Pori, Finland. Feed rates to the pilot furnace were generally about 2 t/h. In addition higher rates were tested for short periods. Oxygen enrichment varied between 75 and 100 %.

The pilot tests succeeded very well. Blister copper with the sulphur contents of as low as 0.04 % were produced and the process control was demonstrated to be relatively easy.

When combined with data on matte granulation, milling and Outokumpu´s previous experience in designing flash furnaces, the pilot test work confirmed that the Outokumpu-Kennecott Flash Converting process was ready for commercial application.

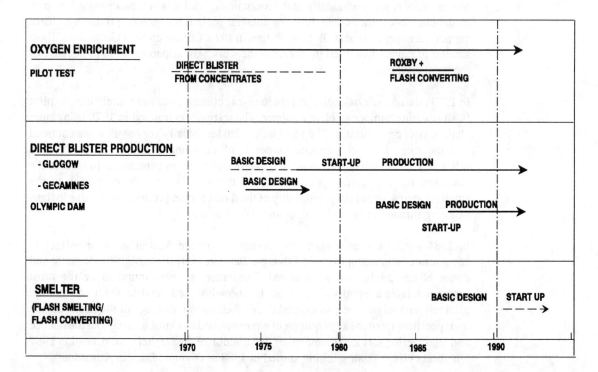

Figure 2. Development of Flash Converting

3

WHY REPLACE PEIRCE-SMITH CONVERTER BY FLASH CONVERTING ?

History of converting

Adoption of a side-blown vessel for the conversion of molten copper mattes to blister copper occurred in the early 1900´s. The first 13´*30´ PEIrce-Smith converter, still the industry standard today, was constructed towards the end of the 1910s.

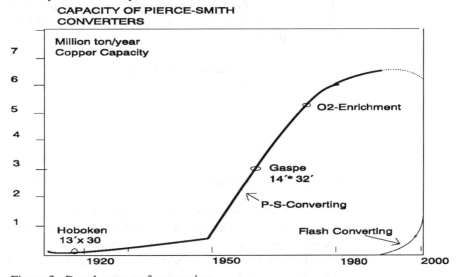

Figure 3. Development of converting

The siphon converting technique was developed in the 1930´s. The benefits of the Hoboken converter and TBRC (top blown rotary converter) compared to Peirce-Smith converter are lower fugitive emissions and more continuous operation (higher on-line-availability). This is true in theory, because the converter is operated at reduced pressure and the vessel can be blown and charged at the same time. Although the Hoboken and TBRC techniques have fulfilled these requirements only partially, they have or they have had industrial applications at the following smelters: Hoboken, Glogow I, Paipote, Caraiba, Inspiration and Rönnskär.

Development of the conventional converting process has been mainly directed at equipment technique. The target of these improvements has been to increase the capacity and to maximize the on-linc-availability. The new modification of Peirce-Smith converter, 15´*35´, was introduced at the beginning of 1960´s. The Gaspe punching machine was put into practice in the late 1960´s.

Today´s requirements for converting

New smelting processes, like flash smelting, are presumed to utilize the reaction heat of the concentrate in smelting and minimize the sulphur dioxide emissions. By increasing the matte grade in the smelting step both of these purposes have been achieved. Because of lower reaction heat of the high grade matte it was necessary to adopt oxygen enrichment in converter blowing air which occurred in the 1970s.

196

Requirements for a modern converter operation can be summarized as follows:

- higher oxygen enrichment

- ability to treat high grade matte

- smooth gas flow to sulphur fixation plant

Sulphur recovery

Sulphur recovery from the converter gases started in the 1960s and has gradually become more general. However, conventional converting faces the following problems as to the sulphur fixation:

- large process gas volumes owing to low oxygen enrichment (max. 30 %) and extensive dilution with air

- intermittent gas flow because of the batch operation

- large volumes of fugitive gases generated in the ladle transfer.

All these factors increase production costs of sulphuric acid.

Since now the development in the copper industry has mainly concentrated at primary smelting processes in order to reduce operating costs and improve sulphur fixation. Nowadays production of high grade matte is generally accepted and the following step will be the adoption of a new converting process. Conventional converting will be displaced by new technique, which is ready to meet the tight requirements for sulphur emissions.

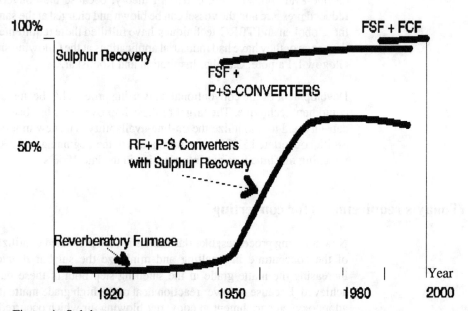

Figure 4. Sulphur recovery

Trends in smelter development

During the past twenty years the development of copper smelting has aimed at higher smelting capacities by adoption of oxygen enrichment in process air. At the same time the air protection aspects have become more and more important.

Smelting capacity has been increased in different ways depending on the smelting method:

A) Reverberatory furnace smelting:
- oxy-fuel burners
- installation of CMT (Teniente Modified Converter) or Noranda reactor to create additional capacity
- various modifications of reverb furnaces

B) Flash smelting and its modifications
- use of higher oxygen enrichment in process air
- equipment improvements

In reverb furnaces oxy-fuel burners offer a cheap alternative to raise the capacity. Although the copper content of matte rises a little, the furnace is still able to treat converter slag without considerable copper losses.

Smelting capacity can be elevated by a CMT or Noranda reactor which is operated parallel with a reverb furnace. In this case part of the concentrate is processed to white metal resulting in a higher copper and magnetite contents in converter slags compared to the earlier practise. This causes difficulties in slag processing in the concentrate smelting unit and increases the copper content of waste slag. As a consequence separate slag treatment becomes necessary.

INCREASE OF SMELTING CAPACITY

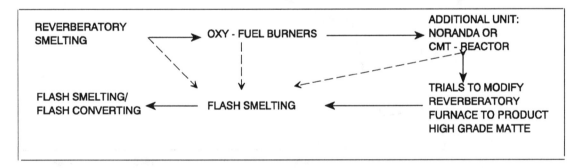

SULPHUR RECOVERY

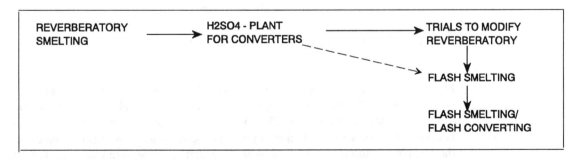

Figure 5. Trends in copper smelting technology

The process modifications described above do not improve the sulphur recovery neither from the economic nor air protection point of view. Fluctuating large gas volumes with low SO_2 content means unavoidably high capital and operating costs in gas cleaning and sulphur recovery plant.

Reverb modifications (Sprinkle, Contop) tested at the following smelters Morenci, El Paso, Palabora and Chuquicamata have aimed at elevation of smelter capacity and production of SO_2 gas suitable for sulphur fixation. The results have not corresponded to the expectations. Furnace cooling and flue dust generation have been the major problems. Flue dust amount can be reduced by means of a conical reaction shaft, but according to Outokumpu´s experience, it causes difficulties in equipment technique. The conical reaction shaft has been tested at the Harjavalta smelter already in the 1950´s and later in Japan.

Nowadays production of high grade matte in the smelting step is widely accepted. From the very beginning production of high grade matte by using optimum or maximum amount of technical oxygen has been typical to the flash smelting. In this way the greater part of the sulphur is recovered to continuous low volume gas stream. Production costs of sulphuric acid from this kind of gas are significantly lower than from converter gas.

In the course of time matte amount in flash smelting has been reduced by increasing the copper content of matte as shown in Fig.6.

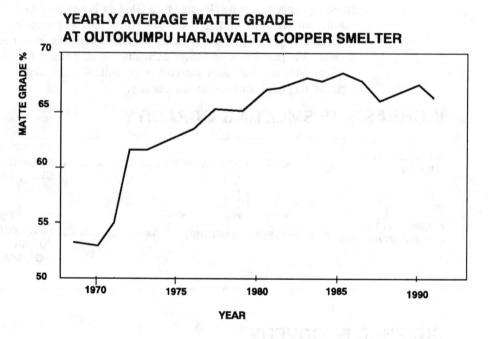

Figure 6. Matte grade at the Harjavalta smelter

Flash smelting furnace is fully developed to produce high grade matte with a copper content of 70-79 %. Conventional converting faces serious problems in the treatment of mattes containing more than 70 % of copper. Very skilled operators are needed because it is necessary to use higher oxygen enrichment in the blowing air and because the slag blow is short. One of the drawbacks is that 15-20 % of the matte is charged to the converter in solid state as revert.

Concentrate burner

Earlier a general comprehension was that when producing high grade matte in the smelting step impurity contents of blister copper exceed the acceptable levels. In this respect the knowledge has partially become obsolete. Outokumpu is developing a new type of concentrate burner by which it is possible to control more accurately the circumstances in the reaction shaft (temperature, oxidation etc.). In this development work the willingness of Magma´s personnel to conduct industrial scale tests has been of great importance. From the preliminary test results it can be concluded that the new concentrate burner has potential to answer all expectations. It is most probable that with the new burner Outokumpu will on a commercial scale arrive at good results which correspond to the theoretical ones presented by the designers of flash smelter modifications.

Table 1. Theoretical effect of temperature on impurity distribution

Impurity	Temperature	Distribution, %		
		Matte	Slag	Gas
Pb	1600 K	3	62	35
	1900 K	1	9	90
Zn	1600 K	10	70	20
	1900 K	2	8	90
Sb	1600 K	15	80	5
	1900 K	<5	10	85
As	1600 K	1	3	96
	1900 K	<1		>99
Ag	1600 K	99	1	0
	1900 K	79	1	20

Flexibility

The Flash Smelting - Flash Converting method increases the flexibility of blister copper production. Smelting and converting are not interlinked with each other. The processes can be operated independently which improves the on-line availability of the smelter. The process conditions can be optimized so that the best possible elimination for harmful elements is achieved. This process concept offers the following options to eliminate harmful elements:

- flue dusts can be circulated back to the concentrate smelting furnace

- impure copper matte can be stored and fed gradually to the converting furnace

- possibility to use different type of slags (lime, silica)

The ability of this process combination to eliminate impurities is equivalent or even better than that of flash smelting - Peirce Smith converting

The process generates continuous low volume of SO2 gas which is suitable for elemental sulphur production.

Automation

Batch converting has required special professional skill based partially on visual observation and auditory perception.

The Flash Smelting - Flash Converting process can be operated autogenously by using oxygen or oxygen enriched air. Temperature and copper content of matte, sulphur content of blister etc. can be determined in advance. Appropriate temperature, pressure and flow rate sensors and computer controlled systems contribute to smooth operation resulting in optimum production conditions.

4
FLASH CONVERTING AND PORTUGAL´S FLASH SMELTER PROJECT

Outokumpu Engineering Contractors Oy has performed a comprehensive feasibility study of the economies of the Portuguese copper smelter project. The process is based on the Flash Smelting-Flash Converting technique because it meets the air protection requirements and improves the project economy. By selecting the above mentioned concept capital costs can be reduced considerably. The capital cost of the flash converting unit is low. The expensive converter aisle is replaced by the small solid matte mill, dryer and flash converting furnace.

The flash converter operation doesn´t require so many operators as Peirce-Smith converting. Relining of the flash converting furnace is needed only after a few years´ operating period which means savings in expensive refractory materials maintenance work. Also energy consumption in flash converting is lower compared to the conventional converting. The relative capital and operating costs are presented in the following table.

Table 2. Capital and operating costs of converting

	FLASH SMELTING PEIRCE-SMITH CONVERTING	FLASH SMELTING FLASH CONVERTING
TOTAL INVESTMENT	100	85-90
TOTAL OPERATING COSTS	100	89
LABOR	100	88
POWER	100	96
MAINTENANCE AND REPAIR	100	85

It is generally believed that Flash Converting decreases copper production costs by 5 cents per pound of copper compared to the conventional technique.

Gas flow from the flash converting furnace is continuous and rich in SO_2 which causes considerable savings in the sulphuric acid plant. An example of the savings with the anode copper production of 150 000 MTPY is shown in Table 3.

Table 3. Savings in sulphuric acid production.

	FLASH SMELTING PEIRCE-SMITH CONVERTING	FLASH SMELTING FLASH CONVERTING
Copper Production	150,000 MTPY	150,000 MTPY
Sulphuric Acid Production	2,000 TPD	2,000 TPD
Sulphuric Acid Plant,desig.	3,000 TPD	2,000 TPD
SO2- Content after dilution	9 - 12%	20 - 30%
Investment Costs	88 MUSD	60 MUSD
Operating Costs	100 kWh/ton	53 kWh/ton
Saving in Power Consumption (5c/kWh)		31,020 MWh/a 1,551,000 USD/a = 0.46 c/lb Cu
Saving in Investment of Acid Plant (15%, 20 years)		1.29 c/lb Cu
TOTAL SAVING IN ACID PRODUCTION		1.75 C/LB BLISTER

Savings in capital and operating costs increase up to 2 cents per pound of copper, if the period of repayment is 20 years, interest rate 12 % and price of electric power 50 USD/MWh.

Since now the Flash Converting process has no commercial scale application but, as mentioned before, the blister flash furnace at Olympic Dam is very close to the flash converting unit. The furnace at Olympic Dam was designed to produce 55 000 MTPY copper from concentrate. In a later stage this furnace will be used as the flash converting furnace to produce 100 000-150 000 MTPY of blister copper depending on matte grade. An interesting comparison can be made with the Rio Tinto Minera copper smelter at Huelva, Spain.

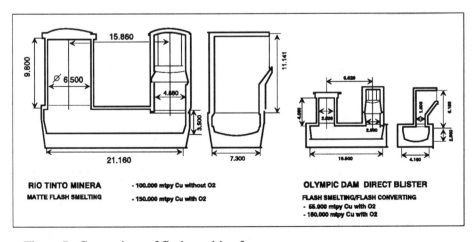

Figure 7. Comparison of flash smelting furnaces

At Huelva the flash furnace produces high grade matte which is converted to blister copper in two or three Peirce-Smith converters. This plant has a production capacity of 100 000 to 150 000 MTPY blister. When two converters are blowing simultaneously the gas volume is about 150 000 Nm^3/h. An equivalent capacity flash converting furnace would have a gas volume of 10 000 to 13 000 Nm^3/h. The reduction in gas volume is about 90 % and the concentration of SO_2 is about 70-80 % compared to 5-12 % in Peirce-Smith converting.

5
CONCLUSIONS

In the copper smelting industry development work has mainly been directed to the primary smelting step so that the modern smelting methods utilizing the benefits of oxygen are gradually replacing the obsolete processes. The converting of copper mattes to blister copper has seen few technical advantages since the adoption of Peirce-Smith converters.

Considering increasingly stringent environmental and hygienic regulations together with requirements for production cost cuts the relevance of Peirce-Smith technique can be questioned today.

The flash converting process, jointly developed by Kennecott and Outokumpu, overcomes the disadvantages associated with Peirce-Smith and other converting options. Because flash converting is a continuous process which is able to utilize tonnage oxygen and produce a small volume of high strength SO_2 gas, considerable savings are achieved in capital and operating costs. Simultaneously sulphur emissions are minimized. The benefits of flash converting become pronounced when high grade matte for converting is supplied by the Outokumpu Flash Smelting process.

At the moment Outokumpu is just about to start to design the commercial scale flash smelting-flash converting process which as a reference plant will be of great importance.

Bibliography

1   J. Asteljoki et al., "Flash Converting", <u>Proceedings of the Fifth International Flash Smelting Congress</u>, ed. M. Murtoaro, (Helsinki, Finland, 1986), 13-43

2   J. Asteljoki, M. Kytö, "Alternatives for Direct Blister Copper Production", (Paper presented at the 114th SME-AIME Annual Meeting, New York, 1985)

3   Y. Anjala et al., "Outokumpu Flash Smelting in Copper Metallurgy - the Latest Developments and Applications" , (Paper presented at the **Copper 87** Conference December 87, in Vina del Mar, Chile)

4   Y. Anjala et al., "The Role of Oxygen in the Outokumpu Flash Smelting Process", (Paper presented at the 28th CIM Annual Meeting, Winnipeg, Manitoba, 1987)

5   P. Hanniala, "Copper Smelting Technology, How Outokumpu Sees Flash Smelting Advancing", (Paper presented at the Metal Bulletin Conference, Santiago, April 1990)

6   K.J. Richards et al. "A New Continuous Copper Converting Process", (Paper presented at the SME-AIME Annual Meeting, San Fransisco, California, November 1983)

7   J. Asteljoki, "Flash Converting - Continuous Converting of Copper Mattes", <u>Journal of Metals</u>, May 1985

8   J. Asteljoki, M. Kytö, "Minor Elements Behaviour in Flash Converting", (Paper presented at the 115th AIME Annual Meeting, New Orleans, 1986)

9   D. Rodolff et al., "Review of Flash Smelting and Flash Converting Technology",(Paper presented at the 115th AIME Annual Meeting, New Orleans, 1986)

10   T.P.T. Hanniala, J.S. Sulanto, "The Development Trends of the Outokumpu Flash Smelting Process for the Year 2000", (Paper presented at the 118th AIME Annual Meeting, Las Vegas, Nevada, February-March 1989)

11   Flash Smelting at 40, Ed. R. Wyllie, <u>E&MJ</u>, October 1986

12   D. Rodolff et al., "Outokumpu Flash Smelting an Update and Retrofit Considerations", (Paper presented at the AIME Annual Meeting, Tucson Arizona, December 1984)

13   T. Mäkinen, M. Kytö, Pilot and Mini-pilot Tests of Flash Reactions, <u>Flash Reaction Processes</u>, ed. D.G.C. Robertson, H.Y. Sohn and N.J. Themelis, The Center for Pyrometallurgy, University of Missouri-Rolla, MO, 69-98

14   Pyromet or Hydromet, Ed. K.R. Suttill, <u>E&MJ</u>, May 1990

15   J. Asteljoki, H.B. Muller, "Direct Smelting of Blister Copper - Pilot Flash Smelting Tests of Olympic Dam Concentrate", <u>Extractive Metallurgy 87</u>, (IMM Publications, London, 1987)

16   T.J.A. Smith et al., Oxygen Smelting at the Olympic Dam Project, (Paper presented at the 28th CIM Annual Meeting, Winnipeg Canada, 1987)

17   Kenneth R. Coyne, " Metals technology slowly advances", (Paper presented at the Capital Metals and Materials Forum Dec. 14. 1989 in Washington)

Minor-element behaviour in copper-making

H. Y. Sohn
Hang Goo Kim
University of Utah, Salt Lake City, Utah, USA
K. W. Seo
University of Utah, Salt Lake City, Utah, USA
(at present, Korea Institute of Science and Technology, Seoul, Korea)

SYNOPSIS

The behavior of minor elements in bath and flash smelting processes for coppermaking has been investigated. Particular attention has been paid to the conditions of high oxygen enrichment in the case of bath smelting and converting. A computer simulation has been carried out to predict the distribution behavior of Pb, Zn, Bi, Sb, and As in recently proposed copper smelting and converting processes with submerged oxygen injection. The technique of stepwise equilibrium modeling was used. The predicted minor-element behavior under high oxygen enrichment was compared with that in the Noranda process (normally 42% oxygen enrichment in smelting and 24% in converting) and in the reverberatory process using air. The effects of various operating parameters such as temperature, the final matte grade in smelting, the initial matte grade in converting, and the O/Fe ratio in slag on the behavior of minor elements are discussed.

In the case of flash smelting, the minor-element behavior is calculated based on a comprehensive mathematical model incorporating the turbulent fluid dynamics of a particle-laden gas jet, heat and mass transfer, and chemical reactions. The volatilization of As, Sb, Bi, and Pb was computed, and experiments were carried out for Sb and Pb in a laboratory flash furnace. Satisfactory agreement between the predicted and measured results was obtained. The degrees of volatilization of minor elements were computed as functions of position and matte grade. With the comprehensive mathematical model developed in this work, a computer simulation can be readily performed to predict the minor-element behavior under various different operating conditions.

The submerged oxygen injection system[1,2,3] developed for modern steelmaking has a potential for application to coppermaking. In this system, tonnage oxygen is injected with a shielding gas, such as nitrogen or natural gas, into a molten bath through submerged injectors extending through the refractory lining of the reactor. The injectors and refractories are protected by surrounding the central oxygen stream with a shielding gas. One aspect which should be addressed is the behavior of minor elements under submerged oxygen injection which involves a much higher oxygen potential and smaller gas volume. The important effect of oxygen enrichment has been reported in previous studies.[4-7] The present study extends the calculation to predict the impact of oxygen technology on the behavior of minor elements in copper smelting and converting processes.

In a flash-smelting process, as particles travel down the reaction shaft within the turbulent jet, they exchange momentum, mass, and energy with the surrounding gas. During this exchange, the elimination of minor elements to the gas phase also takes place. This study extends the comprehensive mathematical model developed by Hahn and Sohn[29,30] to the prediction of minor-element behavior inside the flash-furnace shaft during copper flash smelting incorporating the thermodynamic analysis of Chaubal.[11] In this new model, the minor-element behavior is described in combination with turbulent transport phenomena, chemical reactions, and thermal radiation in a flash-furnace shaft.

This study, therefore, deals with the behavior of five of the most undesirable minor elements in coppermaking, i.e., Pb, Zn, Bi, Sb, and As. Three major volatile forms (M, MS, and MO) for Pb and Zn and four forms (M, M_2, MS, and MO) for Bi, Sb, and As are taken into account.

COMPUTATIONAL PROCEDURE

Model basis

The basic assumptions used in computing the distribution behavior of minor elements in copper smelting and converting in a bath reactor with submerged oxygen injection are as follows:

(1) Autogenous operation.
(2) Steady-state smelting and time-dependent batch converting.
(3) The feed is melted and separated into two phases, matte and slag, as soon as it enters the furnace, and all of the minor elements in feed are dissolved in the condensed phases prior to the volatilization.[8]

(4) A system consisting of copper, matte, and slag phases is under equilibrium conditions.[8]

(5) 100% oxygen efficiency.

(6) All natural gas injected as shielding fluid is assumed to completely burn to CO_2 and H_2O.

The behavior of minor elements in a flash smelter can be described by considering the following[4,11]:

(1) Sulfur removal from the concentrate particles,

(2) The variation of gas and particle temperatures in the reactor shaft,

(3) The elimination of minor elements to the gas phase in the reactor shaft, and

(4) The distribution of the minor elements between the molten phases in a particle.

The basic premise of the volatilization model for flash smelting is that at the surface of the molten particle the partial pressures of the minor-element species are those at equilibrium. Transport of minor-element species to gas is then described by external mass transfer.

Thermodynamic data compilation

The thermodynamic data used in this study were carefully selected from the previous studies[4,8-18] after close comparison between the computer results and the commercially observed data[5,19-28] under several operating conditions in order to obtain more reliable results.

Basic volatilization model

The computer simulation of the thermodynamic distribution behavior of minor elements was performed based on the previous modeling work of Chaubal and Nagamori.[4,8,9,11] Since the basic derivation is presented in their articles, only a brief description of the appropriate equations and newly developed equations will be discussed here.

The effective total pressure of all the minor-element containing species is given by[8]

$$P_M^e = P_M + 2P_{M_2} + UP_{M_uO_v} + WP_{M_wS_Y} + \dots \quad (1)$$

For steady-state bath smelting, the mass balance of minor elements is given by

$$\dot{m}_{fd}[\%M]_{fd} = \dot{m}_{mt}[\%M]_{mt} + \dot{m}_{sl}[\%M]_{sl}$$
$$+ \dot{v}_M A_M/0.224 \quad (2)$$

where $\dot{m}_{fd}$, $\dot{m}_{mt}$, and $\dot{m}_{sl}$ are the mass flow rates of feed, matte, and slag, respectively. $\dot{v}_M$ is the volumetric flow rate of M-containing gas, and A_M refers to the atomic weight of element M. Under equilibrium,

$$P_M^e = \frac{\dot{v}_M}{\dot{v}_g} \quad (3)$$

where $\dot{v}_g$ is the volumetric flow rate of washing gas. Using the definition of the distribution coefficient of minor elements between matte and slag,

$$L_M^{III} = \frac{[\%M]_{mt}}{[\%M]_{sl}}, \quad (4)$$

the mass balance equation (2) can be rewritten as follows:

$$\dot{m}_{fd}[\%M]_{fd} = \dot{m}_{mt}[\%M]_{mt} + \frac{\dot{m}_{sl}[\%M]_{mt}}{L_M^{III}} + \frac{P_M^e \dot{v}_g A_M}{0.224}. \quad (5)$$

Therefore, if all the flow rates are fixed, the concentrations of minor elements in matte and slag can be calculated.

For batch-type bath converting, the process is divided into many microsteps. The differential equation which describes the volatilization of minor elements in a microstep in the converting process, is given by[9]

$$\frac{d[\%M]_i}{dV_g} = \frac{-A_M}{0.224} \frac{k_1[\%M]_i + k_2[\%M]_i^2}{W_T} \quad (6)$$

where k_1 and k_2 are functions of temperature, P_{O_2} and P_{S_2}, and are defined as

$$k_1 = \frac{P_M + P_{MO} + P_{MS}}{[\%M]_i} \quad (7)$$

$$k_2 = \frac{P_{M_2}}{[\%M]_i^2}. \quad (8)$$

Here, we have neglected the volatile oxide and sulfide species with u and w in Eq. (1) greater than unity.

The result of integrating Eq. (6) from the initial conditions $[\%M]_i = [\%M]_i^I$ at $V_g = 0$ to final conditions $[\%M]_i = [\%M]_i^F$ at $V_g = V_g$ is given by[8]

$$[\%M]_i^F = \frac{k_i[\%M]_i^I}{G_M k_1 + k_2[\%M]_i^I (G_M - 1)} \quad (9)$$

where

$$G_M = \exp\left[\frac{k_M^* V_g}{0.224 \, W_T}\right] \quad (9A)$$

$$k_M^* = k_1 A_M \quad (9B)$$

The concentration of minor elements in the molten phases after the volatilization at each microstep can be calculated by Eq. (9). The concentration of minor elements before volatilization can be calculated from

$$W_M = 0.01 \, [\%M]_i^I \, W_i + 0.01 \, [\%M]_i^I \, W_j L_M. \qquad (10)$$

For the first stage of converting (slagmaking; $FeS + (1 + y/2)O_2 = FeO_y + SO_2$), i = matte, j = slag, and $L_M = \left(L_M^{III}\right)^{-1}$, and for the second stage (coppermaking; $Cu_2S + O_2 = 2Cu + SO_2$), i = matte, j = blister copper, and $L_M = L_M^I$.

For the <u>flash-smelting process</u>, the modeling equations to describe the overall phenomena of flash smelting were described in the previous studies by Hahn and Sohn.[29,30] The eliminated amounts of minor elements from particles to the gas phases were considered as source terms in the continuity equation. The PSI-CELL technique[29-30] is used in the calculation of the gas-phase concentration of minor elements. At any given instant, the concentration of a minor-element species in the gas phase at the surface of a particle is assumed to be the equilibrium value. The molar rate of transfer of the minor-element species in the gas phase is expressed as

$$\frac{dm_M}{dt} = k_{m,M} S (C_M^e - C_M^b) f_S. \qquad (11)$$

The estimation of the mass-transfer coefficient, k_m, is based on the following equations:

$$Sh_j = 2 + 0.65 \, Re_j^{1/2} \, Pr_g^{1/3} \qquad (12)$$

$$Re_j = \frac{d_j \, |U_g - U_p| \, r_g}{m_g} \qquad (13)$$

$$Pr_g = \frac{C_{p_g}}{m_g k_g} \qquad (14)$$

$$k_{m_j} = \frac{Sh_j \, D_{m_j}}{d_{p_i}}. \qquad (15)$$

The diffusivity was calculated using the Chapman-Enskog equation,[32] and it was assumed that all minor-element-containing species have the same diffusivity as the metallic gas.

From the ideal-gas law,

$$C_M^e = \frac{P_M^e}{RT} \quad \text{and} \quad C_M^b = \frac{P_M^b}{RT} \qquad (16)$$

$$\frac{dm_M}{dt} = \frac{k_{m,M} S f_S}{RT} (P_M^e - P_M^b). \qquad (17)$$

The weight of a minor element remaining in the molten particle at any given instant is described by[4]

$$W_{M,t} = W_{M,0} - m_M A_M. \qquad (18)$$

Once reaction occurs, vigorous internal motion is initiated.[33] Hence it is assumed that the molten phases in the particle are at equilibrium. The concentration in the copper phase is calculated from[4]

$$[\%M]_{Cu} = \frac{100 \, W_{M,t}}{\left[W_{Cu} + \dfrac{W_{mt}}{L_M^I} + \dfrac{W_{sl}}{L_M^{II}} \right]}. \qquad (19)$$

When there is no copper phase, the following equation is used:

$$[\%M]_{mt} + \frac{100 \, W_{M,t}}{\left[W_{mt} + \dfrac{W_{sl}}{L_M^{III}} \right]}. \qquad (20)$$

The sum of the particle pressures of the gaseous species containing minor elements in the small batch is given by

$$P_M^b = \frac{m_M}{n_g} \, P_T. \qquad (21)$$

Equation (15) is now rewritten using Eq. (19) as

$$\frac{dm_M}{dt} = k_{m,M} \, S f_S \left[\frac{P_M^e}{RT_f} - \frac{m_M}{n_g RT_g} P_T \right] \qquad (22)$$

and $m_M = 0$ at t = 0,

where T_f refers to the film temperature. It should be pointed out that the minor-element volatilization and the sulfur removal from the particle occur simultaneously. As the oxidation reaction proceeds, both temperature and composition of the particles change along with P_{S_2} and P_{O_2}.

COMPUTER PREDICTION

Steady-state bath smelting

The simulation was performed for steady-state smelting in a bath reactor with submerged oxygen injection, and the results were compared with those for the Noranda process (normally 42% oxygen-enrichment) and for the reverberatory process which uses air. The basic operating conditions used in this computer analysis, which are taken to be the same as the conditions of the Noranda process, are listed in Table I. The following aspects were predicted for all the minor elements when the submerged oxygen injection was used:

(1) high elimination by slag,
(2) low elimination by gas, and
(3) low total elimination.

The effects of the operating parameters are as follows:

Table I. The basic operating conditions used in the analysis of the behavior of minor elements in the bath smelting process of coppermaking

Feed

Concentrate	(27.5% Cu, 27.0% Fe, 29.9% S, 0.0% O, 2430 TPD)
Precipitate	(80.0% Cu, 4.2% Fe, 0.8% S, 7.61% O, 50 TPD)
Noranda slag	(40.0% Cu, 22.12% Fe, 10.55% S, 6.81% O, 178 TPD)
Converter slag	(3.0% Cu, 49.73% Fe, 3.0% S, 14.69% O, 92 TPD)
Converter dust	(62.57% Cu, 5.18% Fe, 15.26% S, 10.74% O, 18 TPD)
Matte shell	(Final matte composition, 237 TPD)

Average Contents of Minor Elements in the Feed

Pb	0.2%
Zn	1.0%
Bi	0.02%
Sb	0.01%
As	0.14%

Types of Gas Injected

Air injection	21% O_2, 79% N_2
Noranda tuyere	42% O_2, 58% N_2 in smelting
	24% O_2, 76% N_2 in converting
Submerged oxygen injection	Tonnage oxygen (99.5% O_2, 0.5% N_2) with shielding gas; N_2 (30% of tonnage oxygen) or natural gas (5% of tonnage oxygen)
	Natural gas composition
	85.7% CH_4, 7.6% C_2H_6, 1.5% C_3H_8, 0.8% C_4H_{10}, 1.5% N_2, 0.6% CO_2

Other Operating Conditions

O/Fe mole ratio in slag	1.08
Activity of magnetite	0.75
Average temperature	1200°C

(1) An increase in temperature results in an increase in elimination by volatilization and in total elimination (except Sb).

(2) The effect of O/Fe ratio is negligible. A change in this ratio from 1.08 to 1.18 results in less than 2% change in the distribution of minor elements in different phases. Therefore we can expect that the effect of FeO as well as Fe_3O_4 content will be negligible.

Most of the trends described above are consistent with the previous results at lower degrees of oxygen enrichment[6,8] and under flash-smelting conditions.[4] The detailed results of computer simulation for individual minor elements are as follows:

Lead: The behavior of lead is shown in Fig. 1. Lead is largely eliminated by volatilization at a high sulfur partial pressure and by slagging at a high oxygen partial pressure. An increase in temperature leads to a decrease in the slagging of Pb. The total elimination of Pb increases with an increase in the final matte grade owing to the increase in the slagging of Pb. Although an increase in oxygen partial pressure in the injection gas results in a lower total elimination by volatilization and slagging, the effect becomes smaller at a high matte grade.

Zinc: Zinc is predominantly eliminated by slagging as shown in Fig. 2. Therefore, the elimination of Zn is little affected by a decrease in gas volume. An increase in temperature leads to an increase in both the volatilization and slagging of Zn.

Bismuth: The behavior of bismuth is shown in Fig. 3. Bismuth is mainly eliminated by volatilization. Volatilization of bismuth decreases with an increase in the final matte grade or oxygen partial pressure. This results in a decrease in the total elimination of Bi with increasing oxygen content.

Antimony: As shown in Fig. 4, the elimination of antimony by volatilization and slagging, like that of Pb, is strongly dependent on the oxygen and sulfur partial pressures. An increase in temperature leads to a considerable decrease in the slagging of Sb. This results in a decrease in the total elimination of Sb with temperature.

Arsenic: The behavior of arsenic is shown in Fig. 5. Arsenic, like Bi, is predominantly eliminated by volatilization. However, the volatilization of As decreases substantially with increasing oxygen partial pressure. Slagging on

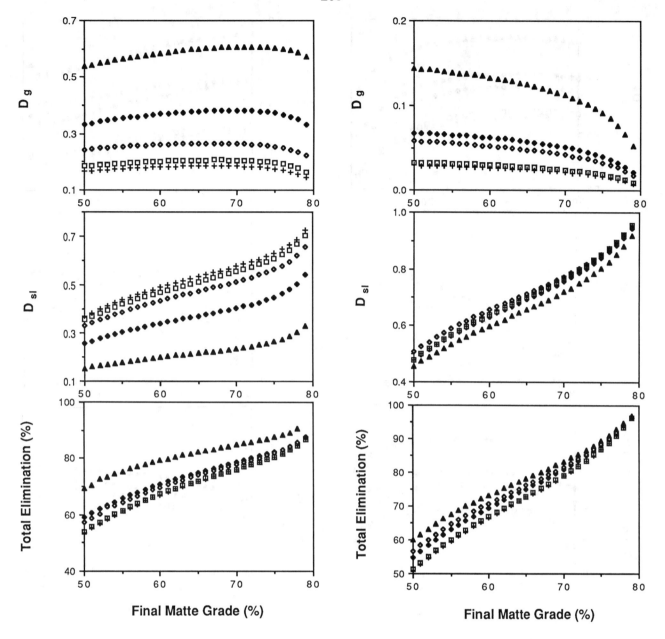

Fig. 1. Elimination of Pb in smelting.

▲ Air, 1200°C
♦ Noranda, 1200°C
□ SOI/N$_2$, 1200°C
◇ SOI/N$_2$, 1250°C
+ SOI/NG, 1200°C

Figure 2. Elimination of Zn in smelting.

▲ Air, 1200°C
♦ Noranda, 1200°C
□ SOI/N$_2$, 1200°C
◇ SOI/N$_2$, 1250°C
+ SOI/NG, 1200°C

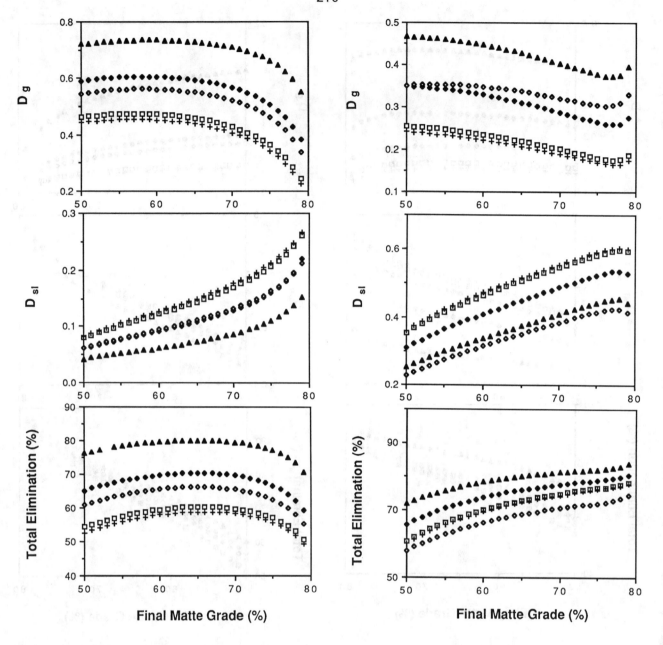

Figure 3. Elimination of Bi in smelting.

▲ Air, 1200°C
◆ Noranda, 1200°C
□ SOI/N$_2$, 1200°C
◇ SOI/N$_2$, 1250°C
+ SOI/NG, 1200°C

Figure 4. Elimination of Sb in smelting.

▲ Air, 1200°C
◆ Noranda, 1200°C
□ SOI/N$_2$, 1200°C
◇ SOI/N$_2$, 1250°C
+ SOI/NG, 1200°C

the other hand increases considerably with increasing oxygen partial pressure and final matte grade. The net result, however, is that, even when a high-grade matte is produced, the total elimination of arsenic in the smelting process decreases with increasing oxygen partial pressure.

Batch-type bath converting

A computer simulation was also carried out to investigate the minor-element behavior for matte converting in a batch-type bath converter with submerged oxygen injection. In this case, a copper matte weighing 10,000 g, containing 0.1 g Pb, 0.5 g Zn, 0.05 g Bi, 0.05 g Sb, and 0.05 g As, is converted to blister copper. Basically the same trends as those for smelting were obtained. For all the minor elements except Zn, the elimination by both volatilization and slagging sharply decreases with an increase in the initial matte grade. However, the volatilization of Zn increases with the initial matte grade, although the slagging of Zn decreases with the initial matte grade. The total elimination of minor elements when submerged oxygen injection is used with nitrogen gas as a shielding gas is shown in Fig. 6. As can be seen, the total elimination of minor elements sharply decreases with an increase in the initial matte grade.

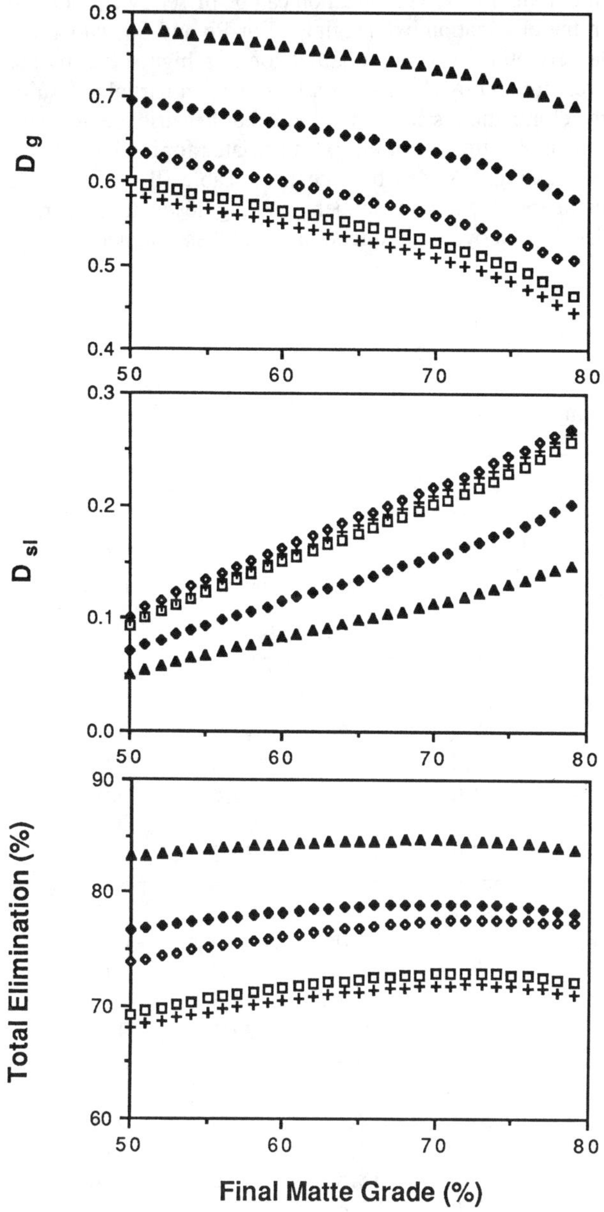

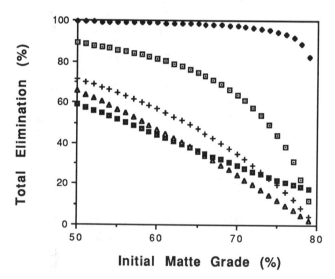

Figure 5. Elimination of As in smelting.

▲ Air, 1200°C
◆ Noranda, 1200°C
□ SOI/N$_2$, 1200°C
◇ SOI/N$_2$, 1250°C
+ SOI/NG, 1200°C

Figure 6. Total elimination pct of minor elements in converting at 1200°C when submerged oxygen injection is used with nitrogen shielding gas.

□ Pb
◆ Zn
■ Bi
+ Sb
△ As

Overall bath-type coppermaking

For a comprehensive analysis of minor-element behavior, the two simulations were combined. In this case, the concentrations of minor elements in the matte charged into the converter were set equal to those obtained from the simulation of copper smelting. The results are shown in Table II.

In general, the elimination of minor elements occurs primarily in smelting, and secondarily in the first stage of converting, as evident in Table II. The computer simulation shows that, when submerged oxygen injection is applied to copper smelting and converting, the elimination of Bi and As, which occurs mainly by volatilization, can be greatly affected because of reduction in gas volume. On the other hand, the elimination of Zn, which is primarily carried out by slagging, is little affected, because the drop in elimination by volatilization can be offset by an increase in the elimination by slagging. For Pb and Sb, which are largely eliminated by volatilization at a high sulfur partial pressure and by slagging at a high oxygen partial pressure, the elimination is somewhat hindered, because the decrease in volatilization is not fully compensated for by the increase in slagging. As can be seen from Table II, a high-grade matte production in copper smelting leads to a marked decrease in overall elimination for all these elements except Zn.

Table II. Distribution of minor elements in entire coppermaking process (1200°C)

| Minor Element | Types of Gas Injected | Final Matte Grade in Smelting (%) | Distribution (%) | | | | | Overall Elimination (%) | [%M] in Blister Copper |
| | | | Smelting | | Converting | | | | |
			Slag	Gas	Slag	Gas	Blister Cu		
Pb	Air	50	15.3	54.2	8.9	19.7	1.9	98.1	0.013
	Noranda	70	40.3	38.3	7.4	7.2	6.8	93.2	0.043
	SOI/N$_2$	50	35.3	18.8	33.7	7.0	5.2	94.8	0.034
		60	47.0	20.4	22.4	4.0	6.2	93.8	0.040
		70	55.5	20.6	13.3	1.8	8.8	91.2	0.054
		75	61.0	19.7	7.6	0.9	10.8	89.2	0.067
Zn	Air	50	45.6	14.4	32.9	7.1	(0.01)	>99.9	0.0004
	Noranda	70	75.7	5.1	16.5	2.7	(0.04)	>99.9	0.001
	SOI/N$_2$	50	48.0	6.3	46.3	2.1	0.3	99.7	0.011
		60	63.9	3.0	31.1	1.6	0.4	99.6	0.013
		70	76.8	2.4	18.6	1.6	0.6	99.4	0.019
		75	85.1	1.8	10.6	1.7	0.8	99.2	0.023
Bi	Air	50	4.2	72.0	1.4	20.3	2.1	97.9	0.001
	Noranda	70	12.9	56.7	1.7	16.7	12.0	88.0	0.007
	SOI/N$_2$	50	8.0	46.2	7.9	19.1	18.8	81.2	0.012
		60	12.2	47.3	5.3	12.6	22.6	77.4	0.015
		70	17.0	42.8	2.8	8.8	28.6	71.4	0.018
		75	20.7	36.4	1.5	7.9	33.5	66.5	0.021
Sb	Air	50	25.2	46.7	9.8	12.9	5.4	94.6	0.0018
	Noranda	70	48.7	28.9	5.4	4.1	12.9	87.1	0.0041
	SOI/N$_2$	50	35.3	25.5	20.5	7.4	11.3	88.7	0.0037
		60	46.5	23.6	13.0	3.9	13.0	87.0	0.0042
		70	54.7	20.1	7.0	1.7	16.5	83.5	0.0052
		75	58.3	18.1	3.7	0.9	19.0	81.0	0.0059
As	Air	50	5.1	78.1	1.3	13.4	2.1	97.9	0.010
	Noranda	70	15.5	63.5	1.7	7.6	11.7	88.3	0.051
	SOI/N$_2$	50	9.4	59.9	6.3	14.3	10.1	89.9	0.047
		60	15.0	56.6	5.0	8.8	14.6	85.4	0.066
		70	20.1	52.8	2.9	4.2	20.0	80.0	0.088
		75	22.9	49.9	1.6	2.0	23.6	76.4	0.103

Flash smelting

A comprehensive mathematical model that combines turbulent transport phenomena of a particle-laden gas jet, chemical reactions, and thermal radiation was developed for the prediction of minor-element behavior inside a reaction shaft of the flash furnace. Experiments were performed with a concentrate feed to which relatively large amounts of sulfides of antimony, bismuth, and lead were added. This was done to increase the accuracy of the chemical analysis for the minor elements. The contents of these elements and the conditions for model predictions are shown in Table III. Experimental data were collected mostly in the middle to lower part of the furnace, because of the difficulty of collection near the burner tip. No experiments using arsenic and bismuth were done.

For the target matte grades of 50 and 70 pct Cu, comparisons between the predicted results and measured results of the fractional elimination of minor elements to the gas phase are shown in Figs. 7 and 8. As can be seen, the experimental data points are rather scattered for the target matte grade of 50 pct Cu. However, for the target matte grade of 70 pct Cu, though the experimental data are some-what scattered, they are in relatively good agreement with the predicted results. The predicted results show that the patterns of elimination of each element are somewhat complicated in the upper section of the furnace. The numerical stability test, by changing the grid spacing in this upper section of the furnace, showed that these complicated patterns of elimination behavior are not the result of numerical instability. The rapid exothermic reactions of sulfide minerals in the particle, a sudden heat loss from the particle to the relatively cold surrounding gases, and the complicated flow and reaction patterns of the sulfide and oxide components of the minor elements appear to make the patterns of elimination rather complicated.

For the elimination behavior in the laboratory furnace shaft with axial feeding of the gas and concentrates through a single-entry burner, the computed results for the laboratory flash-smelting operation with realistic initial contents of each minor element can be summarized as follows:

(1) The elimination of As to the gas phase increases sharply between 0.3 and 0.5 m from the burner. Further than 0.5 m from the burner, it increases rather gradually along the furnace shaft.

Table III. Conditions for laboratory flash furnace experiments for the volatilization of minor elements

Details of the Laboratory Furnace	single entry
Burner tip inside diameter	0.02 m
Furnace inside diameter	0.23 m
Furnace length	1.32 m
Furnace wall thickness	0.02 m
Solid sampler inside diameter	0.02 m
Input Gas Stream	
Linear gas velocity	3.32 m/s
O_2 content	21%
Temperature	303 K
Particle loading	0.697-0.968 (kg particles/kg gas)
Turbulent Intensity at the Inlet*:	0.06
Analysis of Chalcopyrite Concentrate	
Chemical % S	30
% Cu	26
% Fe	27
% H_2O	0.3
% As	0.14
% Sb	0.88
% Bi	1.12
% Pb	1.14
Particle density	4300 kg/m^3
Particle mean diameter	50 μm (for prediction)
Wall Temperature top and side wall	1223 K

*Recommended by Hahn and Sohn[29]

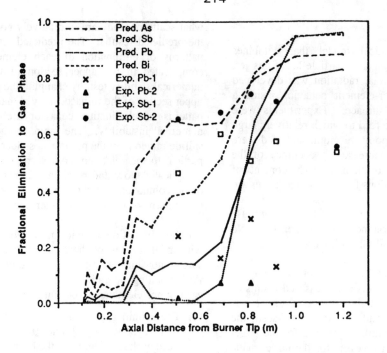

Figure 7. Comparison between predicted and measured results for the fractional elimination of minor elements for target matte grade of 50 pct Cu.

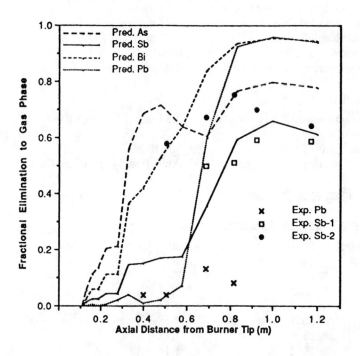

Figure 8. Comparison between predicted and measured results for the fractional elimination of minor elements for target matte grade of 70 pct Cu.

(2) The elimination of Sb to the gas phase increases somewhat gradually along the furnace shaft, and it occurs mostly further than 0.6 m from the burner.

(3) The elimination of bismuth increases sharply at about 0.3 m from the burner, and after this point it increases gradually along the furnace shaft.

(4) At the early stage of the flash-smelting reaction, the elimination of lead is much lower than that of arsenic, antimony, and bismuth, independent of the effect of the target matte grade. More than 0.6 m from the burner, lead elimination increases sharply and exceeds that of other elements.

(5) The elimination of arsenic and bismuth occurs mostly at the initial stage of the reaction of particles just before particles melt completely, and that of lead and antimony, though somewhat different in their elimination behavior, occurs mostly after the complete melting of particles.

The elimination behavior of minor elements in a commercial furnace shaft with typical contents used for the previous predictions was also computed. The result can be summarized as follows:

(1) In the upper section of the furnace, the patterns of elimination behavior are as complicated as predicted in the laboratory furnace.

(2) For the target matte grade of 70 pct Cu, the elimination of all minor elements occurs mostly at about 1.3 m from the burner. However, for the target matte grade of 50 pct Cu, it occurs mostly more than 2.5 m from the burner.

(3) The final elimination of arsenic and antimony is relatively independent of the target matte grade.

(4) Bismuth and lead elimination increases as the target matte grade increases.

(5) The elimination of all minor elements levels off at about 3 m from the burner.

The above predictions at the end of the furnace shaft were compared with the pilot-plant data for the samples collected at the exit of the furnace, as shown in Table IV. Though the predicted results are somewhat higher than the observed values, the table shows a reasonably good agreement between these values.

Table IV. Comparison of observed elimination of minor elements in a pilot plant with the prediction using the conditions of a commercial operation (feed composition: 28% Cu, 25% Fe, 32% S, 0.17% As, 0.01% Sb, 0.01% Bi, 0.01% Pb)

Final Matte Grade, % Cu	Volatilization to the Gas Phase, %							
	Predicted				Observed			
	As	Sb	Bi	Pb	As	Sb	Bi	Pb
50	91	46	72	72	88	42	60-90	70
70	90	48	79	82	77	30	70-90	65

CONCLUSIONS

Computer simulations to predict the behavior of undesirable minor elements such as Pb, Zn, Bi, Sb, and As were carried out for steady-state copper smelting and batch-type copper matte converting in a bath reactor with submerged oxygen injection, and for flash smelting inside a flash-furnace shaft.

The simulation predicted high elimination by slagging due to high oxygen partial pressure and low elimination by volatilization due to reduction in gas volume when submerged oxygen injection was applied to coppermaking. According to the simulations, the elimination of Bi and As can be greatly affected, followed by that of Sb and Pb in decreasing order, and the elimination of Zn is little affected by the increased content of oxygen in the process gas. Therefore, when submerged oxygen injection is used in coppermaking, a feed which contains low amounts of Bi, Sb, and As should be used, and low-grade matte production and high-temperature operation are recommended.

In the case of flash smelting, a comprehensive mathematical model that combines turbulent transport phenomena of the particle-laden gas jet, chemical reaction, and thermal radiation has elucidated the behavior of the minor elements in the furnace shaft during flash-smelting of copper concentrates. Satisfactory agreement between the predicted and measured results was obtained for antimony and lead, especially for the target matte grade of 70 pct Cu. The predicted results show that the patterns of elimination of each element are somewhat complicated in the upper section of the furnace due to the rapid exothermic reactions of sulfide minerals in the particle, sudden heat loss from the particle to relatively cold surrounding gases, and the complicated flow and reaction patterns of the sulfide and oxide components of the minor elements.

ACKNOWLEDGMENTS

The authors wish to acknowledge helpful discussions with Messrs. D. B. George and K. M. Iyer during the course of this work. This work was supported in part by the State of Utah and Kennecott under a grant to establish the State Center of Excellence for Advanced Pyrometallurgical Technology, by the Department of the Interior's Mineral Institutes program administered by the Bureau of Mines through the Generic Mineral Technology Center for Pyrometallurgy under allotment grant No. G1125129, and by the same Mineral Institutes program under allotment grant Nos. G1174149, G1184149, G1194149, and G1104149. A grant of computer time from the Utah Supercomputing Institute, which is funded by the State of Utah and the IBM Corporation, is gratefully acknowledged.

NOMENCLATURE

A_M	atomic or molecular weight of species M
C_p	heat capacity at constant pressure
C	Concentration of gaseous species
D_i	fractional distribution of minor element in phase i
D_M	gas-phase diffusivity of species M
d_p	particle size
f_S	volume fraction of molten particle occupied by sulfides
h	heat-transfer coefficient
k_M^*	volatilization constant defined by Eq. (9B)
k_i	function of temperature, P_{O_2} and P_{S_2}, defined by Eqs. (7) and (8), $i = 1, 2, 3, \ldots$
k_m	mass-transfer coefficient
$L_M^I, L_M^{II},$	
L_M^{III}	distribution coefficient of element M

$$(L_M^I + \frac{[\%M]_{Cu}}{[\%M]_{mt}}, \quad L_M^{II} = \frac{[\%M]_{Cu}}{[\%M]_{sl}},$$

$$L_M^{III} = \frac{[\%M]_{mt}}{[\%M]_{sl}})$$

$[\%M]_i$	pct M in phase i
$\dot{m}_i$	mass rate of flow of i
m_j	time-mean mass fraction of species j
m_M	number of moles of species M contained in the gas of volume V_g
Noranda	Noranda tuyere (42% oxygen enrichment in smelting, 24% in converting)
P_M^0	partial pressure of species M over pure solid or liquid
P_M	partial pressure of species M
P_M^e	effective total pressure of M-containing gases defined by Eq. (1)
P_T	total pressure
R	universal gas constant
Re	Reynolds number
S	surface area of a particle
Sh	Sherwood number
SOI/G	submerged oxygen injector with shielding gas G (N_2 = nitrogen, NG = natural gas)
STP	standard temperature and pressure, 0°C and 1.0 atm
T	temperature in K
t	time
Total Elimination	$100 \times (D_{sl} + D_g)$
V_g	volume of washing gas in liters (STP)
V_i	volume of gas species i in liters (STP)
$\dot{v}_i$	volumetric rate of flow of gas i
W_i	weight of melt i
W_M	weight of minor element M accompany 100 g feed
W_T	$W_T = W_{mt} + W_{sl}/L_M^{III}$ at the first stage of converting and $W_T = W_{Cu} + W_{mt}/L_M^I$ at the second stage of converting

$(\gamma_M^0)_j$	Raoultian activity coefficient of species M at infinite dilution in phase j

Subscripts

b	bulk value
c	concentrate
Cu	copper
f	film
fd	feed
g	gas
M	minor element
m	referring to mass
mt	matte
p	particle
sl	slag

Superscripts

b	bulk value
e	equilibrium
0	value in standard state

REFERENCES

1. Queneau P. E. and Schuhmann R. Jr. Metallurgical process using oxygen. U.S. Patent No. 3,941,587, March 2, 1976.
2. Queneau P. E. Oxygen technology and conservation. Metallurgical Transactions B, vol. 8B, 1977, pp. 357-369.
3. Eacott J. G. The role of oxygen potential and use of tonnage oxygen in copper smelting. In Advances in Sufide Smelting, Basic Principles, vol. 1, edited by H. Y. Sohn, D. B. George, and A. D. Zunkel, Warrendale, Pennsylvania: TMS-AIME, 1983, pp. 583-634.
4. Chaubal P. C., Sohn H. Y., George D. B. and Bailey L. K. Mathematical modeling of minor element behavior during flash smelting of copper concentrates and flash converting of copper mattes. Metallurgical Transactions B, vol 20B, 1989, pp. 39-51.
5. Itagaki K. and Yazawa A. Thermodynamic evaluation of distribution behavior of arsenic, antimony, and bismuth in copper smelting. In Advances in sulfide smelting, basic principles, vol. 1, edited by H. Y. Sohn, D. B. George, and A. D. Zunkel, Warrendale, Pennsylvania: TMS-AIME, 1983, pp. 119-142.
6. Chaubal P. C. Effect of oxygen enrichment on elimination of As, Sb, and Bi during copper converting. Transactions Institute of Mining and Metallurgy, Section C, vol. 98, 1989, pp. C83-84.
7. Sohn H. Y., Kim H. G. and Iyer K. Minor element behavior in copper smelting with submerged oxygen injection. State Center of Excellence for Advanced Pyrometallurgical Technology, Technical Report, University of Utah, Salt Lake City, May 1990.
8. Nagamori M. and Chaubal P. C. Thermodynamics of copper matte converting: Part III and IV. Metallurgical Transactions B, vol. 13B, 1982, pp. 319-329, 331-338.

9. Chaubal P. C. and Nagamori M. Thermodynamics for arsenic and antimony in copper matte converting. Metallurgical Transactions B, 19B, 1988, pp. 547-556.

10. Nagamori M. and Mackey P. J. Thermodynamics of copper matte converting: Part II. Metallurgical Transactions B, vol. 19B, 1988, pp. 567-79.

11. Chaubal P. C. The reaction of chalcopyrite concentrate particles in a flash furnace shaft. Ph.D. Dissertation, University of Utah, Salt Lake City, Utah, 1986.

12. Eriç H. and Timuçin M. Activities in Cu_2S-FeS-PbS melts at 1200°C. Metallurgical Transactions B, vol. 12B, 1981, pp. 493-500.

13. Wahlbeck P. G., Myers D. L., and Truong V. V. Validity of the Ruff-MKW boiling point method. Journal of Chemical Physics, vol. 83, no. 5, 1985, pp. 2447-2456.

14. Nesterov V. N. and Ponomarev V. D. Izvest. Akad. Nauk. Kazakh. S. S. R., Ser. Met., Obogaschchen. i Ogneuporov, 1959, no. 1, pp. 80-84.

15. Bratanov V. and Kunchev K. Zinc Oxide Vapor Pressure Measurement at High Temperature. Rudodobiv Met., vol. 23, no. 3, 1968, pp. 38-41.

16. Arac S. and Geiger G. H. Thermodynamic behavior of bismuth in copper pyrometallurgy. Metallurgical Transactions B, vol. 12B, 1981, pp. 569-578.

17. Chaubal P. C. and Nagamori M. Volatilization of bismuth in copper matte converting — Computer simulation. Metallurgical Transactions B, vol. 13B, 1982, pp. 339-348.

18. Azuma K., Goto S. and Takebe N. Thermodynamic Study of ZnS in copper matte (2nd report). Nippon Kogyo Kaishi, vol. 85, 1969, pp. 935-939.

19. Petersson S., Eriksson S. and Fridfeldt C. Treatment of complex copper concentrates in TBRC at Boliden. CIM Bulletin, vol. 74, no. 832, 1981, pp. 123-127.

20. Minoura J. and Maeda Y. Current operation at KOSAKA Smelter and Refinery. Metallurgical Review of MMIJ, vol. 1, no. 2, 1984, pp. 138-156.

21. Lindkvist G., Nystedt P.-L., and Petersson S. Application of the KALDO Process at the copper smelter of the RONNSKAR works, Boliden Metall AB, Skelleftehamn, Sweden. In Copper Smelting — An Update, edited by D. B. George and J. C. Taylor, Warrendale, Pennsylvania: TMS-AIME, 1981, pp. 41-76.

22. Ohshima E. and Hayashi M. Impurity behavior in the Mitsubishi Continuous Process. Metallurgical Review of MMIJ, vol. 3, no. 3, 1986, pp. 113-129.

23. George D. B., Donaldson J. W. and Johnson R. E. Minor element behavior in copper smelting and converting. In World Mining and Metals Technology, vol. 1, Proceedings of the Joint MMIJ-AIME Meeting, Denver, Colorado, edited by A. Weiss, New York, New York: AIME, 1976, pp. 534-550.

24. Donaldson J. W., Kinneberg D. A., and Themelis N. J. Development of the Kennecott Converter Smelting Process. In World Mining and Metals Technology, vol. 2, Proceedings of the Joint MMIJ-AIME Meeting, Denver, Colorado, edited by A. Weiss, New York, New York: AIME, 1976, pp. 505-533.

25. Edlund S. and Lundquist S. Copper converter practice at the Roennskaer work. In Copper and Nickel Converters, edited by R. E. Johnson, New York, New York: AIME, 1976, pp. 239-256.

26. Johnson R. E., Themelis N. J. and Elterinham G. A. A survey of worldwide copper converter practices. In Copper and Nickel Converters, edited by R. E. Johnson, New York, New York: AIME, 1976, pp. 1-32.

27. Vogt J. A., Mackey P. J. and Balfour G. C. Current converter practice at the Horne Smelter. In Copper and Nickel Converters, edited by R. E. Johnson, New York, New York: AIME, 1976, pp. 357-390.

28. Asahina S., Hosokura K. and Hayashi T. Bismuth recovery from copper smelting dust. In World Mining and Metals Technology, vol. 2, Proceedings of the Joint MMIJ-AIME Meeting, Denver, Colorado, edited by A. Weiss, New York, New York: AIME, 1976, pp. 856-874.

29. Hahn Y. B. and Sohn H. Y. Mathematical modeling of sulfide flash smelting process: Part I. Model development and verification with laboratory and pilot plant measurements for chalcopyrite concentrate smelting. Metallurgical Transactions B, vol. 21B, 1990, pp. 945-958

30. Hahn Y. B. and Sohn H. Y. Mathematical modeling of sulfide flash smelting process: Part II. Quantitative analysis of radiative heat transfer. Metallurgical Transactions B, vol. 21B, 1990, pp. 959-966

31. Sohn H. Y., Seo K. W., Chaubal P. C. and Hahn Y. B. Laboratory studies on the flash smelting of copper concentrate. In Flash Reaction Processes, edited by D. G. C. Robertson, H. Y. Sohn and N. J. Themelis, Center for Pyrometallurgy, University of Missouri-Rolla, Missouri, 1988, pp. 145-166.

32. Bird R. B., Stewart W. E. and Lightfoot E. N. Transport Phenomena, John Wiley & Sons, Inc., New York, New York, 1960, p. 510.

33. How M. E. The Influence of turbulence on the burning rates of fuel suspensions. Journal of the Institute of Fuel, vol. 39, 1966, pp. 150-158.

Electrochemical reductive conversion of chalcopyrite with SO_2

C. A. C. Sequeira
Instituto Superior Técnico, Lisbon, Portugal

SYNOPSIS

The electrochemical reduction of chalcopyrite electrodes was investigated with SO_2 as reducing agent and also with imposed potentials applied externally in the absence of SO_2. The cathodic reduction reactions involved are,

$$2CuFeS_2 + 3Cu^{2+} + 4e^- = Cu_5FeS_4 + Fe^{2+}$$

$$Cu_5FeS_4 + 3Cu^{2+} + 4e^- = 4Cu_2S + Fe^{2+}$$

The anodic oxidation reaction of SO_2 is,

$$SO_2 + 2H_2O = 3H^+ + HSO_4^- + 2e^-$$

During the reaction a defect structure of chalcopyrite, deficient in iron, followed by bornite was observed as intermediate reaction products; and the final product was actually djurleite ($Cu_{1.96}S$). A logarithmic rate expression was derived for broad variations in solution concentration and temperature. Interesting commercial implications are advanced.

In recent years, there has been increasing interest in the development of new copper hydrometallurgical processes to minimize air pollution problems due to low grade SO_2 emissions from conventional smelters. Copper has been produced almost exclusively by smelting and converting methods which achieve high recoveries of copper and precious metals. Although hydrometallurgical processes do not provide clear-cut advantages over existing pyrometallurgical processes, they do offer several possible alternative means for producing copper from sulphide minerals without producing gaseous SO_2. According to Wadsworth[1], for hydrometallurgy to compete it must not only match the smelter economically but also must demonstrate a significant advantage, or quantum jump improvement to become acceptable as a full fledged alternative to smelting. In order to develop a successful process, it is necessary to understand fundamentals of such processes.

Among copper bearing minerals, most research has been conducted with chalcopyrite since it is the most abundant and most refractory of all copper sulphides. A few copper hydrometallurgical processes have been developed and carried to pilot and full scale operation such as the Arbiter process[2], the Clear process[3], the Cymet process[4] and the Sherritt-Cominco[5] process. Apart from direct leaching, conversion of chalcopyrite into a more leachable form may play an important role for these processes. This conversion could be performed by simple heat treatment such as activation with elemental sulphur[6] or roasting[5] in the temperature range of 400°C-550°C to produce covellite or bornite, respectively. And also it has been shown[7] that SO_2 gas can be used as a reducing agent to convert chalcopyrite to djurleite ($Cu_{1.96}S$) in aqueous media.

As was implied by Habashi[8], the cathodic reduction of sulphide minerals has potential importance for hydrometallurgical process since most sulphides are semiconductors with relatively low resistivities. This semiconducting nature of sulphides permits the application of electrochemical techniques developed in the study of metal electrodes. A fairly large amount of research on the electrochemistry of sulphides has been conducted during the last decade, especially for copper sulphide minerals. Baur, Gibbs and Wadsworth[9] studied both anodic and cathodic reactions of chalcopyrite. They showed that chalcopyrite reacts cathodically according to the reaction,

$$2CuFeS_2 + 2e^- = Cu_2S + 2Fe^{2+} + 3S^{2-}$$

In acid solution, the sulphide ion is not stable and reacts with hydrogen ion to form H_2S,

$$6H^+ + 3S^{2-} = 3H_2S$$

Combining the above two equations gives,

$$2CuFeS_2 + 6H^+ + 2e^- = Cu_2S + 2Fe^{2+} + 3H_2S$$

This electrochemical behaviour was investigated further in galvanic reduction of chalcopyrite to chalcocite by Hiskey and Wadsworth[10] and verified later by Nicol[11]. Biegler and Swift[12] also studied the electrolytic reduction of chalcopyrite in acid solutions, confirming the formation of chalcocite with the formation of intermediates such as talnakhite ($Cu_9Fe_8S_{16}$) or mooihoekite ($Cu_9Fe_9S_{16}$) Talnakhite, for example, could form as,

$$9CuFeS_2 + 4H^+ + 2e^- = Cu_9Fe_8S_{16} + 2H_2S + Fe^{2+}$$

and these sulphides are known[13] to tarnish more rapidly than chalcopyrite. Jones and Peters[14] and others[15, 16] studied the anodic behaviour of chalcopyrite with and without oxidizing agent, respectively. Also, it is interesting to note that flotation[17] of sulphides has been shown to be electrochemical in nature.

The purpose of this work was to investigate the electrolytic reduction of chalcopyrite in sulphuric acid media containing cupric and iron ions associated with SO_2 gas.

EXPERIMENTAL

Chalcopyrite electrodes of synthetic material were used in the electrochemical measurements. Synthetic chalcopyrite was prepared by the method described by Dutrizac and MacDonald[12], and modified by Biegler and Swift[12]. The chemical and mineralogical analysis of the massive

product obtained is shown in Table I.

Table I
Analysis of Synthetic
Chalcopyrite

Chem. Element	Anal. Wt%	Mineralogical Anal. Mineral	Wt%
Cu	29.5	$CuFeS_2$	≈85
Fe	32.6	FeS_2	≈ 5
S	36.3	CuS	≈ 1.5
		Cu_5FeS_4	≈ 5
		FeS	≈ 1.5
		Fe_7S_8	≈ 1.0

X-ray diffraction analysis confirmed the presence of minor amounts of pyrite (FeS_2), bornite (Cu_5FeS_4), covellite (CuS), troilite (FeS), and pyrrhotite (Fe_7S_8). The product was shown to be a n-type semiconductor with very low resistivity. Suitable synthetic electrodes were prepared in the form of thin slabs (about 1x1x0.2 cm) from the massive chalcopyrite. The backs of the slabs were coated with a conductive, silver loaded epoxy resin and insulated copper leads attached. The backs and sides of the electrodes were encapsulated in epoxy resin and the front surface abraded flat on emery paper and polished with successive grades of diamond paste to finish with a particle size of 0.25 μm and buffed with 0.1 μm tin oxide.

A typical five-necked flange-topped glass vessel was used for the electrochemical experiments. A platinum counter-electrode compartment (closed with a porous glass frit), Luggin probe, gas inlet, gas outlet and chalcopyrite working electrode were inserted through the necks. The gas inlet tube was positioned near the electrodes to produce both agitation in the electrolyte and a controlled atmosphere. Agitation was further achieved using a magnetic stirrer. Most experiments were carried out in cupric sulphate/sulphuric acid media, molar ratio $Cu^{2+}/CuFeS_2 = 5$, 80°C, 1000 rpm, 1 atm SO_2 and, unless otherwise indicated, these were the conditions used. The solution was purged with sulphur dioxide/nitrogen gas mixtures at a rate of 100 ml min^{-1}. The reference electrode was a saturated calomel electrode (SCE), and all the quoted potentials are with respect to this SCE.

Potentiodynamic polarization curves,

galvanostatic and potentiostatic experiments, and half-cell measurements were performed. Data were obtained using an Oxford Electrodes waveform generator, a PAR 113 preamplifier, a JJ Lloyd xy recorder, a PAR 379 coulometer, and a Keithley constant current source. The sweep rate for the potentiodynamic measurements was 5 mVs^{-1}. Half-cell experiments were carried out in cells of the type:

Cu, C, SO$_2$|H$^+$, Cu^{2+}, Fe^{2+}||
Fe^{2+}, Cu^{2+}, H$^+$|CuFeS$_2$, Cu

to illustrate the electrochemical nature of the reaction involved in this study, and to further examine the effects of several factors on the course of reduction of chalcopyrite electrodes.

RESULTS AND DISCUSSION
Sohn and Wadsworth[7] have shown that the cathodic reduction of chalcopyrite by SO$_2$ in cupric sulphate/sulphuric acid solutions may be expressed as follows:
Cathodic reactions:
$$2CuFeS_2+3Cu^{2+}+4e^-=Cu_5FeS_4+Fe^{2+} \quad (1)$$

$$Cu_5FeS_4+3Cu^{2+}+4e^-=4Cu_2S+Fe^{2+} \quad (2)$$

Anodic reaction:
$$SO_2+2H_2O=3H^++HSO^-_4+2e^- \quad (3)$$

They further claimed that the reaction is electrochemical being affected by factors such as the temperature, the electrolyte agitation and concentration, etc. Present results confirm the overall reaction

$$CuFeS_2+3Cu^{2+}+2SO_2+4H_2O=$$
$$2Cu_2S+6H^++2HSO^-_4+Fe^{2+} \quad (4)$$

which is the sum of the reactions (1), (2) and (3), and give further information on the factors which determine the current efficiency.

Effect of agitation
Figure 1 shows the effect of agitation on the cathodic polarization curve for the chalcopyrite reduction. Without agitation, the curve exhibits a diffusion controlled region at potentials ≈400 mV below the rest potential. This region is characterised by a limiting current

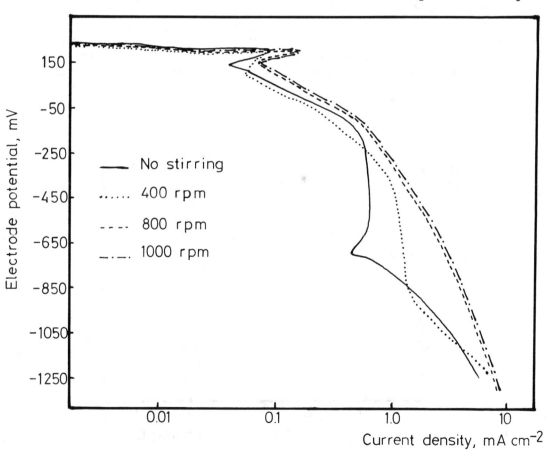

Fig. 1 Effect of stirring on the cathodic polarization behaviour of synthetic chalcopyrite.

density of about 0.5 mA cm^{-2}. With agitation, the same phenomenon appears at much negative potentials and with higher limiting current densities. Above 800 rpm, the cathodic polarization behaviour is independent of agitation and no diffusion limited region is shown. The chalcocite (Cu$_2$S) formed was observed to fall off of the electrode surface for potentials more negative than approximately -250 mV and displayed a classical redissolution region at the potential of the Cu$_2$S/Cu couple for the case of no agitation. Because of the effect of agitation, it was decided to carry out most of the tests at 1000 rpm.

Effect of SO$_2$ pressure

The solubility of gas in an electrolyte can be estimated by the method of van Krevelen and Hoftijzer[18] which relates the Henry's law constant in the solution to that in water at the same temperature. Based on this method we have estimated that the solubility of SO$_2$ in our electrolyte at 1 atm partial pressure of SO$_2$ was about 1.5g moles/l. To control the partial pressure of SO$_2$ in the system, ultra high purity nitrogen was used as a diluent. Therefore electrolytes at 0.1 to 1.0 atm partial pressure of SO$_2$ were tested. The resulting cathodic polarization curves were practically independent

of the SO$_2$ pressure. The reaction rate, expressed in terms of the fraction of Fe^{2+} released (v) into solution, was also determined as a function of the SO$_2$ partial pressure for chalcopyrite/solution systems kept at the rest potential. Again, no variation was observed.

Effect of pH

Cathodic polarization curves obtained for solutions of pH varying from 1.0 to 4.5 have shown that the reduction of chalcopyrite is essentially independent of the hydrogen ion activity. This is in agreement with reaction (3): even if the initial pH is high the forward reaction (3) leads to acid formation re-establishing the high acidity of the medium. The effect of a high initial pH was followed in the half-cell experiments: the current flow between the anode compartment (where reaction (3) took place) and the cathode compartment (where reaction (1) took place) increased when the pH of the anolyte was increased (forward reaction (3) favourable) but then diminished and stabilized along the time as in experiments performed with the 1M H$_2$SO$_4$.

Effect of O$_2$ pressure

Cathodic polarization curves performed in solutions purged with nitrogen and purged with nitrogen/oxygen mixtures showed that

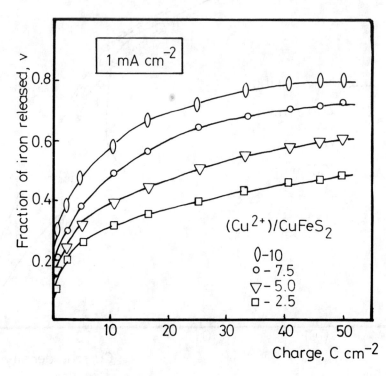

Fig. 2 Galvanostatic curve showing effect of Cu^{2+} addition.

the dissolved oxygen had no effect on the cathodic reduction of chalcopyrite, as it might be expected from analysis of reactions (1) to (3).

Effect of Cu^{2+} and Fe^{2+} concentration

Progressive measurements of the extent of conversion (fraction v of Fe^{2+} released) as a function of initial Cu^{2+} and Fe^{2+} concentrations were performed by galvanostatic means and are illustrated in Figs. 2 and 3, respectively. As the Cu^{2+} concentration increased, the rate of reaction also increased. Figure 3 shows that the reaction rate is retarded with the initial addition of Fe^{2+}, but increased significantly as the reaction proceeds (after passage of a given amount of charge). This "catalytic effect" was also observed by Sohn and Wadsworth[7] during the chemical conversion of chalcopyrite to copper sulphides.

Half-cell experiments carried out in cells of the type

$$Cu, C, SO_2 | H^+, Cu^{2+} || H^+, Cu^{2+} | CuFeS_2, Cu$$
(A)
$$Cu, C, SO_2 | H^+, Cu^{2+}, Fe^{2+} || H^+, Cu^{2+} | CuFeS_2, Cu$$
(B)
$$Cu, C, SO_2 | H^+, Cu^{2+} || H^+, Cu^{2+}, Fe^{2+} | CuFeS_2, Cu$$
(C)

indicated that the current flowing through cell (B) is lower than the current flowing through cell (A), the opposite occurring at cell (C). The current decrease observed in cell (B) may be attributed to the retardation of electrode reaction (3) by the formation of a highly-resistive SO_2-Fe^{2+} complex. The current enhancement observed in cell (C) can be attributed to the lowered activation energy for the surface charge transfer transient process involving both iron and copper according to:

$$Fe^{2+}(aq) + Cu^{2+}(aq) = Cu^+(lattice) + Fe^{3+}(aq)$$

$$Fe^{3+}(aq) + e^- = Fe^{2+}(aq)$$

Cher and Davidson[19] also proposed a similar mechanism to explain the catalytic effect of Cu^{2+} on the oxygenation of ferrous iron in phosphoric acid.

To further examine the effect of Cu^{2+} and Fe^{2+} concentration on the electrochemical reduction of chalcopyrite, cathodic polarization curves were obtained as a function of Cu^{2+}/Fe^{2+} concentration. Figure 4 shows the effect of Cu^{2+} concentration on the cathodic polarization of chalcopyrite in acid media, under stagnant conditions (no stirring). The theoretical rest potential of reaction (1) at standard state is 0.474V but was approximately 0.22V experimentally.

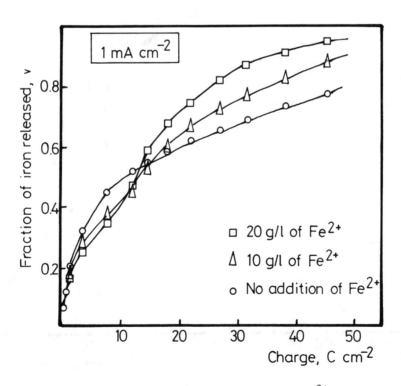

Fig. 3 Galvanostatic curve showing effect of Fe^{2+} addition.

The observed potential may be a mixed potential between the potentials of reaction (1) and the following reaction:

$$CuFeS_2 = CuS + Fe^{2+} + S + 2e^-$$

whose standard potential is -0.006V. This anodic process was proposed by Gardner and Woods[17] in mineral flotation studies and was confirmed using voltammetry to examine the natural flotability of chalcopyrite in acid media. Addition of cupric ion increased both the rest potential and current density in agreement with the leaching data shown in Fig. 4. This figure also shows that at higher copper concentration no diffusion limited region is shown. In other words, the effect of Cu^{2+} on this region is similar to that observed when the agitation is increased (Fig. 1). Therefore, the diffusion controlled region observed in the cathodic polarization curves for chalcopyrite is proposed to be due to the cupric ion depletion at the electrode surface as the reaction proceeds.

Figure 5 shows the effect of Fe^{2+} addition on the cathodic polarization behaviour of synthetic chalcopyrite. As expected from the previous discussion, Fe^{2+} additions greatly affected the current density but had little effect on the rest potential.

Effect of temperature

The effect of temperature on the electrochemical conversion of chalcopyrite to copper sulphides was studied by potentiostatic and potentiodynamic techniques. Figure 6 illustrates the effect of temperature on the potentiostatic reduction of chalcopyrite, performed at -100 mV. The Arrhenius plot shown in Fig. 7 was obtained using the measured slope dq/dt at 40 C cm^{-2}, where q is the amount of charge that passed through the chalcopyrite electrode at -100 mV. This plot indicated an activation energy of 19 Kcal/mole leading to the conclusion that the chalcopyrite electroreduction is controlled by a surface discharge process.

The cathodic potentiodynamic curves for the effect of temperature showed that both rest potential and current density are highly influenced by the temperature increase.

A kinetic analysis of the value of the

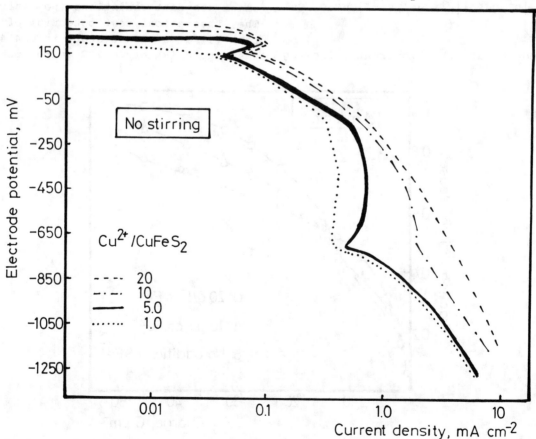

Fig. 4 Effect of Cu^{2+} concentration on the cathodic polarization behaviour of synthetic chalcopyrite.

rate of reaction (expressed in terms of the fraction of iron released, or the current density across the electrode) for broad variations in Cu^{2+}, Fe^{2+}, H^+, SO_2 concentrations and temperature resulted in a general logarithmic rate equation of the type

$$v = v^o \ln (Kt + 1)$$

where v^o is a constant that depends only on the basic pore structure of the chalcopyrite material, and K (expressed in min^{-1}) is the chemical reaction rate constant, which is influenced by temperature and related variables within the system.

Solid products of chalcopyrite reduction
To further examine the reduction of chalcopyrite, X-ray diffractometry of the products, either in situ or scraped from the electrode surface, was used for samples collected after various time intervals under potentiostatic conditions. The imposed potentials varied between +200 mV and -800 mV (vs. SCE). At about 100 mV the solid product of reduc-

tion was generally a defect chalcopyrite intermediate, of the type $Cu_{1+x}Fe_{1-y}S_2$, sometimes mixed with bornite, that prevails for longer times of electrolysis. For potentiostatic runs longer than about 1 hour and potentials in the region of -100 mV the defect structure converged to bornite, suggesting reaction (1) taking place. Below about -200 mV the product is generally bornite mixed with chalcocite (Cu_2S), and djurleite ($Cu_{1.96}S$). At longer times of electrolysis only djurleite is detected forming according to reaction (2). Experimentally metallic copper appears when the potentiostatic run is performed at about -800 mV.

COMMERCIAL APPLICATIONS
From the data presented, the apparent reaction rate of electrochemical conversion of chalcopyrite into more leachable form is relatively fast. This reaction involving SO_2 has an interesting concept to use the energy from sulphur species reaction by intercepting the electrons by means of an electrochemical coupling. Based on reaction (4) as well as the fast leaching behaviour of chalcocite into

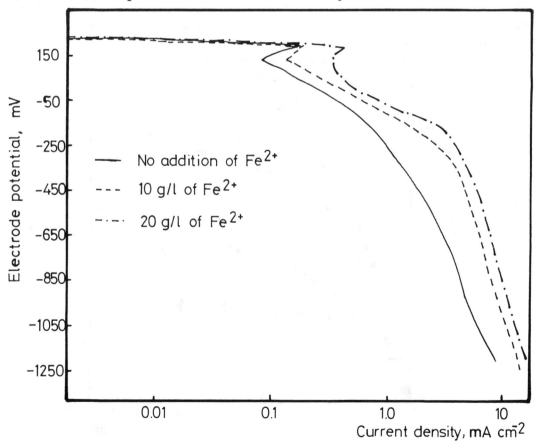

Fig. 5 Effect of Fe^{2+} concentration on the cathodic polarization behaviour of synthetic chalcopyrite.

226

covellite, a new process has been proposed by Sohn and Wadsworth[7] as a possible means to replace cementation and/or
solvent extraction in copper hydrometallurgy. This process would be favoured by
the electrochemical route because the
reaction rate would be enhanced. Therefore the electrochemical conversion may
prove to be an effective means for selective recovery of copper from solution
with concurrent acid generation and
upgrading of copper concentration from
dilute process streams.

Basic research on these lines is being
pursued to explore new ways for the
recovery of base metals. Hydrometallurgical systems have significant problems of
liquid and solid-waste disposals although
they have advantages over pyrometallurgical systems in terms of air pollution
difficulties. As recently mentioned[20],
hydrometallurgy has a place for treating
concentrates which are complex and are
not amenable to smelting. Also, it has a
significant advantage for the low grade
ore amd small scale operations.

CONCLUSIONS
The results of the electrochemical reductive conversion of chalcopyrite with SO_2
in cupric sulphate-sulphuric acid media

indicate the following:
1. For the initial stage of reduction, a defect structure
   of chalcopyrite was observed
   and is probably due to preferential removal of iron
   from the chalcopyrite lattice. Also bornite was shown
   throughout the reaction as
   an intermediate reaction
   product. The final product
   was djurleite ($Cu_{1.96}S$).
2. The presence of a significant amount of ferrous ion
   increased the reaction rate
   dramatically after the
   initial stage of reduction.
   Also, the rate of reaction
   is sensitive to the cupric
   ion concentration.
3. The partial pressure of SO_2
   did not affect the reaction
   rate markedly.
4. Electrolyte agitation has a
   pronounced effect on the
   chalcopyrite electroreduction, being adviseable to
   work at about 1000 rpm.
5. The rate of reaction is
   insensitive to the pH of the
   medium, and to the dissolved
   oxygen.

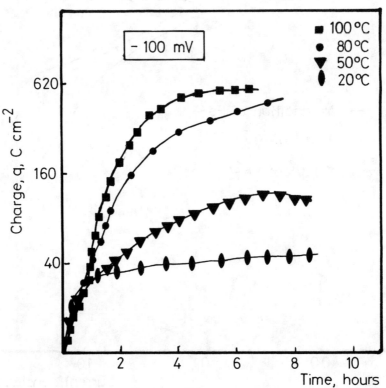

Fig. 6 Effect of temperature on the potentiostatic reduction of synthetic chalcopyrite.

6. The temperature affects markedly the chalcopyrite reduction, being adviseable to work at temperatures of the order of 80°C.

7. The reaction is controlled by a surface discharge process, and a logarithmic rate law seems to describe adequately its kinetics.

8. Commercial application seems possible in specific areas of copper hydrometallurgy.

Acknowledgements

The author wishes to thank Materials Research and Development Corporation, U.S.A., for financial support and research facilities.
Financial support for the preparation of the manuscript is acknowledged to JNICT (Portugal).

References

1. Haas, L. A. and Shafter, J. L. (eds.). Panel discussion on copper hydrometallurgical economics. Journal of Metals, vol. 31, no. 7, July 1979, p. 40-47.

2. Arbiter, N. and Milligan, D. A. Reduction of copper amine solutions to metal with sulphur dioxide. Extractive Metallurgy of Copper, vol. II. New York: The Metallurgical Society of American Institute of Mining Engineering, 1976. p. 974-993.

3. Atwood, G. E. and Curtis, C. H., Duval Corp., U.S. Patents 3,785,944 and 3,879,272 (1974).

4. Ammonia pressure leaching. G. Bolton. In: Hydrometallurgy: Theory and Practice Course Notes. Organised by the Centre for Metallurgical and Process Engineering. Vancouver: The University of British Columbia, 1988. Chapter XIV, 20pp.

5. Swinkels, G. M., Berezowsky, R. M. G. S., Kawulka, P., Kirby, C. R., Bolton, G. L., Maschmeyer, D. E. G., Milner, E. F. G. and Parekh, B. M. The Sherrit-Cominco Copper Process - Part I, II, and III. Canadian Mining and Metallurgical Bulletin, February 1978, p. 1-35.

6. Subramanian, K. N. and Kanduth, H. Activation and leaching of chalcopyrite concentrate. Canadian Mining and Metallurgical Bulletin, June 1973, p. 88-91.

7. Sohn, H.-J. and Wadsworth, M. E. Reduction of chalcopyrite with SO_2 in the presence of cupric ions. Journal of Metals, vol. 32, no. 11, November 1980, p. 18-23.

8. Habashi, F., The electrometallurgy of sulphides in aqueous solutions. Mining

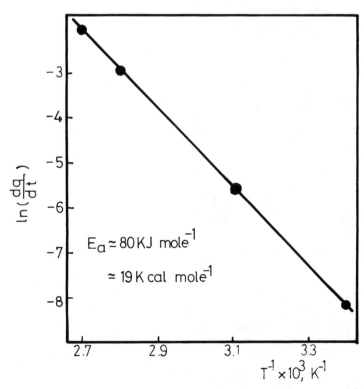

Fig. 7 Arrhenius plot illustrating temperature dependence.
The data in the figure correspond to the data in fig. 6.

Science & Engineering, vol. 3, no. 3, March 1971, p. 3-12.

9. Baur, J.D., Gibbs, H.L. and Wadsworth, M.E.. Initial-stage sulphuric acid leaching kinetics of chalcopyrite using radiochemical technique. Bureau of Mines, RI 7823, 1974.

10. Hiskey, J.B. and Wadsworth, M.E.. Galvanic conversion of chalcopyrite. Metallurgical Transactions B, vol. 6B, 1975, p. 183-190.

11. Nicol, M.J.. Mechanism of aqueous reduction of chalcopyrite by copper, iron and lead. Transactions of the Institution of Mining and Metallurgy, vol. 84, 1975, p. C206-C209.

12. Biegler, T. and Swift, D.A.. The electrolytic reduction of chalcopyrite in acid solution. Journal of Applied Electrochemistry, vol. 6, 1976, p. 229-235.

13. Cabri, L.J. and Hall, S.R.. Mooihoekite and Haycokite, two new copper-iron sulphides, and their relationship to chalcopyrite and talnakhite. American Mineralogist, vol. 57, 1972, p. 689-708.

14. Jones, D.L. and Peters, E.. The leaching of chalcopyrite with ferric sulfate and ferric chloride. Extractive Metallurgy of Copper, vol. II. New York: The Metallurgical Society of American Institute of Mining Engineering, 1976, p. 633-653.

15. Peters, E.. The electrochemistry of sulphide minerals. In Trends in Electrochemistry, eds. by J.O'M. Bockris, D.A.J. Rand and B.J. Welch. New York: Plenum Press. 1977, p. 267-290.

16. Koch, D.F.A.. Electrochemistry of sulfide minerals. In Modern Aspects of Electrochemistry, vol. 10, eds. J.O'M. Bockris and B.E. Conway. New York: Plenum Press, 1975, p. 211-237.

17. Gardner, J.R. and Woods, R.. An electrochemical investigation of the natural flotability of chalcopyrite. International Journal of Mining Processing, vol. 6, 1979, p. 1-16.

18. Dankwert, P.V. Gas-liquid Reaction. New York: McGraw-Hill Book Company, 1970, p. 18-20.

19. Cher, M. and Davidson, N.. The Kinetics of the oxygenation of ferrous iron in phosphoric acid solution. Journal of American Chemical Society, vol. 77, 1955, p. 793-798.

20. Wells, J.A. and Snelgrove, W.R.. The design and engineering of copper electrowinning tankhouses. In Electrometallurgical Plant Practice, eds. P.L. Claessens and G.B. Harris. New York: Pergamon Press, 1990, p. 57-72.

Recycling

Automotive exhaust catalysts: PGM usage and recovery

V. Jung
Degussa AG, Hanau, Germany

Abstract

Different kinds of catalysts are used in chemical processes to accelerate their reactions. One of the most important and well-known examples is the conversion of the pollutants generated by the combustion of fuel to non-toxic products. The significance, supply situation and the recovery processes of the precious metals used as catalysts - platinum, rhodium and palladium (hereafter referred to as PGM) - are described. The pyrometallurgical treatment of spent automotive exhaust catalysts is explained. The newly developed Degussa-process and its first results obtained are mentioned. The problems of logistics are discussed, and a brief view into the future of automotive exhaust catalysts is given.

Introduction

Car exhausts release a number of noxious environmental pollutants into the atmosphere by the combustion of vehicle fuel. Therefore, methods have to be installed to reduce the specific fuel consumption and to convert the exhaust emissions carbon monoxide, uncombusted hydrocarbons and nitrogen oxides into non-toxic substances: carbon dioxide, water and nitrogen. The worldwide demand of industrial catalysts, autocatalysts included, shown in Fig. 1.[1] In 1990 about 34 % of them were used in automotive exhaust catalysts and are expected to increase to about 37 % in 1995.

In 1990 almost 97 % of the newly registered cars in former West Germany were fitted with exhaust gas control systems, 82 % of them with three-way-catalyst.[2]

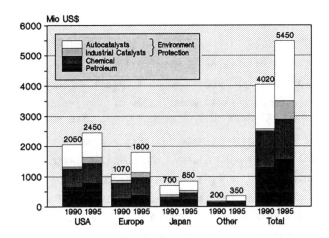

Figure 1 - Demand of Industrial Catalysts

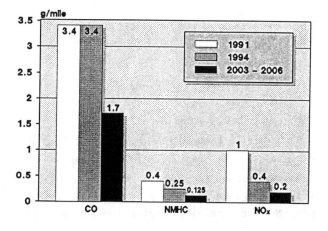

Figure 2 - US-Exhaust Gas Standards

231

The new exhaust gas standards as from 1993, planned for the USA are presented in Fig. 2.[3]
Especially the reduction of NO$_x$ in the future is of great importance in respect to an environmental improvement.

Automotive Exhaust Catalysts

Autocatalysts are based either on ceramic material (honeycomb-type or small pellets varying between 0,3 and 1,2 mm in diameter) or on metallic supports consisting of corregulated stainless-steel foils of an FeCrAl-alloy soldered into a steel can housing. Over the last 15 years the ceramic monolith has emerged as the dominant form of substrate material with still rising tendency, especially in Europe and North America. These honeycombs consist principally of cordierite (2MgO.2Al$_2$O$_3$.5SiO$_2$). The mostly inactive surface of the cordierite material must be coated by an active γ-alumina slurry, the so-called wash-coat, which is dried by a following process.
One or more steps are necessary to apply the precious metals on the substrate.
The catalytyc elements chosen for the purpose of conversion of the car exhaust emissions by means of a three-way-catalyst are platinum and rhodium and, to a lesser extent, palladium. The charge of the precious metals varies about 30-50 g /feet[3].
So 1 to 2 g of precious metal per vehicle is used depending on the type of the car and its engine.

Diesel engines produce small amounts of carbon monoxide, uncombusted hydrocarbons soot and some nitrogen oxides. Recently the use of oxidation catalysts as means of reducing the particulate emissions from diesel engines drew a lot of attention. Carbon monoxide and gaseous hydrocarbons can be removed to a high extent from the exhaust gas by a precious metal based oxidation catalyst.[4] Where the exhaust aftertreatment by filtering devices is concerned, there is a big problem in controlling soot and NO$_x$-production:
higher temperatures burn off the soot but raise, unfortunately, NO$_x$.

The Role of Precious Metals

Various combinations and proportions of the platinum-group metals - platinum, rhodium and palladium - are used in automotive exhaust catalysts. Nowadays catalysts with a platinum/rhodium ratio of about 5 : 1 and approximately 1,5g platinum per kg catalyst have prevailed. Platinum is mainly used for the conversion of hydrocarbons to water and carbon dioxide and acts rapidly because of its fast start-up.
Rhodium, on the other hand, is responsible for converting NO$_x$ to nitrogen by reaction with CO and H$_2$.
The use of palladium has more retreated almost completely into the background since the introduction of the Pt/Rh-based three-way-catalyst. It is less resistant to leaded and sulfur-containing gasoline.
The reduction of vehicle emissions with three-way-catalysts is shown in Fig. 3.[5]

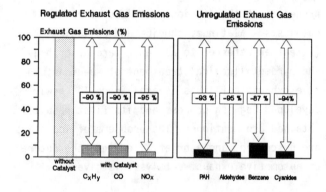

Figure 3 - Reduction of Vehicle Emissions
with Three-Way Catalyst

Market Situation

PGM-Demand

The wide spectrum of applications for the platinum-group metals is shown in Fig. 4.[6] About 114 t of platinum were used in 1990, more than 36% of them for automotive exhaust catalysts. The increase in comparison to 1987 was not very significant (105 t, nearly 35 %). About 84 % of the rhodium was used in automotive exhaust catalysts in the same year. The total demand of rhodium was about 12 t (1987 9,7 t, nearly 75%).

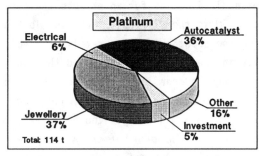

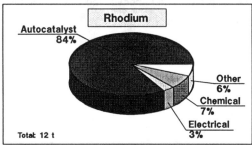

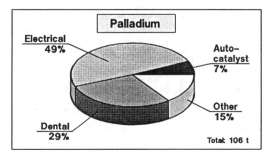

Figure 4 - PGM-Demand in Percentage

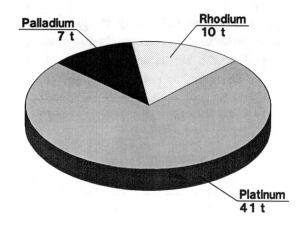

Figure 5 - Proportion of PGM in Autocatalysts

Figure 4 shows the important role of rhodium in spite of the worldwide effort to reduce the quantity of this very expensive precious metal in the production of autocatalysts.

As for palladium a minor application in the automotive exhaust catalysts is evident. Less than 7 % of the total 106 t demand in 1990 is used for this purpose.

The proportion of PGM used in the manufacture of automotive exhaust catalysts is shown in Fig. 5.[6] Up to date more than 340 t of platinum in total have been used worldwide for autocatalysts. The use of palladium will possibly increase if a completely unleaded gasoline network would be established. Then, platinum could partly be substituted by palladium. Although intensive research is being carried out to reduce the rhodium content in autocatalysts, the more stringent exhaust gas standards pending in the USA, and later in Europe as well, will thwart these effects.

Table I shows in detail the PGM-demand in the different countries of the Western World.[6] The amounts of rhodium are not published relative to each country, but in total they are quite well estimated.

Table I. Demand of Platinum, Palladium and Rhodium in 1990 (t)

Country (t)	Platinum		Palladium		Rhodium	
	autocat	total	autocat	total	autocat	total
North America	24.7	16.0	33.8	4.2	4.47	
Japan	57.4	11.3	48.0	2.2	3.73	
Europe + Other	31.5	13.4	24.5	0.7	3.97	
Total	113.6	40.7	106.3	7.1	12.12	10.22

PGM-Supply

Normally, platinum-group metals are found in sulphidic or oxidic copper/nickel - or in chromiteores. The main source of platinum i.e. is the South African Bushvelt Complex with its famous Merensky-Reef and UG2-Reef. The latter contains platinum and rhodium in a ratio of 5 : 1, which is preferred by automotive catalysts manufacturers, but the PGM are difficult to recover because of the high chromium content in this ore.

Table II shows the PGM-supply in 1990.[6] There the dominant role of South Africa is to be seen, where almost 75 % of platinum and about 35 % of palladium are produced. As regards rhodium, no exact data are available so that only the total production of approximately 11,7 t is mentioned.

Table II. PGM Supply in 1990(t)

Country (t)	Platinum	Palladium	Rhodium
South Africa	86.3	37.7	6.37
USSR	21.7	57.4	4.81
USA + Canada	5.7	11.5	0.53
Other	2.0	2.2	0
Total	115.7	108.8	11.71

A considerable increase in platinum production from the extensions of existing mines and more so from the opening of new mines in South Africa is to be expected in the next few years (Tab. III).[5]

Table III - South African Platinum-Production Forecast

Producer	1989	1990	1991	1992	1993	1994
Rusplats	41.07	41.07	42.31	43.40	44.95	44.95
Impala	32.55	33.32	34.10	34.10	35.65	35.65
Westplats	6.97	8.37	8.37	8.37	8.37	8.37
Lebowa	1.86	1.86	3.10	3.72	4.34	4.34
Barplats	0.93	2.94	3.56	4.49	5.27	5.27
Messina	nil	nil	0.93	3.10	4.96	4.96
Northam	nil	nil	1.55	3.10	4.65	6.82
Total Change (%)	83.38 +5.1	87.56 +5.0	93.92 +7.3	100.28 +6.8	108.19 +6.3	110.36 +2.0

PGM in Automotive Exhaust Catalysts

Up to now more than 250 million cars have been on the road worldwide with some form of emission control system. In 1989 nearly 35 million vehicles have been produced, around 25 million or 70 % of which were fitted with three-way-catalysts.[7] One of the main difficulties for an efficient recycling of the PGM is the difficult organisation in collecting the spent automotive catalysts (SAC), especially in the USA. It is estimated that only 70 % of the wrecked cars are really sent back to a workshop, garage or some type of recycling facility; the remaining 30% are deposited elsewhere. This situation will be much better in Europe and Japan. The estimated PGM-recycling rates in the next decade are given in Table IV.[8]

Table IV. Platinum- and Rhodium Supply in Autocatalysts

Year	Element	Total Supply (t)	Supply in Autocatalysts (t)	Supply in Autocatalysts (%)	Recovery from Autocatalysts (t)	Recovery from Autocatalysts (%)
1980	Pt	86.8	2.1	24	0.3	14
	Rh	5.7	0.6	11	0	0
1990	Pt	108.5	46.5	43	7.0	15
	Rh	9.8	7.3	75	0.4	5
2000	Pt	133.3	62.0	47	12.4	20
	Rh	12.7	9.8	77	0.8	8

PGM-Price Movements

The ups and downs in PGM prices are shown in Fig. 6.[9] In August 1987, the platinum quotation at the New York Stock Exchange touched 640 $/oz. The price movement of palladium was less spectacular.

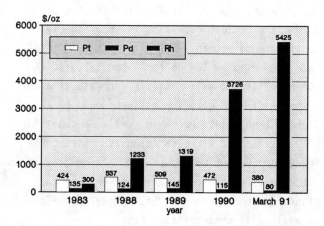

Figure 6 - Average US Dollar Prices of PGM

The rhodium price, however, after having risen from around 300 $/oz in 1983 to 1950 $/oz in 1989, did an unbelievable jump to the 7000 $/oz mark in July 1990. This exposion could be explained by the high demand of rhodium for the automotive exhaust catalysts and additional technical problems in some PGM-refineries. Now the prices are about 400 $/oz for platinum and about 3.000 $/oz for rhodium.

Logistics

The weight of all installed autocatalysts registered in 1988 was about 40.000 t world-wide, 20.000 t of them in USA, 15.000 t in Japan, and only 5.000 t in Europe.[10] The number of vehicles fitted with catalysts is sharply rising in today's Europe. While it has been relatively constant in USA and Japan because of the unchanged legislation in the last years, in Europe more than 7.000 t (a plus of 40 %) were installed in 1990.

There are many problems in collecting the automotive exhaust catalysts in USA because of the wide spread junk yards, scrap dealers, service stations repair shops, and muffler shops. The links in this chain are the collectors, the lot consolidators and the extractors upgrading the low PGM-content to a level at which the material can be sent to a refinery.

In Europe greater control by car companies and refineries will shorten this longthy process. In addition, stronger legislation concerning the wrecking of used or dammaged cars will force cars and spent catalysts as well to be recycled.

Treatment of Automotive Exhaust Catalysts

Because of the high value of the precious metals contained in automotive exhaust catalysts, their recycling is economically very interesting. Stable precious metal prices and relatively large recycling quantities are necessary to justify the complex recycling processes. In addition, the limited natural resources of these metals, rhodium in particular, should be kept in mind.[11] The most important sources of automotive exhaust catalysts for recycling are production rejects, test cars, warranty claims and spent catalysts.

In order to get to the substrate (pellets or monoliths), first of all, the cannings must be removed from the vehicles. In the past these procedures were not carried out by refineries. It takes considerable time to establish a suitable logistic structure to do this work.

In the case of pellets, a sample can be taken directly from the original batch following a careful mixing process, and then submitted to the analytic laboratory for examination.

The preparatory work for monoliths is more complicated because of their inhomogeneous character and size. These materials are first crushed and ground. The resulting product is then mixed in order to homogenise the material and to take a representative sample. In most cases, as well at Degussa-refinery, these precious metal bearing materials are not bought but toll refined for the customer. When the content has been analysed and the customer's approval received, recycling can commence.

Two basic methods of recycling, a hydrometallurgical or a pyrometallurgical procedure, are available:[12]

Hydrometallurgical Methods

In case of dissolving the substrates the dissolution is undertaken by sulfuric acid or in an alkaline medium by means of sodium hydroxide under pressure. The precious metals are contained in the residue which is further leached with chlorine and hydrochloric acid for PGM dissolution.[13] In the alkaline process any silica content remains undissolved and hinders the further processing of the precious metals. It is not advisable to recycle monolithic substrates by these methods because of the conversion of γ-alumina into insoluble α-alumina during a catalyst's lifetime.

On the other hand various methods of dissolving the precious metals with a wide range of precious metal yields are known, e.g. by means of hydrochloric acid and chlorine, hydrochloric acid and nitric acid or hydrochloric acid and hydrogen peroxide. One of the main problems of all these processes is the difficulty in separating the PGM from the nonferrous metals in the dilute solution.

The yields from this process, particularly of rhodium, are unsatisfactory. The negative effects of hydro- metallurgical recycling processes can be summar- ized as follows:

- large quantities of waste water
- leached-out substrates have to be dumped
- precious metal losses
- little use of aluminate liquor/aluminium sulphate

Their advantages are that it works at low temperature, the precious metal content is, in case of a low base metal content, easily monitored (due to the homogeneous solution) and the precipitation processes etc. are easy to carry out.

Pyrometallurgical Methods

In general, pyrometallurgical methods of recycling automotive exhaust catalysts involve smelting the ceramic substrate and simultaneously concentrating the precious metals in a collector metal. Complete scorification of the substrate without losses of the precious metals is of decisive importance in this process.

Especially the high melting point of the alumina pellets (app. 2000°C) is a big problem.Therefore these materials only can be slagged by addition of fluxes or by applying very high melting temperatures. Copper, nickel, lead and iron are generally considered potential PGM-collectors. They are usually selected according to ease of processing and to subsequent wet-chemical stages. The precious metals - platinum, palladium, and rhodium - are separated from the collector metal by sulfuric acid leaching. If copper is chosen, electrolysis can also be used. Compared with hydrometallurgical recycling of automotive exhaust catalysts, the pyrometallurgical method offers far more general advantages:

- high degree of concentration in a metal phase
- high precious metal yield
- can be carried out in nonferrous furnaces (blast furnaces, converters) or special units (i.e. electric furnaces)
- small amounts of by-products and residues

Conventional Melting Process

The temperatures of the furnaces used in the copper-, nickel- or lead-industry are usually about 1300°C and are therefore less suitable for smelting down the ceramic-based automotive exhaust catalysts. They are heated by coke, gas or oil and oxygen-enriched air. The large smelters do not have any problems in adding the above mentioned material. If it represents less than 1 % of the usual total feed, it is without any influence on the smelting process. So, on the one hand, it is a real advantage to treat these precious metal bearing materials in such a large furnace by a well-known metallurgical process with its low melting costs and treatment charges; but, on the other hand, there will be the problem of an effective recovery rate and a high yield of the precious metals, rhodium in particular. The dilution of the PGM is too significant and the amounts of slag occurring in this process are very large. This fact complicates considerably the subsequent concentration processes of PGM-recovery and - refining and leads to higher costs. In this case obtaining the pure metal is a long and arduous way. Therefore this conventional metallurgical process has now been widely abandoned.

Many refineries and non-ferrous metallurgical plants use electric high-temperature furnaces with several advantages:

- compact dimensions and high capacity
- low amounts of slag
- low metal losses
- temperature adapted to specific purpose
- pro-environmental, as very few emissions

The electric furnaces are widely used now for slag-cleaning processes, for the treatment of dust, for the smelting of ore material and many other purposes.[13,14]

A distinction is to be made in a certain respect as to the types of high temperature furnaces for the recycling of automotive exhaust catalysts.

Plasma Furnaces

Plasma furnaces are now in use for the reduction of chromite ores and for processing dusts, automotive exhaust catalysts etc.[16] Fundamental characteristics of the plasma method are high energy density, high temperatures and short smelting times. The high radiation intensity and the the extremely high plasma temperatures of more than 2000°C at the gas/slag interface cause considerable problems for the lifetime of the lining. In addition, the relatively high lead content in certain US autocatalysts caused big problems both for the processing and for the environment. The Iron and Steel Society (ISS) publication from 1987 provides a comprehensive survey of the general use of plasma furnaces and processes. [17]

Electric Furnaces

A specific type of high temperature furnace is the submerged-arc-furnace. In such a furnace the treatment of oxidic materials with high melting points, such as chromite ores, nickel oxides, ore fines etc. seems to be a promising solution. Here, the raw material itself serves as an electrical resistance. The bottom-electrode is represented by the charged material. This process ensures extremely rapid and complete reactions with a high metal yield.[18] The electrical energy is directly transferred to the melt by a conductive bottom lining. The opposing single-electrode is centrally located in the furnace roof. Direct current is normally applied, the polarity depending on different conditions.

Degussa-Electric-Furnace

Degussa obtained a lot of encouraging results by treating precious metal containing material in a lab-scale electric furnace. Consequently, Degussa decided to install a pilot-furnace for the recovery of PGM from automotive exhaust catalysts. The refinery at Wolfgang near Hanau was chosen as the site for this type of furnace, which is used in South Africa for the melting of steel scrap, smelting of pre-reduced iron ore fines and utilized for ferro-alloys and other metallurgical applications (Fig.7).[19,20] This new Degussa-furnace went on stream in late '89.[21]

The power supply of some MW is intentionally oversized to enable Degussa to enlarge the furnace and the throughput if appropriate. A cooling system is installed both for the bottom of the furnace and the power supply system. Electrode holders with pneumatically operated clamps are used.

The furnace shell lining is basic consisting of refractories with high proportions of alumina and MgO. Some nitrogen is blown into the furnace to ensure a neutral atmosphere and to prevent a quick abrasion of the graphite electrode. A removable burner can be used to dry out a new lining, it is moved away, when the melting process starts. A careful preparation of the charge is of great importance. The grain size must be suitable for an undisturbed feeding. No losses via the off-gas should occur. Degussa uses a very sophisticated feeding system based on a continuous, gravimetric or volumetric charging according to the according conditions of the furnace, the actual processing stage and the charge material.

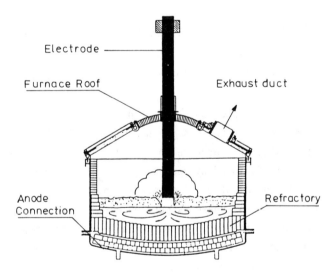

Figure 7 - Submerged-Arc Furnace, ASEA/ABB

Automotive exhaust catalysts require definite amounts of flux to produce a slag that could easily be tapped off. In case of high alumina content, addition of lime in the range of 40 % to 50 % is to be recommended.

If monoliths are to be treated distinctly less quantities of lime and occasionally a pinch of other fluxes might be necessary to avoid a viscous slag (Fig. 8) [22]. Both the PGM-containing auto-catalysts and the flux are weighed and pneumati-cally transported to a mixer unit and then charged into the furnace by a conveyer screw.

This operation is repeated until the PGM-content of the metal is high enough to ensure an econo-mical wet-chemical processing. During the whole process, heat transfer and a good stirring of the melt provides good contact between metal and the precious metals dissolved chemically or physically in the slag. The temperature of the slag is regu-lated by means of the power supply and the feeding system so that no environmental pollution can emerge

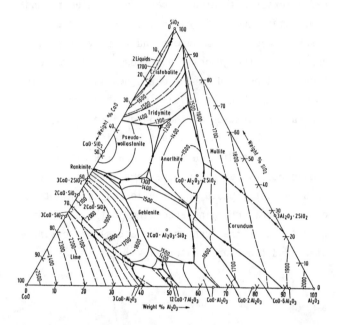

Figure 8 - System Al$_2$O$_3$-SiO$_2$-CaO

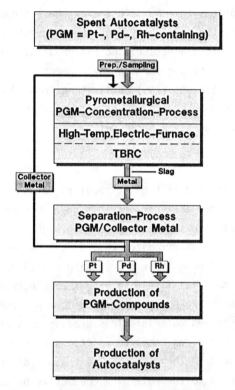

Figure 9 - Flowsheet of the Treatment of Auto-catalysts in an Electric Furnace

Treatment of Autocatalysts in an Electric Furnace

The complete process of treating (spent) automo-tive exhaust catalysts is shown in Fig. 9.[23] The first step of concentrating the PGM in a collector metal is followed by upgrading them in a second metallurgical stage. The separation of the PGM from the metal phase is carried out in the platinum-refinery by wet-chemical means. At least the recovered PGM are used again for the new auto-motive exhaust catalysts.

The metallurgical process begins with the ignition of the electrode. For this purpose a metallic "heel" must be present, or metal sheets have to be given into the furnace for an electrical short circuit and to produce a liquid metallic sump. Then, feed material is charged and smelted down. After waiting for the right moment the tap-off hole is opened and the slag is poured into a slag-pot at the furnace.

A lot of automotive exhaust catalysts were already treated in the Degussa-furnace in the last months. The power-input during the smelting-phase varies about 1.500 kWh/t, the melt itself only needs about 500 kWh/t to keep the temperature. The heat losses are very small. Exact figures do not exist up to now, but we estimate it to be less than 10%. The consumption of the graphite electrode is much higher than in the iron- and steelmaking industry (2 to 3 kg per t steel). However, it lies in the range of about 10 kg per t input.
The slag produced in this electric furnace is low in precious metal content and can easily be dumped or used for other industrial purposes.

239

All temperatures measured on different positions around the furnace shell and near the bottom are monitored and give a continuous information i.e. concerning the process stage, the thickness of the lining and the temperature of the melt. Temperature and pressure of the off-gas are also carefully observed and registered.

The capacity of the actual pilot-furnace should be one hundred tons per year.

However, some more problems have to be solved in the near future. The visual control still must be improved, the feeding system should be totally computerized and the tap-off technology is to be simplified. Of great importance for the smelting process are the quantities and the components of the impurities in the spent automotive exhaust catalysts.

At last a continuous analysis is to be installed for slag and metal to control the process and to interfere at once if necessary, i.e. in changing the flux composition, improving the settling effect of the metalic particles in the slag or in varying the tap-off cycles.

Prospects

The most important application of PGM in the future will certainly be catalysis, the use for automotive exhaust emissions produced by cars, in particular. There will be an increasing public request to avoid or diminuish air pollution. Not only governmental regulations but also personal responsibility are necessary to relieve the environment. If the new stronger emission standards are to be met in the near future the demand of PGM will increase very rapidly. However, the problems of collecting the spent autocatalysts require well-organised logistics. That will be one of the most vital targets of the future.

In technical respect the complete recovery of the PGM from autocatalysts is of great importance in order to preserve the natural sources. Therefore, all efforts are undertaken by many refineries to improve the yield of the precious metals by an effective recovery process. Degussa will hopefully contribute to this aim by treating spent automotive exhaust catalysts economically and pro-environmentally in its new high-temperature electric furnace.

References

1. Degussa AG, Hanau-Wolfgang Forschungs-Pressegespräch Katalyse, December 1989.

2. Statistische Mitteilungen. Kraftfahrt-Bundesamt, no.12, January 1991.

3. E.S. Lox. Katalysator: Entwicklungsstand und -tendenzen. Haus der Technik, Essen: Alternativen zum geregelten Katalysator, Febr. 1991.

4. E.S. Lox, B.H. Engler und E. Koberstein. Diesel Emission Control. Brussels: CAPOC II. Conference, 1990.

5. R. Gerner. Technical and Medical Application of Precious Metals. Trient, Italy: Workshop Precious Metals, 1990.

6. Platinum 1990. Johnson Matthey. London 1991.

7. Verband d. Automobilindustrie e.V. Frankfurt/ Main: Tatsachen und Zahlen, 1990, 54. Folge.

8. Metal Bulletin. 12 March 1990.

9. Degussa AG, Frankfurt/Main. March 1991.

10. Degussa AG, Frankfurt/Main. 1990.

11. Dr. Renner. Platinmetalle. Metalle in der Umwelt. Weinheim: Verlag Chemie, II. 17. 1984

12. V. Jung. Treatment of Automotive Exhaust Catalysts. Cologne: Intern. Symp. Productivity and Technology in the Metallurgical Industries", Sept. 1989, p. 523 - 539.

13. W. Hasenpusch. Vienna, Austria: Symposium "Recycling im Verkehrswesen", Dec. 1989.

14. K.U. Maske. MINTEK, rep. no. M 178, 1985.

15. G. Rath. Der Elektroreduktionsofen - Ein Aggregat für die Nichteisen-Metallurgie. Erzmetall 43, no. 2, 1990, p. 81-85.

16. J. Saville. Recovery of PGMs by Plasma Arc Smelting. New York: 9th Intern. IPMI-Conf., 1985, p. 157 - 168.

17. Plasma Technology in Metallurgical Processing. Grand Junction, CO, USA: Iron and Steel Society, 1987.

18. J.H. Corregan and V. Jahnsen. Electric Arc Furnace Operation. US Patent, no. 4,273, 576, June 1981.

19. S.E. Stenkvist. 12 Years of DC Arc-Furnace Development Leads to US Order. Steel Times, Oct. 1985, p. 480 - 483.

20. ASEA/ABB-Publication, no CH-I 89016 E.

21. Main-Echo, Germany, 14 Febr. 1991.

22. Slag-Atlas. Düsseldorf: Verlag Stahleisen m.b.H., no. ISBN 3-514-00228-2, 1981.

23. Degussa AG. Frankfurt Main. Metal Research Department, 1990.

Recovery of metals from spent catalysts in a DC plasma furnace

D. L. Canham
V. G. Aurich
Billiton Research BV, Arnhem, The Netherlands

SYNOPSIS

The increased use of catalysts in the oil and petroleum industry has resulted in production of significant quantities of spent catalysts comprised of a porous ceramic substrate originally doped with heavy metals (e.g. cobalt, nickel, molybdenum) and further contaminated with heavy metal oxides and sulphides (e.g. nickel, vanadium) during use. These spent catalysts create an environmental threat because of their environmental leachability and pyrophoricity, which together preclude their disposal by landfill. Therefore, considerable effort has been spent in finding an alternative disposal method.

Most activities have centred on some form of immobilisation, in which the toxic components are physically bound in a low permeability medium, or on hydrometallurgical extraction techniques which, because of their complexity and number of process steps, are only viable for particular types of catalyst at relatively high throughputs (ca. 20,000 t/a). Requirement to operate at a large throughput is a disadvantage since it normally requires transport of large quantities of spent catalyst from an extensive catchment area, which itself causes a potential environmental problem. Therefore, a programme has been undertaken at Billiton Research to develop a process which is economically viable at a small scale (ca. 4,000 t/a) and which is flexible enough to process all major types of ceramic substrate based spent catalyst arising in a limited catchment area.

The developed process is carbothermic reduction of the metal oxides present on the spent catalysts in the presence of iron to collect the reduced metal and suitable fluxes to combine the carrier in a fluid slag. Products are an environmentally stable slag with low toxic metals content suitable for either dumping or use as a building material and a mixed ferroalloy containing the cobalt, nickel, molybdenum, etc. which can be sold for its alloy content to a special steel manufacturer or sold as scrap iron. The different uses of the products depend on the composition of the spent catalyst smelted. It is anticipated that the process will be operated in a semi-batch mode in which each different type (or group of types) of spent catalyst is processed to specific products.

To achieve the temperatures and energy densities required for catalyst smelting while minimising the quantities of offgas to be cleaned, the process technology selected was DC transferred arc plasma technology. Results of small scale (30 l.) and large scale (0.2 m^3) trials are presented showing cobalt and molybdenum recoveries to alloy greater than 99% and residual analyses in the slag less than 0.05%.

A preliminary economic analysis for the process indicates that a 4,000 t/a plant will cost approximately U.S.$ 6.2 million. For processing a typical HDS catalyst, a net return of U.S.$. 392 per tonne can be calculated.

INTRODUCTION

Consumption of catalysts in the oil and petrochemicals industry has increased, largely as refiners seek to increase the degree of conversion achieved, but also with greater production and the use of heavier crude oils as refinery feeds. Many catalysts are transition metals (cobalt, vanadium, molybdenum, tungsten, nickel, etc) distributed over the surface of porous ceramic substrates (the carriers), usually in the form of cylindrical extrudates or spheres. During use, these catalysts can become contaminated with carbon, sulphur and heavy metals present in the crude oil. The carbon and sulphur can be removed by regeneration (controlled oxidation), and the metal contaminants can be in part removed by selective leaching (rejuvenation). These processes lead to reusable catalysts having catalytic activity close to that of fresh catalyst. However, after a number of recycling operations the catalyst activity can no longer be sufficiently restored to permit further use. In this state, the catalyst is said to be spent.

Spent catalyst is potentially pyrophoric because of the presence of finely divided metal sulphides and carbon on its surface. Toxic heavy metals on the porous, readily permeable substrate render spent catalysts liable to metal leaching if dumped in conventional disposal sites. Under the Dutch chemical waste regulations (Wet Chemische Afval - WCA), residues are classed as chemical waste if the content of indexed metals exceeds a threshold value listed for each element. In particular, spent catalysts are classed as chemical waste if one (or more) of the molybdenum, nickel or cobalt contents exceed 0.5% by mass. (Future regulations for waste classification may replace a composition limit with a standard leachability limit). Many spent catalysts therefore represent a disposal problem, both for catalyst users and for catalyst manufacturers or suppliers who are increasingly being held responsible since "cradle to grave" tracking of

241

environmentally harmful material has become prominent. In addition, the pyrophoricity and toxicity of the catalysts make it expensive and undesirable to transport spent catalyst over long distances for disposal.

A variety of processes and practices has arisen to dispose of spent catalyst, mostly based on landfilling, physical immobilisation by encapsulation in an inert material (e.g. concrete, pitch) or hydrometallurgical extraction using partial or complete solution. All have advantages and disadvantages, and are in general suitable for a small range of catalyst types. In many cases, the processes employed are economic only at a large scale of operation due to the complexity of the necessary multi-step operations and the associated high capital cost. For example, Cri-Met operates a plant with ca. 20,000 t/a capacity in a preexisting, redundant Bayer processing plant at Braithwaite, Louisiana[1]. This scale of operation can only be justified in areas of high rates of spent catalyst generation (e.g. U.S.A.) and where existing, written-off plant is available.

Pyrometallurgical processes for recovery of base metals from spent catalysts are not common. In some instances, spent catalysts are treated pyrometallurgically by incorporation in the feed for steelworks. In one instance, a process is patented for production of fused alumina from a spent base metal catalyst by a double smelting technique[2]. In the field of precious metal catalysts, many processes exist for recovery of precious metals by a dedicated smelting technique[3-5]. A new process is being developed at Billiton Research in Arnhem (a Shell Research Laboratory) based on DC arc plasma furnace technology, which will allow most types of non-disposable spent catalyst to be processed economically even at a relatively low throughput.

SPENT CATALYST ARISINGS

Spent catalysts under consideration are used in the oil and petroleum refining and chemical manufacturing industries for a variety of processes. Table 1 lists the applications of catalysts used in the industry and Table 2 gives their compositions both as fresh and spent materials.

Table 1   Major industrial base metal catalysts

Process	Application	Active components	Carrier
Distillate hydrotreating	Desulphurisation and olefin saturation	Ni, Mo, (P)	Al_2O_3
	Sulphur removal	Co, Mo, (P)	Al_2O_3
Special hydrotreating	Hydrodesulphurisation	Ni, Mo or W, (P)	Al_2O_3
	Residue desulphurisation	Ni or Co	Al_2O_3
	Residue conversion	Ni, V	SiO_2
	Catalytic cracking	V, Lanthanides	Al_2O_3, SiO_2

Table 2   Typical catalyst compositions

Catalyst	Form	Ni	V	Co	Mo	W	S,P	C
HDS	Fresh	3		3	9			
	Spent	2-3		2-3	4-9		<15	<20
HC	Fresh	3-4		3-4		10-12		
	Spent	3-4		3-4		10-12	4-8	4-6
FCC	Fresh							
	Spent	<0.4	<0.7				5-15	5-15
Residue conversion	Fresh	0.5	1-2					
	Spent	2-6	<15		1-6		5-15	10-30

Table 3 shows the quantity of catalysts available in the U.S.A., Europe and Japan in 1987.

Table 3 Spent catalyst for disposal - 1987

	North America	Europe	Japan
	(Quantity (x1000 t/a))		
Refining			
FCC catalyst	193	24.5	13.6
HDS, HC catalyst	35	2.9	20.2
Other	5	4.7	
TOTAL	**233**	**32.1**	**33.8**
Chemical manufacturing			
Hydrogenation	4.5	3.0	3.8
Ethylene oxide	2.5	n.a.	n.a.
Dehydrogenation	2.1	n.a.	n.a.
Other	18.5	12.0	29.0
TOTAL	27.6	15.0	32.8

An estimate of the current disposal methods for Hydrodesulphurisation (HDS) and Hydrocracking (HC) catalysts in the U.S.A. in 1987 is presented in Table 4.

Table 4   Disposal of spent HDS/HC catalyst in the U.S.A. - 1987 (Total quantity - 35,000 t/a)

Disposal method	Fraction
Dumping (Landfill)	30 - 40%
Immobilisation + Landfill	6 - 20%
Metal reclamation	40 - 55%

Approximately 50% of the spent catalyst generated was treated for metal recovery in modified Bayer plants while the remainder was dumped "as generated" or physically immobilised (in concrete, pitch, etc.) followed by dumping. Table 5 lists 1987 disposal methods for fluidised-bed catalytic cracker (FCC) catalyst in Europe, where it can be seen that the refiners landfill approximately half of their spent FCC catalyst and return the majority of the remainder to the catalyst manufacturer/supplier. Half of the returned material is also landfilled, approximately a quarter used for different catalytic applications and a quarter encapsulated in asphalt.

Table 5 Disposal of spent FCC catalyst in Europe - 1987
(Total quantity - 25,000 t/a)

Disposal method	Fraction	
	(By refiners)	(By catalyst manufacturers)
Dumping (Landfill)	50%	55%
Resold (Cascading)	9%	20%
Filler for asphalt	-	23%
Filler for cement	1%	1%
Returned to supplier	40%	-

Of the catalysts listed, those typically presenting a disposal problem are the HDS, HC and heavily contaminated FCC catalysts. In the past, FCC catalysts have been removed from process when the metal loading on the catalyst has reached approximately 0.5% vanadium, 0.3% nickel. Recent practice has been to use these catalysts further, resulting in a portion of the spent catalyst with a vanadium content of 0.7%. This leads to WCA classification as chemical waste. Although at present only a small proportion of FCC catalyst is classified as chemical waste, the high cost of fresh catalyst is exerting pressure on refiners to use catalysts for longer times, resulting in increased production of high metal bearing spent catalyst. Simultaneously environmental legislation is imposing tighter limits on dumping. These factors together will result in a steady increase in the proportion of spent FCC catalyst that is classified as chemical waste.

For Europe, the expected annual generation in the early 1990s of spent HDS/HC catalyst and of heavily contaminated spent FCC catalyst is approximately 5,000 t/a and 5,000 - 10,000 t/a respectively, although the latter is expected to rise later. From these observations, the quantity of spent catalysts requiring metal recovery in the near future in Europe will be in the range 5,000 - 10,000 t/a. Work for the process considered in this paper has concentrated on developing a plant with comparable capacity.

CATALYST SMELTING PROCESS

The objectives of the catalyst smelting process are:

1) To remove the majority of the heavy metals present by carbothermic reduction and collection in an iron based alloy. Depending on the nature of the metals present in the catalyst, this metal could be sold as a mixed ferroalloy for steel manufacture, sold as scrap iron or further refined for individual metal recovery.

2) To produce a dense slag phase containing the carrier and any residual heavy metals, either as unreduced oxides or uncollected metal. Depending on the nature of the catalyst carrier and the degree of metal recovery, this slag could be used as landfill, as a building material or dumped as industrial waste.

3) To keep atmospheric and liquid emissions during processing within environmentally acceptable limits.

The presence of sulphides on spent catalysts results in production of sulphur dioxide in the furnace offgas.

Recovery of this sulphur dioxide in an environmentally acceptable form is very expensive at the low rates of production anticipated unless a collection facility with spare capacity already exists on the proposed process site. In addition, transport of spent catalyst to the process site is much safer if the sulphides are oxidised prior to movement. Since sulphur disposal capacity exists at most refineries, it was considered appropriate to assume that spent catalyst would be partially oxidised for sulphur removal on refinery sites and that desulphurised (and incidentally decarbonised) spent catalyst would be processed in the DC arc furnace. Preliminary work has therefore concentrated on processing desulphurised catalysts, but planned later work will examine the effects of the presence of sulphur in the process.

To minimise the quantity of slag produced and to maximise its value, it is desirable to minimise the quantity of flux required for smelting. Because of the high melting temperature that such a slag requires, it is necessary to utilise a high intensity furnace technology for the process. To remove the need for a pre-agglomeration step, a process is required in which a substantial proportion of fines can be incorporated in the feed. DC plasma arc furnaces satisfy both of these criteria.

EXPERIMENTAL WORK

Preliminary tests

Preliminary smelting trials were performed in a small DC arc furnace of approximately 30 l. capacity. The furnace is shown schematically in Figure 1. A plasma is generated between a hollow graphite cathode electrode and a submerged partially water-cooled carbon anode. The plasma source can be operated at 2000A current and maximum 300 V potential.

Two types of catalyst were used for the test work to determine the flexibility of the process:

- A commercially available Co/Mo catalyst on an alumina extrudate base.
- A proprietary Ni/V catalyst on a silica sphere base.

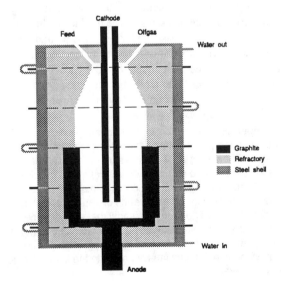

Figure 1
Reactor for small scale tests

Table 6   Product analyses for small scale tests.

	Initial Average slag compositions for Co/Mo/Al₂O₃ smelting tests							
	Catalyst	Test 1	Test 2	Test 3	Test 4	Test 5	Test 6	Test 7
SiO_2	4.6	4.5	4.3	4.8	3.5	1.9	3.0	3.6
Al_2O_3	71.2	63.6	67.2	68.2	62.6	63.0	62.8	64.2
Co_3O_4	3.3	0.42	0.05	0.23	<0.02	0.13	0.05	<0.02
NiO	0.68	0.08	0.02	0.03	0.01	0.03	0.01	0.02
MoO_3	14.2	1.4	0.17	0.67	0.06	0.69	0.17	0.07
P_2O_5	1.2	0.28	0.08	0.16	0.05	0.16	0.12	<0.03
TiO_2	0.14	0.27	0.12	0.14	0.08	0.04	0.03	0.11
Fe_2O_3	0.47	1.5	0.22	1.5	1.7	0.61	1.3	2.5
CaO	0.60	29.2	28.3	25.6	30.9	33.3	33.0	29.2

A series of smelting tests was carried out on approximately 20 kg batches of spent Co/Mo/Alumina catalyst. A typical test consisted of preheating the graphite-lined furnace containing approximately 10 kg catalyst, flux (lime), iron and coke until operating temperature was achieved. Further reaction mix was fed via a screw feeder and drop-tube in the furnace roof into the molten furnace contents. The arc voltage was controlled by controlling the height of the electrode above the slag surface until the plasma became stable during feeding. The furnace power input (current) was adjusted to maintain stable reactor temperatures at approximately 1650 °C. The metal and slag phases were allowed to solidify in the reactor at the end of feeding with power off, although in one test the metal and slag were tapped through a taphole in the reactor side.

During the series of tests, the influences of flux/catalyst ratio and plasma stabilising gas were determined. Additionally, at the end of some experiments iron powder was added to the melt surface to produce a rain of molten iron droplets to act as a collector for small, unsettled metal droplets in the slag.

The composition of the Co/Mo catalyst is shown in Table 6, together with the analyses of slags produced during the test series. Because of metal penetration into the lining, the mass balances for these trials did not fully close. However, from the compositions of the slags produced it is possible to conclude that slags can be produced which are not classified as chemical waste under the WCA regulations. Sufficient slag fluidity can be achieved for metal/slag phase separation with slags containing 28 - 33% CaO. The use of nitrogen (used in test 7) as a plasma stabilising gas does not have significant influence on the process compared to the normally used argon.

While operating with nitrogen, higher voltage drops can be obtained in the plasma than while operating with an argon arc of the same length. This allows more intense heating of the slag surface and results in easier control of the process. Table 7 lists the slag compositions of dip samples taken during a test, showing the influence of settling and iron powder addition. These results show that a settling period of approximately 20 minutes was sufficient to achieve good settling, and that the addition of iron powder as a collector resulted in no significant improvement to the level of toxic metal removal.

Table 7   Settling effects in the slag - Small scale work

Component	Average slag compositions for Co/Mo/Al₂O₃ smelting tests			
	During feeding	End of feeding	After standing (20m)	After iron addition
SiO_2	4.9	5.1	5.0	5.9
Al_2O_3	69.0	67.7	68.5	68.8
Co_3O_4	0.13	0.21	0.08	0.02
NiO	<0.01	<0.01	<0.01	<0.01
MoO_3	0.17	0.35	0.06	0.05
P_2O_5	0.23	0.33	0.20	0.05
TiO_2	0.15	0.15	0.15	0.17
Fe_2O_3	2.0	1.9	1.3	0.94
CaO	25.0	25.8	26.3	26.5

A similar series of trials was carried out using the Ni/V/SiO₂ catalyst. Although different flux/catalyst ratios and different operating temperatures were required, generally similar results were obtained.

It could be concluded that slags that would not be classified as chemical waste under the WCA regulations could be produced from a variety of metal/ceramic catalysts. The versatility of the process was therefore partially demonstrated.

Large scale test

The large scale DC arc furnace utilises the same 600kW power source as the small scale furnace. Figure 2 illustrates the reactor arrangement. The reactor lining is graphite with an insulating backing lining and the bottom anode is a partly water-cooled steel block. All components of the reactor are water-cooled, allowing measurement of heat losses from the reactor. The reactor internal diameter is 0.8 m, and the maximum working depth is 0.4 m giving a working volume of 0.2 m³. The cathode is an axially bored 130 mm diameter graphite electrode. During the trial, argon and nitrogen were used as plasma stabilising gas at different stages.

The furnace was filled with an initial mix of catalyst, lime, coke and flux, together with iron metal to act as collector, and heated with DC arc operation. The purpose of this initial feed material was to protect the furnace lining during preheating. After a molten bath had been formed, the input power was maintained at a level which led to stable reactor

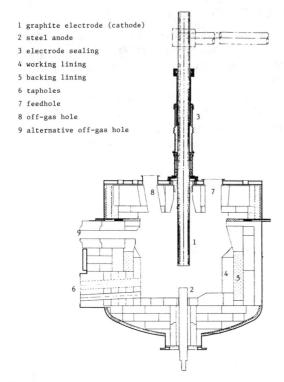

1 graphite electrode (cathode)
2 steel anode
3 electrode sealing
4 working lining
5 backing lining
6 tapholes
7 feedhole
8 off-gas hole
9 alternative off-gas hole

Figure 2
Reactor for large scale tests

temperatures. Instrument problems were experienced in measuring the slag temperatures with the normally employed pyrometer system. Using temperatures measured in the reactor lining (Figure 3), it was estimated that the bulk slag temperature was 1650 - 1700 °C, although the temperature in the vicinity of the plasma was certainly much higher. At this temperature, the power input and reactor heat losses were stable at approximately 120kW.

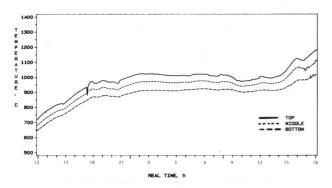

Figure 3
Temperatures in lining during large scale test

Further reaction mix was fed from a bunker via a screw feeder and vertical discharge pipe through the furnace roof. To limit fume emission, a low argon purge was maintained through the feeding system. A feed rate of 30 kg per hour was achieved for 15 h., during which time power input was regulated to keep the furnace temperatures constant. For a

short period, the feed rate was increased to 40 kg per hour with no noticeable effect on plasma operation. However, the increased rate of carbon monoxide formation led to increased temperatures in the offgas to the bagplant since the offgas was burned at the furnace outlet. Feeding rate limitations were therefore controlled by the bagplant capacity, not the furnace power capacity. At the end of this test, slag and metal were tapped from the furnace. Some material was left in the furnace, adhering to reactor walls and floor, etc. This material was not removed from the furnace, since a separate experiment was commenced immediately after the first was completed.

Figure 4 shows the toxic metal analyses of dip samples taken from the slag surface during operation. The period up to 12.00 (real time) corresponds to periods in which the slag was fully molten and during which the reaction mix was being fed continuously. For cobalt, nickel and vanadium, the analysis was below the WCA chemical waste classification limit for almost the entire duration. For molybdenum, the

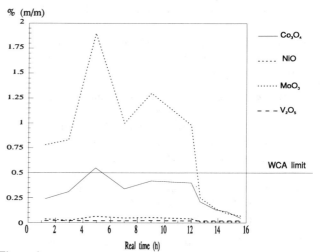

Figure 4
Slag composition during large scale test

average analysis was approximately 1.1%, a factor of more than two above the limit. This indicates that continuous operation would be unfavourable unless long residence times (i.e. large reactor, low throughput) were practical. After 12.00, feeding was ended and the levels of all toxic metals quickly dropped to values around half of the WCA limit. Over the next 90 minutes, the levels dropped further leaving final levels around 10% (maximum) of the WCA limit. This indicated that batch or semi-batch operation would be the favourable operating mode for a commercial plant.

Table 8 shows the compositions of tapped products from the trial.

Conversion of the alloy analyses to metallic components gives an alloy containing 62.7% Fe, 26% Mo, 7.3% Co, 1.3% Ni and % total other metallic impurities. The slag phase is composed of 60.2% Al_2O_3, 30.5% CaO, 4.8% SiO_2 and 5.5% Fe_2O_3. This latter component is not present in the samples of slag taken from the reactor, and almost certainly arises from the iron tube used to open the reactor taphole during oxygen lancing since the quantity present in the tapped slag corresponds with the quantity of iron tube used. The analyses of toxic metals in the slag (Co, Mo, Ni, V) are

Table 8 Compositions and masses of feed materials and products for large scale trial

Analyses (% m/m)	Feed materials			Products		
	Catalyst	Flux	Iron	Slag	Alloy	Dust
SiO_2	4.55	0.20		4.80	0.43	14.57
Al_2O_3	67.90	0.30		60.20	<0.94	17.49
Co_3O_4	3.25			<0.02	9.94	0.74
NiO	0.63		0.25	<0.01	1.65	0.34
MoO_3	13.88			<0.03	39.00	1.07
P_2O_5	1.14			0.19	1.53	1.73
TiO_2	0.15			0.16	<0.02	0.69
MgO	0.06	0.50		0.60	0.02	4.12
Fe_2O_3	0.40	0.15	138.57	5.50	89.52	13.90
Cr_2O_3	0.04		0.18	0.07	<0.01	0.76
CaO	0.21	96.50		30.50	0.12	23.81
MnO_2	0.00		0.95	0.17	<0.03	4.07
Mass (kg)	365.9	146.4	92.0	275.6	73.4	5.2

below the detection limit for the XRF analysis technique employed, and are significantly below the WCA chemical waste classification limit.

An attempt was made to estimate the quantities of slag and alloy remaining in the furnace by closing the mass balance for iron and alumina assuming that the slag and alloy remaining in the furnace had the same compositions as those tapped. This estimate is presented in Table 9 and results in well closed mass balances for cobalt (100%), nickel (80%), and molybdenum (91%), and indicates recoveries to the metal phase of 99% for cobalt, 97% for nickel and 99.6% for molybdenum.

Table 9 Estimated mass balance for large scale trial

Component	Reporting to Slag (%)	Reporting to Metal (%)	Reporting to Dust (%)	Overall Account-ability
SiO_2	94.0	2.4	3.6	124.2
Al_2O_3	99.6	0.0	0.4	100.0
Co_3O_4	0.7	99.0	0.3	99.8
NiO	2.0	97.1	0.9	79.2
MoO_3	0.3	99.6	0.1	91.1
P_2O_5	29.1	67.5	3.4	64.3
TiO_2	94.8	0.0	5.2	131.1
MgO	91.3	0.7	8.0	284.6
Fe_2O_3	17.5	81.9	0.6	100.0
Cr_2O_3	87.9	0.0	12.1	106.6
CaO	98.9	0.1	1.0	89.5
MnO_2	76.6	0.0	23.4	104.7

N.B. Mass balance estimated by assumption that mass balance for iron and alumina are closed by alloy and slag remaining in furnace.

The dust generation is seen to be only 1% of the material fed and shows a small enrichment of manganese (probably insufficient to consider this product as a method for manganese separation). The quantity and composition of dust produced do not present a recycling problem.

COST ESTIMATE

A preliminary economic evaluation has been made for this process, based on processing in the Netherlands on a site with an existing infrastructure suitable for metallurgical processing.

For a throughput of 4,000 t/a, a total capital expenditure estimate of U.S.$ 6.2×10^6 was calculated. For a throughput of 8.000 t/a, the capital expenditure would be increased to U.S.$. 8.5×10^6. (All estimates have +/- 30% accuracy)

For operating expenditure, the costs/revenues vary significantly between different types of catalysts. Two examples are considered:

HDS catalyst

The HDS catalyst considered was a desulphurised catalyst based on nickel and molybdenum as original active constituents. A cost estimate was generated for a process in which 4,000 t/a catalyst would be processed.

It was calculated that the products per tonne of catalyst processed would be 1.18 tonnes slag, containing 34 - 36% CaO, 60 - 64% Al_2O_3 and 0 - 5% SiO_2; and 0.29 tonnes ferroalloy, containing 50% Fe, 11% Ni, 31% Mo and 7% V.

The calculated depreciated operating cost is U.S.$. 800 / tonne, including capital charge, fixed and variable costs.

Offsetting the depreciated operating cost are the revenues obtained from the sale of products and the treatment charge. In calculating the revenues, only 60% of the metal value of the nickel and molybdenum recovered to the ferroalloy is assumed because the alloy is mixed and requires greater control during use in the steel industry. Metal prices used for this evaluation are typical 1987 prices and in particular do not reflect the sharp increase in price of nickel during 1988/1989, resulting in a conservative value for this product. For the slag, it is assumed that the content of toxic metals is sufficiently low that the slag can be used in the building industry, where its composition makes it ideal for use as a high alumina cement.

Using a standard treatment cost for Spent Catalyst treatment and the calculated values of the slag and ferroalloy, a revenue of U.S.$. 1200 per tonne of catalyst processed is calculated.

The overall net revenue from the processing of HDS catalyst is therefore estimated at U.S.$. 400 (1,200 - 800) per tonne of catalyst processed.

FCC catalyst

A similar cost evaluation has been performed for disposal of spent FCC catalyst. Because FCC catalysts are normally rejected with much lower metal loadings (1%) than are HDS catalysts (ca. 20%), the return from ferroalloy sale is very much reduced. Also, the presence of silica in the carrier results in a $CaO/SiO_2/Al_2O_3$ slag which cannot be used for cement manufacture but is suitable for use as ballast material. These factors result in much lower revenue income, requiring a larger throughput to be economically viable. Assuming an exponential extrapolation of the capital expenditure, this process becomes viable at a throughput of 50,000 t/a.

CONCLUSIONS

Existing disposal routes for spent, heavy metal containing catalysts are either environmentally unattractive (immobilisation), require large throughputs (20,000 t/a) to be economically attractive, or are inflexible in terms of the catalyst carrier that can be treated.

Long distance transport of spent catalysts is environmentally unsafe and is being increasingly restricted by legislation.

The expected rate of generation in Europe of spent catalysts which are classified as chemical waste (10,000 t/a) is too low and too varied to utilise existing processes. It is necessary to develop a process which is economically attractive at low throughputs and which is flexible enough to treat a wide spectrum of catalysts.

Carbothermic reduction/smelting of HDS catalysts in a DC plasma arc furnace with lime as flux results in a calcium aluminate slag with indexed metals analyses below the XRF detection limits. This slag would not be classified as chemical waste under the WCA regulations, and is suitable for landfill or as a building material. The molybdenum and cobalt are collected in a mixed ferroalloy which can be sold for its alloy metal content as a special steel additive. Similar results have been achieved for a Ni/V on silica catalyst, demonstrating the versatility of the process.

For a plant capable of treating 4,000 t/a spent catalyst, the capital expenditure is estimated at U.S.$. 6.2×10^6. When processing HDS catalyst, the overall economic evaluation shows a return of U.S.$. 392 per tonne catalyst treated. For processing FCC catalyst, the lower value of the products requires a higher throughput to become economically viable. This is estimated at approximately 50,000 t/a.

Work to date has concentrated on processing desulphurised catalysts. To avoid the cost of this extra process step, further work will consider the influence of sulphur on the process capability.

References

1. Kunzelman, C. CRI-MET's Spent Catalyst Closed-Loop Processing Service. The 1987 CRI-MET symposium on disposal of spent catalysts.

2. Cichy, P. Recovery of heavy metals from spent alumina; Melting, Separation. Patent no. US 4349381 14/09/1982

3. Saville, J. Process for the extraction of platinum group metals. Patent no. EP 0173425 A1 17/06/1985

4. Day, J. Process for the recovery of platinum group metals from refractory ceramic substrates. Patent no. US 4428768 20/10/1981

5. Day, J. Recovery of platinum group metals, gold and silver from scrap. Patent no. US 4427442 24/01/1984

Recovery of rare earths from spent FCC catalysts

B. Alexandre
N. Gérain
A. Van Lierde
Unité des Procédés, Université Catholique de Louvain, Louvain-La-Neuve, Belgium

SYNOPSIS

This paper presents results of laboratory tests performed on spent FCC catalysts. The proposed hydrometallurgical treatment aims to develop a simple chemical process able to extract cerium, praseodymium and lanthanum with high recoveries and in an appropriate form for future treatments in classical metallurgical circuits.

The systematic study of the leaching conditions shows that the temperature is the most important factor in both rare earth recoveries and selectivity towards aluminium. The best results were obtained after one hour of leaching at 50°C with a pulp dilution of 33 % solids by using a 120 g/l H_2SO_4 solution. In such conditions, rare earth recoveries reach about 90 % whereas less than 15 % of aluminium is solubilized. Rare earths are then very selectively precipitated from the leaching solution at pH 3 to double sulfates $RE_2(SO_4)_3.Na_2SO_4.xH_2O$ with a Na_2SO_4 solution. The rare earth precipitation yield is higher than 95 % and the coprecipitation of aluminium is lower than 5 %. This precipitate is then transformed by NaOH to hydroxides of high purity while the Na_2SO_4 solution is regenerated. After drying, the final precipitate contains 32 % Ce ; 22.3 % Pr ; 17.7 % La ; 0.5 % Al ; 0.04 % Na and 0.7 % S. After oxidation at 200°C, this latter can be retreated by leaching with a 18.3 % HCl solution. 96 % of the cerium remains in a 78.4 % CeO_2 residue while at least 85 % of the lanthanum and the praseodymium goes into the solution from which they can be precipitated by one of the numerous well known methods.

INTRODUCTION

In recent years, many studies[1-2] have been carried out in order to recover valuable metals from various spent catalysts, particularly petroleum hydrotreatment catalysts[3] and automobile catalytic converters.[5-6] In the first case, the interesting metals are molybdenum, cobalt, nickel, vanadium and tungsten, while platineous metals are of course valuable elements of exhaust catalysts. For these cases, some industrial plants have already been built.[2-4-7-8-9] Many other spent catalysts are produced since there are many other sorts of catalysts. Among these, FCC catalysts, definitively spent after several cycles of regeneration treatment[10-11], constitute an interesting raw material for rare earths production. Indeed, their grades are often similar to those of raw ores. Moreover, the potential tonnage of these catalysts is more than ten times higher than that of HDS catalysts. Hydrometallurgical treatment of FCC spent catalysts has been based on flowsheets recommended for rare earths concentrates. However, the operating conditions have to be adapted for an economical treatment of such materials.

The specific aims are :

- to produce rare earth compounds in such a form that they can be reused for manufacturing new FCC catalysts or sold in favorable conditions to plants specialized in the separation of rare earth elements ;
- to be able to treat FCC catalysts of different structures or matrices and diverse rare earth grades.

The present research work only dealt with direct selective leaching and chemical precipitations on a decoked FCC catalyst.

CHARACTERIZATION OF THE CATALYST

The decoked spent catalyst sample treated here consists of an ultrastable synthetic zeolite of REY-type dispersed in an amorphous alumino-siliceous matrix.

According to its chemical analysis given in Table I, it contains about 2 % of rare earth oxides : cerium, praseodymium and lanthanum and minor amounts of conventional contaminates of petroleum : iron, nickel, vanadium, lead and copper.

TABLE I

Chemical analysis of decoked spent catalyst sample

Component	Content (wt.%)	Element	Content (ppm)
SiO_2	58.5	Ni	1296
Al_2O_3	34.2	V	1702
Na_2O	0.66	Fe	911
Ce_2O_3	0.82	Pb	103
La_2O_3	0.46	Cu	86
Pr_6O_{11}	0.58		
H_2O	1.00		

As shown in Table II, the rare earths are distributed fairly evenly between the various size fractions :

all physical pretreatments of the material seem therefore unadapted.

TABLE II

Size distribution of spent catalyst and chemical analysis of size fractions

Size range (μm)	Weight (%)	Ce_2O_3 + La_2O_3 + Pr_6O_{11} Grade (%)	Distribution (%)
+ 149	9.3	1.70	8.7
+ 74 - 149	51.3	1.75	49.5
+ 53 - 74	26.2	1.95	28.1
+ 38 - 53	9.8	1.93	10.4
- 38	3.4	1.74	3.3

SELECTIVE LEACHING OF RARE EARTHS

Acid leaching is the conventional extraction process of rare earths from natural concentrates. After calcination, purified bastnasite flotation concentrates containing about 85 % rare earth oxides are usually leached with concentrated HCl.[12] Monazite concentrates, the other large source of rare earths, are treated using two different processes[12-13] :

- direct leaching at 200°C with concentrated H_2SO_4 ;

- NaOH leaching at 150°C to solubilize the phosphates, followed by an acide leaching of

the rare earths with HCl or concentrated HNO_3.

In all these processes, rare earth extraction recoveries reach about 90 %. These facts drew us to consider the three conventional acids (HCl, HNO_3 or H_2SO_4) with a particular attention for the selectivity problems towards aluminium.

Experimental conditions

The spent catalyst sample has always been leached, as it is, in vessels at a controlled temperature and pulp dilution. The pulp homogeneization is done with magnetic or anchor agitators. In each case, the given leaching time includes the warming of the pulp to the required temperature. At the end of the leaching, the pulp is filtered and the cake is washed with demineralized water at room temperature. Leaching solutions and washing water were always mixed. Extraction recoveries are determined by balance sheets on the solutions and the dried residues.

Selection of the leaching conditions

Some preliminary tests showed that a 50°C temperature, a solution volume/dry catalyst ratio of two and acid concentrations of about 120 g/l could be considered as suitable conditions for further studies.

a. Effect of time

As shown in Table III, one hour of leaching in our general operating conditions gives a 88.5 % rare earth extraction recovery and a depressed aluminium dissolution whatever the type of acid. Multiplying the time by three has very little effect on the rare earth extraction but results in higher aluminium recoveries.

TABLE III

Effect of leaching time on results in the following conditions : acid concentration : 120 g/l, pulp dilution : 2/1, temperature : 50°C

Time (h)	Type of acid			Extraction recovery (%)	
	HCl	H_2SO_4	HNO_3	Rare earths	Aluminium
1	-	x	-	85.5	13.0
2	-	x	-	87.9	14.6
3	-	x	-	88.7	17.0
1	x	-	-	84.9	13.5
2	x	-	-	86.5	16.0
3	x	-	-	88.5	19.0
1	-	-	x	85.0	14.1
3	-	-	x	89.6	18.7

b. Effect of temperature

Figure 1 shows the variation of rare earth extraction recoveries as a function of temperature for a leaching time of one hour with a 120 g/l acid solution at a pulp dilution of two. Temperatures lower than 50°C clearly hinder the rare earth dissolution.

Recoveries indeed decrease from about 85 % to 35 % if the temperature drops to 30°C. On the other hand, all increases of the temperature do not significantly improve the solubilization of rare earths but lead to higher aluminium extractions : at 80°C, they are twice what they were at 50°C. The type of acid does not cause any difference in behavior.

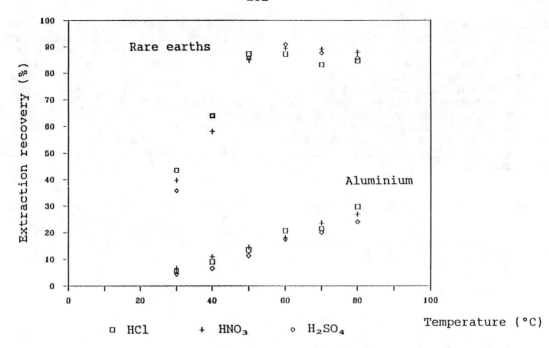

Fig. 1. Variation of rare earth and aluminium recoveries as a function of the temperature (acid concentration : 120 g/l, time : 1 h, pulp dilution : 2/1)

c. Effect of pulp dilution

At 50°C, the effect of pulp dilution on rare earth and aluminium

leaching recoveries is shown in Figure 2 for acid solutions containing respectively 120 g/l H_2SO_4, HNO_3 or HCl and 240 g/l H_2SO_4.

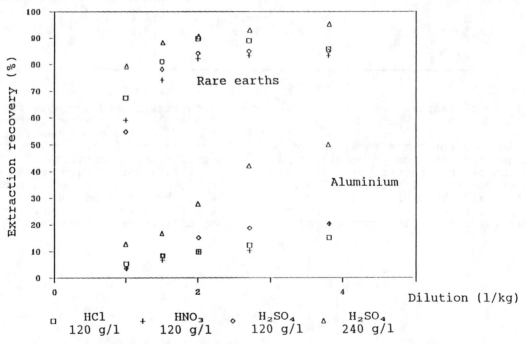

Fig. 2. Variation of rare earth and aluminium recoveries versus the pulp dilution (acid concentration : 120 or 240 g/l, time : 1 h, temperature : 50°C)

With the first solutions, the rare earth dissolution remains constant at its maximum level for a wide range of pulp dilutions, from 2 to 4, whereas aluminium recoveries systematically

increase from about 11.5 to 20 %. At pulp dilutions lower than 2, the rare earth extration clearly decreases to reach only 60 % at a pulp dilution of 1. It can also be seen that an increase of the acid concentration to 240 g/l has very little effect on the rare earth extraction recoveries but improves the aluminium dissolution especially at high pulp dilutions. Therefore the optimal acid addition must be about 240 kg per ton of dry catalyst, for acid concentrations between 120 and 160-170 g/l and dilutions between 1.5 and 2. In such conditions, rare earth recoveries can be as high as 85-87 % after one hour at 50°C. The resulting liquor, after mixing with the residue wash solution (when the ratio wash solution/residue is one), will present the following chemical analysis :

1.1 to 1.3 g/l Ce
0.25 to 0.3 g/l Pr
0.6 to 0.75 g/l La
4.7 to 5.7 g/l Al
15 to 22 g/l free acid.

Sulfuric acid leads to the same results as HCl or HNO_3 and is cheaper. It has therefore been choosen as the only leaching agent for further tests.

TREATMENT OF THE LEACHING SOLUTIONS

Selective chemical precipitation of rare earths

Generally, direct precipitation of rare earths is made with sodium oxalate, carbonate or hydroxide. As a result of selectivity problems towards aluminium or recovery and sometimes filtration difficulties, the rare earths have finally been precipitated from their solutions as a double sulfate of composition $RE_2(SO_4)_3Na_2SO_4.xH_2O$ where RE = Pr, Ce or La. Many tests have been done according to this method. They show that the use of dilute solutions of Na_2CO_3 with a volume/leaching solution ratio equal to 1, leads to the best results. The temperature, the pH and the salt concentrations are also important factors (Table IV). Na_2CO_3 concentrations of 30 to 40 g/l, a final pH of 1 to 3 and temperatures kept at about 85°C in both the mixing and the filtration steps seem to be necessary for precipitating the rare earths with recoveries higher than 95 % while aluminium coprecipitation is limited to 5 %. In fact, a high reaction temperature must be reached because of the respective increase and decrease of the solubility of the aluminium and the double sulfate at rising temperatures.

TABLE IV

Results of the double salt precipitation

Concentration (g/l)		pH	Temperature (°C)	Precipitation recovery (%)			
Na_2CO_3	Na_2SO_4			Ce	La	Pr	Al
20	–	3.55	25	88.5	83.0	85.0	34.0
30	–	3.90	25	95.0	95.0	94.8	87.0
30	–	0.70	25	66.0	70.1	67.3	10.5
30	–	0.90	85	93.7	85.1	90.7	4.0
40	–	1.05	85	95.5	92.5	92.1	6.0
–	40	1.10	85	91.0	90.5	90.9	1.5
40	–	3.35	85	97.0	94.0	93.5	8.0
–	40	2.80	85	95.0	92.0	92.0	2.4

As shown in table IV, the selectivity towards aluminium can be improved by using sodium sulfate rather than sodium carbonate without any loss of rare earth recoveries.

After drying at 100°C, the double sulfate precipitates obtained under such favorable conditions generally has the following chemical composition : 19.5 % Ce_2O_3 ; 10.4 % La_2O_3 ; 13.1 % Pr_6O_{11} and only 0.6 % Al_2O_3.

Recovery of the rare earth oxides

The conversion of the rare earths double salt into hydroxides can easily be done by pulping at 95°C in a caustic soda solution (20-40 g/l) according to the following reaction :

$$RE_2(SO_4)_3Na_2SO_4.xH_2O + 6NaOH = 2RE(OH)_3 + 4Na_2SO_4 + xH_2O$$

As long as the temperature is kept constant, the conversion reachs at least 75 % when NaOH is used in stoechiometric proportions and around 99.9 % when NaOH consumptions are twice as high. After about 30 minutes, the hot precipitate is filtered and washed with boiling demineralized water in order to get rid of a maximum of sodium and sulfide ions. This procedure makes it possible to obtain rare earth hydroxydes concentrates with a composition after drying of 32 % Ce ; 22.3 % Pr ; 17.7 % La and only 0.5 % Al ; 0.04 % Na and 0.7 % S. The collected liquor, which contains less than 0.05 g/l Ce, La, Pr and Al, can be recycled to the precipitation stage without any problems. If 60 kg NaOH per ton of dry catalyst have been used, it is even possible to form the double salt without any other addition of reagent.

Separation of cerium from lanthanum and praseodymium

The separation of cerium from global rare earth oxides concentrate can be done by applying the method of Swaminathan and al[14]. This process consists of :

* an oxidation by air of the oxides at 200°C for at least six hours resulting in a rise of the cerium oxidation state to 4 ;

* the leaching of the calcinated oxides in a HCl solution.

After calcination, the precipitate of rare earth hydroxides becomes a brown calcine with the following chemical composition :

CeO_2 : 43.61 %
La_2O_3 : 22.58 %
Pr_6O_{11} : 22.43 %
Al_2O_3 : 0.96 %
SO_3 : 0.63 %
Na_2O : 0.02 %
H_2O : 3.76 %

This calcine is leached in presence of HCl with a pulp dilution of 1 l/kg. After 4 hours of leaching at 70°C with the pH maintained at 3 by continuous additions of small amounts of a 18.5 % HCl solution, excellent results are already achieved. More than 96 % of the cerium remains insoluble while between 86 and 91 % of the lanthanum and praseodymium dissolves. Table V shows the chemical composition of the lanthanum solution and of the cerium cake obtained after washing with a HCl solution at pH 3 (dilution 0.5 l/kg) and drying at 120°C.

TABLE V

Characteristics of the cake and the solution produced by the leaching of the rare earth oxides precipitates

Component	Cake* content (%)	Solution** content (g/l)
CeO_2	78.9	1.65
La_2O_3	4.3	155.50
Pr_6O_{11}	5.9	193.73
Al_2O_3	0.8	4.0
SO_3	1.15	0.0
Na_2O	0.01	0.11
Cl	5.18	-
H_2O	3.51	-

* Weight : 55 %
** 1.3 l/kg dry oxides precipitates

The praseodymium and the lanthanum can later be precipitated by adding strong ammonium hydroxide and ammonium chloride solution at 50°C.

CONCLUSIONS

This study shows how rare earths can easily be extracted from decoked FCC spent catalysts by direct acid leaching with hydrochloric, nitric and sulfuric acids. Maximum selectivity in the dissolution of cerium, lanthanum and praseodymium in regard to aluminium can be achieved after 1 hour of leaching at 50°C with 240 kg sulfuric acid per ton of dry catalyst. In such conditions, rare earth extraction recoveries of 85 to 87 % can be obtained with an aluminium solubilization lower than 14 %.

The liquor produced can be treated with 30-40 g/l Na_2CO_3 or Na_2SO_4 solutions. By keeping the pH between 1 and 3 and the temperature at 85°C in both the precipitation and the filtration steps, recoveries of about 95 % for rare earths and only 5 % for aluminium can be achieved. The resulting double salt can be transformed to hydroxides by NaOH while the Na_2SO_4 is regenerated. The separation of the cerium from the other lanthanides can be done by calcinating the hydroxides at 200°C under air atmosphere for six hours and leaching the calcine at 70°C with a hydrochloric solution. After about six hours of leaching at pH 3, at least 85 % of the praseodymium and the lanthanum can be dissolved. 96 % of cerium oxides remain in the residue which has the following chemical composition :

CeO_2 : 78.40 %
La_2O_3 : 4.63 %
Pr_6O_{11} : 6.25 %
Al_2O_3 : 0.85 %
SO_3 : 1.15 %
Na_2O : 0.01 %
Cl : 5.18 %
H_2O : 3.51 %

References

1. B.W. Jong and R.E. Siemens. Proposed methods for recovering critical metals from spent catalysts. Recycle and Secondary Recovery of Metals, 1985, p. 477-488.

2. G. Parkinson, S. Ushio, M. Hibbs and D. Hunter. Recyclers try new ways to process spent catalysts. Chemical Engineering, 02/1987, p. 25-30.

3. P.G. Raisoni and S.G. Dixit. Leaching of Co and Mo from a Co-Mo/ -Al_2O_3 hydrodesulphurization catalyst waste with aqueous solutions of sulphur dioxide. Minerals Engineering, volume 1, n° 3, 1988, p. 225-234.

4. Z.R. Llanos, J. Lacave and N.G. Deering. Treatment of spent hydroprocessing catalysts at Gulf Chemical and Metallurgical Corporation. Preprint n° 86-43, 1986, Society of Mining Engineers, Aime.

5. F.K. Letowski and P.A. Distin. Platinum and Palladium recovery from spent catalysts by aluminium chloride leaching. Recycle and Secondary Recovery of Metals, p. 735-745, Proceeding of the Symposium Florida, Aime, 12/1985.

6. S.A. Bonucci and D.P. Parker. Recovery of PGM from automobile catalytic converters. Precious Metals : Mining, Extraction and Processing, p. 463-481. Conference Proceeding, Aime, Annual meeting Los Angeles CA, 27-29/02/1984.

7. C.W. Bradford and S.G. Baldwin (Johnson Matthey). Recovery of precious metals from Exhaust catalysts. Brit. Pat. 1517270 (20/05/75).

8. P. Cichy (Kennekot). Recovery of precious metals from spent alumina-containing catalysts. Europ. Pat. 48823 (07/04/82).

9. W.A. Millsap and N. Reisler. Cotter's New Plant diets an spent catalysts and recovers Mo, Ni, W and V products. Engineering and Mining Journal, volume 179, n° 5, 1978, p. 105-107.

10. C. Marcilly and J.M. Deves. Les contaminants métalliques des catalyseurs de FCC. Institut Français du Pétrole. Cinétique et Catalyse. Rapport IFP n° 33462, 9/1985.

11. F.S. Elvin. Regeneration and reuse of spent cracking catalysts. Symposium on Spent Catalyst. Handling presented at the March 1987 Meeting of the American Institute of Chemical Engineers. New Orleans Louisiana.

12. Y.A. Attia. Extractive metallurgy and refining of Terbium : a review in precious and rare metal technologies. Proceedings of a Symposium on Precious and Rare Metals, Albuquerque NM, 6-8/04/1988, Edited by A.E. Torma and I.H. Gundiler. Process Metallurgy, volume 5, 1989, p. 473-495.

13. Y.W. Miao and S.S. Horng. Decomposition of Taïwan local black monazite by hydrothermal and soda fusion methods. Proceedings of a Symposium on Rare Earths, Extraction, Preparation and Applications, 27/02-02/03/89, Edited by R.G. Bautista and M.M. Wong. The Minerals, Metals and Materials Society, 10/1988, p. 195-206.

14. T.V. Swaminathan, V.R. Nair and C.V. John. Stepwise hydrochloric acid extraction of monazite hydroxides for the recovery of cerium lean rare earths, cerium, uranium and thorium. Op. cit. p. 207-212.

Vibration technique for recovery of non-ferrous, rare and noble metals from secondary raw materials

A. Fedotov
USSR Engineers' Academy, Leningrad Technical University, Leningrad, USSR
G. Denisov
USSR Engineers' Academy, Leningrad Technical University, Leningrad, USSR and
'Mekhanobr' Institute, Leningrad, USSR

SYNOPSIS

The most efficient and ecologically safe method of recovering metals from recycled materials is the vibratory method. This method consists in selective liberation of non-homogeneous phases along intergrowth boundaries and structural defects through application of pulsating compression with shear in a volumetric layer with subsequent vibratory size or component classification both in air and water media.

Choice of parameters and regimes of vibratory grinding or attrition is scientifically substantiated, depending on treated material type.

Presented are major technological methods of treatment of a number of recycled materials, containing non-ferrous and rare metals.

Shown is the efficiency of re-recovery of gold from treated electronic wastes through selective regrinding of feed material to optimum size analysis. Analogous technology of platinum recovery from chamotte lining permits to obtain up to 91 % of platinum from refractory final tailings.

Studied is an efficient technology of purification of copper metallurgical slags from oxide component, based upon classification, vibroattrition and crushing of upgraded slags.

Presented is a new method of vibratory regeneration of heat transfer manufactured corundum with separation of dust fraction of nickel catalyst with its subsequent reutilization.

The efficiency of the new method of recovery of valuable components from recycled materials is shown with examples of treatment of materials containing tungsten, tantalum, niobium and other rare metals.

Recycled (or recoverable) materials generally are multicomponent formations with multiphase composition. Their separation to initial inclusions (grains) requires high precision, low power- and metal-intensive methods, conforming to

257

environmental control.

The most efficient method of recovering metals from recycled materials is the vibratory method, it is also the cleanest method in respect to ecology. This method provides for selective liberation of non-homogeneous phases along intergrowth boundaries and structural defects through application of pulsating compression with shear in a volumetric layer with subsequent vibratory size or component classification both in air and water media.

A new approach to rational disintegration, providing for intercrystalline disintegration of heterogeneous media, has been developed at the "Mekhanobr" Institute /1/.

Complete technological effect is achieved only with combination of the following disintegration principles:

- material is subjected to volumetric pulsating load in a relatively thick layer;
- to initiate stresses along intergrain boundaries, pieces of material should be simultaneously subjected to multiple combined load, including elements of shear, bending and torsion;
- loads should be very precise and correspond to structural strength;
- between loading cycles, pieces of material should retain relative mobility for mutual reorientation, timely removal of final size fraction and adjustment of layer density;

- to initiate incipient cracks on material surface, abrasive interaction of material pieces is required;
- relation between the degree and direction of compressive and combined loads, as well as frequency of action upon the material, should be chosen in accordance with material character and technological requirements.

Fig.1 illustrates the scheme of vibratory disintegration of a material.

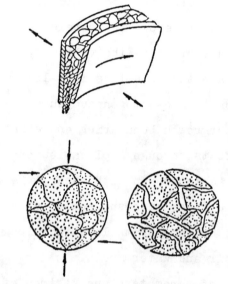

Fig.1. The scheme of selective disintegration of materials.

Practical implementation of the above principles for selective disintegration of multicomponent media can be successfully accomplished by means of vibratory jaw and cone inertia crushers /2/.

This type of machines provides for easy adjustment of force, exerted on the material; effective combination of compression with shear; speed control of final product transportation from crushing chamber; combination of several operations in one apparatus, namely, crushing, grinding, classification, upgrading.

Fig.2 shows the principal scheme of cone inertia crusher. Unbalanced mass vibrator is located on crushing cone stem, to which it is attached by means of a bearing. This design of a crushing machine permits to avoid its break-down, if tramp iron gets into crushing chamber.

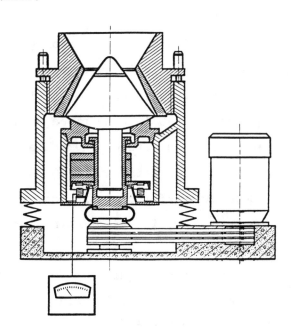

Fig.2. The scheme of cone inertia crusher with unbalanced mass and discharge opening setting sensor suspension.

The developed crushing and grinding equipment does not have any analogues in foreign practice; it was used as a basis for developing new technological processes for treatment of extra strong materials and industrial wastes.

Application of vibratory jaw and

cone inertia crushers in technological flowsheets of treatment of various materials: metals and alloys, slags, hard alloys, electronic boards and other metal-containing materials provides for a high recovery of non-ferrous and rare metals through a sharp decrease of their losses in fine size fractions /2/.

The institutes "Mekhanobr" and VNIIP-Vtortsvetmet have developed a rational technology of treatment of household radio-electronic equipment, that includes two-stage crushing, classification, oversize magnetic separation and undersize electrical separation.

This technology employs cone inertia crusher of KID-300 type (crushing head diameter 300 mm). This crusher provided selective disintegration of radio valve glass with preliminary crushing of feed material to size below 10 mm. This feed size ensures stable operation of the KID-300 crusher with capacity 500 kg/hr.

Magnetic and electrical separation of the crushed product permit to obtain metal concentrate with metal content over 99 %, recovery being 98.4 % /3/.

Retreatment of fire-clay lining through its melting together with nickel raw material is low efficient, since platinum recovery does not exceed 50 %.

Using selective vibratory grinding, the "Mekhanobr" Institute has developed a waste-free technology of gravity concentration of the abovementioned lining material. This technology permits to recover 91 % of platinum into concentrate, containing 10 % of platinum. A flotational method of re-recovery of platinum, ensuring at least 99 % recovery, has been developed at the "Mekhanobr" Institute.

Rational grinding with application of the KID-300 cone inertia crusher permitted to accomplish a full-scale treatment of metal chips into powder at the Leningrad powder metallurgy works for the first time in the Soviet Union.

The fact, that the full-scale technology of high-speed steel PGMS grinding into powder has been brought to a commercial level is proved by:

- the development of an efficient and low-power-intensive process, that practically excludes tungsten losses in burning;

- the development of Specification on the powder, and the pricelist.

Designing of a chip-treating installation, development of processes and their bringing to commercial level in respect of high-speed steel powder production, as well as serial production of articles of this powder, permit to achieve the following technical, economical and practical results at present time:

1 - possibility and expediency of full-scale grinding of chip wastes of various metals and alloys into powder with cone inertia crushers are practically proved;

2 - high efficiency of utilizing and recycling chip wastes and feasibility of low-waste technology are proved;

3 - high economic efficiency of high-speed steel chips grinding into powder with subsequent production of articles from it is proved by economies in power, metal and labour input, owing to rational grinding with new crushers and a high metal utilization factor in blanks, as well as quality increase in final articles.

Introduction of this technology of high-speed steel chips treatment permitted to decrease powder oxidation by the factor of 10 in comparison with the previously used method, the achieved oxidation level being within 0.03 %.

Durability of tools, produced out of this powder, is 3-4 times higher, than that of cast tools.

Technology and equipment for recovery of metals from melting slags have also been developed. This technology was first introduced for copper, silicate nickel and bronze slags. Technological flowsheets for treating various slags are somewhat different. Nevertheless, all these processes are based upon treatment of slags in vibratory jaw crusher with subsequent separation of crushed products through screening. Vibratory jaw crusher permits tramp iron to pass through, it ensures a good

separation of metal and mineral components, owing to high frequencies and acceleration, and it grinds the mineral
fraction. This permits to separate a
major part of pure metal as oversize in
subsequent screening. If necessary, i.
e., with a high metal content in undersize, an analogous process is repeated
in cone inertia crusher, where rounded
grains of metal are rolled into flats
and are also separated through screening.

For certain types of slag, magnetic,
X-ray radiometric or electrodynamic separators are included into flowsheets,
depending on physical-mechanical properties of slags and economic expediency.

The proposed technology permits to
recover up to 90 % of metal from slags.

The developed technology and equipment can be introduced at non-ferrous
smelters, in order to considerably reduce consumption of valuable raw material.

Fig.3 shows the mechanical flowsheet
of the process. A technology for treatment of slags, produced in smelting of
copper in electric furnaces, has also
been developed. Metal fraction is cleaned from oxide component through classification, vibro-abrasive treatment and
crushing of upgraded product.

Another principally new simplified
resource-saving waste-free technology
has been developed for treatment of

tantalum- and niobium-containing wastes
of capacitor production with vibroinertia grinding equipment to obtain tantalum and niobium powders suitable for
recycling in production of hard, high-
-temperature and special alloys, or for
utilization as alloying additions, or in
plasma spraying and other areas. Just
one plant recycles annually 10-15 tons
of tantalum powder.
This technology is also suitable for
other materials and wastes, containing
rare and scarce metals, the economic effect from its introduction amounting to
tens of million of roubles.

Waste-free technology of regeneration of platinum-palladium spent catalysts in cracking of petroleum is also
of interest. As fire-clay lining, spent
catalysts are, as a rule, subjected to
smelting together with nickel raw material. This technology cannot be considered rational, since platinum losses
in slags amount to approximately
200 kg/year.

The combined technology of vibratory-
-mechanical-and-hydrometallurgical upgrading, developed at the "Mekhanobr"
Institute, permits to separate and recycle 85-89 % of regenerated catalyst
in petrochemical industry and to separate platinum-palladium-containing
slags from the remaining 11-15 % and
retreat them into 31-32 % platinum-con-
taining concentrate. So, platinum losses
in waste products are practically

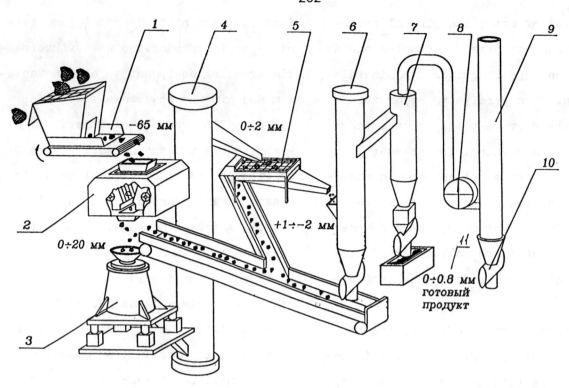

Fig.3. Mechanical flowsheet for treatment of metallurgical slags, moulding sands and ceramic wastes.
1 - Electrovibratory feeder; 2 - Vibratory jaw inertia crusher VSHID 200x1400; 3 - - Crusher KID-450; 4 - Vibrating conveyor Ø 500; 5 - Screen 450Э; 6 - Pneumatic separator CB-1000; 7 - Cyclone ЦН15-650; 8 - Air blower BBD-8; 9 - Filter FBK-60; 10 - Discharge unit; 11 - Final product 0 + 0.8 mm.

eliminated. Besides platinum concentrate, aluminium hydroxide and iron ocher also are marketable products, i.e., a waste-free technology is practically achieved.

A new technology of selective retreatment of recycled resources, containing precious metals, based upon application of vibratory grinding machines, has been developed and introduced. The KID-300 and KID-100 cone inertia crushers are used in a technology, developed for grinding various electronic articles, based on ceramics, before and after pickling. It was established,

that grinding with cone inertia crushers permits to liberate precious metals from ceramic coat and provides for up to 1 % of additional recovery of metals, not recovered before. Treatment of pre-pickled ceramics through grinding with the KID-100 with subsequent re--pickling is a promising technology, since application of three KID-100 crushers can provide for treatment of 60 t of recycled resources per year and will permit to re-recover precious metals, not recovered before, through additional pickling of grinding products.

At present, on the basis of complex

laboratory and full-scale investigations, a classification of recycled materials has been developed in accordance with the principle of macro- and microstructural strength in relation with specific aspects and factors of vibroinertia grinding with consideration of requirements of size analysis and structural homogeneity of final products.

A number of technologies and items of equipment for treatment and regeneration of recycled materials, that exceed the level of the world practice, have been developed and introduced. They are protected by 18 patients.

Under the scientific guidance of the USSR Engineers' Academy a number of programs have been drawn up and are implemented now, aiming at rational treatment of a wide range of recycled materials.

Contracts have been concluded with a number of foreign companies, willing to acquire soviet equipment and "know-how".

environmental control, Leningrad, 1984, 41 p.

References

1. Revnivtsev V.I., Denisov G.A., Zarogatsky L.P., Shishkanov Yu.P. Technology and equipment for vibroinertia selective disintegration of materials of any strength, Leningrad, 1990.
2. Denisov G.A. Treatment of metalcontaining wastes, London, The Institute of Mining and Metallurgy, 1990, 75 p.
3. Denisov G.A., Revnivtsev V.I., Itkin G.Ye. Complex problems of regional

Car scrap recycling towards 2000

W. L. Dalmijn
J. A. van Houwelingen
Delft University of Technology, Faculty of Mining and Petroleum Engineering, The Netherlands

SYNOPSIS

In Europe 13 million cars are produced per annum which will end their economic life cycle with an average age of 10 years. The processing of scrap cars starts with size reduction by a hammermill, or shredder. After size reduction the magnetic fraction is removed by a magnetic separator. The remaining non-magnetic fraction is screened and each fraction is processed in a specific way, adapted to the particle size. In this way it is at present possible to recover metal values, at a high grade and recovery, and reject products with an acceptable contamination level for the environment.

The increased use of plastic, complex and composite materials in the automotive industry will have a negative effect on the recovery of metals and waste production from car scrap unless technology and design for recovery are improved.

For the disposal of different waste products new quality control equipment has to be designed to meet environmental standards.

INTRODUCTION

Car scrap contains a high proportion of valuable materials and the average composition of a car[1] is given in Fig. 1. As can be seen, a proportion of iron and steel has been replaced in course of time by lighter materials, such as aluminium and synthetic materials, in order to improve energy efficiency, to obtain weight reduction and optimize the drag fractor or C_w of the car body.

In 1989 there were about 128 million cars in use in Europe and the forecast for 1994 is 148 million cars [2]. In 1989 about 11.5 million tonnes of car

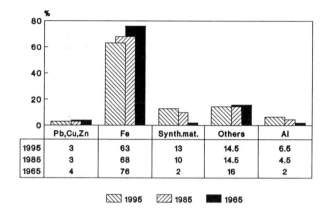

	Pb,Cu,Zn	Fe	Synth.mat.	Others	Al
1995	3	63	13	14.5	6.5
1985	3	68	10	14.5	4.5
1965	4	76	2	16	2

1995   1985   1965

Figure 1 Car composition (Daimler-Benz AG)(1)

scrap were available in Western Europe for processing.

By recycling metals and materials from obsolete cars, resources and energy are saved. When the separation and quality control systems are optimized, metals can be reused in their original application. An example is the aluminium recovered from the non-ferrous fraction of the car shredder operations which is transported in molten state from the V.A.W. plant in Grevenbroich, Germany, to the Volkswagen works at a distance of approximately 250 km. With further improvements in the separation process, the reuse of other secondary metals and materials can be extended.

PROCESSING OF CAR SCRAP

For the recovery of metals and materials from obsolete cars, size reduction and separation is necessary. The general flow sheet of such an operation is shown in Fig. 2. Controlled size reduction is essential to obtain the right degree

FLOWSHEET CARSCRAP

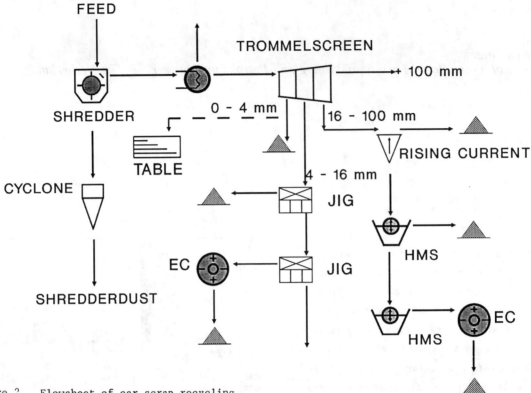

Figure 2   Flowsheet of car scrap recycling

of liberation for optimum grade and recovery in combination with minimum operating costs and metal losses.

During the size reduction the dust and very light material such as foam and textile are removed by air suction. The dust and the other materials are separated by means of cyclones and a wet collection system. This fraction also contains some metal foil and metal dust generated by the size reduction process in the hammermill. In some countries, it is classified as hazardous waste, what makes the shredder operation more expensive and sometimes even impossible.

After liberation, the iron fraction is removed magnetically. Entangled and unliberated material such as copper wires are also recovered with the magnetic fraction. Galvanized steel will of course also report to the magnetic fraction. The zinc concentration in the iron fraction will increase in the future due to improved corrosion protection by galvanizing. This zinc will eventually end in

the dust of an E.A.F. from which it can be recovered by a Waelz process [3].

After removal of the magnetic fraction, the remaining material is screened before the next separations take place. In general, four fractions are obtained by trommel screening in combination with vibrating screens: larger than 100 mm, 16 to 100 mm, 4 to 16 mm, and 0 to 4 mm. The larger fraction is sometimes screened at 65 instead of 100 mm.

PROCESSING OF THE + 65 OR + 100 MM FRACTION

Valuable material larger than 100 mm can be hand picked and the remainder can be land filled. Alternatively, the + 100 mm fraction is reshredded. A different approach has been chosen by Lindemann, where preconcentration and reshredding are an essential part of the dry separation philosophy [4]. For the optimization of grade and recovery of the metal fractions, further liberation of the oversize is essential in order to improve the efficiency of the mechanical separation process.

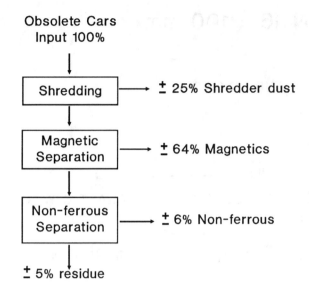

Obsolete Cars
Input 100%

Shredding → ± 25% Shredder dust

Magnetic Separation → ± 64% Magnetics

Non-ferrous Separation → ± 6% Non-ferrous

± 5% residue

Figure 3   Mass balance obsolete cars

CONCENTRATION OF THE 16 TO 65 MM OR 100 MM FRACTION

Depending on the efficiency of the air suction on the shredder, the metal content of the 16 - 65 (mm) fraction is 50 - 60%. With increased costs of land filling, the shredder operators tend to lower the suction thereby tolerating more ballast and lower the metal content of the feed material to the metal concentration plants (Fig. 3). This feed material, which will first go to a density separation, still contains light materials such as wood, plastic foam, textile and plastic foil. Preconcentration by a rising current separator [5] is necessary to avoid ferro silicon losses and to increase through-put of the heavy medium plant. The overflow of the rising current separator contains aluminium sheet material and magnesium due to the combination of shape factor and density. These metals can be recovered by an eddy current separator which is becoming more and more common practice (Fig. 4). The reject fraction from the eddy current separator still contains some metal such as copper wire, stainless steel foil and small metal particles, which constitute a potential problem for land filling.

The underflow of the R.C. separator is fed to the heavy medium plant operating in a ferro silicon slurry for the recovery of the non-ferrous metals. This is performed in two steps. In the first step, light material such as rubber, plastic and composite materials are removed at a density of 2.4,   together with magnesium and some other metals which can be recovered by an eddy current separator [6]. The reject fraction of this separation also contains some metals, with potential problems as mentioned earlier.

In the second step the mixed non-ferrous metals, along with some other materials, are separated at a density of 3.3 for the recovery of aluminium. The light fraction is subsequently treated at a density of 2.7 - 2.8 for the separation of unliberated aluminium mainly contaminated with iron. This can be recycled at the shredder or treated separately, for example in a Coreco smelting furnace. The aluminium fraction still contains glass, copper wire and other unliberated material, in propotions running as high as 10 - 15%. This product can be upgraded with an eddy current separator to more than 99% aluminium with a recovery of over 98% [7]. The rejected product contains a high percentage of copper wire and other metals mixed with plastic such as printed circuits. This product is at present land filled, but, in the near future, other ways will have to be found to recover the metals and other materials.

The sink fraction of the dense medium separation at 3.3 contains the bulk of the heavy metals: zinc, brass, copper, stainless steel, lead, etc. These metals are sold as a mix. A significant part of it goes to Taiwan, where it is hand sorted and remelted. The hand sorting is possible due to the extremely low wages combined with import taxes on primary metals in that country. Treatment plants for that material ar also installed in Europe and the U.S.A.

DEVELOPMENTS OF A METAL DETECTION SYSTEM

A possible solution to recover the metals and other materials from mixed products such as the reject from the aluminium fraction mentioned above is the metal detection system recently developed by S + S company at Schönberg, Germany[8]. This metal detector consists of a row of 10 mm metal detection coils which allow a line scan over 1 m or more. The system is mounted at the lower end of a ramp, at an angle of about 40°, with a width of 1 m and a length of about 60 cm. With a vibrator feeder, the material is fed to the ramp where singulation, and therefore the grade of the

FRACTION 16 - 100 mm

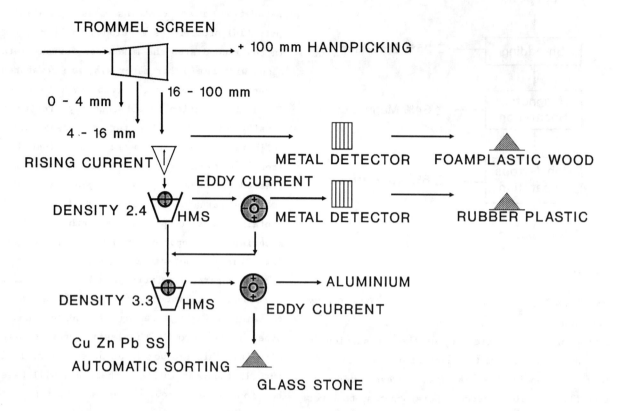

Figure 4   Flowsheet of car scrap recycling

obtained product, is improved by increased speed of the particle flow. Reject materials are ejected by an array of air blasting valves parallel to the detection system, at distances correlated electronically with the delay time to reach the individual blast valves.

An identical type of separator has been developed for photometric sorting in the glass industry. In this case, the metal detection coils are replaced by laser diodes. The results of the first machines are impressive and many applications in the recycling industry will certainly be developed.

ALTERNATIVE DRY PROCESSING OF THE + 16 MM FRACTION.

The non-ferrous metals can also be concentrated by dry mechanical concentration techniques [4]. For this approach, the starting material has to be shredded to less than about 30 mm what will cause some metal losses but will on the other hand improve the liberation and grade of the recovered

metals. The flowsheet in fig. 5 is a combination of air classification, screening, eddy current separation and stoners or air tables. Improvements have to be made for decreasing the zinc content in the aluminium fraction and reducing the aluminium losses to the heavy non-ferrous fraction.

Another process has been developed by Metallgesellschaft [9], a dry separation system for the fraction 16 - 65 mm. It relies on a metal / non-metal separation and the sorting of all metals by the combination of a pulsed laser and an atomic emission spectroscopy. The schematic diagram of this process is given in Fig. 6. Owing to its ability to identify light metals together with a high detection speed, this system is superior to processes based on X ray analysis.

Dry separation processes grow in importance because the water treatment plant and the subsequent disposal of metal sludges can be avoided.

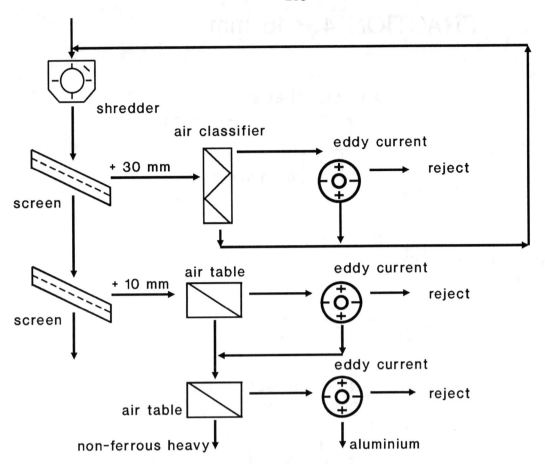

Figure 5   Flowsheet dry separation car scrap

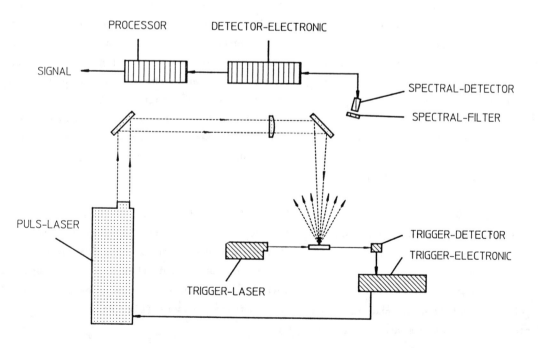

Figure 6   Automated sorting of non ferrous car scrap

FRACTION 4 - 16 mm

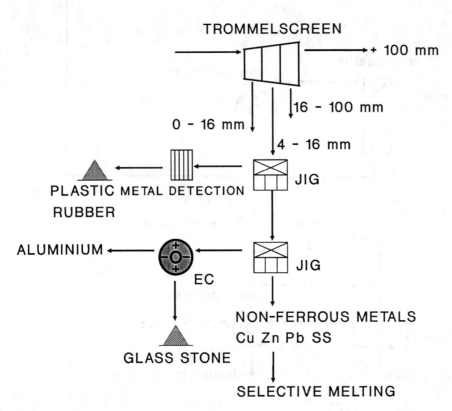

Figure 7   Flowsheet of car scrap recycling

PROCESSING OF THE - 16 MM FRACTION

Because of the lower proportion of metals, typically 15 - 25%, and the increasing sedimentation time with decreasing particle size, this fraction is less attractive from an economic view point for separation by heaavy medium. An alternative for the processing of the 4 - 16 mm fraction is given in the flowsheet of Fig. 7, which is based on two separations by jigging.

The feasibility of this process has been proven by different operators.

In the first jig, the rubber and plastic can be removed as a light fraction [10].

This still contains light metals and thin material which can be removed by eddy current separation in combination with metal detection equipment.

The heavy fraction is fed to a second jig, in which glass, stone and aluminium are separated from other metals. The aluminium from the lighter fraction can be separated by eddy-current techniques, as described earlier. From the remaining heavy metals fraction, lead and zinc can be recovered by means of selective melting [11]. The main advantage of the jig is the simple construction and the high capacity which, depending of the density of the feed, is of the order of 15 tonne/hour per m^2. A drawback is the amount of water necessary and the water cleaning and sludge removal systems.

PROCESSING OF THE - 4 MM FRACTION

For the moment, the fraction smaller than 4 mm is dumped. The processing of this fraction, by means of shaking tables and other mechanical separation methods, is however, under investigation for environmental reasons (fig. 8).

FRACTION 0 - 4 mm

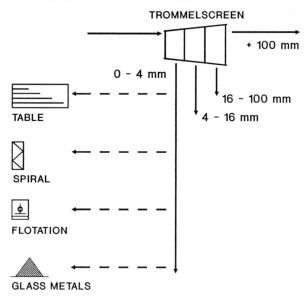

Figure 8   Processing of the 0 - 4 mm fraction

CONCLUSIONS

The processing of car scrap has improved in the last two decades to an extent that recovered metals can often be reused in their original application. The use of plastic and other materials to decrease weight and increase energy efficiency are however going to have a negative effect on the economy of car scrap processing operations. The development of dry separation processes and particle sorting will enlarge future possibilities, whereas new quality control equipment will improve the quality of reject products.

ACKNOWLEDGEMENTS

The authors are indepted to Mrs. E.M. Boons for the preparation of the manuscript and the contributions made by S.P.M. Berkhout and E.J.M. van Poppel.

REFERENCES

1. Voigt B. NE-Metalle im Automobilbau. Metall, 42, Jahrgang Heft 3, März 1988.

2. Nijkerk A.A. Hoeveel auto's vrij voor de sloop? Magazine Recycling Benelux. 24e jaargang nr. 3, juni /juli 1990.

3. Kola R. The processing of steelworks waste. Lead zinc go. The Minerals, Metals and Materials Society 1990.

4. Julius, J. Rückgewinnung von Nichteisenmetallen in Shredderanlagen durch trockensortierung. Metall 38 Jahrgang Heft 8, August 1984.

5. Dean K.C., Sterner J.W., Shirts M.B., Froisland, L.J. Bureau of mines research on recycling scrapped automobiles. United States Department of the Interior Bureau of Mines, Bulletin 684, 1985.

6. Dalmyn W.L., Voskuyl W.P.H., Roorda H.J. Recovery of aluminium alloys from shredded automobiles.

7. Dalmyn W.L., Dreissen H.H., Roorda H.J., Basten A.T. Mechanical separations of non-ferrous metals by means of gravity concentration and eddy-currents.

8. Personal communication with Mr. W. Ott of S + S, Schönberg, Germany.

9. Sattler H.P. Automatic sorting of non-ferrous metals from automobile shredders. Proceedings Recycling of non-ferrous metals, Williamsburg 1990, AIME meeting.

10. Witteveen H.J., Dalmyn, W.L. Concentratie van niet-ijzer metalen uit autoschroot. De Ingenieur nr. 3, March 1989.

11. Weiss K.J., Auer R.Fr., Dalmyn W.L. Selektives Abtrennen verschiedener NE-Metalle aus Gemengen. Metall, 36. Jahrgang Heft 4, April 1982.

Sekundäraluminium im Automobil— Einsatz und Rückgewinnung

H. Orbon
Metallwerk Olsberg GmbH, Essen, Deutschland

Mit einer Sekundäraluminiumproduktion von ca. 1,6 Mio Tonnen ist die Europäische Gemeinschaft der größte Sekundäraluminiumhersteller der Welt, gefolgt von Japan mit rund 1,03 Mio t und den USA mit ca. 1,03 t. Mengenmäßig dürfte die UdSSR, worüber leider keine Produktionszahlen vorliegen, ebenfalls in der Spitzengruppe liegen. Innerhalb Westeuropas weist Deutschland mit ca. 560.000 Jato in 1990 die höchste Produktion auf.

Tabelle 1

1990	Sek.-Al t/a	Primär-Al t/a	Anteil Sek.-Al / Gesamtproduktion %
EC	1,6 Mio	2,3 Mio	41
Japan	1,03 Mio	0,03 Mio	97
USA	1,03 Mio	4,05 Mio	20
GERMANY	0,56 Mio	0,72 Mio	44

Mit einem Anteil von etwa 60 % ist die Automobilindustrie Hauptverbraucher von Sekundäraluminium.

Wenn heute über den Einsatz eines Werkstoffes im Fahrzeugbau gesprochen wird, dann spielen Energie- und Recyclingaspekte neben Festigkeits- und Korrosionseigenschaften eine sehr wichtige Rolle.

Für Aluminium kann bei ganzheitlicher Energiebetrachtung festgestellt werden, daß es sich um einen Werkstoff mit positiver Energiebilanz handelt. Für diese Feststellung ist folgendes zu berücksichtigen:
Die Bewertung des Energieverbrauches darf sich nicht ausschließlich auf die Herstellung des Produktes beschränken. Dem Energieaufwand für sämtliche Prozeßstufen von der Gewinnung des Rohstoffes bis zum fertigen Produkt muß die mögliche Energieersparnis während der Nutzungsdauer - im Vergleich zu Wettbewerbsmaterialien - gegenübergestellt werden sowie auch der Energieinhalt durch die stoffliche Verwertung nach Gebrauch.

Beim Recycling von Aluminiumprodukten erspart das Aufbereiten und Einschmelzen zum Sekundärmetall bis zu 95 % der Energie, die für die Primärherstellung aus Bauxit erforderlich war. Wichtig ist ebenfalls eine Betrachtung der Energiequellen, aus denen Primäraluminium gewonnen wird, da sich Öl, Kohle, Nuklear- und Wasserenergie in ihren Wirkungsgraden erheblich unterscheiden, was einen direkten Einfluß auf die Höhe des Primärenergiebedarfes hat. So hat Stromerzeugung aus Kohle z.B. nur einen Wirkungsgrad von 33 %, bei Einsatz von Wasserkraft erzielt man dagegen Wirkungsgrade um 90 %. Wasserkraft ist mit über 60 % die weltweit am meisten eingesetzte Energiequelle zur Herstellung von Primäraluminium.

Die vorstehenden Aussagen sind nachfolgend auf der Basis von Unterlagen der Aluminium-Zentrale, Düsseldorf, graphisch dargestellt:

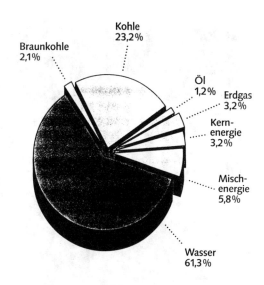

Energiequellen zur Primäraluminiumherstellung in der westlichen Welt 1988

Bild 1

273

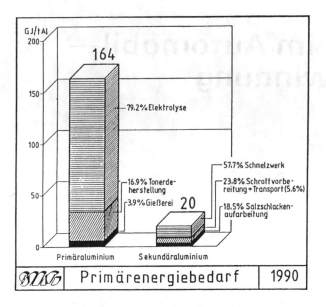

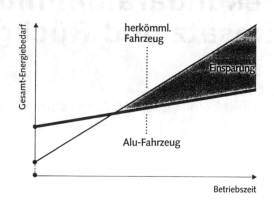

Positive Gesamt-Energiebilanz am Beispiel Fahrzeug

Bild 2

Bild 3

Der Einsatz von Sekundäraluminium im Automobil erfolgte bisher fast ausschließlich in Form von Gußlegierungen. Es kann davon ausgegangen werden, daß über 8o % aller im Fahrzeugbau eingesetzten Gußwerkstoffe auf Sekundärmetallbasis hergestellt wurden. Bekannte Beispiele für Gußteile aus Sekundäraluminium sind: Getriebe- und Kurbelgehäuse, Zylinderköpfe, Ansaugrohre sowie Komponenten des Brems-, Lenk- und Kühlungssystems; Beispiele, die in den nachfolgenden Abbildungen dargestellt sind: (11 Dias VAW, VDS-Druckgußwettbewerb)

Das vorrangige Motiv für einen Einsatz von Aluminium im Automobil ist neben seinen guten Recyclingeigenschaften die Gewichtsersparnis mit dem Ziel der Treibstoffersparnis. Durch die Verwendung von derzeit ca. 5o kg Aluminium pro Fahrzeug anstelle spezifisch schwererer Materialien reduziert sich der Benzinbedarf für alle zur Zeit in Deutschland laufenden Fahrzeuge um rund 1 Milliarde Liter pro Jahr.

Neben Gußlegierungen finden aber auch Aluminiumwalzfabrikate in Form von z.B. Karosserie- oder Kühlerteilen sowie Aluminiumprofile Anwendung im Automobilbau.

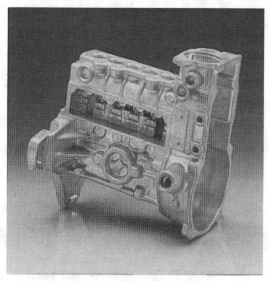

Bild 4

Gehäuse für Einspritzpumpe
Druckguß, 2,1 kg

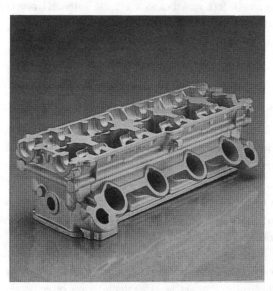

Bild 5

Zylinderkopf
Kokillenguß, 17 kg

Bild 6
Aufnehmergehäuse
Druckguß, 0,94 kg

Bild 7
Hinterachsgehäuse
Sandguß, 6,8 kg

Nach US-amerikanischen Angaben von Doehler-Jarvis stieg
der Aluminiumanteil im Automobil von 65 lbs in 1976 auf
12o lbs pro Fahrzeug in 1989. Es trat also nahezu eine
Verdoppelung ein. Für das Jahr 2ooo wird dort mit einem
durchschnittlichen Aluminiuminhalt von 18o lbs pro PKW
gerechnet. Für Europa ist mit ähnlicher, wenn nicht
höherer Zuwachsrate zu rechnen.

Wie sich unter Recyclinggesichtspunkten der Wert eines
Altautomobils ändert, wenn man das Fahrzeug des Jahres
1988 mit einem "Aluminium-intensiven" auf der einen Seite
und einem "kunststoffintensiven" Auto andererseits vergleicht, zeigt die nachfolgende Abbildung, die einem
Bericht des Vereins Deutscher Ingenieure entnommen wurde.

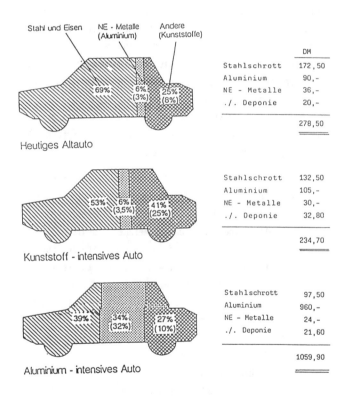

	DM
Stahlschrott	172,50
Aluminium	90,-
NE - Metalle	36,-
./. Deponie	20,-
	278,50

Heutiges Altauto

	DM
Stahlschrott	132,50
Aluminium	105,-
NE - Metalle	30,-
./. Deponie	32,80
	234,70

Kunststoff - intensives Auto

	DM
Stahlschrott	97,50
Aluminium	960,-
NE - Metalle	24,-
./. Deponie	21,60
	1059,90

Aluminium - intensives Auto

Materialzusammensetzung und Wertinhalt von Altautos
(für 1t Auto; Metallpreise 1. Sept. 1988, Deponie 80.- DM/t)

Gemessen am jährlichen Verbrauch in Deutschland beträgt
der Anteil von Sekundäraluminium zur Zeit etwa 35 %.
Diese Zahl sollte jedoch nicht mit der tatsächlichen
Recyclingrate verwechselt werden. Betrachtet man die
einzelnen Anwendungsgebiete separat und dazu ihre spezifische Nutzungsdauer, so sind erheblich höhere Recycling-
quoten festzustellen.
Wie die nachfolgende Abbildung zeigt, liegt der Rücklauf
von Altschrotten aus dem Verkehrsbereich bei über 9o %.

Bild 8

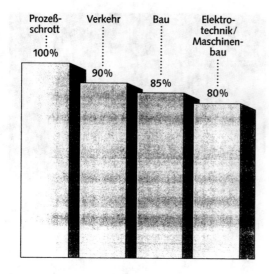

Prozeß-schrott 100% Verkehr 90% Bau 85% Elektro-technik/Maschinen-bau 80%

Recycling-Raten für Prozeßschrotte und einzelne Anwendungsbereiche 1990

Bild 9

Hauptgußwerkstoffe, die Sekundäraluminiumhütten für den Einsatz im Automobil liefern, sind Legierungeh des Typs AlSiCu. Kolbenlegierungen sind meist Spezialleglerungen der Kolbenhersteller; moderne Motorengehäusekonstruktionen benutzen eutektische und übereutektische Werkstoffe des Typs AlSiCu; Gußwerkstoffe für Räder sind meist sehr reine AlSi-Legierungen, die auf Reinaluminiumbasis aufgebaut sind. Auch wenn Gußwerkstoffe im Automobil nicht ausschließlich auf Sekundärrohstoffbasis aufgebaut sind, sind sie aus Recyclinggesichtspunkten wichtig. Bei der Verschrottung des Automobils tragen sie zur qualitativen Aufbesserung des rückgewonnenen Aluminiuminhaltes bei.

Auf einen Punkt sollte hierbei in aller Deutlichkeit hin-gewiesen werden, die Technologie einer Sekundäraluminium-hütte, die dem Stand der Technik entspricht, ermöglicht durch gezielte Aufbereitung und schmelzmetallurgische Verarbeitung ein Werkstoffrecycling ohne Qualitätseinbuße. Der gelegentlich zu hörende Einwand, Aluminiumschrott sei nur durch Verschneiden mit Primär- oder Rein-Aluminium qualitativ gleichwertig recyclierbar, ist nicht zutreffend. Wenn z.B. aus einem verschrotteten Aluminium-Zylinderkopf durch Einschmelzen nicht direkt wieder die entsprechende Zylinderkopflegierung herstellbar ist, so liegt die Ur-sache hierfür nicht in einer verfahrensbedingten Qualitäts-verschlechterung. Vielmehr verlangt die Gießerei von der Sekundärhütte eine reinere Basislegierung zur Herstellung des Zylinderkopfes, um eine Verarbeitungsspanne für gießerei-interne Kreislaufverunreinigungen zu haben. Diese Tatsache spiegelt sich in den meisten Normvor-schriften für Gußlegierungen wider, die für die Zu-sammensetzung von Ingots oder Flüssigmetall bei be-stimmten kritischen Elementen stärker eingeschränkte

Toleranzen vorschreiben als für die Zusammensetzung des aus der entsprechenden Legierung herzustellenden Gußteiles.

Wenn bisher ausgeführt wurde, daß Sekundäraluminium nahezu ausschließlich in Form von Gußlegierungen im Automobil eingesetzt ist, so muß darauf hingewiesen werden, daß moderne Aluminiumrecyclingwerke unter Ein-beziehung aufwendiger Aufbereitungs-, Sortier-, Schmelz- und Kontrolltechniken durchaus in der Lage sind, Recy-cling auf dem Qualitätsstandard von Knetwerkstoffen durchzuführen und dies in vielen Fällen auch bereits praktizieren. Befürchtungen, Knetlegierungen würden durch Recycling auf das "niedrigere Niveau" einer Massen-Sekundärgußlegierung "abgewertet", sind völlig unbe-rechtigt.

Da im Rahmen dieser "First European Metals Conference" an anderer Stelle in einem speziellen Vortrag über die Technologie des Aluminiumrecyclings berichtet wird, möchte ich mich hier darauf beschränken, lediglich das Verfahrensschema einer modernen Sekundäraluminiumhütte darzustellen, das im wesentlichen in vier Bereiche aufge-teilt werden kann: Aufbereitung, Schmelzen, Qualitäts-kontrolle, Reststoff-/Abfallbehandlung.

Wenn man über Aluminium im Automobil und sein Recycling spricht, kann man das nicht ohne Hinweis auf die all-gegenwärtige Umweltproblematik tun. Aufgrund seiner charakteristischen Eigenschaften, wie gutes Korrosions-verhalten, niedrige Dichte, gute Recyclierbarkeit u.ä., trägt Aluminium wesentlich dazu bei, den Gebrauch eines Automobils immer umweltfreundlicher zu gestalten. Im öffentlichen und vor allem im politischen Bewußtsein setzt sich jedoch immer mehr durch, daß ein Produkt nicht nur in seinem Gebrauch umweltfreundlich sein sollte, sondern daß auch seine Herstellung und die zu seiner Herstellung benötigten Rohstoffe unter Umwelt-gesichtspunkten zu bewerten sind. Es ist klar, daß nach deutschen bzw. europäischen Umweltvorschriften hergestelltes Aluminium teurer sein muß als Material, das ohne oder mit unzureichenden Maßnahmen hergestellt wurde. Leider ist die Angebotsmenge von Sekundäralu-minium, das ohne Umweltschutzmaßnahmen hergestellt wurde, auf dem europäischen Markt beachtlich und im Preisangebot absolut wettbewerbsverzerrend. Hier sind gesetzliche und Kontroll-Maßnahmen unbedingt erforder-lich.

Es dürfte für den Anwender von Recyclingmaterial, also auch für die Automobilindustrie, durchaus imagefördernd sein, wenn er in Zukunft darauf achtet, daß sein Produkt nicht nur umweltverträgliche Anwendungsbedingungen bietet, sondern auch unter Einsatz umweltverträglich hergestellter Rohstoffe produziert wurde. Es wäre zu wünschen, wenn

Herstellung von Sekundär-Aluminium

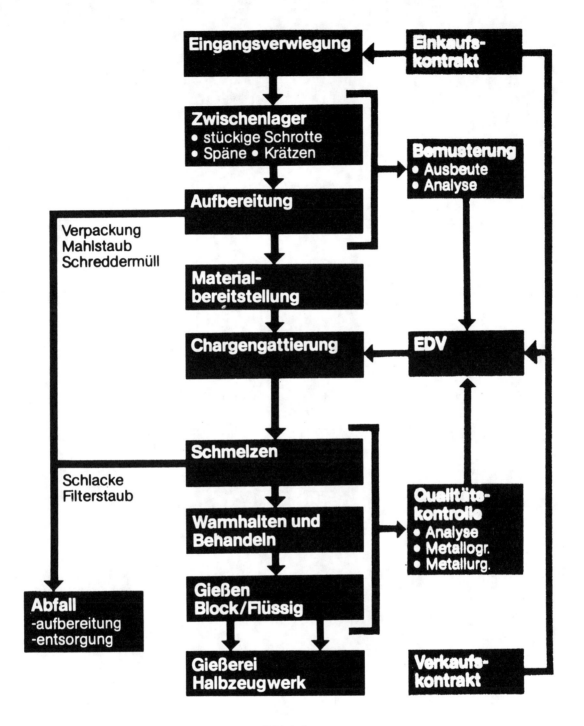

Bild 1o

mit dem Angebotspreis für einen Rohstoff, so auch für Sekundäraluminium, der kontrollierbare Nachweis für eine umweltfreundliche Rohstoffherstellung abgegeben werden müßte.

Soweit meine Ausführungen zum Sekundäraluminium im Automobil, für deren Zusammenstellung ich dankenswerterweise auf Unterlagen der VDS Vereinigung Deutscher Schmelzhütten und der Aluminium-Zentrale e.V., Düsseldorf, zurückgreifen konnte.

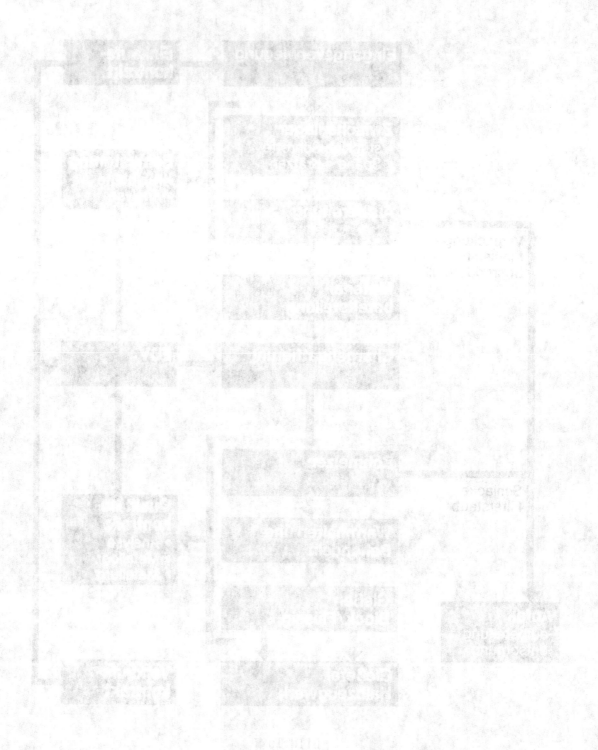

Les fours tournants dans l'industrie européenne de la récupération des métaux non-ferreux

M. Rousseau
Stolberg Ingenieurberatung GmbH, Neuss, Allemagne

RESUME

Les types de fours tournants utilisés pour le recyclage des métaux non-ferreux, sont le plus souvent des fours à tambour avec garnissage basique. L'analyse des échanges de chaleur explique pourquoi le rendement thermique des fours tournants est supérieur à celui des fours à reverbère. La prédominance des fours tournants, par ex. pour le recyclage du plomb, du cuivre ou de l'aluminium, restera assurée pour de nombreuses années encore, en particulier grâce à l'utilisation de brûleurs avec air de combustion enrichi et/ou préchauffé ou bien de brûleurs à oxygéne.

ABSTRACT

The type of rotary furnace currently most employed in the recycling of non-ferrous metals consists of furnaces lined with basic refractories. The analysis of heat transfer explains why the thermal efficiency of rotary furnaces is higher than those of reverberatory furnaces. The rotary furnaces will continue to predominate, e.g. for lead, copper or aluminium recycling, for many years to come especially with regard to the utilization of burners with oxygen-enriched air and/or preheated air or even with oxy-fuel burners.

INTRODUCTION

La présente communication a pour objet de traiter des principaux types de fours rotatifs utilisés dans les opérations métallurgiques de récuperation des métaux non-ferreux. Elle se limitera aux fours ayant pour but d'obtenir différentes phases liquides où les métaux ad hoc sont concentrés et dont les caractéristiques sont indiquées au tableau I et à la figure 1. Ils travaillent pour la plupart de façon discontinue. Les fours de ressuage constituent en cela une exception car opérant en continu et n'étant généralement pas garnis de réfractaires. Les fours privilégiant les réactions en phase solides (p. ex. fours Waelz pour les poussières d'aciérie) et ceux équipés de lances ou de tuyères et travaillant principalement comme convertisseurs (fours de type Kaldo TBRC utilisés p. ex. par Boliden, Metallo-Chimique et Chemetco) ne seront pas considérés par la suite.

Les fours tournants garnis de réfractaires sont particulièrement bien adaptés à l'industrie de la récuperation (voir Tab. II) eu égard à leur capacité, leur simplicité d'utilisation, leurs bons rendements thermiques (p. ex. 30 à 35 %) et métallurgique; toutes choses permettant une adaption rapide aux conditions du marché (dans cette optique les opérations discontinues représentent plutôt un avantage). Ceci explique l'utilisation prédominante en Europe de ces types de fours, au contraire p. ex. de l'Amérique du Nord utilisant encore nombre de fours réverbères (n'ayant des rendements thermiques que, p. ex. de 15 à 20 %). A ceux-ci sont toutefois substitués de plus en plus de fours tournants comme par.ex. chez Cominco (Trail, Canada) en 1988 (ϕ 3,5 m x 3,5 m) ou bien chez Masters Metals (Cleveland, E.-U.) en 1985/87 (ϕ 3 m, ϕ 3, 6 m et ϕ 4 m avec ratio de forme 1, 2 à 1,3).[4]

Les fours tournants (voir fig. 1) sont équipés de brûleurs à fuel ou à gaz situés sur l'avant du four (à tambour long) ou au contraire à l'arrière (four à tambour court). Ce dernier brûleur (type lance) est soumis à de hautes contraintes thermiques (refroidissement nécessaire) mais a l'avantage de permettre le chargement du four, brûleur en marche. Le chargement s'effectue normalement par goulotte ou avec des machines spécifiques (p. ex. avec auges de chargement).

TAB. I: PRINCIPAUX TYPES DE FOURS TOURNANTS UTILISES DANS LES INDUSTRIES DE RECUPERATION [1]				
Désignation	Ratio de forme L : φ	Dimensions habituelles (m)	Exemple d'utilisation pour le recyclage du plomb	
			charges	production
Four à tambour court	1,0 à 1,5	φ = 2 à 4 L = 2 à 5	7 à 20 t	5000 à 20000 tpa
Four à tambour long	1,5 à 3,0	φ = 1 à 4 L = 2 à 6	0,5 à 10 t	500 à 15000 tpa
Four de ressuage	3 à 11	φ = 0,5 à 1,5 L = 3 à 6	continu	15000 à 35000 tpa

N.B.: L = longueur, φ = diamètre

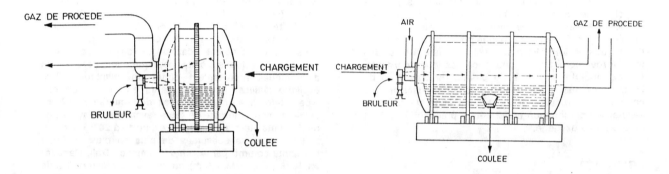

FOUR A TAMBOUR COURT   **FOUR A TAMBOUR LONG**

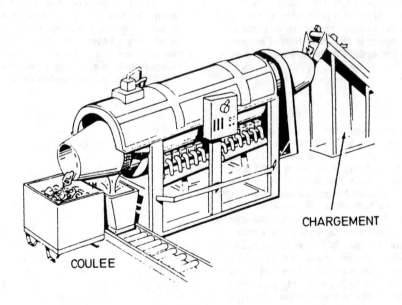

FOUR DE RESSUAGE

FIG. 1:   VUES DE DIFFERENTS FOURS TOURNANTS [2, 3]

TAB. II: **PROPRIETES INTRINSEQUES DES FOURS TOURNANTS GARNIS DE REFRACTAIRES**

AVANTAGES:
-     Opération en atmosphères controlées: neutre, réductrice, oxydante, par là:
-     Selectivité métallurgique, par là:
-     Obtention de scories assez pauvres en métaux et/ou absorbant certaines impuretés (S, Cl)
-     Possibilité d'obtenir plusieures phases liquides (p. ex. une matte, un speiss),
-     Volatilisation sélective partiellement possible,
-     Large éventail de granulométrie des matières premières: des fines aux fragments,
-     Pertes par radiation minimalisées,
-     Excellent rendement thermique, et par là:
-     Profil de température très régulier, et par là:
-     Usure uniforme des réfractaires,
-     Faible volume spécifique de gaz de procédé,
-     Excellente capacité spécifique de traitement,
-     Utilisation d'agents réducteurs bon marché,
-     Homogénéisation de la charge (p. ex. réglage de la rotation de 0,1 jusqu'à plus de 3 t.p.m.) et par là:
-     Contact intensif entre la charge et les additifs (réducteurs, scorifiants),
-     Scories en équilibre avec le métal <u>et</u> la phase gazeuse,
-     Simplicité d'utilisation: en opération, pour la mise en route et l'arrêt et par là:
-     Adaptation rapide des opérations aux conditions du marché
INCONVENIENTS:
-     Construction plus compliquée (que les fours à réverbères) surtout pour les fours basculants,
-     Consommation assez importante en réfractaires,
-     Température de travail limitée à environ 1200 - 1300 °C (rarement plus),
-     Surtout adapté au traitement de matières peu humides,
-     Température relativement haute des gaz de procédés, et de par là:
-     Génération assez élevée en poussières,
-     Difficulté à charger des blocs volumineux,
-     Capacité par charge relativement faible (p. ex. même pour les fours Thomas, pas plus de 70 à 100 t)

MECANISMES DES ECHANGES DE CHALEUR

Les échanges de chaleur, dans des zones éloignées de la flamme, entre charge solide et gaz ont été récemment étudiés en détail et modellisés pour les fours tubulaires (avec brûleur à air ou air enrichi) [5]. Leurs mécanismes (voir Fig. 2), en grande partie applicables aux fours étudiés ici, sont les suivants.

La phase gazeuse transmet par radiation et convection, de l'énergie aux surfaces exposées de la charge et des parois, tandis qu'un échange de chaleur a lieu ensuite entre ces surfaces. Chaque partie de la paroi parvient régulièrement (Cf mouvement de rotation) en contact avec la charge, assurant (par rapport aux fours à sole) un transfert additionnel d'énergie, principalement par conduction. La surface de la charge est constamment renouvelée; les matières chauffées migrent sous la surface, où il s'établit des gradients de température, c.à.d. des transferts de chaleur dans la charge, et des circulations secondaires. En conséquence, les différences de température (p.ex. 30 à 90 K pour le brûleur à air et jusqu' à plus de 200 K pour ceux à oxygène) dans les zones de contact entre la charge et la paroi sont relativement faibles; ceci évite ou limite la surchauffe locale des réfractaires. Pour tous les types de fours les pertes de chaleur par l'enveloppe représentent 15 à 20 %. L'isolation des réfractaires peut les réduire, mais (spécialement dans le cas des brûleurs à oxygène) doit être utilisée avec prudence pour éviter toute surchauffe.

Le transfert de chaleur de la flamme à la charge, via les gaz de procédés, est différent selon qu'il s'agisse de brûleurs à air ou à oxygène. Avec l'air (flamme longue mais plus froide), dans la partie supérieure de la charge, prédomine le transfert par radiation (p. ex. 60 - 80 %) [6, 7] du fait de la génération des produits de combustion (CO_2, H_2O) et ce dans les régions proches de la flamme (p. ex. de 0,5 à 0,8 fois le diamètre du four) [8]. Les transferts par convection (primaire et secondaire) vers la charge viennent en seconde position. Ce type de flamme longue permet d'obtenir une flamme dite inversée ("reverse flamme") particulièrement propice pour les fours à tambours courts [9].

L'augmentation de la température de flamme par préchauffage de l'air et/ou son enrichissement en oxygène (p. ex. 10 K par 0,3 à 0,5 % O_2) permet d'accroître la productivité des fours jusqu'à un certain niveau. Au delà, seule l'introduction des brûleurs à oxygène pur permet un accroissement drastique. Pour cela, 2 à 2,2 m^3 d'oxygène sont nécessaires pour la combustion complète de 1 m^3 de gaz naturel ou de 1 kg du fuel.

L'utilisation des brûleurs à oxygène procure de nombreux avantages intrinsèques (voir Tab. III) prépondérant si l'oxygène disponible est assez bon marché, mais peut poser des problèmes d'environnement (auxquels il est possible de remédier). Les températures supérieures à 1300 °C conduisent en effet à une plus haute concentration de NO_x (même si l'émission totale est réduite). Enfin les combustibles eux mêmes contiennent de l'azote, de sorte qu'ils génèrent p. ex. la moitié du NO_x formé [10]. La température de flamme (et donc la formation de NO_x) peut être réduite par l'introduction de H_2O [11], de gaz porteur de CO_2 (p. ex. recirculation partielle de fumées) [7], par le contrôle de l'oxygène mis à disposition (p. ex. moins de 2 à 4 % O_2) via la mesure de son taux dans les fumées, ou si le procédé le permet par des conditions plus réductrices (cf CO).

Le passage de brûleurs à air aux brûleurs à oxygène pur doit être préparé en détail, en considérant en particulier leur position dans le four (modification éventuelle) et les implications sur l'environnement.

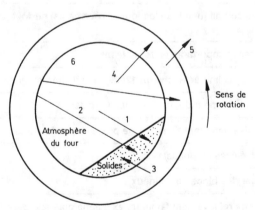

(1) Des gaz de procédé à la charge exposée (Radiation + Convection)
(2) Des parois exposées à la charge exposée (Radiation)
(3) Des parois recouvertes à la charge (Radiation + Convection + Conduction)
(4) Des gaz de procédé aux parois exposées (Radiation + Convection)
(5) De l'enveloppe à l'extérieur (Radiation + Convection)
(6) Des parois exposées aux parois exposées (Radiation)

FIG. 2: MECANISMES DES ECHANGES DE CHALEUR (VALABLE DANS UNE ZONE ELOIGNEE DE LA FLAMME DU BRULEUR) 5

TAB. III	INFLUENCE DE L'UTILISATION DES BRULEURS A OXYGENE PUR
AVANTAGES	
-	Economies de combustibles (par ex. 50 % ou plus),
-	Rentable pour oxygène bon marché (p. ex. prix du kg de fuel au moins deux fois celui du m^3 d'oxygène),
-	Accroissement de la productivité (p. ex. 10 à 30 %),
-	Réduction drastique des gaz de procédé (60 % ou plus) et corrélativement,
-	Dans une installation existante: gaz avec plus long temps de séjour (cf rendement thermique) et vitesse plus faible (cf poussières),
-	Scories plus pauvres,
-	Accroissement du point de rosée (cf durée de vie des matériaux de filtration),
-	Diminution des émissions totales (poussières, NO_x, SO_2),
-	Excellente combustion des hydrocarbures,
-	Haute température de la flamme: contrecarre une éventuelle formation de dioxine.
INCONVENIENTS	
-	Prix de l'installation de chauffe (2 à 3 fois plus elevé),
-	Prix de l'oxygène suivant le lieu,
-	Réfractaires pouvant être soumis à de plus grandes contraintes thermiques,
-	Accroissement possible des concentrations en NO_x et SO_2

REFRACTAIRES [12, 13]

Les fours tournants possèdent pour la plupart un garnissage basique ou tout au moins peu acide (avec quelques exceptions encore, comme dans l'industrie du cuivre: four Thomas, four de perchage).

Le type de réfractaire dépend entre autre des métaux et oxydes présents et des températures de travail p. ex. pour les rebuts cuivreux 1100 à 1350 ° C, plombifères 1000 à 1100 ° C, mais moins de 1000 °C (jusqu'à 750 - 850 ° C) pour les rebuts alumineux.

Les plus basiques sont utilisés pour la récupération du plomb: principalement magnésie (avec plus de 80 % MgO) ou magnésie-chrome (de 55 % à 80 % MgO et jusqu'à p. ex. 16 % Cr_2O_3). Diverses chamottes sont utilisées en plus pour la récupération du cuivre et surtout de l'aluminium (p. ex. environ 40 % Al_2O_3).

EXEMPLES D'UTILISATION DES FOURS TOURNANTS

Plomb

Les fours rotatifs de dimension moyenne (p. ex. diam. 3 x 4 m) et équipés de brûleurs à air permettent de traiter environ de 50 à 150 tpj de rebuts ou plus. De telles installations (voir Tab. IV) sont décrites en détail dans la littérature technique [1, 13], et deviennent de plus en plus complexes en ce qui concerne les équipements de protection de l'environnement. Ceci est illustré par la consommation des fours (diam. 3,6 x 5 m) à l'usine de Villefranche (Métaleurop) de 45 kWh/t Pb pour la réduction contre 105 kWh/t Pb pour la filtration, et par la figure 3 (BSB-R, R.F.A.).

Cuivre

Les fours à tambour et basculants (dits fours Thomas) sont très fréquemment utilisés [13] p. ex. à la Norddeutsche Affinerie (N.A.), Hambourg, R.F.A., [16]. Leur capacité varie de 8 t jusqu'à 20 à 40 t en moyenne (c.à.d. env. diam. 3 x 6 m) ce qui peut correspondre à 15000 - 20000 tpa de matières.

Les résultats de l'introduction des brûleurs à oxygène pur, pour la refonte de rebuts cuivreux, dans les fours à anodes de la M.H.O. (Belgique) expriment bien la tendance pour tous les types de fours tournants: réduction du temps de refonte (env. 60 %), de la consommation d'énergie (env. 30 %) et corrélativement accroissement de productivité (env. 80 %) [17].

TAB. IV: FOURS TOURNANTS POUR LA RECUPERATION DU PLOMB ET DES MATIERES PLOMBIFERES [1,10,14,15]

Paramètres	Four φ 3 m x 4 m (charge: 10 - 12 t de matières)		Four φ 3 m x 5,3 m (charge: 30 t de matières)[3]		
	air	oxygène pur	air	air enrichi (22,2 % O_2)	oxygène pur
Fuel lourd	110 kg/h ou moins de 80 kg/t charge	37 kg/h ou moins de 40 kg/t charge	238 kg/h ou 55 - 65 kg/t matières	209 kg/h ou 40 kg/t matières	114 kg/h
Electricité	110 kWh/t charge	50 kWh/t charge	n.c.	n.c.	n.c.
Réfractaires	8,5 kg/t charge	6 kg/t charge	n.c.	n.c.	n.c.
Porteur d'oxygène	air	oxygène pur	2500 Nm^3 air/h	1900 Nm^3 air/h	200 Nm^3 O_2/h
Gaz de procédé	21000 Nm^3/h	9000 Nm^3/h	5000 à 7000 Nm^3/h	n.c.	n.c.
Poussières	193 kg/h	57 kg/h [1]	n.c.	n.c.	n.c.
Durée de vie des filtres (Poly-acrylnytrile)	env. 12 mois	+ 20 % [2]	env. 12 mois	n.c.	n.c.
Scories	n.c.	0,7 unité Pb en moins	n.c.	n.c.	n.c.
Productivité	5,7 charges/j	5,3 charges/j	n.c.	n.c.	n.c.

1) 100 % en dessous de 1 Micron
2) moins de 5 mg/Nm^3 après filtration
3) retraitement d'écumes du raffinage de plomb

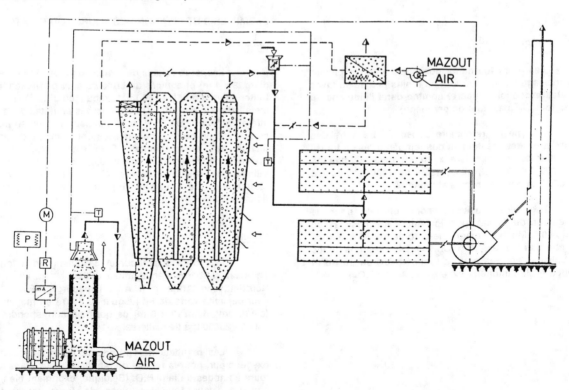

Fig. 3   Four rotatif court; arrangement typique du four, refroidisseur et filtre (Approx. à l'échelle)[2]

285

Aluminium

Les matières fines et fragmentées sont généralement refondues sous couche de sels (KCl, NaCl, evtl. NaF) dans des fours à tambour long (capacité 5 à 20 t). Des fours moyens de 10 à 12 t (capacité de fusion env. 1 t/h) peuvent traiter jusqu'à 12000 tpa de rebuts [18].

Une amélioration notable (voir Tab. V) peut être aussi obtenue par préchauffage de l'air. Les résultats obtenus avec des brûleurs spéciaux (pourvus de récupérateurs céramiques couplés), pour les fours circulaires de 12 t (capacité de fusion env. 3,5 t/h) d'Aluminium Corp. (Dolgarrog, Royaume-Uni) sont significatifs à cet égard. Les concentrations en NO_x sont par là réduisibles de 70 à 80 % [19]. L'enrichissement en oxygène permet par contre d'augmenter notablement la capacité de traitement des fours tournants: par ex. + 30 % pour une consommation de 20 m^3 O_2/t Al [20].

TENDANCES FUTURES ET CONCLUSION

De plus en plus de métaux non-ferreux doivent être récupérés à partir de rebuts complexes et d'alliages. Les nouvelles méthodes de préparation permettent jusqu'à un certain point une récupération sélective, mais les fractions obtenues doivent être retraitées en général par voie pyrométallurgique [21].

Les fours rotatifs garnis de réfractaires sont particulièrement bien adaptés à ces travaux complexes de fusion/réduction (voir Tab. II) ou différentes matières secondaires (assez fragmentées) peuvent être traitées grâce à une large gamme de températures et d'atmosphères de four. Seuls des rebuts à grande valeur et/ou volumineux peuvent justifier l'utilisation d'autres fours (réverbères, plus encore fours à induction).

Pour les matières métalliques à bas point de fusion et/ou relativement pures où seule une refonte est nécessaire, les fours rotatifs de ressuage (à chauffage direct ou indirect) peuvent constituer une alternative comme p. ex. pour des plaques d'accumulateurs au plomb (les crasses et oxydes devraient être toutefois retraités en fours classiques).

Pour les matières très complexes et/ou à valeur importante, des réacteurs fixes de type SIRO-SMELT/ISASMELT devraient occuper progressivement des segments bien particuliers du marché [22, 23, 24].

Toutefois les fours tournants sont restés, malgré la concurrence de certains procédés utilisant des réacteurs spécialisés, un des outils essentiels du recyclage des métaux non-ferreux. La grande flexibilité de ces fours ainsi que l'utilisation d'un air de combustion enrichi en oxygène et/ou préchauffé ou bien de brûleurs à oxygène, devraient assurer, pas seulement en Europe, leur pérennité pour de nombreuses années encore.

TAB. V: FOUR POUR LA FUSION DES REBUTS DE FABRICATION DE L'ALUMINIUM [19]		
Paramètres	Brûleur à air	Brûleur à récuperateur incorporé
Capacité du brûleur	3,9 MW	2,3 MW
Température		
* dans le four	maxi 1100 ° C	maxi 1100 ° C
* des fumées	maxi 1200 ° C	maxi 200 ° C
* de l'air de combustion	30 ° C	maxi 1000 ° C
Consommation de combustible	1500 kW/t (fuel)	920 kW/t (gaz naturel)

286

Littérature

1. Rousseau M., Lead Scrap Processing in Rotary Furnaces: A Review, 5th Int. Recycling Congress (IRC), Berlin, 1986, pp 1721 - 1727
2. Melin A., M. Rousseau, Recyclage du plomb: Les techniques industrielles actuelles, Acta Tecnica Belgica, ATB-Metallurgie, Vol. XXII No. 4, pp 229 - 235
3. Rousseau M., K. Boeger, Recovery of Zinc and Aluminium from Secondaries by Sweating, TMS/AIME (Etats-Unis), Technical Paper A 88-7
4. Jaeck M.L., Primary and Secondary Lead Processing, Proc. Int. Symp. CIM, 28th An. Conf., Trail, Canada, Aug. 1989, pp 106
5. Barr P.V., J.K. Brimacombe, A.P. Watkinson, Heat-transfer Model for the Rotary Kiln with Pilot Plant Verification, Pyrometallurgy '87, IMM (London), Sept 1987, pp 53 - 90
6. Gorog J.P., T.N. Adams, J.K. Brimacombe, Regenerative Heat Transfer in Rotary Kilns, Met. Trans. 8, Vol. 13 B, June 1982, pp 153 - 163
7. Knauber K., Moderne Erdgas-befeuerte Industriebrenner, G.W.I., Bd. 36, 1987, H. 4, pp 226 - 229
8. Gorog J.P., J.K. Brimacombe, T.N. Adams, Radiative Heat Transfer in Rotary Kilns, Met. Trans. 8, Vol 12 B, March 1981, pp 55 - 70
9. Fritsch W.H., Technisch-wirtschaftliche Auslegung von Industrie-Brennern, G.W.I., Bd. 32, 1983, H.5, pp 194 - 201
10. Lamm K.F., Use of oxy-fuel burners, Productivity and Technology in the Metallurgical industries, TMS/GDMB, 1989, pp 473 - 481
11. Hanzlik K., Wie sinnvoll ist O_2-Anreicherung?, G.W.I., Bd. 34, 1985, H. 9, pp 370 - 374
12. Granitzky, K-E, Auswahl des Feuerfesten Baustoffes, 1 Met. Sem. der G.D.M.B., 1975, pp 143 - 168
13. Rousseau M., K. Boeger, Processing of Complex NF-Secondaries in Rotary Furnaces: A Review, 5th Recycling Congress (IRC), Berlin, 1986, pp 1721 - 1727
14. Deininger L., G. Fuchs, Neue Dreflammofenanlage, Erzmetall, 33, Nr. 5, 1983, pp 226 - 229
15. Deininger L., Einsatz von Sauerstoffbrennern, GDMB-Bleifachausschuß 1984, 10 pages
16. Ann., Kupferverarbeitung der Norddeutschen Affinerie, Erzmetall 39, 1986, Nr. 10, pp 518 - 519
17. Lombeck K., Neuere Entwicklungen beim Schmelzen von Cu-Schrotten, Int. Recycling Symp. der Metallgesellschaft AG, Nov. 1986, Budapest, 18 pages
18. Schlug K., Das neue Al-Umschmelzwerk Ronshofen, Metall, H. 11, Nov. 1980, pp 1052 - 1053
19. Jasper H.D., Moderne Feuerungstechnik für Öfen, 3. Jahrestagung der VDI-GEI, Darmstadt, Febr. 1986, 19 pages
20. Bücker W., Sauerstoff zur Steigerung der Schmelzleistung bei Aluminium, O_2-Anwendung in der Pyrometallurgie, 10. Met. Sem. G.D.M.B., 1980
21. Rousseau M., A. Melin, The Processing of Non-Magnetic Fractions from Shredded Automobile Scrap; A Review, Ressources Conversation and Recycling, 2, 1989, pp 139 - 159
22. Barett K.R., Lead Smelting Technologies into the 90's, ILZIC Conf., New Delhi, Nov. 88, pp 25.1 - 25.11
23. Anon., Billitons's New Arnheim Furnace, Metal Bulletin, 26.02.1990
24. Hartley C.J., Zinnrecycling mit dem Sirosmelt-Verfahren, Metall, H.9, Sept. 1990, 873 - 875

The TBRC as a unit with a promising future for secondary copper plants

H. Bussmann
Hüttenwerke Kayser Aktiengesellschaft, Lünen, Germany

ABSTRACT

Plants for copper winning from secondary raw material in their conventional structure are operating with the aggregates blast furnace, converter, refining furnace, and refining electrolysis. Between the single operation steps there are necessarily considerable material reverse runnings which cause a loss of metal and additional energy.

It showed that regarding the sphere of blast furnace, converter, and refining furnace the situation can be considerably improved by use of a TBRC (Top Blowing Rotary Converter). At the same time the emissions are reduced considerably, as by operation of the reactor in a close casing not only the process gases but also the secondary emissions from various charging operations will be totally captured. Alternatively the aggregate can do the reduction work of the shaft furnace, the oxidation work of the converter, and the refining work of the anode furnace with a high effectiveness and a lot of advances.

Materials as computer scrap can be charged continuously by a charging device. Therefore, huge quantities of them can be used without any problems regarding keeping the air clean.

Regarding converter work besides other advantages especially the good adjustability of temperature and slagging practice have to be remarked.

As in the secondary metallurgy the materials used are stored and sampled in small units, high efficient aggregates for charge processing are to be preferred and promise a higher economy.

Therefore, with the TBRC an aggregate is available which due to its flexibility can be adjusted to permanently changing input relations and which, moreover, due to its kind of construction and its kind of operation is able to considerably improve the environmental situation.

The recovery of copper from secondary raw materials is carried out in the following steps: reduction, converting, fire refining, and electrolytic refining. Secondary raw materials are brought to the suitable processing step according to their chemical composition. The reduction processes are mainly carried out in blast furnaces, and the metal from the blast furnace is blown together with alloy scrap in scrap metal converters. During this process the slag arising is returned to the reduction unit. The converter copper is refined

copper anodes are refined electrolytically into cathode copper. This common kind of process shows that there are considerable material recycles before the final end product cathode copper is

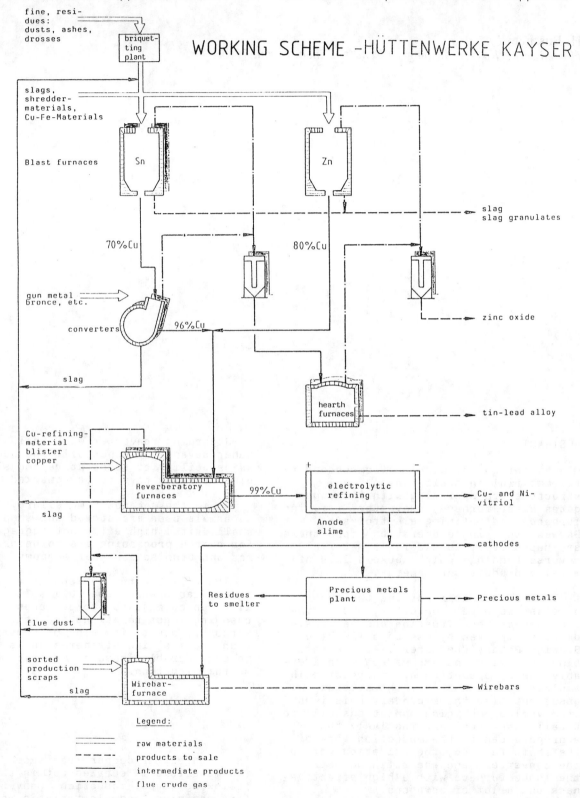

WORKING SCHEME -HÜTTENWERKE KAYSER

fine, residues: dusts, ashes, drosses

briquetting plant

slags, shreddermaterials, Cu-Fe-Materials

Blast furnaces

Sn

Zn

slag
slag granulates

70%Cu

80%Cu

gun metal bronce, etc.

converters

96%Cu

zinc oxide

slag

hearth furnaces

tin-lead alloy

Cu-refiningmaterial blister copper

Reverberatory furnaces

99%Cu

electrolytic refining

Cu- and Ni-vitriol

slag

Anode slime

cathodes

flue dust

Residues to smelter

Precious metals plant

Precious metals

sorted production scraps

Wirebarfurnace

slag

Wirebars

Legend:

raw materials
products to sale
intermediate products
flue crude gas

further in the anode furnace, and during this operation copper refining scrap is added - the slag from this process is also returned to the blast furnace. The

reached. Besides energy requirements for melting and reduction these material recycles cause metal losses by passing the units several times.

Hüttenwerke Kayser has carried out calculations, tests, and planning for easier processing with material returns as low as possible.

In non-ferrous metallurgy the TBRC (Top Blown Rotary Converter) has been used in recent years. The TBRC is a rotating and tilting furnace with burning and blowing injector lances: it looks like a concrete mixer.

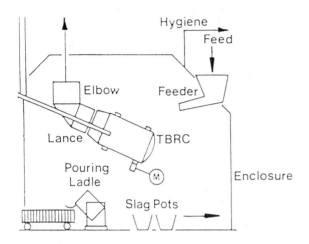

Fig. 2 - TBRC arrangement

In the blast furnace reactions are slow and incomplete due to an incomplete mixing. Due to the excellent mixing effect of the TBRC the diffusion paths have been reduced, therefore, the attainment of equilibrium is accelerated.

Also, when using the TBRC for converter work and refining work an accelerated and complete reaction is achieved due to good mass transfer.

By using oxygen both for heating and as reaction gas the required fuel and the quantities of waste gas are reduced. If the TBRC is working in a hood which closely surrounds it, the process waste gases as well as the secondary emissions which occur upon feeding and emptying can all be captured. This means a considerable reduction of environmental pollution compared with the usual processing.

The plant's flexibility is considerably improved by the fact that the TBRC can do the blast furnace's reduction work as well as the converter's and anode furnace's refining work, by carrying out the oxidation process in the same vessel.

For a blast furnace working effectively on reduction work a preparation of the material is necessary by mixing, crushing, agglomeration, and sieving. If the reduction work is done in the TBRC this is largely unnecessary. The mixing is done by the TBRC's turning. Coarse material gets dissolved in the moving melt, and fine material does not matter, as there is not a feed column with gas flow. By using melted materials, as for example, liquid slags the energy for remelting is saved. The charge still can be influenced while running by addition of flux and reduction agents. By this the distribution of the elements accompanying copper can be influenced favourably for the following process steps.

By working step by step, removing the metal under the slag layer, the elements can be parted according to their reduction reaction. Moreover, low final contents in the slags can be achieved without producing high amounts of metal with high iron content.

All these processes are helped by the intensive mixing during rotation, which allows the fluid-solid and fluid-liquid diffusion stages to be of minimal length, producing rapid and complete reactions.

For reduction work in the TBRC all solid reduction agents can be used. For carbon containing materials no special requirements have to be met regarding stability, piece size or ash content. Therefore, the cheapest reducing materials can be used. Also if using iron as reduction agent one only has to take care that the brick lining does not get damaged due to the piece size. Moreover, too big pieces unnecessarily prolong the reduction times.

All these facts show that regarding reaction kinetics, requirements for feed material, and energy utilization of the heat contents of combustible feed material as metallurgical reactor the TBRC is superior to the blast furnace.

In oxidation work as necessary upon converting and refining, one also has to point out the favourable reaction mechanism. Due to the rotation of the reactor there is always a good mixture of the material, which leads to a uniform concentration and new surfaces for the gas-melt-reaction. This is especially important for reactions which do not produce gases, as in the metallurgy of copper secondary materials.

By using oxygen there are high reaction velocities and minimum losses of heat recorded, therefore, for processing it is necessary to add only small amounts of energy besides the oxidation energy available. The oxygen necessary for oxidation can be utilized optimally, as the melt which wets the vessel's

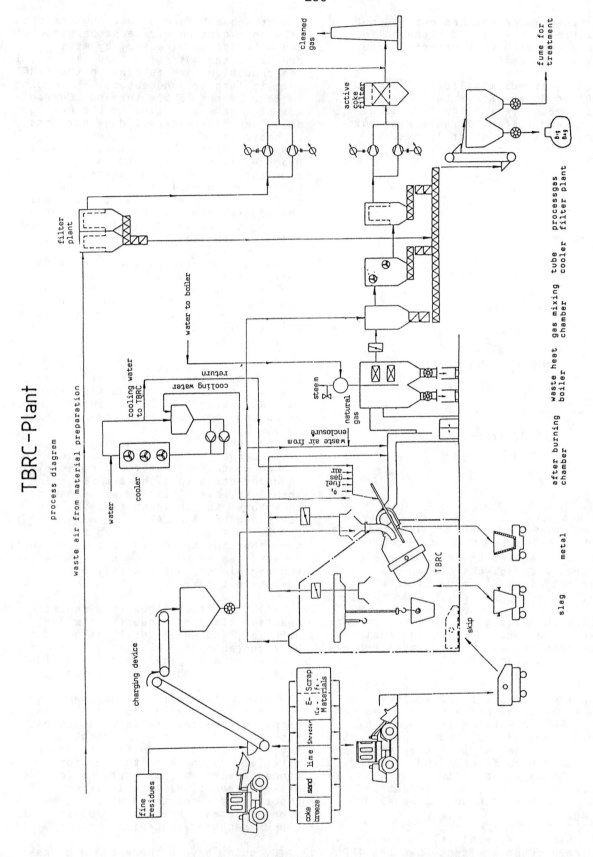

TBRC-Plant

process diagram

walls can react afterwards with the furnace atmosphere.

By dosage of oxygen addition and fuel feed the reaction temperature can be controlled for the respective refining step. This is not possible with the usual scrap converter. With the TBRC, however, one has the possibility to influence the metallurgy in such a manner that a certain selectivity upon slagging occurs.

Lead and tin, for example, can be slagged at low temperatures while the zinc gets eliminated as metal vapor. Especially these separations are very important for secondary copper plants because the metals accompanying the copper have to be removed and recovered economically, and without adverse ecological effects.

Naturally with this unit the quantities of waste gas are low if it is working with an oxyfuel burner. Attachments of combustible material in the feed can be burnt so that their burning heat is used for processing if the burner is fed overstoichiometrically with oxygen. With a suitable charging device working continuously, a dosed addition of metal containing plastic material is possible; the burning energy set free can be a big part of the process heat giving more favourable costs. Moreover, bigger quantities of these materials could be used while keeping the air clean without any larger problems.

Working with such a unit, therefore, is also favourable for the concept of waste management of car wrecks. Car wrecks should largely be disassembled, whereby most of the recovered parts should be reusable. The disassembled plastic parts should at least be suitable for reuse.

Nevertheless, it is assumed that a combustible, metal-containing fraction will be obtained which should go to energetical and metallurgical utilization. With its offgas treatment system, the TBRC offers ideal conditions to use this fraction.

For the command of all process steps the offgas treatment system should be designed in that way as to allow total capture and follow-up treatment of the gases. According to Hüttenwerke Kayser's planning, this can be achieved in a reliable way.

The process offgases and offgases from a hood under which the charging operations are effected are conveyed to a combustion chamber for afterburning where a complete burn and destruction of detrimental halogen-containing compounds is achieved at high temperatures. A constantly high temperature is maintained in this chamber by means of an auxiliary burner. The heat content of the gases is used in a waste heat boiler, so that their temperature at the outlet is 450° C. By mixing the air drawn out of the casing it is possible to cool them to about 200° C. This shock-cooling prevents recombination of dioxines and furanes. After further cooling by a heat exchanger, the offgas is brought to the required temperature for cleaning with a cloth-filter. There, the metal-bearing dusts of the metallurgical processes are separated from the gases. They are led to reutilization for the recovery of zinc, lead, and tin.

By means of a fixed bed filter with active coke, installed after the cloth-filter, it will be possible to meet the future limits for dioxines and furanes of 0.1 ng toxicity equivalent.

Secondary raw materials for copper recovery are usually delivered in small lots, sampled and stored separately.

For continuous processes, mixtures have to be made at great expense, enabling process steps to be carried out uniformly over a long period of time. As it allows specially suitable processing of the material supplied, a smaller, high-efficient unit for individual charges can lead to better metallurgical results and, therefore, increase the profitability of the plant.

The TBRC which was introduced in steel metallurgy at the end of the 50's and proved only little appropriate there, was used since mid of the 70's first to recover lead from secondary raw material with great success. Then followed its application in tin and copper metallurgy for the recovery from ore concentrates and also from residues. It is used with great success precisely in case of widely varying operational conditions. Tests by Hüttenwerke Kayser have shown that the favourable metallurgical conditions described are to be expected. Thus, for the recovery of copper from secondary raw materials, a reactor is available which allows metallurgical processes to be carried out in a way better adapted to the input.

Therefore, as described above, the operation of a TBRC represents not only a metallurgical progress, but also a large step towards the improvement of the environmental situation.

Lead and copper recycling in the Boliden Kaldo

Lennart Hedlund
Boliden Contech AB, Skelleftehamn, Sweden

SUMMARY

In 1976 the Lead Kaldo Plant was started up at the Rönnskär Smelter of Boliden. The Lead Kaldo was developed to treat complex dusts from the copper smelter. Over the years the process was modified and production increased.

In the early 1980s the stocks of dusts were consumed, and other materials tested. This included leach residues, lead batteries and concentrates. Copper scrap with high organic content was also successfully smelted.

In 1978 a Copper Kaldo Plant was built for smelting of copper concentrates. It was found that large amounts of scrap could be included in the charge, due to good heat efficiency and mixing in the Kaldo

In 1989 the Copper Kaldo was closed down because of lack of concentrates and refinery capacity and the Lead Kaldo was modified for smelting of lead concentrates.

Significant for the Boliden Kaldo Processes are high productivity, low energy consumption, low emissions to environment and superior in-plant hygiene.

INTRODUCTION

The Rönnskär Smelter of Boliden Metall AB is situated in the north of Sweden, 800 km north of Stockholm. The operations began in 1930 with treating of the very complex ore from the Boliden mine. The development has led to treatment of complex concentrates with precious metals, lead, zinc clinker, sulphuric acid and liquid SO_2 as important by-products. Since 1945 lead also has been produced in a separate lead-circuit.

Copper Smelting

The copper plant consist of an electric furnace, fed by calcine from a fluid-bed roaster and dried secondary materials. The slag is treated for recovery of zinc, lead and copper in a slag-fuming furnace. The matte is converted in Pierce-Smith converters. The tankhous has a capacity of 105 000 tons a year.

The copper concentrates treated contain considerable amounts of lead and zinc and hence 8 000 tons of lead-bearing by-products are produced every year. In order to treat these very complex materials a Lead Kaldo Plant was erected in 1976.

Lead concentrate smelting

Smelting of lead concentrate started up in a separate lead-line 1945. The concentrate were smelted directly in an electric furnace and refined in kettles ref.2. The electric furnace was closed down 1989, due to environmental legislation. Today, lead concentrate are smelted in the Lead Kaldo Plant.

TREATMENT OF LEAD-BEARING DUSTS

Lead-bearing dusts from the copper converters and the clinker furnace were in the 1960s sold to other lead producers. Penalties for arsenic, antimony, cadmium and halides (table 1) soon made the sales unprofitable.

Table I. Compositions of materials treated in the Lead Kaldo Plant

	Dust from copper converters %	Dust from zinc clinker plant %	Residue from zinc clincer leaching %
Pb	43,1	50,0	48,9
Zn	11,1	14,7	5,0
As	3,3	7,4	2,2
Sb	0,22	0,43	0,95
Sn	0,81	1,21	1,96
Cu	0,6	0,1	0,12
S	9,9	6,4	9,2
Cl	0,61	1,46	0,06
F	0,022	0,7	0,1
Cd	0,73	0,42	

It was necessary to develop a process of our own to treat these materials.

In 1972 the first trials were conducted of the Mefos 5-ton Kaldo furnace in Luleå. The tests indicated that the Kaldo furnace (a top-blown rotary converter) was a suitable unit for the smelting of these materials.

Plant description

The dusts were stored in a separate building and, via daybin and chargebins, charged to the furnace by drag chains. Fluxes, iron and coke are charged from separate bins above the furnace.

The furnace is totally enclosed by a ventilated hood. Inside diameter of the furnace is 3,6 m and the length is 6,5 m. During operation the furnace inclination was first 22 degrees, later on changed to 28 degrees to give larger working volume. Maximum rotation speed is now 15 RPM.

Heat is supplied by an oxygen/oil burner in a watercooled lance.

The offgas hood is water cooled and reinforced by a basic castable.

For the gas cleaning a Venturi scrubber is used, operating at 2 000 mm pressure drop. Venturi water flows to a thickener for settling. The sludge is filtrated, and the filter cake dried in a oil heated drum dryer and then returned to the process.

Lead is cooled in ladles before pouring to kettles and cast into 2,5 ton blocks from the kettles. (Figure 1.)

After several test-campaigns the full scale process was designed as shown (figure 2).

Pelletized dust, granulated slag from the fuming furnace, soda, lime and dross from the lead kettles were charged batchwise to the furnace. Heat is supplied by an oxygen/oil burner. During the melting the lead sulphates decomposes to lead oxide and SO_2. After melting coke was charged continuously for the reduction of the lead oxide in the slag. When a lead content of less than 1 % was reached, the slag was skimmed into ladles and transferred to the fuming plant. The lead was collected in a kettle. When 60-90 tons of lead were collected in the kettle, the lead was pumped to the Kaldo furnace for refining. In pilot scale it was easy to get a selective oxidation of tin to a dry dross that was pulled out of the furnace before the oxidation was continued to produce a fluid arsenic-antimony dross.

Process modification

The time and labour consuming oxidizing refining method was discontinued quite soon. Instead iron was added to form a iron-arsenic-speiss. The fluid speiss was easy to skim off to a ladle and could be discarded after solidification. Instead of collecting lead in a kettle and pumping it back, it was collected in the furnace and refined after 2-3 heats, then tapped in a ladle.

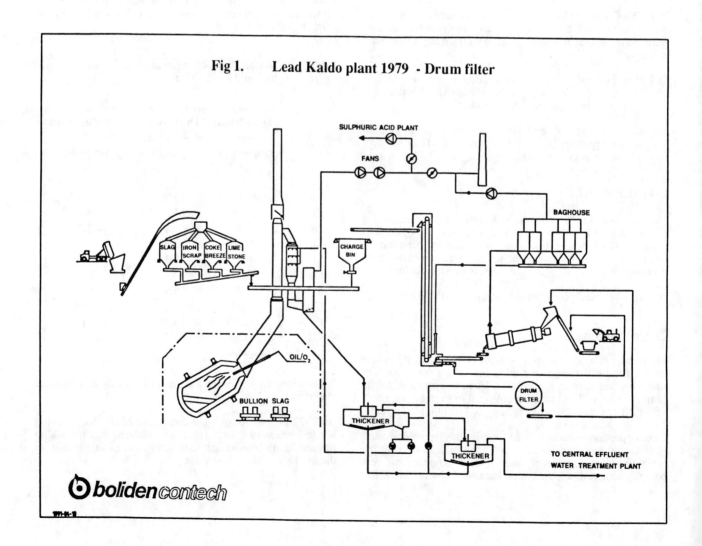

Fig 1.   Lead Kaldo plant 1979 - Drum filter

Fig 2. Flowsheets for processing of lead dust

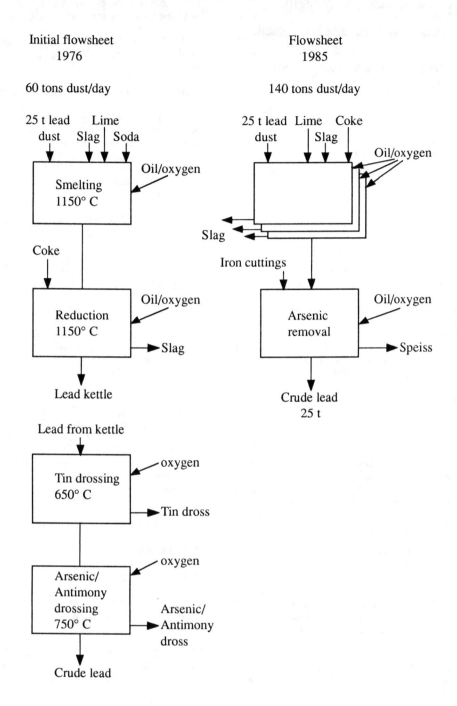

Initial flowsheet
1976

60 tons dust/day

25 t lead dust   Lime
Slag   Soda

Smelting
1150° C

Oil/oxygen

Coke

Reduction
1150° C

Oil/oxygen

Slag

Lead kettle

Lead from kettle

Tin drossing
650° C

oxygen

Tin dross

Arsenic/
Antimony
drossing
750° C

oxygen

Arsenic/
Antimony
dross

Crude lead

Flowsheet
1985

140 tons dust/day

25 t lead dust   Lime   Coke
Slag

Oil/oxygen

Slag

Iron cuttings

Arsenic
removal

Oil/oxygen

Speiss

Crude lead
25 t

Minor modifications of the slag composition made it possible to operate without the expensive soda. (Table II)

Table II. <u>**Products from the Lead Kaldo Plant.**</u>

	Slag	Crude Lead	Speiss
Pb	1,5	96-97	3,5
Zn	15		
Fe	22		50
As	0,5	<0,1	30
Cu	<0,1	<0,1	12
Sb	<0,1	0,7	
Bi	<0,1	0,3-1,5	
SiO_2	25		
CaO	23		

The pneumatic feed of coke for the reduction was unreliable, and batchwise charging of coke was introduced.

The most serious remaining problem was foaming slag. The reaction between coke and lead oxide evolves gas, and if the viscosity of the slag is too high, bubbles will be trapped in the slag and cause foaming. During melting the fayalite slag from the fuming plant oxidized to magnetite. This was avoided by adding the slag at the end of the melting period.

Now the process ran smoothly, except for the wet gas cleaning system.

Variations in the raw materials caused variations in the dust and thus the pH in Venturi water was unstable. At the end settling conditions in the thickener and properties of slurry were almost out of control. The rotary drumfilter could not take this, and large amounts of lead were lost with filtrate and overflow water. A change to press filter was a big improvement. It does not operate continuously, but it always gives a good separation and a good filter cake.

During mid -82 the refractory wear increased. This coincided with higher lead content in the dust treated. Tests were done to prereduce the charge to bring down the lead content of the slag. Quite surprising it was found out that all the coke could be added at the start of the charge. The reduction would start before melting, and when the melt was fluid, it was also reduced. This method was also a lot faster than the previous 2-step process.

The theory was that the melting should be oxidizing to avoid PbS-evaporation, which had been experience during the pilot plant tests. When one old truth was found to be false, others were to be tested. How should the burner be run? Oxidizing, neutral or reducing? The answer was unexpected: Oxidizing burner is best for reduction. The melting time became shorter and oil consumption lower without increased coke consumption.

The refractory wear was caused by the refining step. At the beginning sponge iron was used, but later it was exchanged for cheaper iron cuttings with only 70 % metallic iron, the rest oxide. The metal reacts very fast with the arsenic in lead, and forms a fluid speiss at temperatures below 1000°C. On the other hand the iron oxide, mixed with some remaining slag, needed very high temperatures to melt. This overheating was very harmful for the lining. First improvement was not to melt all the contents of the furnace. It was sufficient to add the iron and rotate the furnace while heating for a few minutes. The speiss could be skimmed and the unmolten slag was left in the furnace.

Later it was found out that the iron could be added together with the rest of the charge. Speiss was formed simultaneously with the lead, slag and speiss were tapped together.

In 1982 processing of leach residues from the leaching of zinc clinker was started. No major changes in the process were needed.

SMELTING OF LEAD CONCENTRATES

After some years of operation most of the stored lead dusts were consumed. Melting lead concentrates was a natural option. During the pilot plant tests at Mefos prior to the erection of the Lead Kaldo Plant, a process for smelting lead concentrates was demonstrated. In 1981 some full scale tests with concentrates were conducted and in 1982 several thousand tons were smelted in a bigger campaign.

In 1989 the plant was equipped with an offgas boiler and pneumatic concentrates feeders. (Figure 3.)

The old electric furnace plant was closed, and concentrate smelting moved to the Lead Kaldo. Smelting of lead dust was less profitable so the dusts were stored.

The process has two steps: melting and reduction. The melting takes place when dry concentrate is fed through a lance to a nozzle where it is mixed with oxygen and air. The heat of oxidation is in most cases sufficient to melt the concentrate and fluxes added. In order to get low sulphur content in the lead some lead must be oxidized resulting in about 30 % lead in the slag.

After "melting" the lead in the slag is reduced with coke breeze. The heat is provided by oxygen/oil burner. It was found out that highest melting capacity was achieved when concentrates were not molten prior to the reduction. Coke is added to the solid material at a temperature of 900-1000°C and reduction starts in solid phase. When the slag is fluid reduction is finished.

LEAD BATTERY SMELTING

Pilot plant tests in the 1970s indicated that lead batteries could be autogenously smelted in Kaldo. This was now to be tested in full scale in the Lead Kaldo plant.

Batteries and lead sulphates (to simulate circulating dust) were charged directly to the furnace. Oxygen was introduced via the burner lance.

First experience was that it was more difficult to burn modern plastic-case batteries than the ebonite-type of the 1970s. The plastic evaporated, and burned in the offgas hood.

On the other hand, the temperature inside the furnace was low, so dust losses also were low. When all parameters at last were optimized, it was found out that smelting and reduction took place simultaneously. Partial combustion of the plastics in the furnace created a very reducing atmosphere, so oxides and sulphates in the paste were reduced to lead. Almost no slag was formed, as only slag component was aluminium-silicate from some ebonite-type batteries still appearing in the scrap.

Only problem was that the ventilation filter at the Lead Kaldo Plant wasn't equipped with a spark arrester, so charging had to be very carefully done.

Fig 3.   Lead Kaldo Plant 1991 - Filter press

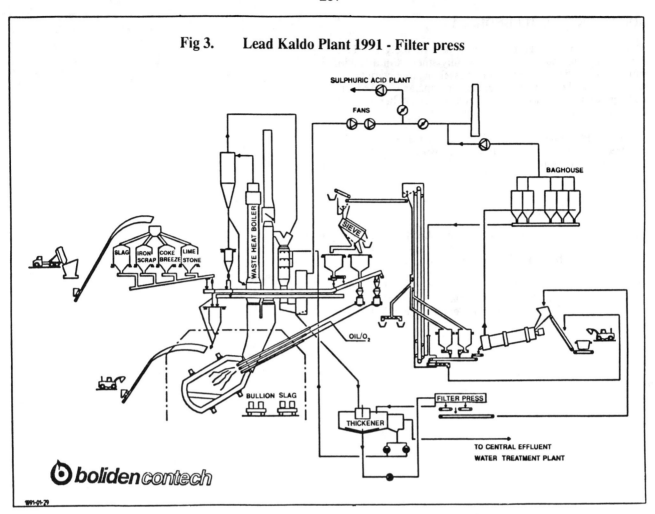

BURNING OF COPPER SCRAP

Boliden Metall has traditionally always processed all domestic low-grade copper scrap. Organic material has to be removed from cables and electronic scrap before further processing.

In the early 1970s the scrap was burned in a stationary furnace. However, the removal of organics was not perfect. In addition the dust and gas emissions were too large. The authorities demanded new equipment with lower emissions to be built.

In this situation it was decided to test burning of scrap in the Lead Kaldo Plant. The results were very encouraging. The rotation of the furnace makes a perfect combustion possible. The organics are burned with oxygen in the furnace and the gases afterburned with secondary air in the water cooled hood at a high temperature. The highly effective Venturi takes care of the chlorine and dust and thus the emissions are well below the legislated values.

The slag can be discharged, and the metal is transferred in molten state to the Pierce-Smith converters of the copper smelter.

In 1982 the furnace hood was extended to accommodate a fully enclosed skip hoist for charging of scrap to the furnace.

Burning of scrap was first tested in 1980 and today it contributes to about 100 days of the production year.

COPPER CONCENTRATE SMELTING

In 1978 the Copper Kaldo Plant was started up. Ideas to produce blister copper were abandoned, because of problems to keep heat balance and a reasonable converting speed.

Best concept was to smelt concentrates with oxygen to a rich (60-65 % Cu) matte. The matte would then be transferred to the Pierce-Smith converters for conversion to blister copper. Copper content of the slag was about 1 %, so the slag was discarded. It was found out that the excess of heat created when smelting with pure oxygen could be used to smelt large amounts (25 % of the charge) of reverts and scrap. Good mixing by rotation ensured that the metals dissolved in the matte. The cold charge could be of any size as long as it would go through the furnace mouth, and moisture content was not critical. Fine-grained concentrates or concentrates with high carbon content, that created problems when roasted and smelted in the electric furnace, could be used in the Kaldo with no problems.

Copper scrap smelting

Smelting of low-grade copper scrap is also possible in the Kaldo. Such a process would use an oxygen/fuel-burner for smelting and an air/oxygen lance for converting of the "black copper". The metallurgy is described in ref.2.

298

BOLIDEN KALDO TECHNOLOGY

The Kaldo processes have proven to be extremely flexible. Almost any material can be successfully smelted in a Kaldo furnace. Even in the very same vessel a large variety of materials can be treated in separate campaigns. Based on Boliden Kaldo technology, a secondary copper smelter is under construction in Italy, and a lead smelter in Iran.

A small Kaldo furnace for smelting of anode slime is included in a precious metals plant, which will be constructed by Boliden for KGHM in Poland.

References

1. H.I. Elvander; The Boliden Lead Process, Symposium on "Pyrometallurgical Processes in Non-Ferrous Metallurgy", Pittsburgh 1965.

2. D.S Flett et.al "Oxygen refining of black copper in a top blown rotary furnace". Copper 87 vol 4:Pyrometallurgy of Copper page 425-439.

Minor Metals

Purity and long-term stability of 8-hydroxyquinoline-based metal extractants

G. Haesebroek

MHO—a Division of ACEC–Union Minière, Research Department, Olen, Belgium

SUMMARY

8-Hydroxyquinoline derivatives such as Kelex ® 100 from Schering AG are well known commercial extractants for the recovery of Gallium from Bayer process aluminate solutions. Despite their high extraction power and outstanding selectivities, they suffer from a limited resistance to oxidation, which in the long run, leads to diminishing physical and chemical performances.

In 1985, MHO set up an extensive research program aimed at improving the long term stability of Kelex ® 100 when contacted with aqueous caustic solutions.

The study showed that an increase of the active component content of the extractant results in a substantial improvement of the resistance of Kelex ® 100 against oxidation, thus avoiding pronounced deleterious effects on the composition and performances of the Kelex ® 100 containing solvent when used in contact with concentrated aqueous caustic solutions in the presence of air.

Eventually, this work led to the development of a new Kelex ® 100 type (named Kelex ® 100 S) which is now commercially available.

INTRODUCTION

Industrial solvent extraction experience proved that the long term stability of the solvent phase is of primary importance to control the process economics. A poor stability increases reagent consumption. In most cases the degradation products also tends to accumulate in the organic phase which leads to depressed physical and chemical performances, decreased equipment capacity and eventually lower profitability.

This is why process engineers try to avoid using extractants with limited chemical stability.
However less stable products are sometimes used because of other unique properties. Commercial 8-Hydroxyquinoline derivatives are good examples of such extractants.

Their outstanding metal chelation properties have already been shown in the early seventies [1]. Their usefulness for the recovery of Gallium from Bayer process aluminate solutions was demonstrated in the late seventies [2] and industrial application followed soon.

Despite these encouraging results, the sensitivity of this kind of molecule to oxidation when contacted with strong caustic solutions limited further developments.

Inert gas blanketing and reduction of the double bond in the unsaturated side chain by hydrogenation both help to limit the oxidation of 8-Hydroxyquinoline derivatives [3].

Even with both these improvements, the oxidation leads to unfavourable ratios of active compound to oxidation products in the solvent phase after a few years of industrial use. This is due to the high solubility of the oxidation products in the organic solvent.

Commercial 8-Hydroxyquinoline derivatives are relatively impure products. Kelex ® 100 from Schering contains only some 75 vol % of active 8-Hydroxyquinoline derivatives [4].

MHO and Schering were both interested in increasing the stability of such compounds. A joint research program started in 1985. The key results of this study are reported in this paper.

EXPERIMENTAL

Long term stability tests

These tests consisted of stirring equal volumes of aqueous and organic phases together over a long period of time ranging from 1000 up to 3000 hours.

At different time intervals, solvent samples were taken and analysed for their total and active extractant concentrations (Gas chromatography (GC) analysis and Cu loading capacity measurement according to Schering 's standard procedures).

Temperature, volumes and gas flow rates above the mixture were monitored and controlled.

<u>Organic phase</u> (Solvent)

Two different solvents were studied. They contained 10 vol % of Kelex ® 100 (as delivered), 25 vol % of Isodecanol and 65 vol % of Escaid 110 (from EXXON Chemicals).

These solvents differed only in the purity of the Kelex ® 100 component. Two qualities prepared by Schering were tested. Their properties are given in table I.

Table I

Kelex ® 100 type (extractant)	Active material content [1]-vol%	Cu loading capacity (g Cu/10g Kelex)
77% [2]	77	0,79
90+% [3]	92	0,97

(1)   Determined by GC analysis.

(2)   Commercial quality at the beginning of the study.

(3)   Improved purity.

<u>Aqueous phase</u>

Aqueous solutions of NaOH, prepared from demineralized water and a technical grade 50% NaOH solution, with concentrations varying from 40 to 100 g NaOH per litre.

<u>Parameters</u>

Beside the purity of the Kelex ® 100 extractant, we also looked at the influence of the alkalinity of the caustic aqueous solution and of the oxygen content of the gaseous atmosphere above the aqueous/organic mixture on the stability of Kelex ® 100.

RESULTS

The stability of the extractants is measured by following the Cu loading capacity of the extractant-containing solvent as a function of the total contact time between aqueous and organic phases.

Both solvents were prepared with equal concentrations of extractant (Kelex ® 100). Due to the different purities of the extractants, the initial active component concentration of the considered solvents slightly differs (7,7 and 9,2 vol %). Therefore the active component concentration of both the solvents were normalised to 100 at time 0 in the following graphs (fig. 1 - 3). This must be taken into account in the further discussions.

<u>Purity of the extractant</u> (Kelex ® 100)

Figure 1 clearly shows the positive influence of a purification of the commercial Kelex ® 100 extractant on its stability in caustic conditions.

In the chosen operating conditions, the loss of activity of the reagent is almost twice as much with the common 77 % quality as it is with a 90+ % quality.

It must be emphasized that these tests have been developed in order to obtain substantial degradation rates. Obviously, a direct extrapolation to industrial practice is not straight forward. In continuous operations, the solvent phase is recycled at a low frequency (determined by solvent hold-up and flow rate).

Furthermore, in each of its cycles, the solvent is intimately contacted with the caustic solutions only for limited periods of time in the mixers of the sections where caustic

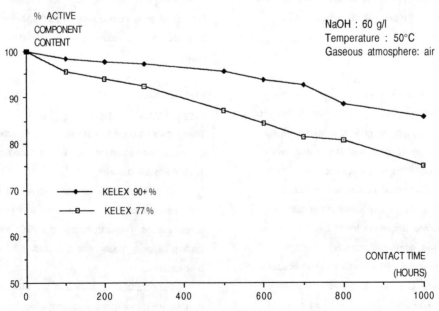

Figure 1 : Influence of the purity of Kelex ® 100 on its stability.

solutions are added. The degradation rate in industrial plants would therefore be much lower than the rate measured in our standard test.

Alkalinity of the caustic aqueous phase
Figure 2 shows that at concentrations to 60 g/l (1,5 N) sodium hydroxide, the oxidation of Kelex ® 100 (90+ %) remains

limited. Further increase of the alkalinity accelerates the degradation process. At 100 g/l (2,5 N) NaOH, high oxidation rates are measured. These results confirm that the purification of Kelex ® 100 S depresses its reactivity towards oxidizing agents. They also show that a careful choice of the operating conditions is important for maximizing the performances of Kelex ® 100.

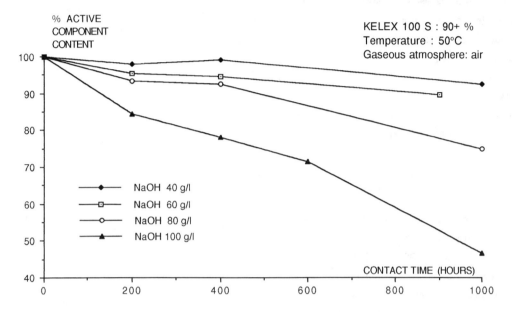

Figure 2 : Effect of the NaOH concentration on the stability of Kelex ® 100 90+ %.

Oxygen content of the gaseous atmosphere
Figure 3 shows that a reduction in the oxygen content of the gaseous atmosphere above the solvent aqueous phase mixture leads to a pronounced decrease of the oxidation rate. It is important to note that the oxidation rate is quite independent of

the O_2 concentration below 5 vol % O_2. This is very interesting, because it demonstrates that it is not necessary to make the equipment completely gastight to obtain satisfactory gains. This technique can therefore easily be used in industrial plants at low costs.

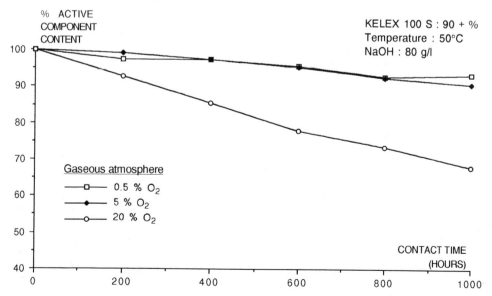

Figure 3 : Effect of the oxygen content of the gaseous atmosphere above the solvent/ aqueous caustic solution mixture on the stability of Kelex ® 100 90+ %.

304

CONCLUSIONS

The long term stability of 8-Hydroxyquinoline derivatives contacted with aqueous caustic solutions can be dramatically increased by increasing its purity. Schering AG was able to reduce the content of inactive components in its Kelex ® 100 reagent from ± 23 vol % down to ± 8 vol %. The purer product, now commercially available as Kelex ® 100 S, proved to have a better oxidation resistance in caustic solvent extraction conditions. Kelex ® 100 S has also other advantages [5] as a metal extractant. We conclude that it is a very valuable alternative for other commercially available 8-Hydroxyquinoline derivatives.

ACKNOWLEDGMENT

The author wishes to thank the MHO-Management for having given the permission to publish this paper and other people of MHO involved in this study and in the preparation of this paper.

REFERENCES

1. W.M. Budde & al - USP 3637711 - 25/1/72 - Beta-alkenyl substituted 8-Hydroxyquinolines.

2. A. Leveque, J. Helgorsky - ISEC '77 - Proceedings - Vol. 2 - The recovery of Gallium from Bayer Process Aluminate Solutions by Liquid-Liquid Extraction - p. 439 - 442.

3. D.L. Gefvert, H.J. Richards - WO 82/01369 - 29/4/82 Stabilization of substituted 8-Hydroxyquinoline Hydrometallurgical reagents.

4. G.P. Demopoulos, P.A. Distin - Hydrometallurgy 1983, 11 - On the structure and composition of Kelex ® 100 - p.389 - 396.

5. A. De Schepper, G. Haesebroek, A. Van Peteghem - USP 4942023 - 17/7/90 - MHO a division of ACEC-Union Minière - Metal extraction process with substituted 8-Hydroxyquinolines.

Improved technology for In, Ge and Ga recovery in an electrolytic zinc plant

Tian Runcang
Guangzhou Research Institute of Non-ferrous Metals, Guangzhou, China

SYNOPSIS

A study on stripping Ge in synergistic
extraction has been carried out aimed at
the inprovements in the author's previous work "New technology for In, Ge,
Ga recovery in an electrolytic zinc
plant." In the original technology, HF
was used as the stripping agent. Although
the stripping rate was high, a series of
difficulties occured in enrichment and
purification for the next step. For this
reason, stripping with liquid ammonia was
studied and was put into operation in
China in 1986. With this method, some
drawbacks in the original process were
overcome but other serious problems
appeared in production. In this paper,
emphasisis laid on a new stripping agent
AN-64 (supplied by Xingang Chemical Plant,
Guangzhou) and two alternative routes
for recovery of Ge from stripping liquid
have been proposed. Not only can the new
method overcome the drawbacks of the HF
stripping process, such as toxicity, high
corrosivity, difficulty in recovery of Ge
from the stripping liquid and low in
stripping rate using ammonia as stripping
agent, but it can retain all individual
advantages, such as the high stripping
rate in the HF stripping process and non-
toxic, low corrosivity, does not need a
second enrichment and easy to recover Ge
from the stripping liquid using ammonia.
It is a successful method and a significant improvement on the previous technology.

INTRODUCTION

Since the publication of "New technology
for In, Ge, Ga recovery in electrolytic
zinc plants"[1] and its industrial application in one of China's smelters in 1985,
the difficult problem of recovering In,
Ge, Ga by complete extraction in H_2SO_4
has been solved industrially. However,
it is still rather difficult to strip Ge
in the synergistic extraction system as
the extract formed by Ge and D2EHPA plus
YW100 (supplied by shenyang Mineral Processing Reagent Factory) is very stable.
In the previous study, two kinds of
stripping agents for Ge were used in production or pilot-scale.[1-2]

In the first method, 2N HF solution
was used as the stripping agent for Ge.
In spite of a high stripping rate, the
stripping solution is toxic, strongly,
corrosive containing a large amount of
fluorine ions and the flowsheet of the
process is lengthy, because a second
extraction enrichment has to be followed.
Moreover, the products could not be
obtained directly from the stripping
solution byhydrolysis. A certain amount
of Al salt had to be added to fix fluorine before a chlorinating distillation.
However, after the addition of Al salt,
the viscosity of the solution would
increase, causing difficulty in distillation.

In the second process, 10% $NH_3 \cdot H_2O$ +
1% $(NH_4)_2SO_4$ was used as the stripping
agent, this was non-toxic, fluorine-free
and employed a simple flowsheet (not
requiring a second extraction enrichment).
But the Ge stripping rate was as low as
56-60%, leading to an accumulation of Ge
in the organic phase. As a result, the
continuety of extraction in production
was impeded.

In order to improve and perfect the

new technology of recovering In, Ge, Ga in electrolytic zinc plants, special research for stripping Ge has been made, aiming at studying and finding out a new Ge-stripping agent. This agent should be high in stripping rate, non-toxic, low corrosivity, cheap and easy to get as well as no harmful effects on the next procedure. After selection and testing, AN-64 has been found to meet all the abovementioned requirements. Not only can it overcome the drawbacks that HF or $NH_3 \cdot H_2O + (NH_4)_2SO_4$ possess as stripping agents but it also retains their advantages.

EXPERIMENTAL

The experiment was conducted in a 60 ml glass separatory funnel. The Ge-stripping agent used was AN-64. Ge-rich organic phase was prepared by Ge-bearing feed (provided by smelter) extracting with 20% D_2EHPA + 1.25% YW100-kerosine. After mixing for a given time in an oscillator, it was settled to separate into two layers. The aqueous phase was run out to analyze its Ge concentration. According to analytical results, stripping ratio O/A and Ge concentration in Ge-rich organic phase, the stripping rate for Ge could be calculated. The benzfluorenone colorimetric method was used for Ge analysis.

Main factors affecting Ge stripping

(a) Effect of AN-64 concentration on Ge stripping
Testing conditions:
Ge-concn. in Ge-rich organic phase
	0.34g/l,
ratio	O:A=3
mixing time	5min.
temp.	room temp.

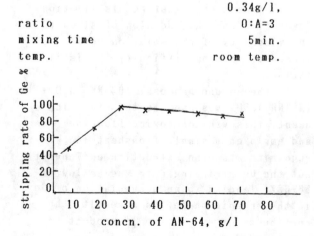

concn. of AN-64, g/l

Fig.1.   AN64 concentration dependency on Ge-stripping

It can be seen from Fig.1 that when AN-64 concentration is in the range of 25-35g/l, a peak value will appear in the curve of Ge-stripping. When the concentration is 27g/l, the stripping rate of Ge reaches its maximum -- 97%; when the concentration is less than 18g/l, the stripping rate is lower -- less than 75%; when the concentration is higher than 44g/l, not only does the stripping rate reduce but emulsification appears. Therefore, it is recommendable to take the concentration of AN-64 as 27g/l.

(b) Effect of stripping-equilibrium time on Ge-stripping
It is shown in Fig.2 that stripping equilibrium can be reached within 3 min. In order to ensure enough time for the reaction to reach equilibrium, the time can be taken as 5min.
Testing conditions:
Ge concn. in Ge-rich organic phase
	0.48g/l
ratio	O:A=2
AN-64 concn.	27g/l
temp.	room temp.

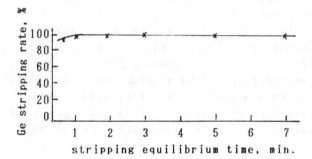

stripping equilibrium time, min.

Fig.2 Equilibrium time versus Ge-stripping rate

(c) Effect of phase ratio (O/A) on Ge-stripping
Testing conditions:
Ge concn. in Ge-rich organic phase
	0.47g/l
AN-64 concn.	27g/l
equilibrium time	5min.
temp.	room temp.

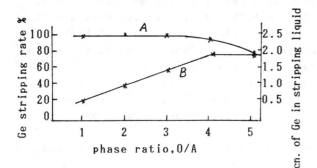

Fig.3 Relation between phase ratio and Ge-stripping
  A - stripping curve
  B - Ge-concn. in stripping liquid versus phase ratio

It is shown in Fig.4 that when the stripping ratio O:A=1 to 4, Ge-stripping rates are all as high as 98% or more and Ge-concn. in the stripping liquid is increased with the increase of ratio. However, when the phase ratio reaches over 4, a downward inflection point appears in Curve A, which indicates reduction of stripping rates; while Curve B tends to be flat, indicating saturation of Ge in the stripping liquid. Hence, when the concn. of Ge-rich organic phase is around 0.47g/l, it is suitable to choose 4 as phase ratio.
(d) Effect of Ge-concn. in Ge-rich organic phase on Ge-stripping
Test conditions:

AN-64 concn.   27g/l
equilibrium time   5min.
ratio   O:A=1 or 4
temp.   room temp.

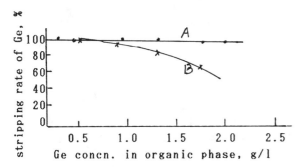

Fig.4 Relation between Ge concn. in Ge-rich organic phase and Ge-stripping rate
  A - ratio O:A=1
  B - ratio O:A=4

Results in Fig.4 indicates that at different phase ratios (the other stripping conditons are fixed), with the enhancement of Ge concentration in Ge-rich organic phase, Curve A falls gradually and Curve B drops more rapidly, illustrating that by means of adjusting the phase ratio or controlling Ge-concentration in organic phase, a high stripping rate can be achieved and a given Ge-concn. can be ensured. For example, in one Chinese smelter, the Ge concentration in the Ge-rich organic phase is usually around 0.42-0.52g/l, the phase ratio can be chosen as 4; whereas for some other smelters, Ge concentration in Ge-rich organic phase is higher(1.8-2.3g/l), then the phase ratio can be taken as 1.

Test on circulating the organic phase
After the organic phase is extracted and stripped, whether it can be recycled in use or not determines the feasibility of applying AN-64 as a new stripping agent. For this reason, circulation of the organic phase was examined in the test and the results are shown in Fig.5.
Testing conditions:
aqueous phase   real plant feed
Ge-concn.   0.083g/l
organic phase 20% D2EHPA + 1.25% YW100 -
 kerosine (supplement YW-100 each time)
ratio   O:A=1:5
equilibrium time   5min
temp.   room temp.

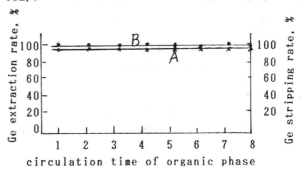

Fig. 5 Circulation time of organic phase versus Ge-extraction and Ge-stripping
  A - Ge extraction curve
  B - Ge stripping curve

Results from Fig.5 show that Ge-extraction rate for each circulation is in the range of 94 - 96%(average of 95%) and Ge-stripping rate is about 96 - 98% (average of 97%). After 11 times of extraction - stripping circulation, both extraction rate and stripping rate of Ge fluctuated within the allow range, not showing any tendency of declination.

Technological route for recovery of Ge from stripping solution

(a) Evaporation enrichment -- hydrolysis to precipitate Ge (process 1)

When AN-64 is used as the stripping agent, the Ge concentration in the stripping solution usually averages 1.88g/l. If Ge is precipitated by hydrolysis directly from this stripping solution, the hydrolysis rate will not be high since Ge concentration in the solution is not high enough. Generally, evaporation enrichment is applied to raise the Ge concentration in the stripping solution. But if evaporation causes any loss of Ge, this process should not be used. For this reason, special tests have been conducted to confirm that evaporation enrichment will not cause any Ge loss.[3]

Normally, Ge concentration in the enriched solution after evaporation is 9.3g/l. In order to recover Ge from the solution, 50% $NH_3 \cdot H_2O$ is added for neutralization, the end point of which is controlled between pH 8.8 - 9.1. Then the neutralized solution is filtered and the precipitate is washed. Finally, the product is dried and crude GeO_2 containing 25% or more Ge is obtained. Recovery of this process is 96.3% (from stripping solution to crude GeO_2).

(b) Evaporating enrichment -- chlorinating distillation (process 2)

In consideration of the economic effect, it is possible that some smelters may be interested in producing GeO_2 of high purity. So the second technological route for recovery of Ge has been studied, namely, the evaporating enrichment -- chlorinaing distillation. In this process, enriched solution, obtained by evaporation, and concentrated hydrochloric acid are added separately into the chlorinating distillation flask. After mixing, the total acidity of the solution is at the range of 8 - 10N. The solution is put on an electric hot plate to be heated, when the temperature reaches 78 - 82°C, Ge will be distilled out in the form of $GeCl_4$, which is then absorbed by water and hydrolyzed to GeO_2 of higher purity. When the temperature is suddenly raised to 100°C, the heating and distillation is stopped. The residual liquid from distillaion is analyzed, Ge concentration in which is less than 0.01g/l. Recovery of this method is over 97% (from stripping solution to GeO_2 of higher purity).

Analysis of economic and technical indexes

Comparison between AN-64 as stripping agent (is new process) and HF or $NH_3 \cdot H_2O$ + $(NH_4)_2SO_4$ as stripping agent is shown in Table 1.

The data listed in Table 1 show that when AN-64 is taken as stripping agent, the average Ge-stripping rate is 97%, approaching the index of HF -- 98%, but 37 - 41% higher than the rate with $NH_3 H_2O$ plus $(NH_4)_2SO_4$. Moreover, when AN-64 is applied, the unit consumption of raw materials is the lowest, 313.7 RMB/Kg Ge.

Table 1 Comparison of economic and technical indexes between the new process and the old one

process	stripping agent	Ge-stripping rate %	Ge hydrolysis	recovery % chlorinating distillation	unit consumption of raw materials RMB/Kg Ge
new	AN-64	97	94.8	97	313.7
old	HF	98	/*	97	415.2
	NH H O + (NH) SO	56-60	55	55	341.0

* Hydrolysis cannot be used here.

CONCLUSION

It is a significant improvement on the
original process to use AN-64 as the
stripping agent, which not only can
guarantee the feasibility of the new
process but also reduce the cost of GeO_2
production. Not only are the drawbacks
existing in each of the two stripping
agents used in the original process
overcome but it retains all the advantages of the other techniques. In addition, two technological routes for
recovering Ge from the stripping solution
are offered so that different smelters
can decide, depending on their own conditions.

References
1. Tian Runcang, International Mineral
Processing and Extractive Metallurgy.
Kunming, China, 1984, 10, P. 615.
2. Li Shuzhen, Tian Runcang, Chen
Xinglong, Rare Metals, 1988, 3, P. 299.
3. Chen Xinglong, Tian Runcang, Li
Shuzhen, Metallurgy of Non-ferrous Metals,
1990, 5, P. 9.

Production of molybdic trioxide by high-temperature oxidation of molybdenite in a cyclone reactor

Igor Wilkomirsky

Department of Metallurgical Engineering, University of Concepción, Concepción, Chile

ABSTRACT

A process to obtain high-purity molydic oxide by oxidizing molybdenite concentrates in a cyclone reactor is being developed. Dry molybdenite concentrate and oxygen are fed continuously into a cyclonic reactor where a fast oxidation takes place at 1200-1600ºC, a temperature at which gaseous MoO_3 is directly formed. The off gases, containing SO_2, oxygen and gaseous MoO_3, are quenched with cool air to 500-550ºC to condense the molybdic oxide, which is then recovered in high temperature filters. The gases are further washed in a venturi scrubber to recover the rhenium volatilized during the oxidation of molybdenite.

Impurities, such as copper and iron are collected in the lower section of the reactor in the form of a slag, which contains molybdenum as copper and iron molybdates. This slag is treated separately to recover the molybdenum.

The molybdic oxides produced, as a fine dust with particle size less than 5 microns, contain from 98 up to 99.5% MoO_3.

INTRODUCTION

Molybdenum trioxide has been produced by controlled roasting of molybdenite concentrates in multiple hearth furnaces,which,although a well developed technology, requires a high grade molybdenite concentrate since few impurities can be volatilized during roasting.

Fluidized-bed roasting has several advantages over multiple hearth furnace in terms of output per unit hearth area, control of the process and energy requirement, but it also requires a high-grade molybdenite.

In both roasting alternatives,temperature is kept below 650ºC to avoid sintering produced by MoO_3 crystal formation from gaseous phase and molten MoO_3.

In the new process that is being developed (1) advantage is taken of the high vapour pressure of the MoO_3 by performing the reaction above 1100 ºC with oxygen or oxygen-enriched air to generate the gaseous oxide,which is further condensed to obtain a high-grade product of MoO_3.

MOLYBENITE OXIDATION

Oxidation of molybdenite disulphide (molybdenite) above the boiling point of molybdic oxide (1152 ºC) generates gaseous MoO_3 according to the overall reaction.

$$MoS_{2(s)} + 3.5\ O_{2(g)} \rightarrow MoO_{3(g)} + 2\ SO_{2(g)}$$

This reaction is strongly exothermic, generating -136.8 kcal at 1200ºC.

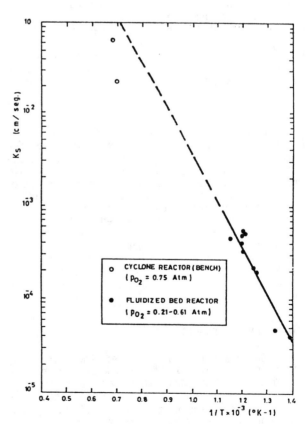

Figure 1: Reaction rate of molybdenite oxidation.

Molybdic oxide has one of the highest vapour pressures of the heavy metal oxides. Above the melting point of MoO_3 (795ºC), it can be estimated by the relationship (2)

$$Log\ p_{moO_3} = -\frac{11,820}{T} - 7.04\ log\ T + 30.44 (mm\ Hg)$$

By performing the reaction of oxidation above the boiling point of MoO_3, the formation of lower molybdenum oxides, such as MoO_2, Mo_4O_{11} and MoO_{26} can be minimized, since no diffusion control is established through the solid MoO_3 product layer and interparticles that could displace the thermodynamic equilibrium from MoO_3 to other lower oxides, as occurs in multiple hearth fur - naces, and to a lesser extent, in fluidized bed reactors.

The reaction rate constant follows approximately an exponential function with temperature, as can be observed in Figure 1. At 500 ºC it is about 1.1×10^{-4} cm/seg, increasing to 6.2×10^{-4} at 600 ºC and to 8×10^{-2} at 1100 ºC. This represents an oxidation rate over 120 times faster for a cyclone reactor operating at 1100ºC compared with a fluidized bed reactor at 600ºC(4).

At temperatures prevailing in multiple hearth furnaces and fluidized bed reactors it appears that the initial reaction control mechanism is the chemical reaction at the interface, followed by a mixed control, which is progressively replaced by a diffusional control through the MoO_3 layer growing outward.

For reactions temperatures above the melting point of MoO_3, the dense molten phase coats the unreacted core of MoS_2 decreasing even further the diffusion of oxygen to the reacting interface. As temperature increases over 800ºC, volatilization kinetics of MoO_3 becomes faster up to 1195ºC, where the boiling point is reached. At this temperature chemical reaction is extremely fast, and the control of the reaction becomes predominantly the oxygen transfer to the reacting surface of MoS_2.

The combined heat, mass and momentum transfer in cyclone or flash reactors plays a complex interaction at temperatures above 1195ºC. Temperature instability at particle level due to the unsteady heat generation and dissipation rates increases particle temperature up to 600-700 ºC above the nominal reaction temperature, the condition under which a combined heat and mass control is established. In a cyclone reactor an additional resistance mechanism appears also to exist due to the molten slag phase that can entrap reacting molybdenite particles. This phenomenon it is not present in a flash-type reactor.

EXPERIMENTAL RESULTS

Early attempts in 1975 at the University of Concepción (3) indicate that a cyclone reactor could be an alternative to conventional roasting technology. These initial studies and later developments showed that a high grade MoO_3 could be produced in this type of reactor.

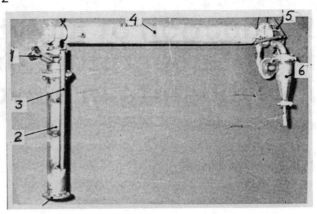

Figure 2: Bench scale cyclonic reactor
1. inlet pipe. 2. cyclonic reactor
3. oxygen inlet. 4. condenser. 5. low speed dust collector. 6. Product cyclone.

The experimental bench scale cyclone used in some of the early versions is shows in Figure 2. The reactor has a long barrel with a single tangential entry where a suspension of dry molybdenite and oxygen enriched air was injected. Additional inlets allow the introduction of oxygen or air along the cyclone barrel to complete the reaction. Under ideal conditions, a molten layer of $CuO-SiO_2$ slag should form, discharging through the apex of the cyclone while the gaseous MoO_3 together with the gaseous products (SO_2 and O_2) exit the reactor via the vortex to be cooled in a water jacketed condenser. Fine MoO_3 crystals growing along the condenser were continuously removed by means of a low rotating device. The solid MoO_3 was further separated from the gases in a conventional cyclone.

Due to the constrains imposed by the construction materials of the reactor, the upper reaction temperature was about 1150ºC, just below the boiling point of MoO_3.

Molybdenite concentrates used ranged from 0.5 up to 2.5% copper (Table I), although the concentrate used in bench test contains below 0.5% copper on average.

Table I. **Molybdenite Concentrates Used (Wt.%)**

Mo	S	Cu	Fe	Ca	Insol.
52-56	25-29	0.5-2.5	0.8-1.3	0.1-0.2	5.8

In operating the bench scale units, several problems arise, mainly in the cyclone reactor itself due to the formation of accretions and blockage of the feed inlet. The high viscosity of the molten phase and its slow movements along the reactor wall did not allow a continuous operation for more than a relatively short period of time.

Condensed MoO_3 as a very fine powder (-400 mesh) could contain variable amounts of impurities as well as an unreacted MoS_2. Purity varies from 85 up to + 99% of MoO_3, depending on several operational variables such as gas/solid ratio, gas velocity inside the reactor, reaction temperature and particle size. Figure 3 shows the influence of the reactor temperature on the purity of MoO_3 collected in the condenser.

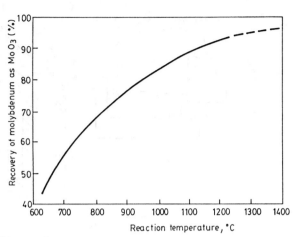

Figure 4: Recovery of molybdenum as MoO_3 from MoS_2 as a function of temperature.

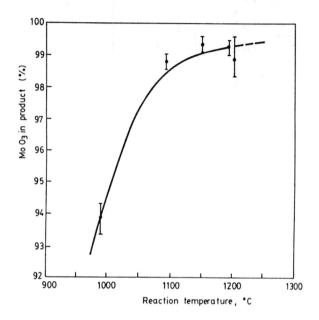

Figure 3: Influence of the reaction temperature on the purity of MoO_3 obtained.

At temperatures above 1100 ºC the MoO_3 formed contains from 96.3 up to 99.7%, copper sulphate and silica being the main contaminants.

Recovery of molybdenum as MoO_3 condensed product from molybenite showed a strong dependence on temperature . (Fig.4). At temperatures below about 800ºC, the semi-molten residue collected from the cyclone reactor contains large amounts of unreacted molybdenite, while at higher temperatures (above 900ºC), most of the molybdenum is found is the slag in the form of copper and iron molybdates.

Formation of $CuMoO_4$ increases rapidly with temperature. At 650 ºC in 5 minutes of reaction time, less than 40% of the copper present has been reacted with MoO_3, while at 900ºC reaction is completed in less than 2 minutes (Fig.5).

The inversion temperature for thermal decomposition of copper molybdate is about 1600 ºC, which can be obtained by keeping the molten bath at this temperature by means of burners. The slag formed can also be treated by acid leaching to recover the molybdenum by SX.

Based on the laboratory and bench scale results, a pilot unit is being commisioned to develop both the concept of the cyclone and flash oxidation of molybenite with oxygen.

The schematic view of the reactor and ancillary equipment is shown in Figures 6 and 7.

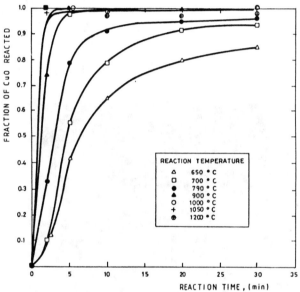

Figure 5: Formation of copper molybdates from CuO and MoO_3 in a static bed.

The reaction shaft enables the operation of either a cyclone reactor, as shown in the figure, or a flash vertical burner. Maximum operating temperature will exceed 1700 ºC.

Dry molybdenite is fed with oxygen enriched air. Technical oxygen(95%) is added to the cyclone, preheated to 200-300ºC. Steam can be added to enhanced the volatilization of molybdenum trioxide as $MoO_3 \cdot 3 H_2O$

Slag is collected in a crucible, which is kept molten by means of burners. Off gases containing the gaseous MoO_3 are quenched with air in a condenser and the fine MoO_3 is recovered in a sintered metal filter of 0.5 microns. Off gases from the filter at 380-400ºC are water scrubbed to solubilize the rhenium contained.

The cyclone reactor is refractory lined, with water-cooled walls.

The pilot unit is fully instrumented.

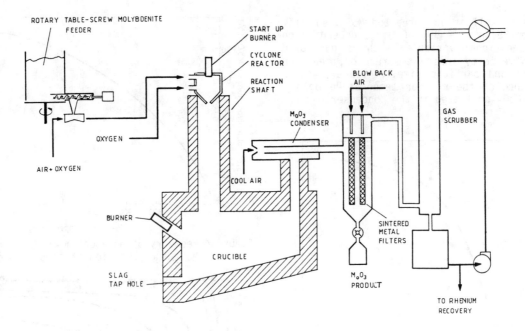

Figure 6: Schematic diagram of a cyclone/flash pilot unit
for high temperature oxidation of molybdenite.

Figure 7: Pilot plant set up.
1.Cyclone reactor. 2.reaction shaft. 3.crucible
4.Condenser. 5.Hot filter. 6.Pneumatic feeder.

BIBLIOGRAPHY

1. Wilkomirsky I.,Petit-Laurent H. and Reghezza A., "Production of high purity molydic oxide from molybdenite". Chilean Pat. 37.285, (28 Aug. 1990).

2. Kelly, K.K., U.S. Bureau of Mines Bull.383, (1935).

3. "Direct production of high purity molybdic oxide from molybdenite". University of Concepcion, Research Grant Nº 2.10.14 (1975).

4. Wilkomirsky I., Brimacombe J.K. and Watkinson P."Kinetics of the oxidation of molybdenite" Trans.IMM, Vol.86, Nº 3, p C16-C22,(1977).

ACKNOWLEDGEMENTS

The authors wish to thank H.Petit-Laurent, A.Reghezza,M.E. Alarcón, Dr.R.Padilla and J.Gacitua for their collaboration. Thanks are also extended to FONDECYT, Chile, for the Research Grant Nº 89-0723.

Trends in the development of processes for recovery of rare metals

A. A. Titov
I. F. Poletayev
V. A. Krokhin
A. A. Schelkonogov
State Institute for Rare Metals, Moscow, USSR

ABSTRACT

An effective method to process the complex rare earth titanoniobates is their chlorination with chlorine gas in presence of the carbon reductant.

The chlorination methods offer good opportunities for recovery of rare metals from secondary raw materials.

Solid-state methods for processing concentrates of zirconium and rare-earth metals offer a more rational use of mineral resources.

INTRODUCTION

The paper is concerned with some of the most recent trends in the practice of recovery of rare and rare-earth metals from primary and secondary raw materials.

Chlorination processes for the metallurgical recovery of rare metals (niobium, tantalum, titanium, zirconium) and their compounds have widespread use in the USSR. Chlorination methods are characterized by high parameters of recovery of metal values, by a comparatively simple separation of elements in the course of the process and by a possibility to carry out deep purification of chlorides and to obtain from them high-purity ductile metals and chemical compounds.

The chlorination methods offer good opportunities for recovery of rare metals from secondary raw materials.

Research that is under way in this field gives hope for further progress.

Solid-state methods for processing concentrates of rare metals have seen further development. When applied to the production of chemical compounds of zirconium and rare-earth metals, these methods offer a more rational use of mineral resources and greater environment safety.

Chlorination

Chlorination methods constitute the basis of the technologies for processing layered titanium-tantalum-niobium raw minerals containing rare-earth.[1-2] Chlorination is effected in the melt of chloride salts. In the process of chlorination, a finely divided concentrate in a mixture with coke forms a molten bath containing molten chlorides of sodium, calcium and rare-earth elements.[3]

A concentrate having the following chemical composition has been subjected to chlorination: TiO_2 - 38.6, Nb_2O_5 -8.8, rare-earth metals - 31.2, Ta_2O_5 - 0.5, Fe_2O_3 - 1.48, SiO_2 - 3.2, Al_2O_3 - 2.0, ThO_2 - 0.71 (wt %).

Chlorides of rare-earth metals, sodium and calcium formed in the process of chlorination build up in the melt and are periodically withdrawn from the system. Niobium, tantalum, titanium, aluminium and iron chlorides, carbon

315

dioxide gas and carbon monoxide are directed in the gaseous state to a condensation system for fractional condensation.

To purify chlorides of desired metal values from iron and aluminium use was made of the salt purification method based on the difference in the thermal stabilities of complex chlorides of niobium, tantalum, titanium, iron and aluminium.

As iron and aluminium chlorides react with potassium and sodium chlorides, thermally stable compounds are formed - for instance, $K(Na)Fe(Al)Cl_4^-$ and, as a result, a gaseous mixture of refractory metals chlorides is purified from iron aluminium.

In the process of chlorination of titanium niobates of rare-earth metals, apart from pentachloride, niobium oxychlorides are also formed. This fact is demonstrated by thermodynamic calculations and tahes place in practice.[2] At a temperature of $1000^{\circ}C$ the fraction of niobium oxychloride reaches 50%.

For subsequent separation and deep purification of Nb and Ta chlorides the need arises for conversion of Nb oxychloride to Nb pentachloride, the latter being a sufficiently convenient chemical compound to be amenable to purification by physical and chemical methods. This conversion process is based on a chemical reaction between Nb oxychloride and carbon tetrachloride.

Consequently, to produce Nb and Ta in the form of pentachlorides, use is made of three main chemical reactions, namely

$$2Nb_2O_5 + 6Cl_2 + 3C \longrightarrow 4NbOCl_3 + 3CO_2$$
$$Fe(Al)Cl_3 + KCl \longrightarrow KFe(Al)Cl_4$$
$$2NbOCl_3 + CCl_4 \longrightarrow 2NbCl_5 + CO_2$$

The chlorination approach to processing titanium niobates represents a more rational approach to the problem of Nb/Ta separation. The rectification method is particularly promising in this

respect, since this reactant-less waste-free process makes it possible to separate Nb pentachlorides from Ta pentachlorides and to carry out their subsequent full purification.[4]

Practical implementation of this separation method is limited by the highly corrosive working medium, which leads to considerable corrosion of metal values and makes impossible the use of tray-type rectification columns.

These technical difficulties were overcome by using packed-type rectification columns with a diameter of 300 mm. The main working surfaces of such columns are protected with a special coating.[4] The operating parameters of condensation are selected so as to enable the discharge of distillate (i.e. Nb and Ta pentachlorides) from the rectification column in the form of powder.

The essential novelty of the above-described process flow-sheet for separation of niobium and tantalum resides in the use of rectification, since hitherto extraction was used for this purpose.

Effective methods have been developed to enable Nb and Ta pentachlorides to be converted to such chemical compounds as find wide acceptance in engineering, such as niobium and tantalum oxides, lithium niobates and tantalates, lead niobates, lead magnesia niobates niobium and tantalum carbides.[5]

The majority of the above numerated chemical compounds are synthesized via intermediate compounds - chloroxoethylates[4] - in accordance with the following flowsheet:

Chloroxoethylates $\xrightarrow{\text{Heat treatment}}$ Oxides, niobates, tantalates, carbides.

The above products contain impurities in amounts ranging from 10^{-2} to 10^{-5}wt%.

Another field of application for chlorination techniques for processing titanoniobates of rare-earth metals is envisaged for the first stage of the leaching treatment of rare-earth metals with nitric acid. A solid residue resulting from the leaching treatment contains Ta, Nb and Ti. To recover these elements the residue is subjected to chlorination. This treatment offers the advantages of lower chlorine and reductant consumption rates and a lower discharge of chlorine containing effluents.

Of basic importance is the fact that the chlorination temperature has been reduced from 1000°C to 850°C, the lower process temperature giving considerably milder requirements for the material used for manufacturing the chlorination reactor. Its service life, although operating exposed to attack by corrosive molten media at 850°C, is prolonged by a factor of three.

Rare-earth metals are concentrated in nitric acid solutions from which they are separated and purified by extraction

Fig.1 illustrates a promising method for the recovery of Nb, Ta and rare-earth metals from are concentrates.

Processing secondary raw materials

Recovery of Nb and Ta from secondary raw materials constitutes an important source of these metals. Considerable quantities of Nb are contained in the wastes resulting from the production of electrolytic capacitors, superconducting alloys, hard-alloy products, niobates of alkali metals and alkaline-earth metals.

To process spent niobium capacitors a method has been developed based on a reaction between the material of spent capacitors and chlorine, in an aqueous medium, at a pH 6, in the presence of variable-valency metal chlorides. Impurities (iron, copper, nickel, manganese) are solubilized, while niobium and tantalum do not react with chlorine and remain in a solid residue. The process of chlorinating niobium-containing wastes is a multi stage technique, variable-valency iron ions serving as chlorine carriers. It is this fact that explains the possibility to chlorinate copper, nickel and iron at relatively low temperatures. The above process

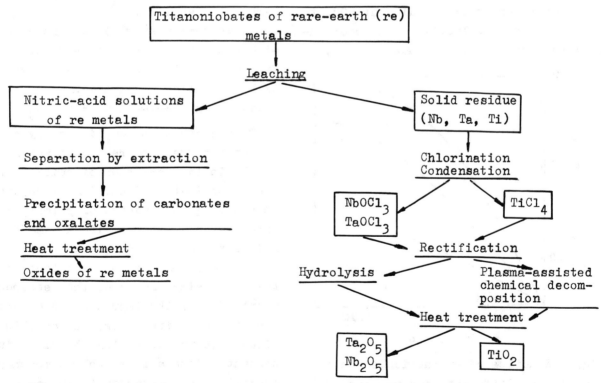

Fig.1. Flowsheet for processing titanoniobates of rare-earth elements

proceeds in accordance with the following equations:

$$2Fe + 6HCl \longrightarrow 2FeCl_3 + 3H_2$$

$$Ni(Cu) + 2FeCl_3 \longrightarrow Ni(Cu)Cl_2 + 2FeCl_2$$

$$2FeCl_2 + Cl_2 \longrightarrow 2FeCl_3$$

At the final stage, the Nb/Ta concentrate thus formed is subjected to additional hydrometallurgical treatment followed by chlorination, in which case the chlorination degree reaches 90% at 500°C. If chlorination is not preceded by chemical treatment, the chlorination degree at the same temperature does not exceed 50%. The lower chlorination degree under the same process conditions is explained by the presence of contaminants forming high-boiling chlorides. The latter create an obstacle to mass transfer of chlorine and lower the extent of chlorination.

The hydrochlorination method is also suitable for recovery of rare metals from discarded superconducting alloys containing 25% Nb, 25% Ti and 50% Cu. As such waste material is treated with chlorine gas in an aqueous medium at 60°C to 95°C, copper is solubilized in accordance with a kinetic curve shown in Fig.2.

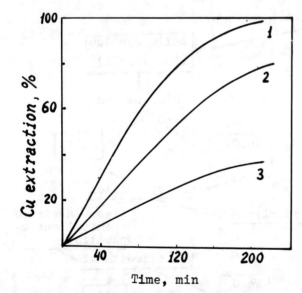

Fig.2. Kinetics of extraction of Cu to solution at different temperatures (°C): 1 - 95; 2 - 80; 3 - 65.

The Nb- and Ti-containing residue, after hydrochlorination and drying, is subjected to chlorination with chlorine gas to form $NbCl_5$ and $TiCl_4$. When chlorination is carried out at 700°C for 60 minutes, the degree of conversion is 97% by weight.

Fig.3 shows the sequential order in which process operations for Nb and Ta recovery from wastes are carried out. Significant possibilities for the recovery of rare metals from wastes are apparent.

Sintering Zr with CaCO_3

The Soviet zirconium industry is mainly based on processing zircon. Zircon concentrates containing at least 65% ZrO_2 are used for further processing by the Soviet industry. Latest research into primary decomposition of zircon with lime-salt mixtures made it possible to develop and partially to master on a commercial scale new processes offering superior economic parameters and a higher degree of protection against market instability.[6-8]

As is common knowledge, zircon reacts with calcium carbonate at a molar ZrO_2:$CaCO_3$ ratio of 1:1 in accordance with the following equation:

$$ZrSiO_4 + CaO \longrightarrow ZrO_2 + CaSiO_3$$

This process is complicated by a number of intermediate reactions which result in the formation of acid-soluble zircon silicate and calcium zirconate. In studying the process of primary decomposition of zircon with a lime/chloride mixture at a $CaCO_3$:ZrO_2 ratio of 0.1 - 0.6 it was established that with the increasing amount of the decomposing agent ($CaCO_3$) the degree of decomposition of zircon increases, and as this figure reaches 90 to 92% the zirconium content in the acid soluble compounds present in sinter products increases appreciably. A rise in the process temperature, a longer sintering time and

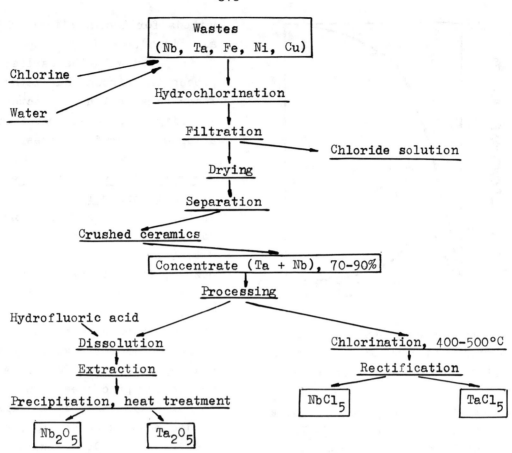

Fig.3. Schematic diagram of process flowsheet for recovery of Nb and Ta from wastes

the use of fluxing agents contribute to lower the zirconium content in the form of these compounds in the sinter products.

The final product (ZrO_2) of this reaction contains 0.2 - 0.3% CaO an amount that rapidly increases as the zircon decomposition degree exceeds 95%. These phenomena associated with the nonequilibrium nature of the solid-phase process make it impossible to achieve complete decomposition of zircon resulting in the formation of ZrO_2 alone. If zircon is fully decomposed, there is invariably formed a noticeable amount of zirconates, which ultimately leads to zirconium losses, since purification is normally conducted with acid solutions.

Moreover, when zircon is fully decomposed, the reaction product contains an elevated amount of CaO (up to 3%). When the zircon decomposition degree is 90 to 94%, it becomes possible to obtain

a product with minimal zirconium losses (1 to 1.5%) and with a minimal CaO impurity content (0.25 to 0.35%). In the latter case, ZrO_2 contains an admixture of silica (3 to 5%) in the form of zircon. This product is fairly usable for the production of refractories and abrasives.

Fig.4 shows a typical kinetic picture of the zircon decomposition process using $CaCO_3$ in the presence of $CaCl_2$. To purify ZrO_2 from the bulk of $CaSiO_3$, sintered zircon cakes ground to a grain size of 150 m were subjected to leaching using various acids.

It was established that the best purification results are attained by using two-stage acid leaching, namely: the first leaching stage is carried out cold with 5 - 6% HCl (8 to 10% HNO_3), while the second leaching stage is carried out under heating conditions with 10% HCl or 15 - 18% HNO_3.

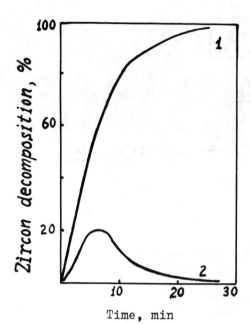

Time, min

Fig.4. Kinetics of process of zircon decomposition at 1250°C: 1 - zircon, 2 - part of dissolved zirconium.

In the first leaching stage the bulk of $CaSiO_3$ passes into solution, while in the second leaching stage the impurities Fe, Al and a portion of the Si are solubilized.

As a result, a product is obtained containing impurities in the following quantities, (wt %): 0.25-0.35 CaO; 0.06-0.10 Fe_2O_3; 0.08-0.15 Al_2O_3; 0.10--0.15 TiO_2; 3.00-5.00 SiO_2.

The second purification stage can be also carried out cold if use is made of mixtures of acids (HCl + HF) or (HNO_3 + + HF). The latter mixture makes it possible to obtain a by-product $Ca(NO_3)_2$ formed after neutralizing the solutions with lime.

ZrO_2 resulting from the first leaching stage was used as starting material for the production of other zirconium compounds via the process stage of producing basic zirconium sulphate.

Systematic research into the solubility of zirconium products in various acids made it possible, in general, to develop a process flowsheet for the production of zirconium compounds from zircon concentrate using lime/salt

methods for decomposition of raw material (Fig.5).

Compared with the earlier methods for processing zircon, our method features a lower consumption of reactants, a versatility of process operations and the absence of harmful wastes.

Recovery of rare-earth metals from phosphate minerals

In the USSR, rare-earth concentrates containing such minerals as yttrosynchisite (Y, Ca)FCO_3 $CaCO_3$, xenotime, monazite, yttrofluorite with a total content of rare-earth metals of up to 6-8%, are used as one of the sources of yttrium-group rare-earth metals (REM).

Methods for recovery, purification and separation of REM are based upon extraction processes. However, in order to solubilize REM, it is first necessary to decompose REM-bearing phosphate minerals (YPO_4, $CePO_4$). Fluorocarbonate and fluoride minerals are readily soluble in acids. Until recently, the only effective method for primary decomposition of phosphate concentrates was sintering of concentrates with soda ash.

In spite of its high efficiency, this method is often complicated by partial fusion of the feedstock. Therefore, depending on the specific composition of each concentrate it is always necessary to make a proper choice of the specific conditions for decomposition.

New research aimed at elucidating the mechanism of action of various calcium and sodium compounds upon REM-containing phosphates[9-10] made it possible to establish that this reaction with soda ash follows an exchange-type mechanism.

$$2LnPO_4 + 3Na_2CO_3 \longrightarrow Ln_2O_3 + 2Na_3PO_4 + 3CO_2$$

However, because of the instability of REM carbonates the reaction shifts fully to the right. For normal running of this process it is necessary that at the temperature of from 600 to 750°C soda ash be help in liquid state, i.e.

it is necessary to have a flux remaining intact throughout the entire process (e.g. a eutectic mixture of NaCl + + Na_2SO_4).

Novel reactants were devised capable of effectively decomposing REM phosphates.

te-based solid solution, both acid-soluble.

Hence, the above-described complex research into decomposition of fluorophosphate REM concentrates and into leaching of cakes became the basis for development of an up-to-date method for

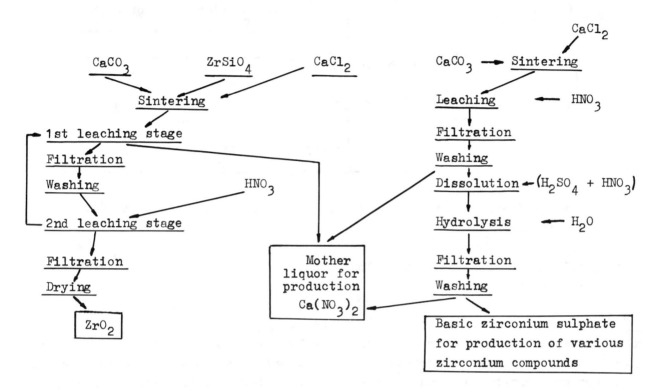

Fig.5. Generalized process flowsheet for production of zirconium compounds.

In the presence of Na_2SO_4 as flux, REM phosphates form within the temperature range 900 to 950°C addition products with Na_3PO_4, whereas within the range 800 to 900°C such addition products are formed with NaF. Such addition products are readily dissolved in nitric acid and therefore these sodium salts can be used as decomposing agents.

The most remarkable feature of this research is the fact that a possibility was created for an exchange process between REM phosphates and fluorite. At elevated temperatures (950 to 1000°C) in the presence of liquid phase the following reaction tahes place:
$$LnPO_4 + CaF_2 \text{---} Ca(Ln)F_2 + Ca_5(Ln)(PO_4)_3F$$
and as a result REM pass to a fluorite--based solid solution and a fluoroapati-

processing REM concentrates (Fig.6).

The method illustrated by Fig.6 is based upon the decomposition of concentrates at 700 to 800°C with a sulphate/soda mixture (which is a waste of aluminium production). If fluorite impurity is present in the concentrate in a sufficient quantity the mixture $Na_2SO_4 + Na_2CO_3$ can be used in a small amount, but in that case the process temperature should be raised up to 950 to 1000°C.

The main advantage offered by this method is that it affords a guaranteed effective decomposition regardless of the mineral composition of concentrates. Apart from economic advantages, the low consumption rate of reactants rules out the formation of sodium silicates

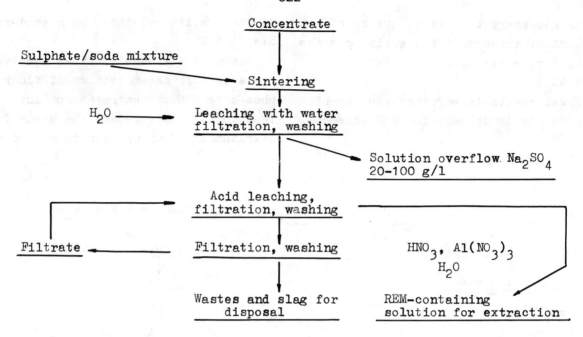

Fig.6. Schematic flowsheet of a process for decomposing fluorophosphate REM concentrates.

where by filtration of nitric-acid pulp slurries is facilitated.

CONCLUSION

The chlorination method for the recovery of rare metals features high process parameters and permits the production of various chemical compounds. The chlorination method constituted the basis of a number of techniques for the production of ultrapure substances for various technical purposes. The efficiency of this method was confirmed for recovery of niobium and tantalum from wastes. The metallothermic reduction of chlorides opens up a possibility for producing metals more economically.

In order to produce zirconium and REM chemicals, effective methods were developed for their recovery. These methods are based on hydrometallurgical processes and on processes of solid-phase interaction of reactants. Farther progress in this direction in processing REM -containing materials makes it possible to lower material losses and considerably to reduce discharge of harmful substances to the atmosphere.

References

1. Morosov I.S. Uses of Chlorine in Metallurgy of Rare and Non-Ferrous Metals. Moscow: Nauka Publishers, 1966. 250pp.
2. Nazarov Yu.N., Titov A.A. Tsvetniye Metally J., 1987. N 6, p.57-58.
3. Krokhin V.A., Solyakov S.P., Mal't-sev N.A. GIREDMET Transactions, vol.24, 1969. p.153-160.
4. Titov A.A., Karavainy A.I., Mal't-sev N.A. Tsvetniye Metally, no.3, 1988. p.56-58.
5. Titov A.A. Fundamentals of Processes for Producing Special-Purity Niobium Pentachlorides and Oxides, TSNIITSvetmet Research and Development Institute for Economic Studies and Information Services, 1986. p.48.
6. Poletayev I.F., Krasnenkova L.V., Smurova T.V. Tsvetniye Metally, vol.12, 1988. p.56-58.
7. Poletayev I.F., Krasnenkova L.V., Berestiuk A.S. Tsvetniye Metally, vol.9, 1983. p.67-68.
8. Poletayev I.F., Smurova T.V., Kras-

nenkova L.V. GIREDMET Transactions. Moscow, 1987. p.97-102.

9. Chuvilina E.L., Poletayev I.F., Zimina G.V. J.Inorganic Chemistry, vol.34, no.5. 1989. p.1274-1280.

10. Baryshnikov N.V., Chuvilina E.L., Poletayev I.F. J.Inorganic Chemistry, vol.28, no.5. 1983. p.1303-1308.

La lixiviation chlorurante d'alliages Fe–Si: une voie d'avenir dans la production du silicium?

F. Margarido
M. H. Bastos
Centro de Valorização de Recursos Minerais (CVRM), Departamento de Engenharia de Materials, Instituto Superior Técnico, Lisbon, Portugal

RESUME

L'obtention industrielle du silicium, au four eléctrique, est caracterisée par une grande consommation d'énergie, ce qui met à l'ordre du jour la recherche de nouveaux procédés, moins exigeants du point de vue énergétique.

On peut envisager le dévelopment de la voie hydrométallurgique de purification d'alliages Fe-Si comme un outil important pour la reconversion et diversification de l'industrie du silicium, dans la mesure où il devient possible d'opérer à des températures nettement inférieures (le degré de pureté des matières premières étant aussi moins élevé).

On a effectué des essais de lixiviation chlorurant, portant sur des alliages Fe-Si à 75% Si.

Les essais ont été établis suivant la méthodologie des plans factoriels ce qui a permis, outre l'étude des paramètres fondamentaux de la lixiviation, l'analyse de l'influence de la composition structurale des alliages.

L'étude de l'influence de la structure a été efectuée par comparaison des résultats obtenus en plusieurs plans factoriels d'essais identiques (du type 2^k, c-a-d à deux niveaux et k facteurs) utilisant des alliages structurellement différents.

Le traitement statistique des résultats a permis l'obtention des modèles mathématiques empiriques, qui ont mis en évidence la sélectivité de l'agent lixiviant utilisé (HCl ou HCl + $FeCl_3$) vis-à-vis des impuretés les plus significatives dans les alliages (Fe, Ca, Al).

Ainsi, la lixiviation du calcium et de l'aluminium est due à l'acide chlorydrique, l'action du chlorure ferrique étant toujours nocive.

Par contre le chlorure ferrique est l'agent lixiviant par excellence du fer.

Le rapport établi entre les rendements de la lixiviation des impuretés et la composition structurelle des alliages montre qu'on ne peut atteindre le degré de purification voulu, sauf si on parvient à atteindre une structure adéquate des alliages pendant le procédé industriel de fabrication.

INTRODUCTION

Le Si joue actuellement un rôle important, dans le contexte industriel, vu l'étendue de leurs applications: on l'utilise dans la préparation des alliages métalliques, pour l'industrie des silicones, la production de semi-conducteurs et des fibres optiques.

Malgré cette conjoncture globalement favorable, l'industrie du Silicium métallurgique (teneur égale où supérieure à 98%) traverse de grandes difficultés, qui sont en rapport avec les caractéristiques du procédé traditionnel de production en four eléctrique: une consommation d'énergie fort élevée et une grande exigence en ce qui concerne le degré de pureté des matières premières.

Le développement de la voie hydrométallurgique de purification d'alliages Fe-Si peut être envisagé comme un outil important pour convertir et diversifier l'industrie du silicium, dans la mesure ou il devient possible d'operer à des températures nettement inférieures et avec des matières premières ayant aussi un niveau de pureté moins elevé.

L'application de la méthodologie des plans factoriels aux essais de lixiviation chlorurante portant sur des alliages Fe-Si à 75% Si - par l'étude systématique de l'influence des paramètres fondamentaux de la lixiviation dans la solubilisation des principales impuretés (Fe, Ca, Al), en rapport avec la composition chimico-structurale des alliages - a pour but l'établissement des conditions d'optimisation de l'opération.

L'étude de l'influence de la composition chimico-structurale des alliages dans le rendement de solubilisation des impuretés est effectué par l'analyse comparée des résultats obtenus dans les mêmes conditions de lixiviation avec des alliages structurellement différents.

Les conclusions se rapportent au procédé industriel à suivre, dans le but de la production du silicium "métallique".

PROCÉDÉS DE PURIFICATION DU SILICIUM PAR LIXIVIATION ACIDE

La lixiviation est une des opérations utilisées, depuis longtemps, dans la purification du silicium "métalique". Les premiers travaux remontent aux années 30, époque où Becker et Tucker ont préparé du silicium à 99%[1].

Actuellement, on peut envisager deux niveaux dans l'utilisation des procédés hydrométallurgiques: ceux qui portent soit sur le Si avec une teneur $\geq$ 98%, soit sur des alliages de Fe-Si (avec une teneur en Si $\geq$ 60%). Dans le premier cas on obtiendra le Si avec une pureté de l'ordre de 99,9% (convenable pour des applications dans les domaines de l'énergie solaire et de l'électronique) tandis que dans le deuxième on ne parvient pas à dépasser 99,2% en Si (le produit étant utilisable dans l'industrie chimique).

PURIFICATION D'ALLIAGES Fe-Si

La lixiviation d'alliages Fe-Si a pour but l'élimination des impuretés présentes, par l'action des acides chlorhydrique, fluorhydrique, sulphurique où nitrique, suivant la réaction générale[2]:

$$Si,(xFe,yAl,zCa)_{(s)} + KH_mA_{(aq)} \rightarrow Si_{(s)} + xFeA_{\frac{\alpha}{x}(aq)} + yAlA_{\frac{\beta}{y}(aq)}$$

$$+ zCaA_{\frac{\gamma}{z}(aq)} + K_{\frac{m}{z}}H_{2(g)} \qquad (1)$$

Il n'y a pas beaucoup d'études sur la production du Si à partir d'alliages de Fe-Si. Dans la recherche bibliographique effectuée, on trouve seulement deux brevets norvégiens qui utilisent, d'ailleurs, la même technologie sur des alliages Fe-Si différents (respectivement avec 60% Si[3] et 90 à 94% Si - procédé Silgrain[4].

Dans les deux cas on utilise un réacteur continu, à contre-courant, la lessive constituant le flux ascendant. Le rendement de la lixiviation des impuretés dépend du type et concentration des phases intermétalliques, dans les alliages de Fe-Si, qui sont solubles.

PLAN DE TRAVAIL FACTORIEL. METHODE DE LA SURFACE DE REPONSE

Le plan factoriel d'essais établit un ensemble d'expériences, dont on prétend mesurer la réponse (variable dépendante), qui correspond à toutes les possibilités de combinaison des variables de contrôle (facteurs).

L'étude de tous les facteurs, simultanément, est basée sur la supposition qu'ils opèrent indépendamment[5,6], condition qui doit être confirmée par l'analyse des résultats.

La méthode de la surface de réponse (MSR)[7] permet aussi bien la détermination du modèle mathématique qui s'ajuste le mieux aux résultats obtenus - à travers des tests d'hypothèses de signification appropriés aux paramètres du modèle - comme de l'ensemble optimale des niveaux des facteurs expérimentaux, qui correpondent au maximum (ou minimum) de la valeur de la réponse. En tous cas, même si on ne peut pas accéder aux meilleures valeurs de la variable réponse, le MSR permet d'atteindre une connaissance plus profonde du procédé etudié.

On considère l'analyse factorielle du type 2^k

par l'établissement de plans factoriels à deux niveaux, dont les facteurs (en nombre de k) peuvent prendre seulement deux valeurs - correspondantes aux niveaux inférieur et supérieur (définis par rapport à un niveaux "standard", équidistant de ces valeurs extrêmes).

L'utilisation de variables codifiées de façon homogène, au lieu des variables originales, rend plus facile la construction des plans factoriels d'essais. La formule de codification qui définit les variables codifiées, x_i[8] est la suivante:

$$x_i = \frac{2X_i - (X_{iB} + X_{iA})}{X_{iA} - X_{iB}} \qquad i = 1,2,...,k, \quad (2)$$

étant

X_{iB} = niveau inférieur de la variable indépendante X_i

X_{iA} = niveau supérieur de la variable indépendante X_i

Dans cette formule on utilise la notation ± 1 pour les niveaux des variables codifiées, ie, le niveau supérieur (+1) et inférieur (-1).

On peut donc construire la matrice du plan factoriel d'essais qui représente l'ensemble des différentes possibilités de combinaison des niveaux des variables codifiées. Chaque ligne de la matrice spécifie la combinaison de niveaux des divers facteurs qu'il faut adopter, et maintenir fixes, pour chaque essai.

La réplique d'essais dans le niveau "standard" conduit à l'estimation de la variance de l'erreur expérimentale (V_e), ce qui correspond à une estimation de la dispersion du procédé d'observation.

L'ensemble des valeurs de la variable de réponse Y, obtenu expérimentalement et concernant les différentes combinaisons des facteurs, est ensuite analysé et développé statistiquement.

CARACTÉRISATION DE LA MATIÈRE PREMIÈRE

Analyse chimique

On fait figurer dans le Tableau I les résultats du traitement statistique correspondant à six analyses efféctuées sur deux alliages Fe-Si a 75% Si (lots 1 et 2), portant sur les éléments qu'on veut lixivier.

L'écart-type relatif, concernant les analyses effectuées, est dans les limites admises (2%), sauf pour le Fe (dans le lot 1) où cet écart est légèrement plus élevé (2,2%).

Analyse microstructurale

L'étude microstructurale[9] a été effectuée sur des échantillons des deux lots, ce qui a mis en évidence l'existence de deux phases principales - Si et α-Fe$_{1-x}$Si$_2$ - outre des phases CaAl$_2$Si$_{1,5}$, Al-Fe-Si, Ca-Al-Si-Fe et, sporadiquement, CaSi$_2$[2].

La phase Ca-Al-Si-Fe, designée par Caalsifer[2] est isotypique de la phase α-Fe$_{1-x}$Si$_2$, qui représente environ 50% de l'alliage Fe-Si à 75% Si.

Tableau I - Composition chimique partielle des alliages

Element \ Echantillon	Valeurs Statistiques	Lot 1	Lot 2
Fe	Moyenne (%)	21,44	18,30
	Ecart-type	0,46	0,24
	Ecart-type relatif (%)	2,2	1,3
Al	Moyenne (%)	1,75	1,74
	Ecart-type	0,04	0,02
	Ecart-type relatif (%)	2,0	1,0
Ca	Moyenne (%)	0,90	0,91
	Ecart-type	0,01	0,01
	Ecart-type relatif (%)	1,2	1,0

En ce qui concerne la phase β-FeSi$_2$, elle est uniquement décelable dans le lot 1.

PROCÉDÉ EXPÉRIMENTAL

Les lots des alliages Fe-Si utilisés pour les essais de lixiviation, après fragmentation et criblage préalables, avaient une gamme de granulométrie fixée entre 10-22,4 mm.

Ces lots ont été ensuite soumis à une lixiviation acide, avec de l'acide chlorhydrique et du chlorure ferrique.

Les réactions chimiques de solubilisation des principales impuretés sont les suivantes:

$$Me_{(s)} + nHCl_{(aq)} \rightarrow MeCl_{n(aq)} + \frac{n}{2} H_{2(g)} \qquad (3)$$

$$Me_{(s)} + nFeCl_{3(aq)} \rightarrow MeCl_{n(aq)} + nFeCl_{2(aq)} \qquad (4)$$

avec:

    n = 2 pour Me = Fe ou Ca
    n = 3 pour Me = Al

Pour rendre compte de l'action simultanée de l'acide chlorhydrique et du chlorure ferrique sur la solubilisation des impuretés en question on a admis comme réaction globale de la lixiviation, pour chaque élément, celle qui résulte de la somme algébrique des réactions considérées ci-dessus.

L'acide chlorhydrique, outre son action lixiviante, doit maintenir le pH du milieu suffisamment bas ($\leq$ 2) pour empêcher la précipitation des hydroxydes des cations résultant de la solubilisation des impuretés[4].

La durée des essais, thermostatisés à 102°C ($\pm$ 2°C), a été constante (= 4h)· Aucune agitation mécanique n'a été utilisée au cours des expériences, puisque le dégagement d'hydrogène a été considéré suffisant pour garantir l'agitation de la pulpe.

Les essais ont été établis, suivant la méthodologie de l'analyse factorielle, en deux plans à deux niveaux et trois facteurs: le rapport liquide/solide, la concentration d'HCl, la concentration de FeCl$_3$. 6 H$_2$O, respectivement designés par x_1, x_2 et x_3.

Aussi avons-nous dû effectuer les huit essais nécessaires, dans chaque plan, à l'étude de ces trois facteurs, et encore quatre essais dans le niveau "standard", pour estimer l'erreur expérimentale.

La matrice du plan factoriel d'essais, où on représente, les combinaisons des niveaux codifiés des facteurs étudiés, est indiquée dans le Tableau II.

Tableau II - Matrice des niveaux codifiés du plan factoriel d'essais - 2^3

Essais	Facteurs		
	x_1	x_2	x_3
1	-1	-1	-1
2	+1	-1	-1
3	-1	+1	-1
4	+1	+1	-1
5	-1	-1	+1
6	+1	-1	+1
7	-1	+1	+1
8	+1	+1	+1
9 à 12	0	0	0

Les niveaux "standard" des facteurs x_2 et x_3 correspondent aux concentrations d'HCl et FeCl$_3$ qui ont été calculées à partir de leur stoechiométrie respective dans les réactions globales de lixiviation des trois impuretés, et tenant compte de la composition chimique des alliages utilisés dans chaque plan factoriel.

Pour le 1er plan 2^3 l'indication des facteurs, et de leurs valeurs dans les niveaux respectifs pour les deux lots, est resumée au Tableau III.

Tableau III - Facteurs et niveaux respectifs dans
le 1er plan factoriel (2^3).

Lot	Facteurs	Niveau			Unités
		Inférieur (-1)	"Standard" (0)	Supérieur (+1)	
1	x_1	6:1	8:1	10:1	ml/g
	x_2	10	24	38	g/l
	x_3	0	170	340	g/l
2	x_1	6:1	8:1	10:1	ml/g
	x_2	8	22	36	g/l
	x_3	0	150	300	g/l

DISCUSSION DES RÉSULTATS

Les rendements d'extraction obtenus pour le Fe,le Ca et l'Al dans le lot 1 sont indiqués au Tableau IV.

Tableau IV - Rendements d'extraction (%) du Fe,
Ca et Al dans le lot 1(1er plan factoriel)

Essais	Rendements d'extraction (y)%		
	Fe	Ca	Al
1	0,19	53,50	17,49
2	0,36	55,83	15,71
3	0,43	58,50	22,37
4	0,78	55,28	24,14
5	2,34	52,17	15,17
6	3,26	49,72	14,86
7	4,69	51,00	13,89
8	10,72	61,81	20,43
9	2,14	56,56	15,09
10	2,60	57,11	18,97
11	3,39	54,56	16,91
12	2,60	59,00	20,23

L'analyse effectuée après estimation de la variance de l'erreur expérimentale montre que celle-ci est très grande pour le Fe et l'Al, tout en étant acceptable pour le Ca. Ce problème peut -être dû aux analyses chimiques, étant donné le grand facteur de dilution exigé par le haut niveau des concentrations de ces éléments dans les lessives finales. L'hétérogénéité structurale de l'alliage du lot 1 peut aussi contribuer à l'erreur expérimentale constatée.

L'estimation des coefficients des polynômes de régression[2] a été effectuée avec un programme de régression linéaire multiple qui détermine, par l'application du test des hypothèses de signification (Test t-Student, Test Fisher-Snedecor) les effets à conserver dans le modèle de régression des résultats expérimentaux, pour un certain niveau de signification α préalablement choisi.

L'analyse de ces effets montre que, pour le Fe et l'Al, tous les effets principaux et d'interaction des facteurs sont très significatifs (α=0,01), ce qui justifie qu'ils soient considérés dans le modèle de régression.

Par contre, pour le Ca - et dans da gamme retenue pour la variation des valeurs des facteurs -

- aucun effet n'est significatif à moins de l'erreur expérimentale, ce qui veut dire qu'on ne saura pas distinguer la variabilité de la réponse , entre niveaux,de celle qui est dûe à l'erreur expérimentale. C'est pourquoi il n'y a pas lieu de retenir le modèle de régression.

Les modèles de régression qui correspondent au Fe et Al sont les suivants:

$$Y_{Fe} = 2,828 + 0,934x_1 + 1,309x_2 + 2,406x_3 + 0,661x_1x_2 + 0,804x_1x_3 + 1,144x_2x_3 + 0,616x_1x_2x_3 \quad (5)$$

$$Y_{Al} = 17,984 + 0,778x_1 + 2,2x_2 - 1,92x_3 + 1,3x_1x_2 + 0,78x_1x_3 - 1,128x_2x_3 + 0,413x_1x_2x_3 \quad (6)$$

Pour tester le niveau de correspondance entre ces modèles polynômials du troisième ordre et la variabilité observée de la réponse,on a réalisé des tests d'homogénéité des variances de l'erreur expérimentale et des résidus. On a verifié que les modèles mathématiques proposés pour le Fe et l'Al présentent un niveau de signification de 5%.

Le niveau d'ajustement obtenu pour ces modèles, à travers le coeficient de corrélation multiple (R^2), est proche de 100% .

On résume, au Tableau V, les rendements d'extraction obtenus pour le Fe, le Ca et l'Al dans le lot 2.

L'erreur expérimentale estimée, dans le plan d'essais effectués sur le lot 2, est très grande pour les trois impuretés, ce qui peut être justifié par l'ordre de grandeur du facteur de dilution utilisé dans les analyses chimiques. Il faut signaler que le lot 2 ne présente pas l'hétérogénéité structurale verifiée dans le lot 1.

Les résultats de l'estimation des coefficients des facteurs étudiés - et le niveau respectif de signification - montrent que, pour le Fe, les effets des facteurs x_1, x_3 et ceux correspondant à leur interaction sont les seuls significatifs.

Pour le Ca, aucun effet principal, ou d'interaction des facteurs, n'est significatif dans l'intervalle de variation étudié.

En ce qui concerne l'Al, seul le facteur x_2

a un effet significatif.

Tableau V - Rendements d'extraction (%) du Fe, Ca et Al,dans le lot 2(1^{er}plan factoriel)

Essais	Rendements d'extraction (y)%		
	Fe	Ca	Al
1	0,22	48,21	14,31
2	0,43	54,53	16,24
3	0,46	62,72	21,64
4	0,69	60,03	23,13
5	2,29	47,23	14,57
6	7,63	48,49	15,37
7	4,81	50,36	16,81
8	8,77	58,52	17,96
9	3,97	54,29	16,32
10	3,97	63,08	21,61
11	4,58	67,03	23,22

Une fois les modèles de régression élaborés pour les trois impuretés, et réalisés les tests d'homogénéité des variances de l'erreur expérimentale et des résidus respectifs, on peut conclure que, pour un niveau de signification de 5%, seul le modèle du Fe est ajusté, son degré d'ajustement étant de 94%.

Par contre, les modèles proposés pour le Ca et l'Al sont inacceptables comme représentation des résultats expérimentaux, étant donné que leurs niveaux d'ajustement sont, respectivement, de 55% et 45%.

Le modèle de régression du Fe est le suivant:

$$YFe = 3,274 + 1,218x_1 + 2,713x_3 + 1,108x_1x_3 \quad (7)$$

Les résultats de l'application du 1er plan factoriel montrent des rendements de solubilisation insuffisants pour les impuretés.

Dans le but de procéder à l'optimisation globale de la lixiviation de ces impuretés - Fe, Ca, Al - on a effectué l'analyse des coefficients qui affectent les facteurs x_i, à fin de déterminer le(s) effet(s) prédominant(s) et le sens de leur variation. On a constaté que les facteurs significatifs, et donc retenus dans les modèles explicatifs de la variation du rendement d'extraction des impuretés, ne sont pas les mêmes pour les deux alliages étudiés. Toutefois, l'analyse soigneuse de tous les coefficients qui affectent ces facteurs mène à la conclusion que le sens de leur variation est le même, dans les lots considérés.

L'analyse systématique de l'influence des variables sur le rendement de solubilisation des impuretés montre, pour les différents modèles mathématiques élaborés, que:

1 - Pour la variable x_1 (rapport L/S)

L'accroissement du niveau de ce facteur a un effet significatif sur les rendements d'extraction du Fe, surtout quand la concentration de FeCl$_3$ est au niveau supérieur (figs. 1 et 2). Dans l'alliage 1, cet effet est plus fort quand l'HCl est au niveau supérieur (fig. 1.a).

Pour les rendements d'extraction de l'Al, obtenus dans l'alliage 1, l'accroissement de L/S est nuisible si les concentrations d'HCl et de FeCl$_3$

sont au niveau inférieur, étant très favorable quand ces facteurs sont au niveau supérieur(fig. 3).

Dans l'alliage 2, l'obtention de modèles concernant la lixiviation de l'Al et du Ca n'a pas été possible, situation décrite auparavant, et qui correspond aussi au cas du Ca, dans l'alliage 1.

2 - Pour la variable x_2 (concentration d'HCl)

L'accroissement de la concentration d'HCl favorise, toujours, le rendement d'extraction du Fe (fig. 4) et, considéré séparément, celui de l'Al (fig. 3). Toutefois, si on prend la concentration de FeCl$_3$ au niveau supérieur, et la rélation L/S au niveau inférieur, la solubilisation de l'Al sera contrariée par l'augmentation de la [HCl].

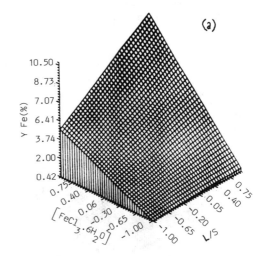

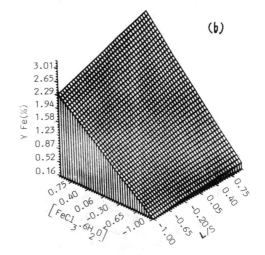

Fig. 1 - Surface de réponse pour le Fe, modèle 5-1^{er}plan factoriel 2^3, alliage 1; a) x_2 est au niveau +1; b) x_2 est au niveau -1.

Même sans modèle on a verifié un effet favorable de [HCl](1) dans la lixiviation du Ca.

---

(1) Conclusion obtenue par l'analyse des coefficients qui affectent les facteurs.

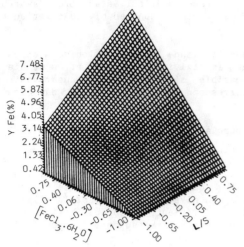

Fig. 2 - Surface de réponse pour le Fe, modèle 7-
- 1er plan factoriel, alliage 2.

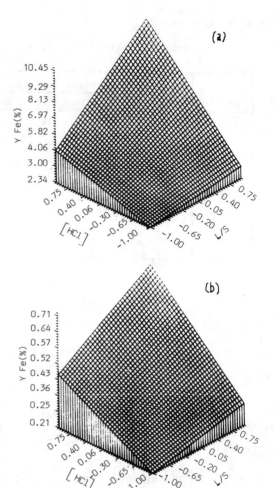

Fig. 4 - Surface de réponse pour le Fe, modèle 5
- 1er plan factoriel 2³, alliage 1;
a) x_3 est au niveau +1; b) x_3 est au
niveau -1.

3 - Pour la variable x_3 (concentration de
$FeCl_3$. $6H_2O$)

L'accroissement de la concentration de
$FeCl_3$. $6H_2O$ agit favorablement sur la solubilisation du Fe (fig. 1 et 2), et négative-
ment sur celles de l'Al et du Ca[1].

Pour l'élaboration du deuxième plan factoriel
d'essais, et tenant compte de l'effet contradic-
toire de $[FeCl_3]$ dans la lixiviation globale des
impuretés, on a maintenu les mêmes valeurs pour
les niveaux des facteurs x_1 (rapport L/S) et x_3
(concentration de chlorure ferrique), tandis que
les niveaux de la concentration d'HCl (x_2) ont
été augmentés de façon à ce que l'actuel niveau
inférieur coincide avec le niveau supérieur du
premier plan.

Les valeurs adoptées pour chaque niveau des
facteurs, sont indiqués au Tableau VI.

Les résultats obtenus pour la lixiviation de
l'alliage 1 sont résumés au Tableau VII.

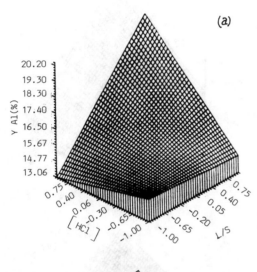

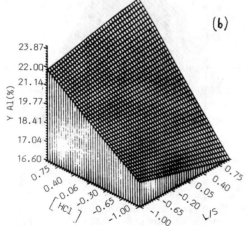

Fig. 3 - Surface de réponse pour l'Al, modèle 6 -
1er plan factoriel 2³, alliage 1;
a) x_3 est au niveau +1; b) x_3 est au niveau -1.

---

(1) Conclusion obtenue par l'analyse des coefficients qui affectent les facteurs.

Tableau VI - Niveaux des facteurs dans le 2ème plan
factoriel, utilisés pour les alliages 1 et 2.

Lot	Facteurs	Niveau			Unités
		Inférieur (-1)	"Standard" (0)	Supérieur (+1)	
1	x_1	6:1	8:1	10:1	ml/g
	x_2	38	79	120	g/l
	x_3	0	170	340	g/l
2	x_1	6:1	8:1	10:1	ml/g
	x_2	36	73	110	g/l
	x_3	0	150	300	g/l

Tableau VII - Rendements d'extraction (%) du Fe,
Ca et Al, pour le lot 1 (2ème plan factoriel).

Essais	Rendements d'extraction (y)%		
	Fe	Ca	Al
1	0,43	58,50	22,37
2	0,78	55,28	24,14
3	4,30	56,92	46,42
4	5,05	62,64	46,09
5	4,69	51,00	13,89
6	10,72	61,81	20,43
7	7,81	54,92	20,97
8	10,42	49,31	21,09
9	4,69	54,56	15,82

L'estimation des nouveaux coefficients des facteurs, et l'analyse respective de signification, montrent que: l'effet du facteur x_3 est très significatif pour la lixiviation des trois impuretés, tandis que le facteur x_2 l'est seulement pour l'Al; l'interaction des facteurs x_2 et x_3 est significative pour la lixiviation du Ca et de l'Al, alors que l'interaction des trois facteurs l'est pour le Ca.

Les tests d'homogénéité des variances de l'erreur expérimentale et des résidus, dans les modèles élaborés pour les trois impuretés, ont révélé qu'ils sont - pour le Ca et l'Al - très significatifs ($\alpha=0,01$) et ajustés (le niveau d'ajustement est de 88%). Le modèle pour le Fe, bien que significatif, n'est pas ajusté ($R^2= 62\%$): la variation globale enregistrée n'est expliquée que dans une faible mesure.

Les modèles de régression qui décrivent le rendement d'extraction du Ca et de l'Al sont, pour l'alliage 1, les suivants:

$$YCa = 56,104 - 2,038x_3 - 1,795x_2x_3 - 3,170x_1x_2x_3 \qquad (8)$$

$$YAl = 25,691 + 6,718x_2 - 7,830x_3 - 4,783x_2x_3 \qquad (9)$$

Les résultats des essais de lixiviation de l'alliage 2 sont repris dans le Tableau VIII.

L'analyse de signification des facteurs montre que, pour le Fe, les facteurs x_1, x_2, x_3 et l'interaction des facteurs x_1, x_3 et x_2, x_3 sont très significatifs tandis que pour le Ca c'est le facteur x_3 et pour l'Al les facteurs x_2, x_3 et leur interaction.

L'ajustement effectué pour les modèles élaborés montre qu'ils sont très significatifs et ajustés ($R^2 = 99\%$) pour le Fe et l'Al, tandis que pour le Ca il est significatif mais non ajusté ($R^2=75\%$).

Tableau VIII - Rendements d'extraction (%) du Fe,
Ca et Al dans le lot 2 (2ème plan factoriel)

Essais	Rendements d'extraction(y)%		
	Fe	Ca	Al
1	0,46	62,72	21,64
2	0,69	60,03	23,13
3	4,58	63,79	50,41
4	5,34	64,01	51,75
5	4,81	50,36	16,81
6	8,77	58,52	17,96
7	5,61	47,64	19,39
8	9,16	49,18	15,80
9	3,97	58,24	20,72

Les modèles de régression du Fe et de l'Al sont les suivants:

$$YFe = 4,821 + 1,063x_1 + 1,245x_2 + 2,160x_3 +$$
$$+ 0,815x_1x_3 - 0,948x_2x_3 \qquad (10)$$

$$YAl = 26,401 + 7,226x_2 - 9,621x_3 - 7,121x_2x_3 \qquad (11)$$

L'analyse de l'influence des variables des modèles mathématiques, dans le rendement de solubilisation des impuretés, met en évidence les effets suivants:

1 - Pour la variable x_1 (rapport L/S)

L'accroissement du niveau de ce facteur est très significatif pour les rendements d'extraction du Fe (alliage 2), quand le facteur x_3 ($[FeCl_3]$)est au niveau supérieur; n'est pas significatif si celui-ci se trouve au niveau inférieur (fig. 5). L'augmentation du niveau de ce facteur est encore favorable à la lixiviation du Ca, quand les niveaux des facteurs x_2 et x_3 sont opposés (fig.6).

2 - Pour la variable x_2 (concentration d'HCl)

L'accroissement de la concentration d'HCl a un effet très significatif sur le rendement d'extraction du Fe (alliage 2) quand le facteur x_3 est au niveau inférieur, et peu significatif si x_3 est au niveau supérieur (fig. 5). Il est favorable à la lixiviation du Ca et de l'Al si le facteur x₃ est au niveau inférieur (figs. 6 à 8).

concentration d'HCl. En effet, la tendance globale de l'accroissement de la concentration de FeCl₃ est de contrarier la solubilisation du Ca, ce qui n'est pas le cas dans les essais 4 et 8 (dans le 1ᵉʳ plan factoriel) et 2 et 6(dans le 2ème plan).

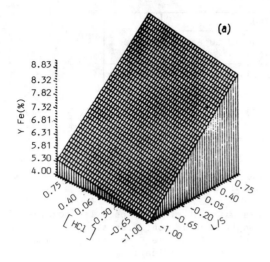

Tableau IX - Synthèse de la matrice des niveaux codifiés(1)

Essais : 1 → 8: L/S ↑ ; HCl↑ ; FeCl₃ ↑

5 → 4: L/S ↑ ; HCl↑ ; FeCl₃ ↓

7 → 2: L/S ↑ ; HCl↓ ; FeCl₃ ↓

3 → 6: L/S ↑ ; HCl↓ ; FeCl₃ ↑

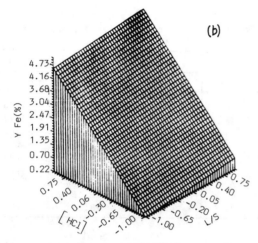

Fig. 5 - Surface de réponse pour le Fe, modèle
10 - 2ème plan factoriel 2³, alliage 2;
a) x₃ est au niveau +1; b) x₃ est au
niveau -1.

3 - Pour la variable x₃ (concentration de
FeCl₃ . 6H₂O)

Le chlorure ferrique favorise la lixiviation du Fe (fig. 5) et est nuisible pour le Ca et l'Al.

Outre le traitement statistique des résultats, qui à conduit à l'élaboration des modèles, on a fait aussi une représentation graphique directe, qui permet une interprétation plus immédiate. Ceci est une méthode plus expéditive et qui a l'avantage de permettre la représentation graphique simultanée de différents plans factoriels, élaborés à partir de la même matrice de niveaux codifiés(Tableau IX), comme on peut voir dans les figures 9 et 10.

L'analyse des figures montre une tendance d'évolution des effets des facteurs étudiés, quand on compare les plans factoriels entre eux, semblable pour chaque impureté, dans les deux alliages.

La seule exception apparait dans la lixiviation du Ca, dans l'alliage 1, quand on analyse l'effet du FeCl₃, en conservant le niveau supérieur de L/S pour n'importe quelle valeur de la

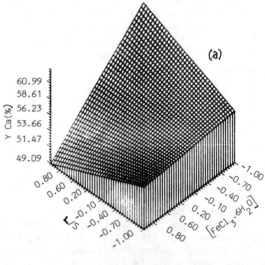

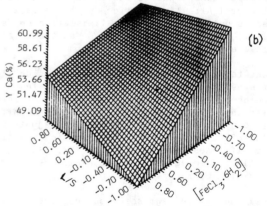

Fig. 6 - Surface de réponse pour le Ca, modèle8-
- 2ème plan factoriel 2³, alliage 1 ;
a) x₂ est au niveau +1; b) x₂ est au
niveau -1.

La solubilisation du Ca est favorisée par l'augmentation de la concentration d'HCl, ce qui est aussi le cas de l'Al. La lixiviation de cette impureté est contrariée par l'augmentation de la concentration du FeCl₃. D'ailleurs ce facteur favorise exclusivement la solubilisation du Fe.

_____

(1) Voir Tableau II.

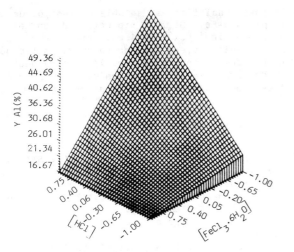

Fig. 7 - Surface de réponse pour l'Al, modèle 11-
- 2ème plan factoriel 2³, alliage 2.

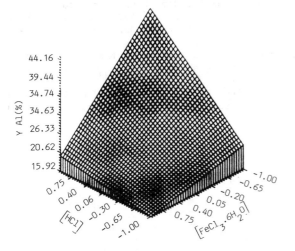

Fig. 8 - Surface de réponse pour l'Al, modèle 9 -
- 2ème plan factoriel 2³, alliage 1.

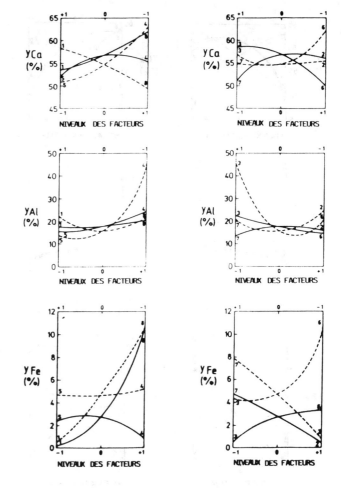

Fig. 9 - Rendements d'extraction en fonction des
niveaux des facteurs (alliage 1).

Légende ———— essais du 1er plan factoriel
---- essais du 2ème "   "

L'anomalie verifiée peut-être en rapport avec
l'hétérogéneité structurale de l'alliage 1, qui
présent un coefficient de variation plus grand
dans sa composition chimique(1).

On sait que la phase CaSi$_2$, sporadiquement dé-
tectée dans les deux alliages, est soluble. Le
fait qu'elle a une plus grande incidence dans
l'alliage 1 peut justifier les concentrations
plus élevées du Ca dans la lessive finale.

L'étude structurale du solide, après lixiviation, a conduit à l'identification du Si et
α-Fe$_{1-x}$Si$_2$ comme phases insolubles. Les phases
qui constituaient initialement ces alliages -
CaAl$_2$Si$_{1,5}$, Al-Fe-Si, Caalsifer, CaSi$_2$ - ont été
léssiviées.

_____

(1) - Voir Tableau I

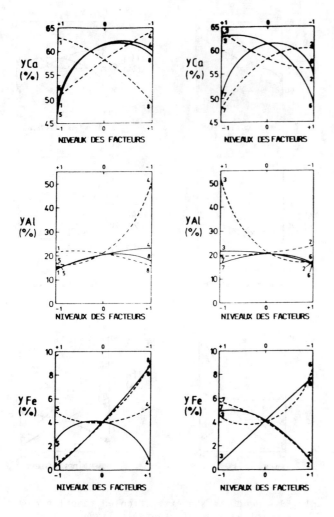

Fig. 10 - Rendements d'extraction en fonction des
niveaux des facteurs (alliage 2)

Légende ——— essais du 1er plan factoriel
——— essais du 2ème "   "

En synthèse, on peut conclure que:

- le rapport liquide:solide (L/S) n'est pas un
  facteur significatif dans la solubilisation des
  impuretés signalées;

- l'HCl est l'agent lixiviant, par excellence, du
  Ca et de l'Al;

- le $FeCl_3$ est l'agent lixiviant principal du Fe;

- les pertes de Si par dissolution sont insigni-
  fiantes, étant donné que cet élement métallique
  est hautement réfractaire à l'action du milieu
  acide utilisé;

- l'alliage 1 réagit de façon semblable à l'allia
  ge 2, à l'exception des anomalies verifiées dans
  la lixiviation du Ca, justifiées par un plus
  grand coefficient de variation dans sa composition chimique initiale;

- la composition structurale de ces alliages sem-
  ble ne pas être la mieux ajustée à l'obtention
  de rendements élevés pour la lixiviation globale des impuretés.

- la phase Caalsifer est soluble et isotypique de
  la phase $\alpha\text{-}Fe_{1-x}Si_2$: la production de une allia
  ge synthétique qui favorise la présence de la
  phase Caalsifer, au lieu de sa phase isotypique
  (représentant environ 50% de l'alliage) peut
  rendre rentable la production de Si de pureté
  élevée par voie hydrométallurgique.

Bibliographie

1. Margarido, F. et Bastos, M.H. (1988). Silicates Industriels, 53, 1-2, 11-19.
2. Margarido, F.. A Importância da Composição Es-
   trutural na Lixiviação de Ligas Fe-Si, Thèse
   Doctorat. IST/UTL (1989).
3. Christiania Spigerverk, (1972). Brevet Français 7115882.
4. Kolflaath, J.A. (1957).Brevet USA 2923617.
5. Cochram, W.J. et Cox, G.M. (1957). Experimental Designs. 2nd ed., John Wiley & Sons, N.Y..
6. Kempthorne, O. (1983). The Design and Analysis
   of Experiments. Robert Krieger, Florida.
7. Box, G. et Draper, N. (1977): Empirical Model-
   -Building and Response Surfaces. John Wiley &
   Sons, Nova York.
8. Khuri, A. et Cornell, J. (1987). Response Surfaces. Designs and Analysis. Marcell Dekker ,
   Nova York.
9. Margarido, F. et Figueiredo, M.O. (1988). Materials Science and Engineering, A104,249/54.

Recovery of vanadium from slags by sulphiding

K. Borowiec
Warsaw University of Technology, Institute of Solid State Technology, Poland
(at present, QIT–Fer et Titane Inc., Sorel, Canada)

SYNOPSIS
Molten vanadiferous titania slag containing 2.0 wt % V_2O_3 was sulphided with elemental sulphur with an addition of MgO under reducing conditions at around 1600°C to form iron-vanadium rich sulphide. Sulphided slags after fine grinding were subjected to gaseous sulphation at 450°C with $SO_2 + O_2$ mixture to obtain water soluble $VOSO_4$. Water leaching at room temperature gave 88.2 % vanadium extraction calculated on the basis of the vanadium content in the slag. A slightly lower degree of extraction was obtained by sulphiding the slag with $S_2 + C$ at 1200°C and then sulphation with water leaching.

INTRODUCTION
More than 80 % of the world's vanadium production is recovered from titanomagnetite ores. Titanomagnetites consist of a titanium-rich magnetite i.e.Fe_3O_4-$Fe_2TiO_{4(ss)}$, some ilmenite i.e.$FeTiO_3$-$Fe_2O_{3(ss)}$ and gangue minerals. Vanadium distribution between coexisting magnetite and ilmenite phases show some scatter, but the magnetite phase is usually richer in vanadium, where V^{+3} replaces Fe^{+3} as shown by Schuiling et al.[1] Vanadium in these ores amounts to between 0.2 - 0.9 % and can be recovered.

These ores can not be smelted in an iron blast furnace because of the high content of TiO_2, and the highly reducing conditions which exist in such furnaces. There is extensive reduction of Ti^{+4} which gives a mushy high-melting point slag which may contain titanium carbides and nitrides. These ores can be smelted in electric reduction furnaces where less reducing conditions can be achieved. This gives a pig iron (hot metal) which contains a certain amount of vanadium. By blowing the pig iron in a converter, a slag rich in vanadium is obtained. This slag can be used for production of ferrovanadium or V_2O_5.

Vanadium can be extracted from titanomagnetite ores by roasting with sodium salts without preliminary smelting to pig iron. Often, the ore is first upgraded magnetically. This not only increases the feed grade, but also rejects a major amounts of gangue constituents such as SiO_2 and silicates which could increase the consumption of sodium salts. Successful salt roasting requires extensive oxidation to convert vanadium from its V^{+3} to its leachable V^{+5} form. In most studies reported, Na_2CO_3 roasting has generated the best vanadium water leach recoveries when compared with Na_2SO_4 or NaCl. Sodium vanadate is leached by hot water (70°C), and the residual Na_2O content in the range 0.7 to 1.3 wt % always remains in the leached pellets which causes difficulties in the subsequent iron-making process.

As it was mentioned above, the smelting of vanadiferous titanomagnetites or ilmenites under rather reducing conditions results in the extensive reduction of vanadium into a high-carbon pig iron melt. Bong-Chan and Kruger[1] have shown that extensive reduction of titania slags can remove, at the very most, 50 % of the vanadium content of the slag. They also found that when FeO content of titania slag is reduced to 1 - 2 wt % the most of the silicon, manganese and chromium contents are reduced into the metal. At the same time the Ti^{+4} in the slag is reduced completly to Ti^{+3}. This extensive reduction of titania slag requires a large increase in the electric energy consumption, up to unacceptable level of 2000 kWh/t slag.

Smelting of titanomagnetite or ilmenite ores under less reducing conditions, produce a low-carbon iron, and a low melting slag with 8 - 10 wt % FeO, containing nearly all the vanadium, chromium and manganese values. This slag may subsequently be subjected to a refining process whereby the vanadium can be recovered.

One refining process is by sulphiding or hydrosulphiding according to the reactions;

$$(MeTi_2O_5)_a(Ti_3O_5)_{1-a} + (a-\tfrac{1}{4})S_2 = aMeS + (3-a)TiO_2 + (a-\tfrac{1}{2})SO_2 \qquad (1)$$

$$(MeTi_2O_5)_a(Ti_3O_5)_{1-a} + aH_2S = aMeS + 2aTiO_2 + (1-a)Ti_3O_5 + aH_2O \qquad (2)$$

where Me represents easily sulphided metals i.e. Fe, Mn, V or Cr.
In reaction (1) carbon may be used as a reducing agent instead of Ti^{+3} from the slag.

It was shown by Borowiec et al.[3] that by sulphiding titania slag with Na_2S melt under moderately reducing conditions iron and manganese were transferred to the sulphide phase, wheras vanadium and

and chromium were retained in the ferrous pseudobrookite phase. The purpose of the present investigation was to study to what extent vanadium can be extracted from a molten and solid titania slag by sulphiding under more reducing conditions.

THEORETICAL CONSIDERATION

The reduction process of V_2O_3 from slag to Fe-V melt at 1600°C is governed by the equilibrium of the reaction 3)

$$3Fe_{(1)} + V_2O_{3(1)} = 3FeO_{(1)} + 2V_{(1)} \quad K_3 \ll 1 \quad (3)$$

The equilibrium constant of this reaction was calculated to be $K_3 = 3.3 \cdot 10^{-10}$. For calculation the standard state of Fe and FeO was chosen to be in liquid form and for V and V_2O_3 to be pure solid form. As the equilibrium constant is much less than one, the recovery of vanadium requires a very low FeO content in the slag, which will lead to a high content of Ti^{+3} in the slag with correspondingly high slag melting temperature and a large increase in the electric energy consumption. Even at strongly reducing condition, the recovery of vanadium in Fe-V-C bath will in any case be incomplete.

Borowiec [4] has investigated the Fe-V-S system at 1100°C and found that the equimolar mixture of FeS + V after annealing in an evacuated sealed silica tube for 24 hr has been converted completly to the VS + Fe mixture. It means that the equilibrium of the reaction:

$$FeS_{(1)} + V_{(s)} = Fe_{(1)} + VS_{(s)} \quad K_4 \gg 1 \quad (4)$$

is shifted strongly to the right which indicates that vanadium has higher affinfor sulphur than iron. Using the Gibbs free energy for formation of VS given by Burylew et al. [5] gave the equilibrium constant for the reaction (4) at 1600°C $K_4 = 1.4 \cdot 10^3$. The data estimated recently by Pei [6] for ΔG^0 formation of VS gave the value of $K_4 = 3.6 \cdot 10^2$. The standard states of the components were chosen as indicated in reaction (4).

The combination of reaction (3) and (4) gives reaction (5) for which the Gibbs free energy at 1600°C was calculated to be 39.2 kcal which gave the equilibrium constant $K_5 = 2.7 \cdot 10^{-5}$ i.e. several orders of magnitude higher than for reaction (3).

$$2FeS_{(1)} + Fe_{(1)} + V_2O_{3(s)} = 3FeO_{(1)} + 2VS_{(s)} \quad 1 > K_5 \gg K_3 \quad (5)$$

This suggests that a better recovery for vanadium from the slag should be possible by sulphiding under moderately reducing conditions than by reduction to Fe-V-C melt.

The practical sulphiding of molten titania slag which consist of two immiscible liquid phases i.e. ferrous pseudobrookite with a general formula: $(FeTi_2O_5)_a(MgTi_2O_5)_b(V_2TiO_5)_c(Al_2TiO_5)_d$- $(Ti_3O_5)_e$ and the glassy silicate: $(Mg,Ca,Al,Fe,Ti)SiO_3$ can be accomplished by using oxidizing or non-oxidizing sulphiding agents. Distribution of the different elements between the three solid-phases i.e. ferrous pseudobrookite with a higher content of MgO and Al_2O_3 than before sulphiding i.e. $(MgTi_2O_5)_f$- $(Al_2TiO_5)_g(Ti_3O_5)_h$ and the sulphide $(Fe,V,Ti)_{1-x}S$ and the glassy silicate will depend, among other things, on the degree of reduction of the slag. In order to study this distribution some basic studies were made on synthetic slags.

BASIC STUDIES

Different starting sample compositions with about 35 - 45 wt % TiO_2 and with different content of MgO and S were smelted to give a 80 % TiO_2 titania slag and sulphide corresponding to what may come from smelting/sulphiding of vanadi-ilmenites or titanomagnetites. Vanadium content was kept constant at about 3.0 wt % V_2O_3. The sulphur was introduced as FeS for the experiments with a low sulphur content. For the higher sulphur content, MgO was replaced by MgS.

The starting mixtures were divided into three series: A) the samples with a low MgO content, B) the samples with a high MgO content, C) the samples with high MgO and Al_2O_3 contents. The representative compositions of each group are presented in Table I.

In order to obtain various degree of reduction some of the TiO_2 was replaced on an equimolar basis, with metallic titanium powder in each group of the samples. The mixtures of chemicals were compressed into briquettes and were heated to fusion in glassy carbon crucibles in a stagnant atmosphere of purified argon. The samples with the lowest degree of reduction were completely molten at around 1480°C, with increased reduction the melting temperature rose to about 1620°C. The samples were brought to about 10 - 20°C above their melting temperature, kept there for 5 min, and allowed to cool rapidly in the furnace. During the smelting a number of simultaneous reactions took place:

$$5 TiO_2 + Ti = 2 Ti_3O_5$$
$$2 FeS + Fe + V_2O_3 = 3 FeO + 2 VS$$
$$MgS + FeO = MgO + FeS$$
$$FeO + Ti_3O_5 = Fe + 3 TiO_2$$

and the corresponding equilibria was established.

In spite of the glassy carbon crucibles being rather inert some carbothermic reduction undoubtedly also occured, as shown by gas evolution from the melt, and from an increased gas pressure in the closed system. There appeared also to have been some sulphur losses by volatilization. For these reasons the degree of reduction of the slag, and the sulphur content, could only be determined

Table I. The representative compositions of the starting mixtures before melting.

Compo-nents wt %	Series A		Series B		Series C	
	Low S	High S	Low S	High S	Low S	High S
MgO	1.00	–	8.00	–	8.00	–
MgS	–	2.40	–	17.50	–	17.50
Al_2O_3	3.00	3.01	3.00	2.74	12.00	10.95
TiO_2	45.00	41.09	43.00	35.59	40.00	32.86
Ti	–	2.40	–	2.19	–	2.19
V_2O_3	3.00	3.01	3.00	2.74	3.00	2.74
FeS	43.50	43.58	38.50	35.14	32.50	29.66
SiO_2	4.00	4.01	4.00	3.65	4.00	3.65
CaO	0.50	0.50	0.50	0.45	0.50	0.45
Total	100.00	100.00	100.00	100.00	100.00	100.00

by analizing the slag/mixture. After cooling, the samples were examined by X-ray diffraction and microprobe analysis.

Melting of the samples with low and high contents of MgO produced a four phase combination, i.e. 1) pseudobrookite phase, $(MeTi_2O_5)_x(Ti_3O_5)_{1-x}$, where Me = Mg,Al,Fe,V, 2) sulphide phase (Fe,V,Ti)S, 3) glassy silicate, and 4) metallic iron phase. The samples with a high content of MgO and Al_2O_3, in addition to these phases gave also, the spinel $MgAl_2O_4$.

The compositions of these phases with the same degree of reduction of the slag which is indicated by a similar content of FeO in the pseudobrookite phase i.e. 1.20, 0.98 and 1.15 wt % FeO for the series A, B and C, respectively, are presented in Table II.

As seen the V_2O_3 contents in the silicates and also the V contents in the metallic phases were found to be very small for all series of experiments. The interesting thing about the vanadium recovery is a considerably lower content of V_2O_3 in the pseudobrookite phase (M_3O_5) obtained in the series B, i.e. for the samples with a high content of MgO. It appears from microprobe analysis that as the content of FeO in the pseudobrookite phase is decreased from 6 wt% to 1 wt% the composition of the M_3O_5 phase in the samples of series B was shifted gradually to the magnesium titanate i.e. $(MgTi_2O_5)_x(Ti_3O_5)_{1-x}$ solid solution as a result of removal of FeO and V_2O_3. As the pseudobrookite phase occured in all samples, and as all samples contained the metallic iron phase, the FeO content of the pseudo-

Table II. Distribution of elements between different phases obtained after melting of high sulphur content samples in glassy carbon crucibles.

Compo-nents wt %	Series A		Series B		Series C		
Oxide	M_3O_5 phase	Silicate	M_3O_5 phase	Silicate	M_3O_5 phase	Silicate	Spinel
MgO	6.75	5.41	12.42	18.13	5.85	11.84	27.46
Al_2O_3	5.46	25.91	3.77	24.99	6.13	24.07	64.52
TiO_2	85.95	7.64	83.72	5.57	86.62	6.19	5.93
V_2O_3	1.82	0.29	0.51	0.16	2.24	0.18	0.54
FeO	1.20	8.86	0.98	6.43	1.15	7.43	0.49
SiO_2	0.07	43.80	0.04	36.13	0.05	41.52	0.02
S	n.d	0.79	n.d	1.10	n.d	0.90	0.01
CaO	0.13	5.25	0.12	4.34	0.11	4.98	0.07
Total	101.38	97.95	101.56	96.85	102.15	97.11	99.04

Ele-ment	Sulphide phase	Metallic phase	Sulphide phase	Metallic phase	Sulphide phase	Metallic phase
Ti	3.86	0.07	5.11	0.12	4.81	0.23
V	5.91	0.09	14.45	0.04	5.02	0.11
Fe	46.02	98.84	38.19	98.49	46.60	98.26
S	40.98	0.16	40.62	0.12	41.54	0.17
Ca	0.04	0.01	0.06	0.01	0.04	0.01
Mg	0.01	0.03	0.02	0.04	0.02	0.01
Total	96.82	99.20	98.45	98.80	98.03	98.79

338

brookite phase was chosen as a measure of the oxygen potential for the samples. Figs. 1 and 2 shows the distribution ratio for vanadium defined as the ratio of (V wt% in MS)/(V_2O_3 wt% in M_3O_5) between the sulphide and the pseudo-brookite phase as a function of the FeO content of the latter for the samples with a low and a high content of sulphur, respectively.

Fig. 1. Vanadium distribution between sulphide MS and pseudobrookite M_3O_5 phase for samples with a low sulphur content (solid lines). Vanadium distribution ratio for the induction furnace experiments is shown by a dashed line.

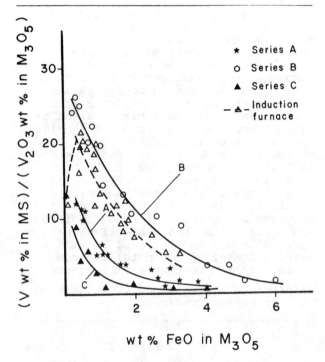

wt % FeO in M_3O_5

The most interesting feature of these Figures is the formation of vanadium rich sulphides in a wide range of the degree of reduction of slag for samples with a high MgO content. This is indicated by high distribution ratio on both Figures, although this effect is more pronounced for samples with a higher sulphur contents. Vanadium distribution ratios between the sulphide and the glassy silicate were found to be one order of magnitude higher than between the M_3O_5 phase and the MS phase. These results suggest that sulphiding of molten vanadiferous titania slag should be done under reducing conditions with MgO added.

SULPHIDING OF MOLTEN TITANIA SLAG

Sulphiding was done in an induction furnace on batches of about 1.5 kg of slag at around 1600°C. Titania slag with the bulk composition given in Table III, i.e. similar to Tyssedal

Fig. 2. Vanadium distribution between sulphide MS and pseudobrookite M_3O_5 phases for samples with a high sulphur contents.

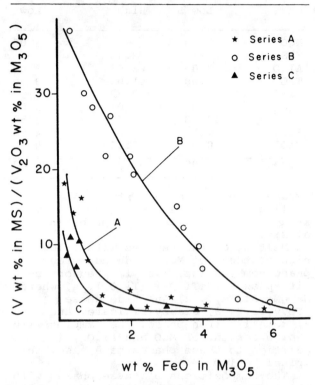

wt % FeO in M_3O_5

slag was remelted in a graphite crucible. Various degrees of reduction were obtained by various holding times at smelting temperature. During smelting further reduction of FeO to metallic iron and Ti^{+4} to Ti^{+3} took place. Elemental sulphur was added by means of a graphite "plunger" and 120 g of MgO as micro-pellets was added to the molten slag. During remelting, due to extensive reduction a certain thickening of the slag was observed and after the sulphur/MgO addition the slag once more became quite fluid. The slag/sulphide mixtures with different degrees of reduction were poured into cast iron ingot molds and were cooled and then analyzed by X-ray diffraction and microprobe analysis.

Similar to the synthetic mixtures most of these samples consist of a four phase combination i.e. the pseudobrookite phase M_3O_5, sulphide phase MS, glassy silicate, and metallic iron. The overreduced slag for which FeO content in the M_3O_5 phase was found to be below 0.2 wt%, in addition, to these phases gave also the tagirovite M_2O_3 phase with the V_2O_3 content in the range 1.1-1.4 wt%. The V_2O_3 content of the M_3O_5 phase ranged from 0.35 to 1.09 wt%.

As the solubility of V_2O_3 in glassy silicate is very low and the M_3O_5 phase is the major phase of the slag, the distribution ratio for vanadium between the MS phase and the M_3O_5 phase can be use as a measure of degree of vanadium extraction. Vanadium distribution ratio

Table III. The bulk composition of vanadiferous titania slag.

Components	wt %	Components	wt %
TiO_2	76.2	MgO	7.1
FeO	7.3	V_2O_3	2.0
SiO_2	4.8	CaO	0.5
Al_2O_3	1.2	Other	0.9

for larger scale experiments is shown by
a dashed line on Fig. 1. It is apparent
that distribution ratio increases with
decreasing FeO content in the same way
as for series B, i.e. for the synthetic
mixtures with high MgO content. For the
most reduced experiments, corresponding
to 0.15 wt % FeO in the M_3O_5 phase the
distribution ratio decrease from about
20 to 12. This is due to the formation
of tagirovite M_2O_3, which decrease the
vanadium content of the MS phase from
12 - 16 wt % to 8 - 9 wt %.

FURTHER TREATMENT OF THE SLAG

It was clear from the microscopic investigation that a major part of the vanadium occured as the MS primary and secondary grains which adhere strongly to
the silicate and pseudobrookite phase,
and the grinding gave essentially trans-
granular cracks. Therefore it was not
possible to obtain a concentrate significantly enriched in the sulphide grains
by using both magnetic separation or
flotation.

The leaching test with boiling HCl
and H_2SO_4 gave only a small amount 5 -
10 % of vanadium dissolved in these
acids, although markedly more iron has
been extracted than vanadium.

It is well known that vanadyl sulphate, $VOSO_4$, has very high solubility
in cold water. It suggests that the extraction of vanadium from the sulphide
grains can be achieved via formation of
$VOSO_4$ by gaseous sulphation. Treatment
of the sulphided slag under high SO_2 and
O_2 potentials in the temperature range
of thermodynamic stability for $VOSO_4$,
i.e. below 480°C will lead to transformation of (Fe,V,Ti)S solid solution to
the corresponding sulphates. Titanyl
sulphate, $TiOSO_4$, will decompose to TiO_2
during water leaching after sulphation.
Borowiec et al.[3] have shown that alkaline earth oxides from the M_3O_5 phase
and glassy silicate can be transformed
to the sulphates above 750°C by using
SO_2 and air. It can be expected that
during sulphation at 450°C, the M_3O_5
phase and glassy silicate remains unreacted.

For the extraction of vanadium from
these vanadium rich sulphides grains,
the fine ground slag to - 200 mesh was
subjected to gaseous sulphation with
$SO_2 + O_2$ in order to obtain the water
soluble $VOSO_4$. These experiments were
carried out on the slag with 1 - 3 wt%

of FeO in the M_3O_5 phase at 450°C. The
10 gram sample of slag was placed in a
ceramic boat in a 24 mm diam. quartz
tube furnace and was heated-up to 450°C
under N_2 flow, and then the $SO_2 + O_2$
equimolar mixture was admitted for varying periods of time. After sulphation,
the samples were leached with water at
room temperarure for 1 hr. The leach
solution were analyzed for vanadium content. The degree of vanadium extraction
was calculated on the basis of the vanadium content in slag, and increased from
74 % for 0.5 hr of sulphation, to 86.5%
for 1 hr, and 88.2 % for 2 hr of sulphation, respectively. Extension of sulphation time beyond 2hr did not improve
the degree of extraction. Sulphation of
MgO and CaO from the M_3O_5 and glassy
silicate at such low temperature was
negligible.

SULPHIDING OF SOLID TITANIA SLAG

The experiments were carried out on the
same slag, of which the composition is
given in Table III. The ground slag
- 200 mesh was mixed with 3 wt % coal
and sulphided with sulphur vapour. Nitrogen was used as a carrier gas for the
sulphur vapour by passing its through a
liquid sulphur bath at 400°C. All runs
were made with 10 gram of slag for 0.5,
1, 2 and 4 hr at 1200°C. After sulphiding for these periods, the furnace
temperature was decreased to 450°C and
then the equimolar mixture of $SO_2 + O_2$
was admitted for 2 hrs in order to allow
oxidation of the vanadium-rich MS grains
to water soluble $VOSO_4$.

The analysis of leach solutions indicate that this procedure gave a slightly lower degree of vanadium extraction
then the previous one i.e. sulphiding
of molten slag and gaseous sulphation.
By extending the sulpiding time from
0.5 hr to 1 hr and 2 hr, the degree of
vanadium extraction was increased from
65.6 % to 71.3 % and 77.5 %, respectively. A further extention to 4 hr gave
78.4 % of vanadium recovery.

Sulphiding with C + S_2 is the endothermic and extra energy is necessary.
Thus, one can visualize an industrial
process when briquetted slag + coal is
treated with sulphur vapour in an elec-
trically heated furnace, at 1100-1200°C,
whereupon the briquettes are treated
with SO_2 + air or oxygen at decreasing
temperature,i.e. below 480°C, for the
highest possible conversion of vanadium-
rich sulphides grains into vanadyl sulphates. Manufacture of sulphur vapour
and handling of S_2 represents technical
difficulties in this process. This problem can be avoided by using sulphiding
of solid slag with Na_2SO_3 + C mixture
at 1100 - 1200°C in reducing conditions.
During sulphiding the formation of the
solid solution of heavy metals with sodium (Fe,Na,V,Ti)S is expected. The extraction of vanadium from sulphides can

be done by gaseous sulphation at the
same conditions as described above or
by acid leaching. The vanadium recovery
in the range 65 - 75 % obtained in the
preliminary experiments indicated that
this procedure can be practical.

Acknowledgments

The present research was partly made at
the Department of Metallurgy, NTH-
Trondheim, Norway, and partly at the
Warsaw University of Technology, Poland.
The microprobe analysis were made with
the assistance of Bard Totdal of the
Physics Department, NTH-Trondheim.

References

1. Schuiling R.D. and Fennstra A. Geo-
chemical behaviour of vanadium in Iron-
Titanium Oxides. Chem. Geology, vol. 30,
1980, p. 143-150.
2. Bong-Chan and Kruger J. Upgrading of
ilmenite and titania slag by DC-ARC
smelting. Extractive Metallurgy and
Materials Science Part 2, Sept. 21-24,
1987, Editors: Li Songren, Jin Zhangpeng
and Z. Yong Jian. Symposium organised
by CSUT and TUC in China.
3. Borowiec K., Rosenqvist T., Tuset J.
and Ulvensoen J.H. Synthetic rutile
from titaniferous slags by a pyrometallurgical route. In: Pyrometallurgy 87,
p. 91-119, 1987, The Institution of
Mining and Metallurgy.
4. Borowiec K. Sulphidization of solid
titania slag. Scand. J. of Metallurgy
in press.
5. Burylew B.P., Mishin P.P. and
Tseitlin M.A. Thermodynamics of the process of desulphurizing liquid and solid
steel by vanadium. Fiz-Khim. Issled.
Metall. Protsessov (Sverdlovsk) vol. 12,
1984, p. 57-62.
6. Pei B. Sulphidization of rare metal
(Cr,V,Ti) oxides. Ph.D. thesis.
Norwegian Institute of Technology,
Trondheim, 1989.

Electrolytic reduction of Eu(III) in acidic chloride solutions

T. Hirato
H. Majima
Y. Awakura
Department of Metallurgy, Kyoto University, Kyoto, Japan

ABSTRACT

The method of electrolytic reduction is deemed to be well-suited for the selective reduction of Eu(III) to Eu(II) and subsequent precipitation of highly pure $EuSO_4$. To more fully understand fundamental aspects of the electrolytic reduction of Eu(III), cathodic polarization characteristics of Eu(III) on a titanium electrode were examined in aqueous solutions of $EuCl_3$ and $EuCl_3$-HCl. The effects of cathode potential, catholyte agitation speed, $EuCl_3$ concentration, temperature, etc. were investigated. The reduction of Eu(III) started at a cathode potential of about -0.6 V vs SHE. The plateau current for the reduction of Eu(III) was reached at around -1.2 V vs SHE. A further decrease in cathodic potential resulted in a decrease in the reduction current. The plateau current shown by the cathodic polarization curve exhibited a half order dependence on the stirring speed of electrolyte and first order dependence on the $EuCl_3$ concentration. The apparent activation energy was 15.7 kJ·mol^{-1}. These results suggest that the plateau current given by the cathodic polarization curve is diffusion-limiting in the electrolytic reduction of Eu(III).

Batch-type electrolytic reduction of Eu(III) was investigated using a bipolar electrolytic cell, which consisted of a titanium cathode, a platinum anode and an anion exchange membrane. The effects of pH, reduction current, catholyte flow rate, temperature and Eu(III) concentration on the reduction rate of Eu(III) and on current efficiency were investigated. It was found that a current efficiency of over 0.6 could be achieved by keeping the pH of the electrolyte constant at 2, in order to avoid hydrolysis of Eu(III).

INTRODUCTION

In recent years, the importance of rare earth metals has increased in the field of functioning materials, such as magnetic materials, superconducting materials and so on, which are recognized as the supporting elements of modern technology. However, the physical and chemical properties of these rare earths are similar, and thus their mutual separation is generally very difficult. Although the solvent extraction technique has been employed in the separation and concentration of rare earth elements, the separation of these elements is still difficult.

Rare earths are present inherently as trivalent ions in aqueous solution, some of which can be reduced to divalent ions. Standard redox potentials M^{3+}/M^{2+} of several rare earth metal ions are tabulated in Table I[1,2]. The redox potentials shown in brackets has not been confirmed.

Table I Standard redox potentials of rare earth metal ions

Reaction	E°/ V
$Eu^{3+} + e = Eu^{2+}$	-0.43
$Tm^{3+} + e = Tm^{2+}$	(-1.5)
$Yb^{3+} + e = Yb^{2+}$	-1.15
$Sm^{3+} + e = Sm^{2+}$	(-1.55)

As is obvious in this table, the reduction potential of Eu(III) to Eu(II) is larger than that of the other redox systems. Thus Eu(III) can be selectively reduced more easily than the other trivalent rare earth ions.

Industrial processing of Eu(III) involves reduction to Eu(II) by zinc powder or zinc amalgam, and the Eu(II) thus formed is precipitated as sulfate by means of a fractional crystallization method[3]. However, it is necessary that the zinc introduced during reduction of Eu(III) is removed from the resultant solution. Recently, Lu et al. reported the electrolytic reduction of Eu(III) using a graphite electrode[4]. Furthermore, the electrolytic reduction of U(VI) to U(IV) was studied in the authors' laboratory using a titanium cathode of a high hydrogen overvoltage[5-8]. Since the reduction was achieved with high efficiency using a titanium cathode, the reduction of Eu(III) is also expected to be efficient.

In the present study, to obtain better fundamental understanding of the electrolytic reduction of Eu(III) using a titanium cathode, cathodic polarization characteristics of the reduction of Eu(III) were studied. Batch-type electrolytic reduction of acidic $EuCl_3$ solution was also investigated, using a bipolar electrolytic cell.

EXPERIMENTAL PROCEDURES

Materials and reagents

Aqueous $EuCl_3$ solution used in this study was prepared as follows: Eu_2O_3 of 99.99 % purity, which was supplied by Nippon Yttrium Co. Ltd., was dissolved in aqueous HCl solution, and then recrystallized as $EuCl_3 \cdot xH_2O$. A mother liquor of $EuCl_3$ was prepared by dissolving the recrystallized $EuCl_3$ in deionized water. The concentration of Eu(III) in the mother liquor was determined by means of an EDTA titration method using xylenol orange as an indicator. The concentration of Eu(II) was determined by redox titration using potassium dichromate. The Eu(III) concentrations and pH of the $EuCl_3$ solutions used in the electrolytic reduction experiments were adjusted with the $EuCl_3$ mother liquor and diluted HCl solution. Unless otherwise stated, the Eu(III) concentration of $EuCl_3$ solution used in this study was $0.1 \ kmol \cdot m^{-3}$. All other inorganic

chemicals used in this study were commercially supplied and of reagent grade. Deionized water with a specific resistivity of 5×10^5 ohm cm was used in the preparation of aqueous solutions.

Experimental apparatus

The experimental set ups used in this study are essentially the same as those reported in previous papers[6,7]. Fig.1 shows the apparatus used for the determination of cathodic polarization characteristics. A rectangular vessel was divided into two compartments by an anion exchange membrane fixed at the center of the vessel. The size of the titanium cathode was 1.5 cm x 1.5 cm, while that of the platinum counter anode was 5.0 cm x 1.0 cm. The cathode compartment was filled with acidic $EuCl_3$ solution, and the anode compartment with $0.1 \ kmol \cdot m^{-3}$ HCl solution. The catholyte solutions were agitated by means of a magnetic stirrer throughout the electrolysis. A three electrode method was employed for polarization measurements during the electrolytic reduction of Eu(III). A AgCl-Ag electrode

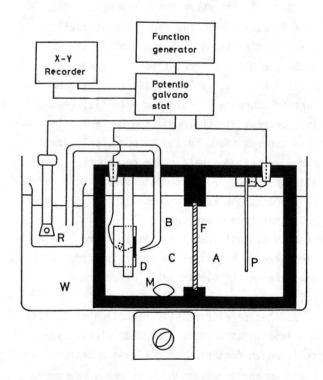

Fig.1 Schematic illustration of the experimental apparatus used for polarization measurement of the electrolytic reduction of Eu(III): (A) Anodic compartment, (B) Agar salt bridge, (C) Cathodic compartment, (D) Ti cathode, (F) Anion selective membrane, (P) Pt anode, (R) Reference electrode, (M) Magnetic stirrer, (W) Thermostated water bath.

filled with 3.3 kmol·m^{-3} KCl solution was used as the reference electrode. All potentials measured in this study are reported as a function of the standard hydrogen electrode (SHE) instead of the 3.3 kmol·m^{-3} KCl AgCl-Ag electrode. An agar salt bridge containing saturated potassium chloride was used as a liquid junction between the reference electrode vessel and the cathode compartment. The cathode potential was controlled by a potentiostat with a function generator, operating at a constant scanning rate of 1 mV·s^{-1}.

Fig.2 shows the batch-type electrolytic cell used for the reduction of Eu(III). The electrolytic cell is bipolar, consisting of titanium electrodes, each plated on one side with platinum. The titanium side was used as a cathode, and the platinum side, as an anode. The area of each electrode was 5 x 10 cm^2. Anion exchange membranes (Tokuyama Soda Company Ltd., ACH-45) were placed alternating with electrodes to separate the cell into cathode and anode compartments. The distance between the membrane surface and each electrode surface was 1 cm. The electrolytic cell was connected to a galvanostatic D.C. source to carry out the electrolytic reduction of Eu(III) at a constant current. 700 ml of anolyte, containing 0.1 kmol·m^{-3} HCl, and 700 ml of EuCl$_3$ catholyte were circulated through each compartment using a separate pump. The flow rate

of catholyte was kept constant at 3.3 l·min^{-1}. The electrolyte reservoirs were immersed in a thermostated water bath. The temperature was maintained at 298 K, unless otherwise stated. The pH of catholyte was controlled by additions of HCl with an autotitrater. N$_2$ gas was bubbled in the catholyte to avoid oxidation of Eu(II) by O$_2$ in the atmosphere. Chlorine gas formed in anolyte was removed by absorption towers containing 1 kmol·m^{-3} KI and 1 kmol·m^{-3} Na$_2$SO$_3$.

EXPERIMENTAL RESULTS AND DISCUSSION

Cathodic polarization curve of a titanium cathode in aqueous EuCl$_3$ solution

Cathodic polarization curves were determined for aqueous 0.1 kmol·m^{-3} EuCl$_3$ solutions using an agitation speed of 400 min^{-1} at the following temperatures: 288, 298, 308 and 318 K. The results obtained are depicted in Fig.3. The reduction current was found to increase with elevations in temperature. It is also clear in this figure, that the current begins to rise at a cathode potential of around -0.6 V vs SHE, and reaches a plateau at around -1.2 V vs SHE. This plateau does not continue for long, as the current decreases eventually with reductions in cathode potential. An increase in the current due to hydrogen evolution at the cathode is observed at a potential of around -1.6 V vs SHE. It is noteworthy that a considerable change in the color of the titanium cathode surface was detected after the polarization experiment. The

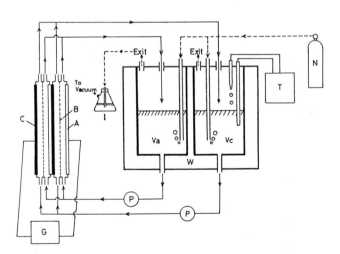

Fig.2 Schematic illustration of the experimental apparatus used for batch-type electrolytic reduction of Eu(III): (A) Pt anode, (B) Anion selective membrane, (C) Ti cathode, (P) Pump, (G) Galvanostat, (W) Thermostated water bath, (Va) Vessel for anolyte, (Vc) Vessel for catholyte, (I) Cl$_2$ gas absorber, (N) N$_2$ gas cylinder, (T) Autotitrater.

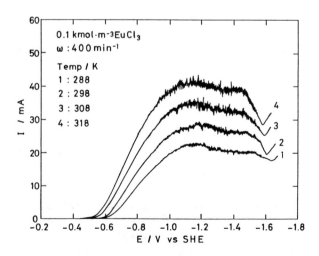

Fig.3 Effect of temperature on the cathodic polarization curve of aqueous EuCl$_3$ solution

surface product was analysed by ESCA. The energy spectra of Eu and O exhibited peaks at 137.3 and 532.2 eV, respectively. The energy spectra are very similar to those of Eu_2O_3. These findings suggest that a film, which inhibits reduction, formed on the cathode surface by hydrolysis of Eu(III) at a low cathode potential.

The current values of the plateau portion of the polarization curves are plotted against temperature in the form of an Arrhenius plot, as shown in Fig.4. The curve is essentially linear, and the activation energy estimated from the slope of the straight line is 15.7 kJ·mol^{-1}. This value suggests that mass transfer controls the electrolytic reduction of Eu(III) in aqueous solution[9-11] under the experimental conditions which determine the plateau.

The effect of agitation speed of catholyte on the plateau current is shown in Fig.5. In this figure, values of the current are plotted against the square root of agitation speed. There exists a linear relation, passing through the origin, between the value of the plateau current and the square root of agitation speed. According to Levich, the theoretical rate of mass transfer in the solution boundary layer adjacent to the surface of a solid under forced convection is pro-

portional to the square root of agitation speed[12]. Fig.6 depicts the linear relationship between the plateau current and EuCl$_3$ concentration. Judging from these results, the plateau

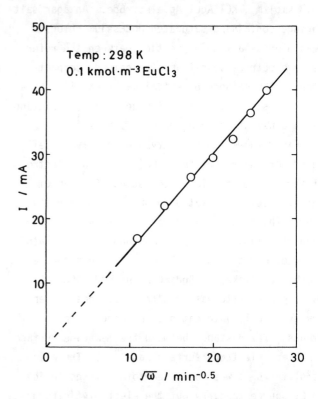

Fig.5 Relationship between the plateau current of the cathodic polarization curve of aqueous EuCl$_3$ solution and the square root of agitation speed

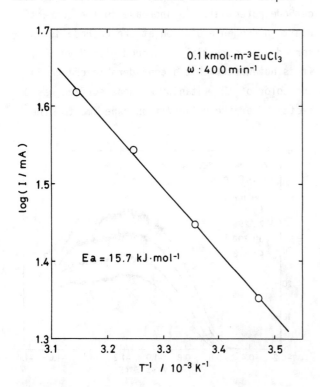

Fig.4 Effect of temperature on the plateau current of the cathodic polarization of aqueous EuCl$_3$ solution

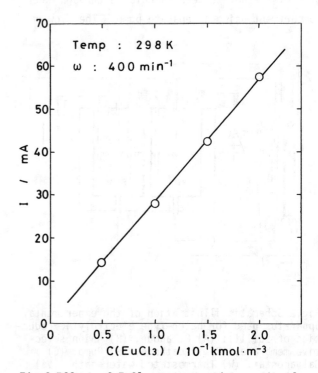

Fig.6 Effect of EuCl$_3$ concentration on the plateau current of the cathodic polarization curve of aqueous EuCl$_3$ solution

current of the cathodic polarization curve is a diffusion-limiting current in the electrolytic reduction of Eu(III).

Cathodic polarization curve for the aqueous solution system, $EuCl_3$-HCl

In industrial solvent extraction processes for the separation and concentration of rare earths, aqueous solutions containing rare earth salts acidified with HCl or HNO_3 are generally subjected to treatment. Therefore, cathodic polarization curves for aqueous $EuCl_3$-HCl solutions are of interest.

Curve 1 in Fig.7 shows the cathodic polarization curve obtained with aqueous 0.1 $kmol \cdot m^{-3}$ $EuCl_3$ - 0.1 $kmol \cdot m^{-3}$ HCl solution. For the sake of comparison purpose, cathodic polarization curves obtained in aqueous 0.1 $kmol \cdot m^{-3}$ $EuCl_3$ solution and in aqueous 0.1 $kmol \cdot m^{-3}$ HCl solution, are depicted by Curves 2 and 3, respectively, in the same figure. Comparison of Curve 1 and Curve 2 shows that the addition of HCl to the $EuCl_3$ solution slightly decreased the cathode potential for initiation of Eu(III) reduction. Also, there is no longer a distinct plateau current, or diffusion-limiting current in the Eu(III) reduction.

Since the cathodic polarization curve, which was determined in 0.1 $kmol \cdot m^{-3}$ HCl aqueous solution, is believed to correspond to hydrogen gas

evolution at the titanium cathode, the difference between the currents shown by polarization curves 1 and 3 should correspond to the reduction current of Eu(III) in aqueous 0.1 $kmol \cdot m^{-3}$ $EuCl_3$ - 0.1 $kmol \cdot m^{-3}$ HCl solution. This difference in currents is illustrated by Curve 4 in Fig.7, which shows that the electrolytic reduction of Eu(III) can proceed in an aqueous solution of $EuCl_3$-HCl. However, the reduction current decreases at a cathode potential lower than -1.0 V vs SHE, when the evolution of H_2 gas becomes significant.

Batch-type electrolytic reduction of acidic $EuCl_3$ solution

Electrolytic reduction of Eu(III) was performed in solutions of different pHs using the bipolar cell shown in Fig.2. The time variation of the concentration of Eu(II), which was produced by the electrolytic reduction, is shown in Fig.8. The broken line indicates the theoretical time variation of the Eu(II) concentration, which was derived assuming that a current efficiency of unity. As is clear in this figure, there is no significant difference between the reduction rates at pH 2, and at pH 1. The reduction of Eu(III) was almost complete in about 4 h. The current consumed by the reduction of H^+, which was calculated from the amount of HCl added by the autotitrater during the experiment, showed good agreement with the current calculated from the difference between the reduction rate curve obtained experimentally and that predicted theoretically. This suggests that a possible side

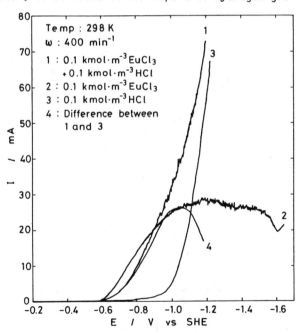

Fig.7 Cathodic polarization curves in various aqueous $EuCl_3$ solutions

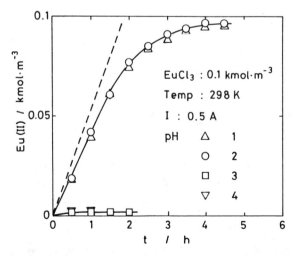

Fig.8 Effect of pH of catholyte on the electrolytic reduction of Eu(III)

reaction is the reduction of H$^+$. On the other hand, the reduction of Eu(III) was limited at pH 3 and pH 4. At these higher pHs, it was observed that the cathode surface was coated with a white product. It is believed this product formed by the hydrolysis of Eu(III), because the pH of the solution in the vicinity of the cathode surface was raised by H$^+$ reduction. Therefore, in the experiments which followed, the pH of the catholyte was maintained at a level of 2 throughout the electrolysis.

Fig.9 shows the effect of electrolytic current on Eu(III) reduction. The times required for complete reduction of Eu(III) were 5 h at 0.25 A and 4 h at 0.5 A. The reduction rate of Eu(III) at 0.75 A decreased after 1.5 h due to a violent

evolution of H$_2$ gas. Fig.10 shows the mean current efficiency of the electrolysis depicted in Fig.9. Although the mean current efficiency decreased as the reduction progressed, it remained above 0.6 at the end of the electrolysis at 0.25 A. Thus, in the batch-type electrolytic reduction, excess current was not always desirable, from the viewpoint of current efficiency.

The data shown in Fig.9 are replotted in Fig. 11, according to the first order rate law. In Fig.11, the logarithm of the ratio of Eu(III) concentration at time, t, to the initial concentration is plotted against the reduction time. In the electrolysis at 0.5 A, a linear relationship remains throughout the reduction. This result indicates that the reduction rate is proportional to the concentration of Eu(III), and suggests that the reduction of Eu(III) proceeded under the condition of a diffusion-limiting current. On the other hand, for the electrolysis at 0.25 A, the initial reduction rate was less than the limiting current. After 3 h of reduction, when the Eu(III) concentration became lower, the reduction proceeded at the limiting current and exhibited the same slope as that obtained at 0.5 A. At 0.75 A, the reduction of Eu(III) proceeded initially at the limiting current; however, after 1.5 h of reduction, the reduction rate became smaller. This may be attributed to a decrease in cathode surface area available for Eu(III) reduction. The cathode surface is covered with H$_2$ gas

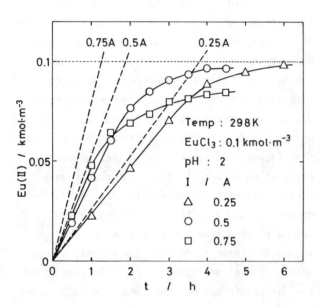

Fig.9 Effect of reduction current on the electrolytic reduction of Eu(III)

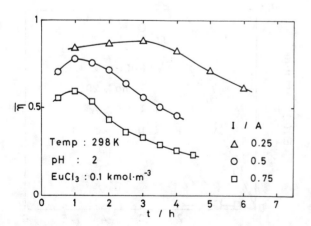

Fig.10 Mean current efficiency of the electrolytic reduction of Eu(III) using different reduction currents

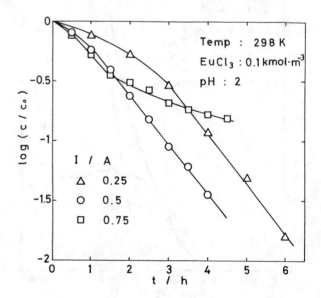

Fig.11 First order rate law plot of the data shown in Fig.9

more significantly at a higher current density.

Electrolytic reduction experiments were performed at a constant current of 0.5 A using aqueous 0.1 $kmol \cdot m^{-3}$ $EuCl_3$ solution as catholyte, under a flow rate of catholyte ranging from 3.3 to 1.5 $l \cdot min^{-1}$. The reduction rate of Eu(III) decreased dramatically with a decrease in flow rate. This result can be understood from the decrease in the diffusion limiting current density due to the increase in thickness of the diffusion layer.

The effect of temperature on the electrolytic reduction of Eu(III) was examined at 298, 308 and 318 K, using aqueous 0.1 $kmol \cdot m^{-3}$ $EuCl_3$ solution as catholyte and maintaining the reduction current at 0.5 A. Although the reduction rate increased with increasing temperature, the effect of temperature was not significant under these experimental conditions. This is reasonable, since an increase in temperature causes an increase in the reduction rate of H^+, as well as that of Eu(III).

The results of the electrolytic reduction of $EuCl_3$ solution containing different concentrations of $EuCl_3$ are shown in Fig.12. Although the reduction rate of Eu(III) decreased gradually with the progress of reduction, almost complete reduction was achieved, in each case, after about 4 h. The mean current efficiencies at the initial stage of electrolysis were around 0.4 and 0.2 for 0.05 and 0.025 $kmol \cdot m^{-3}$ $EuCl_3$ solutions, respectively, and these efficiencies decreased with the progress of the electrolysis. The first

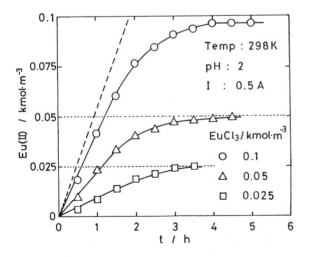

Fig.12 Effect of the $EuCl_3$ concentration of catholyte on the electrolytic reduction of Eu(III)

order rate law plots of the data for 0.05 and 0.025 $kmol \cdot m^{-3}$ $EuCl_3$ are almost identical to that of 0.1 $kmol \cdot m^{-3}$ $EuCl_3$ shown in Fig.11. These findings suggest that the electrolytic reduction proceeds under the condition of the limiting current. The low current efficiency observed at low $EuCl_3$ concentration was caused by current consumption by the evolution of H_2 gas.

CONCLUSIONS

To more fully understand fundamental aspects of the electrolytic reduction of $EuCl_3$, the polarization characteristics of electrolytic reduction of Eu(III) using a titanium cathode were examined. Batch-type electrolytic reduction of Eu(III) was also investigated using a bipolar electrolytic cell, in which a titanium plate was used as the cathode. The principal results obtained are as follows:

1. The cathodic polarization curve for aqueous 0.1 $kmol \cdot m^{-3}$ $EuCl_3$ solution exhibited an increase in electric current, at the start of the electrolytic reduction of Eu(III) to Eu(II), at around -0.6 V vs SHE. The plateau current for the reduction of Eu(III) was observed at around -1.2 V vs SHE, but further lowering of the cathode potential resulted in a decrease in the reduction current. The electrolytic reduction of Eu(III) could proceed in aqueous 0.1 $kmol \cdot m^{-3}$ $EuCl_3$ - 0.1 $kmol \cdot m^{-3}$ HCl solution, although the reduction current decreased at a cathode potential lower than -1.0 V vs SHE, when the evolution of H_2 gas became significant.

2. The value of the plateau current exhibited by the cathodic polarization curve of Eu(III) increased with an increase in agitation speed of catholyte and with the concentration of $EuCl_3$. A half order dependence on agitation speed and the first order dependence on $EuCl_3$ concentration were demonstrated. The apparent activation energy for the plateau current was found to be 15.7 $kJ \cdot mol^{-1}$. These findings suggest that the plateau current is a diffusion-limiting current in the reduction of Eu(III).

3. In the batch-type electrolytic reduction of Eu(III) using a bipolar electrolytic cell, the effects of pH, reduction current, catholyte flow rate, temperature, and Eu(III) concentration on

348

the reduction rate of Eu(III) and on current
efficiency were investigated. It was found that
complete reduction of aqueous 0.1 $kmol \cdot m^{-3}$ $EuCl_3$
solution with a current efficiency of over 0.6
could be achieved by keeping the catholyte pH at
2, in order to avoid hydrolysis of Eu(III).

ACKNOWLEDGMENT

Financial assistance from The Kawakami
Memorial Foundation to T. Hirato is greatly
acknowledged.

REFERENCES

1. W.M. Latimer; "Oxidation States of the
   Elements and their Potentials in Aqueous
   Solutions", 2nd Ed., Prentice Hall, New York
   (1952), p.294.
2. L.B. Aspery, B.B. Cunningham; "Progress in
   Inorganic Chemistry", Vol.11, Edited by F.A.
   Cotton, Interscience, New York (1960), p.267.
3. C.G. Byown and L.J. Sherrington; Journal of
   Chemical Technology and Biotechnology, 1979,
   Vol.29, p.193.
4. D. Lu, J.S. Horng and C.P. Tung; Journal of
   Metals, 1988, Vol.40, p.32.
5. H. Majima, Y. Awakura and S. Hirono; Metallurgical Transactions B, 1986, Vol.17B, p.41.
6. H. Majima, Y. Awakura, K. Sato and S. Hirono;
   Metallurgical Transactions B, 1986, Vol.17B,
   p.69.
7. Y. Awakura, K. Sato, H. Majima and S. Hirono;
   Metallurgical Transactions B, 1987, Vol.18B,
   p.19.
8. Y. Awakura, H. Hiai, H. Majima and S. Hirono;
   Metallurgical Transactions B, 1989, Vol.20B,
   p.337.
9. M. Eigen, W. Kruse, G. Maass and L. de Maeyer;
   Progress in Reaction Kinetics, 1964, Vol.2,
   p.287.
10. M. Anbar, Z.B. Alfass and H. Bregman-Reisler;
    Journal of the American Chemical Society,
    1967, Vol.89, p.1263.
11. B. Cercek and M. Ebert; Journal of Physical
    Chemistry, 1968, Vol.72, p.766.
12. V.G. Levich; "Physicochemical Hydrodynamics",
    Prentice Hall, New York (1962), p.70.

Environment

Best available technology— a viewpoint on the development and application of the concept to the European non-ferrous industry

Tony Connell
Britannia Refined Metals, Northfleet, England
Robert Maes
MHO—a Division of ACEC–Union Minière, Hoboken, Belgium
José Poncet
Asturiana de Zinc, Avilés, Spain

SYNOPSIS

The implementation within the European Community of the concept of Best Available Technology (BAT) was enacted in June, 1984, with the promulgation of Directive 84/360 on the combating of air pollution from industrial plants.

This Directive requires Member States to introduce a system of prior authorisation for the operation of new industrial plants or substantial alteration of existing plants to ensure that the best available technology not entailing excessive costs (BATNEEC) is applied. Furthermore, it enables the European Council to establish emission limit values based on the use of BATNEEC, taking into account the nature, quantities and harmfulness of the emissions concerned.

In 1988 the European Commission initiated a pilot exercise with the aim of establishing a framework for the exchange of information between Member States and the Commission on BAT, together with corresponding emission limits achieved. Seven production sectors were chosen, including heavy metals from non-ferrous industrial plants.

Technical Working Committees were formed for each sector to exchange and collate the information and to produce a technical note for presentation to the Commission and to national government experts. This information will constitute a technical basis for the Commission to prepare directives on specific industrial sectors and annexes defining limit values of emissions.

The authors of this paper, who were members of the Heavy Metals Technical Working Group, have endeavoured to give an insight into the development of the technical note and of its importance and implications to the industry.

A description of progress and pursued objectives is given. The practical application of the concept to the non-ferrous sector is discussed and future trends are outlined.

THE COUNCIL DIRECTIVE

Directive 84/360/EEC,[1] which was adopted on June 28, 1984, aims at the prevention or reduction of atmospheric pollution from industrial plants.

Article 3 of the Directive requires that the exploitation of specified plants, including installations for the production and transformation of metals, should be submitted to prior authorisation of national authorities in all the Member States of the Community. Such authorisation may only be granted when the authority has ascertained that the best available technology (BAT) is used, provided that the measures applied do not entail excessive costs (NEEC).

Article 7 states that the Member States and the Commission must exchange information concerning pollution reduction or prevention, technical processes and equipment, air quality and emission limit values. As a result of this exchange, the Commission may establish emission limit values based on BATNEEC, taking into account the nature, quantities and harmfulness of these emissions.

The concept of BAT was not defined in the Directive and neither were the criteria used to assess "proven" BAT. Nor was the definition of NEEC given : for instance, was it capital as well as operating costs ? However, the following interpretation of BAT progressively received acceptance : it is the technology or group of technologies which, through performances, was adequately demonstrated to be the best commercially available as regards the minimization of emissions to the atmosphere, providing that it was proved to be economically viable when applied to the relevant industrial sector.

A further provision of the Directive is that the Member States follow the evolution of BAT and, when proven, if necessary, designate the new BAT by imposing appropriate conditions on plants that had received authorisation to operate, again without entailing excessive costs. New technologies may thus become BAT and the associated emission values may become the values considered for future authorisation. The BAT concept is thus inherently of an evolutionary nature.

PRELIMINARY DEVELOPMENTS OF THE BAT CONCEPT

The practical implementation of Directive 84/360 apparently encountered considerable difficulties,[2] amongst others owing to the important effort required to define the relevant technologies. Therefore, the Commission realized that the best approach would be through Article 7, requiring an exchange of information between Member States, and initiated in 1988 what was termed a pilot exercise with the aim of establishing a framework for this exchange of information in order to prepare technical notes defining BATs, associated emissions, etc. A BATNEEC Exchange of Information Committee was set up and charged with the definition of priorities, and Technical Working Groups were established to draft technical notes related to the following production sectors :

1. Benzene
2. Cement manufacture
3. Ammonia production
4. Nitric acid
5. Sulphuric acid
6. Toxic waste incineration
7. Heavy metal emissions from non-ferrous plants

The Technical Working Groups consisted of national government experts and industrial experts from the Member States. In the case of the heavy metal emissions from non-ferrous plants, the group was coordinated by an environmental consultant appointed by the Commission and the action of the industrial experts was coordinated by Eurometaux, the European federation of non-ferrous industries. Volunteer experts presented themselves for drafting parts of the technical note.

Different meetings of the Technical Working Group for non-ferrous plants took place, with the following progress :
- March, 1989 : The proposed structure of the technical note was presented to the Working Group by the coordinator and the format agreed at the first meeting. As illustrated, Table 1 gives the contents of the technical note in its final form as established in April, 1991.
- May, 1989 : Parts of the technical note already drafted by volunteer experts were discussed and potential authors nominated by Eurometaux for specific undrafted parts were introduced.
- September, 1989 : A general discussion of all the documents that were produced took place to enable the coordinator to prepare a "final report" by November, 1989.

SOME REMARKS MADE BY THE INDUSTRY ON THE IMPLEMENTATION OF THE BAT CONCEPT

Independently of the activities of the Working Group at the three meetings described above, some comments on the implementation of the BAT concept had already been made in the Environment, Health and Safety Committee of Eurometaux.[4] The new approach to environmental questions was welcomed, albeit with a few concerns about its applicability to the non-ferrous metals industry, such as the fact that each plant applies processes with specific characteristics that are difficult to extrapolate in general terms. The opinion was also expressed that the concept should be extended to all sectors of the environment in a "multi-media" approach. Questions were raised about the industry being forced to completely reassess its processes, the

353

Table 1 : Contents of technical note for heavy metal emissions from non-ferrous industrial plants[3]

PREFACE

PART A: GENERAL ASPECTS

   AI   Introduction

   AII   Legal Provisions of Member States
           Relevant for Plant Authorisation

   AIII   Processes and Installations Included in
           this Technical Note

   AIV   Heavy Metals And Their Compounds Emitted
           by Non-Ferrous Industrial Plants

   AV   References to Part A

PART B: HEAVY METAL EMISSIONS CAUSED BY
          NON-FERROUS INDUSTRIAL PLANTS AND
          THEIR CONTROL

   BI   Introduction

   BII   Emissions of Heavy Metals Caused by
           Non-Ferrous Industrial Plants

   BIII   Air Pollution Control Technology

   BIV   Cross-Media Aspects

   BV   Monitoring of Emissions

   BVI   References to Part B

PART C: CONSIDERATIONS WITH REGARD TO PARTICULAR
          PROCESSES

   CI   Introduction

   CII   Lead Works

   CIII   Copper Works

   CIV   Zinc Works

   CV   Tin Works

   CVI   References to Part C

PART D: ANNEXES

   DI   Examples of the Performance Characte-
           ristics of Dust Collectors

   DII   Example of Emissions Figures of
           Primary Lead Production

   DIII   Plant Authorisation Procedures in
           EC Member States

   DIV   Recommended emission limits for
           lead-, copper-, and zinc-works

354

resulting costs and the way in which these would be assessed in different countries.

Similar concerns were also expressed by the authors of this paper during the Technical Working Group meetings and these were notified to the Environment Department (DGXI) of the Commission. For example, competition between processes for a lowering of emissions by a few per cent was considered futile. Furthermore, there were differences of opinion regarding processes applying for BAT that were not unanimously considered as "proven" technology.

"PLENARY" MEETINGS FOR EXCHANGE OF INFORMATION BETWEEN COMMISSION AND NATIONAL EXPERTS

Following the establishment of "final versions" of the technical notes for the different production sectors meetings took place in the framework of the BATNEEC Exchange of Information Committee, bringing experts from the Commission and from national Governments together, under the chairmanship of a representative of DGXI responsible for BAT activities; industrial experts were invited as observers. The following progress was achieved :

- December, 1989 : Most of the established technical notes were agreed, but reservations were expressed for the heavy metals sector, where the large number of processes mentioned and problems underlined by industry were recognized to need further consideration. It was also concluded that a firm legal ground for BAT activities should be established by appropriate Council legislation.
- February, 1990 : Requests to the Council to formally set up the BAT Committee and to define the legal status of the technical notes were proposed. Owing to the complexity and variety of processes listed in the technical note on heavy metal emissions, some reservations were expressed on making the total BAT technical notes legally binding and it was suggested that they should remain of an informative nature and be completed by a separate section defining the standards that would be binding and enforceable.

REMARKS MADE BY EUROMETAUX ON BEHALF OF THE EUROPEAN NON-FERROUS INDUSTRY

Following a request of the Chairman at the meeting of February 1990, Eurometaux presented some serious concerns to the Commission that it considered should be addressed in the follow-up meeting for the heavy metals sector. They summarized the main difficulties that had been encountered in the implementation of the BAT concept for the sector.

1. It was clear from the descriptions brought together in the technical note that the production processes of non-ferrous metals were very varied and had a very large number of unit operations that were dependent upon specific economic and social conditions and possible feed materials. The aim was to designate a process for the production of a given metal as the best available technology and this therefore appeared irrealistic; even if one concentrated on the unit operations, attention had to be paid to the entire processing circuit in which they were linked in order to treat given feed materials.

2. Primary feed materials were, in essence, mineral and thus variable in both form and composition. Recycled materials were also extremely varied in shape and combination. Metallurgical processes, particularly in Europe, often combined both types of feed materials (for the sake of environment protection), with corresponding implications on the choice and performance of unit operations.

3. With regard to environmental protection, emissions at the stack for a given process were important, but not the only polluting factor to be taken into consideration. Fugitive emissions were also important, but not directly related to the selected process. The overall environmental impact, including all effluents, residues and waste products, as well as recycling capabilities, should therefore be considered in the assessment of BAT.

4. Futile competition between processes on the basis of their emissions should be avoided by

coupling a given reduction of the particular emission (air, water, waste, etc.) with a fair appreciation of the corresponding reduction of harmfulness and with the price that has to be paid for it, as referred to in article 8 of the Directive.

5. One had to be aware of the tendency for some companies or groups to outbid their competitors by pushing their own process in order to favour their engineering divisions. There was consequently a risk of distorting the information released, particularly about new processes, on how they achieve the required performances and how far they may be considered as proven, reliable and flexible technology. It was therefore important to define the parameters of proven technologies for each specific plant or equipment relative to the process.

6. The assessment of costs associated with the application of a process or unit operation would always be complicated by local circumstances, such as the cost and availability of energy, specificity of feed materials or implications of the overall processing circuit in which the introduction of the process or unit operation was considered.

7. Notwithstanding the above remarks, it was considered that the BAT action was fruitful, at least by the information that it would bring together and the inventory of processes that it would establish. The belief was expressed that the action should be extended in a positive direction by giving the produced documents a certain enforcement in order to progressively reach harmonization of the environmental standards for the non-ferrous industry all over the EC.

"RESTRICTED" MEETING AT EUROMETAUX'S OFFICE

An informal discussion to exchange views on Eurometaux's reservations took place with the reponsible representative of DGXI and the coordinator. This meeting was very constructive and probably played an essential role in the future decisions which were going to be taken in relation with the implementation of the BAT concept.

The industry convinced that it was not merely manoeuvring to delay enforcements but that fundamental obstacles were encountered that would have to be cleared in order to enable the Directive to be properly and effectively enacted. The principle of "pilot exercise" being practised with the establishment of these first technical notes was fully realized and it is not surprising that some major interventions were made by the non-ferrous sector, to which environmental questions were particularly sensitive.

The workload required to define all the relevant technologies was recognized and time would be given to perfect what had already been achieved. Practical difficulties to define emissions associated with given processes were realized. For the delicate point of "proven" technology, Eurometaux was requested to contribute to the development of a definition. A climate of mutual appreciation was established, inspiring confidence that a way would be found to apply the Directive with the necessary flexibility, taking into account the complexity of involved processes, the variety of feed materials and the diversity of local situations. The hope was expressed that a completed version of the technical note on heavy metal emissions would be finalized in the course of 1991.

THE CONCEPT OF PROVEN TECHNOLOGY

The feeling of the authors of this paper is that no authority can be conceived that could settle out of hand whether a given technology is proven or not. The decision can only be established by some form of jurisprudence coupled with past experience. Therefore, it was decided that a statement should be developed to define the minimum criteria that a technology should meet to be considered as best available technology. After consultation of all Eurometaux' members, this statement became the following:

"In order to be taken into consideration as best available technology, a technology must be proven i.e. the technology must have been sufficiently operated during a reasonable period of time, according to the complexity of the process, at a stated capacity, processing the various feedstock materials intended, so that appropriate assessments can be made of recognised performance criteria and results recorded in an appropriate form, in terms of reliability, consistency of environmental performance and economic factors, which can be proved and controlled, if need be."

It was also felt that minimum criteria for technology evaluation should include details on essential operating data, such as :
- a description of the process and the equipment and/or plant in which it is practiced, with mention of capacities
- physical and chemical characteristics of process materials
- the period under evaluation and the degree of availability of equipment and/or plant
- process cross media aspects
- a technical description of environmental controls for emissions to air, water and waste, and their operational performance, including ambient air quality
- economic factors.

This list should be considered as a recommendation of minimum information to be released on processes applying for BAT. The list was also established with the intention of providing support to companies being constrained by authorities requiring them to apply technologies that in their view, were not proven for the conditions of the intended application. It may be stressed, in this connection, that the interpretation of these principles remains quite intractable, as exemplified by the case of a well established process whose transposition to slightly different conditions has recently resulted in significant start-up difficulties.[5]

FINALIZATION OF THE TECHNICAL NOTE ON HEAVY METAL EMISSIONS

A last round of consultations was organized in February, 1991, with the aim of finalizing the technical note. The Technical Working Group was split into sub-groups, dealing separately with lead, zinc and copper, which, in fact, agreed more or less unanimously with their respective

decisions. A suggestion had been made beforehand by the coordinator, based on a proposal from the German Umweltbundesamt, to define emission limits associated with metallurgical processes on a general TA Luft-type base[6] and to seek agreement on a draft document established along this line. After consultation of Eurometaux' members, this procedure was accepted by the industrial experts. It presented the advantage of overcoming the difficulty of collecting information on emissions for each individual process. However, it did not exactly follow the approach recommended by the Directive, but did appear as an acceptable initial solution.

Discussions took place in an excellent atmosphere. Most proposed amendments to the suggested emission limits were unanimously approved. Furthermore, the comments of industry on proven technology were not only accepted for inclusion within the technical note but would also be used as a reference to all industrial sectors technical notes.

FUTURE TRENDS

At the end of the meetings of February, 1991, the responsible representative of DGXI summarized the future orientations of the BAT activities as follows[7] :
- The technical notes that had been approved unanimously by the experts of the Technical Working Groups would be presented to the Commission and to national government experts for final agreement.
- The Commission would organize at least two meetings per year of the Exchange of Information Committee in order to approve technical notes, improve their quality, define further work programmes and decide upon updating of technical notes whenever necessary.
- Approved technical notes would be published by the Commission as EUR documents.
- New technical notes would not contain emission limits but only a description of the BAT, including relevant operating and investment costs and associated emission values.
- The technical notes would not be binding documents for the Commission, neither for the Mem-

ber States nor for the industry. They would constitute a technical basis for the Commission to prepare directives on specific industrial sectors and annexes defining emissions limits.

- The annexes would be inserted in a more general directive prepared in the framework of an integrated approach for industrial pollution.

It was intimated that this integrated approach would consider the overall environmental impact in the assessment of BAT.

These new DG XI guidelines were received very positively by all the attending experts.

References

1. Directive du Conseil du 28 juin 1984 relative à la lutte contre la pollution atmosphérique en provenance des installations industrielles (84/360/CEE), Journal Officiel des Communautés Européennes, 16.7.84.

2. Bartaire J.-G., La politique de la Commission des Communautés Européennes en matière de pollution atmosphérique, Pollution Atmosphérique, juillet-septembre 1990, pp. 267-278.

3. Technical Note on Best Available Technology Not Entailing Excessive Costs for Heavy Metal Emissions from Non-Ferrous Industrial Plants, Final Report, April 1991, to be published by the Commission as EUR document.

4. Barbour A.K., Europe on the 1992 horizon - Views of the European non-ferrous industry - Its hopes and concerns, Politiques "Environnement, Hygiène et Sécurité" des Communautés Européennes, Eurometaux, 28-29 septembre 1989, pp. 89-91.

5. Anon., World's largest flash smelter - Under strict environmental demands, Magma meets rules and decreases costs as well, Engineering and Mining Journal, January 1990, pp. C37-39.

6. Technische Anleitung zur Reinhaltung der Luft, TA Luft, Deutscher Wirtschaftsdienst.

7. Bartaire J.-G., New orientations of the BAT activity, Document DGXI/A/3/JGB/kl, 18.1.91

Environmental legislation and advances in tailings disposal technology in North America and Europe

J. P. Haile
Knight and Piésold Ltd, Vancouver, British Columbia, Canada
M. Cambridge
Knight Piésold & Partners, Ashford, Kent, England

SYNOPSIS

California has traditionally led North America in the development and enforcement of environmental legislation. In 1984, the so-called "Sub-Chapter 15" regulations governing discharge of wastes to land was promulgated into law, with prescriptive standard requirements for containment of all mining related wastes. The legislation was based on the primary objective of protection of underlying groundwater and no significant impact on downstream beneficial water usage. While the overall requirements of Sub-chapter 15 appropriately achieve this objective, the prescriptive standards are sometimes inappropriate, and impose unnecessary constraints on the design of waste containment systems. This limitation has been recognised, and current trends in the USA and Canada are focusing on site specific objectives and containment systems.

The development of sub-aerial tailings disposal systems in North America over the last 10 years has played a major role in providing a cost-effective method of achieving or exceeding environmental protection standards. In California, the Jamestown Mine, a 6000 tpd gold mining operation and the first mine to be permitted under the new "Sub-chapter 15" regulations, has a sub-aerial tailings management system permitted as an engineered alternative to prescriptive standards. The facility is designed to achieve a fully drained, stable mass of tailings suitable for immediate reclamation on completion of operations, with separate storage of all process liquids and precipitation. The same design concepts have been proposed for a potential gold processing plant in northern Greece.

In the U.K., environmental legislation has historically developed piecemeal in response particularly to public pressure and to research developments. The Environmental Protection Act recently enacted is intended to draw previous regulations and legislative acts together and to provide a unified approach to all matters concerning environmental pollution. This legislation has, to an extent, been influenced by EC Directives and by the lead given by North America.

In some fields, however, particularly in mining, there has been a joint approach by both Governmental bodies and industrial concerns. The Aberfan disaster, for instance, led to stringent mine waste disposal legislation with regard to safety and stability, though, environmental and pollution regulations were not included. Current environmental legislation seeks to correct this, though local planning agreements and conscientious mine operators have often pre-empted statutory enforcement. Some tailings disposal projects in the UK have therefore been in the van in using state of the art technology to achieve safe and efficient tailings disposal and to minimise environmental pollution.

This paper outlines current North American legislation governing the disposal of mineral processing wastes, including the requirements for tighter controls and total containments, and compares these to European practice. The development and application of sub-aerial tailings disposal technology, particularly with respect to total containment/zero release, is described with case histories and operating experience in North America, the United Kingdom and Europe.

INTRODUCTION

Knight Piésold have been involved in the disposal of a wide range of mine waste products throughout the world for in excess of forty years, and have been responsible for planning, design and construction supervision, and decommissioning of tailings dams during this period. The resulting mine waste disposal experience has been developed against a background of widely varying legislation and often in its absence. The consistent design requirement has been the safe, efficient and environmentally acceptable disposal of waste products. During the last ten years, the mining industry has found itself operating under increasingly stringent environmental legislation, much of which has been developed in response to public concern, inadequate environmental standards and the effects of industrial pollution. The leading role in the increasing tightening of legislation has been North America with the European Community following their environmental lead. As a result mining companies have found themselves operating under strict guidelines with little flexibility during planning and development of new mines due to the often restrictive legislative framework within which they are constrained to work.

The experience of the firm based on worldwide project experience and legislative regimes is that a more pragmatic and flexible approach to waste disposal is feasible without detriment to the environment. This paper compares the current legislative environment on either side of the Atlantic and indicates where such a flexible approach has been successful and provides case histories which show the benefits of this attitude in mine planning and environmental protection.

WASTE DISPOSAL LEGISLATION IN NORTH AMERICA
General Approach

Legislation governing the design of mining and industrial waste disposal facilities throughout North America is generally based on the prevention of any significant impact on beneficial uses of groundwater and surface water downstream of the site. The focus is on the control of liquid wastes, it being assumed that solids wastes can be fairly easily contained. Significant impact on downstream beneficial water usage is variously interpreted to mean no detectable change in any background concentrations, to an allowable change up to prescribed receiving water quality limits. In many cases this requires a zero discharge of any liquids associated with the waste being stored.

The basis for any waste disposal facility design generally requires a comprehensive knowledge of existing or background conditions at the site, the physical and chemical characteristics of the waste to be stored, and site specific containment features or construction materials.

Establishing background conditions requires identification of resources potentially at risk, whether they be water quality, fisheries, soils, air quality, vegetation or the habitat of a rare and endangered plant or animal species. Particular emphasis is required to establish site-specific hydrology, hydrogeology and water quality data over at least one year of monitoring. This information is required to assess potential impacts of the proposed design and to provide actual monitoring data during operations. For existing operations undergoing expansion or modification, background water quality data may be significantly modified by past practices, and the assessment of new designs may be based on improvements to background conditions.

Site specific geotechnical information and containment features, such as natural clays, glacial tills or impervious bedrock, and the extent and possible use of these materials as part of the waste facility design, also form an essential part of the site data relevant to any design. Geotechnical data can often be obtained in conjunction with baseline hydrogeological investigations.

Waste Characterisation

Characterisation of the waste to be stored forms an essential starting point in the design of any containment facility. For mining wastes, some standard tests are usually required to determine the potential for acid generation and whether the material should be classified as a special waste.

Acid generation is the atmospheric oxidation of sulphides, with the ensuing chemical and biological processes resulting in a lowering of pH and leaching of heavy metals into water passing through the material. The prediction of acid generating potential is based on static acid/base accounting methods or more extensive kinetic tests. A static test includes the following:

- Sulphur content determination (%S)
- Calculation of the theoretical acid generating potential (AGP)
- Neutralisation potential determination (NP)
- Calculation of the net neutralisation potential (NNP)

The results of these tests determine if more extensive kinetic tests are required, which involve weathering under laboratory or site conditions in order to establish the potential for and rate of acid generation. One such laboratory test is referred to as the Humidity Cell test. Since the onset of acid generating conditions is rate controlled, the results of kinetic tests can be very important in determining the need for short-term control technology, which may differ from the long-term abandonment plan.

The classification of a waste material as a "special" or "hazardous" waste is generally based on the so-called EP toxicity test, or variations on it. This test involves leaching the waste with acetic acid at pH 5.0, and comparing the concentrations in the leachate to maximum allowable limits. Classification as a special waste initiates a different level of governing review and legislative rules in most jurisdictions.

The results from the above tests often form the basis on which wastes are classified and in many States determine prescriptive standard design details, which may or may not be the most appropriate design for individual site specific conditions. A more rational approach to waste characterisation is to carry out a full analysis of the waste liquid phase to characterise the source term for the facility design, and to include a leach test using carbonic acid at pH 5.5 to simulate long-term leaching with rainfall.

The current best-practice overall testing program for characterisation of the waste materials from the actual process flow sheet(s) is as follows:

- Physical characterisation of the solids to determine geotechnical properties.
- Acid/base accounting on the solids (to test for acid generating potential) and kinetic tests if appropriate.
- Leach tests with acetic acid (required for special waste classification).

- Leach tests with carbonic acid (to simulate long-term leaching with rainfall).
- A full analysis of the tailings liquid phase, and any other liquid waste streams from the process.

Once the waste characterisation data has been developed, it forms the basis, together with baseline environmental conditions, of the prescriptive design or the development of a site specific design to prevent or mitigate environmental impact. How the above data is applied in different legislative arenas is discussed in the following sections.

TABLE 1

WATER QUALITY CRITERIA FOR CALIFORNIA (1)

Parameter	Maximum Contaminant Levels (1)	Water Quality Criteria (3) Drinking Water	Freshwater Aquatic Life
Ammonia			0.02
Arsenic	0.05	0.05	
Barium	1	1	
Beryllium		0.011 - 1.1	
Cadmium	0.010	0.010	0.0004-0.0012
Chloride	250, 500	250	
Chlorine			0.002-0.0010
Chromium	0.050	0.050	0.100
Copper	1.0	1.0	
Cyanide		0.2	0.005
Fluoride		1.4 - 2.4	
Iron	0.3	0.3	1.0
Lead	0.050	0.050	
Manganese	0.050	0.050	
Mercury	0.002	0.002	0.00005
Nickel			
Nitrates, Nitrites	45	10	
Selenium	0.010	0.010	
Silver	0.050	0.050	
Sulphate	250, 500	250	
Sulphide			0.002
TDS (TFR)	500, 1000	250	
TSS			
Zinc	5.0	5.0	
pH		5.9	
Conductivity	900, 1600		6.5 - 9.0

(1) All values in mg/l except pH which is in standard units and conductivity which is in micromhos/cm.

(2) Source: California Administrative Code, Title 22, Environmental Health, Sections 64435 and 64473.

(3) Source: Quality Criteria for Water (USEPA, 1976).

California

The design of any waste management unit in California is regulated by the California Administrative Code, Title 23, Chapter 3, Sub-chapter 15 "Discharge of Waste to Land", which is administered by Regional Water Quality Control Boards. Sub-chapter 15 is an all embracing piece of legislation written primarily for municipal and industrial wastes. It includes some special provisions for the mining industry in Article 7 "Mining Waste Management".

The requisite waste characterisation testwork in California differs from all other States, in that the EP toxicity test is replaced by the California WET (Waste Extraction Test) test, which uses a citric acid leach in contrast to acetic acid. Relevant maximum contaminant levels for the WET test and water quality standards for California are shown on Table 1.

On the basis of the results of acid/base accounting, the WET test, water quality standards and anticipated process reagents, waste materials are classified as follows:

Group A: Hazardous wastes that pose a significant threat to water quality.

Group B: Waste containing non-hazardous constituents with a low risk, but with some potential for water quality degradation.

Group C: Inert material causing no degradation of water quality except for turbidity.

Based on the classification of the waste, Table 7.3 of Sub-chapter 15 provides prescribed liner requirements as reproduced in Table 2. Most non-hazardous waste products would be classified as Group B, and would require a double liner under the entire facility, consisting of an outer liner of 2 feet of clay with a permeability of 1×10^{-6} cm/s or less, an inner liner of clay or synthetic membrane, and a blanket drain between the two to monitor leakage through the inner liner and reduce the head acting on the outer liner. This drain is termed the Leach Collection and Removal System (LCRS).

While attempting to provide complete protection by the provision for double liner systems, these prescriptive standards have some in-built deficiencies which can result in increasing seepage and the creation of a long-term liability problem for the mine operator.

Implicit in the regulations is the notion that the prescribed liner systems will reduce impacts on

TABLE 2

PRESCRIPTIVE STANDARDS FOR WASTE CONTAINMENT IN CALIFORNIA (CAC Title 23, Chapter 3, Sub-chapter 15)

STATE WATER RESOURCES CONTROL BOARD

Waste Group	Waste Management Unit	Geologic Setting	Liner (s) Permeability Value in cm/sec	Leachate Collection and Removal System
A	Waste Pile	per Section 2531 (b)(1) OR single clay liner (1) $\leq 1 \times 10^{-7}$		required
	Surface Impoundment or Tailings Pond	not applicable	double liner, both $\leq 1 \times 10^{-7}$ cm/s outer: clay; inner: clay or synthetic	required (2)
B	Waste Pile	per Section 2532 (b)(1) OR single clay liner $\leq 1 \times 10^{-6}$(1)		required
	Surface Impoundment or Tailings Pond	not applicable	double liner, both $\leq$ 1×10^{-6} cm/s outer: clay or natural permeability (3); inner: clay or synthetic or single replaceable clay liner (4)	required (2)
C	Waste Pile, Surface Impoundment, or Tailings Pond	not applicable	not applicable	not applicable

(1)   Synthetic liner may be used for short-term containment (see Sub-section 2572 (f)(1) of this article).
(2)   Liner and leachate collection an removal system for tailings pond must be able to withstand the ultimate weight of wastes.
(3)   Permeability of $\leq 1 \times 10^{-6}$ cm/sec or natural geologic materials may replace outer liner of double liner system.
(4)   Single clay liner ($\leq 1 \times 10^{-6}$ cm/sec) for surface impoundment, to be removed before last 25 percent (minimum 1 foot thickness) of liner is penetrated by fluid, including waste and leachate.

groundwater to acceptable levels. However, a simple calculation using Darcy's Law to calculate seepage through a 10^{-6} cm/s outer liner under unit hydraulic gradient (for vertical flow under zero head, i = 1.0), shows that for any reasonably sized facility the seepage quantity can be quite large (8.6 m^3/d/ha) and could in fact have a significant impact. Furthermore, by containing all liquid wastes under fully saturated conditions within an inner liner, seepage through the inner liner can be very significant and will persist for a very long time.

California regulations do include, however, an allowance for engineered alternatives to prescriptive standards providing that equivalent or better water quality protection is demonstrated.

TABLE 3

WATER QUALITY CRITERIA FOR BRITISH COLUMBIA (1)

Parameter	Special Waste Regulation (2)	Water Quality Criteria (3) Freshwater Aquatic Life	Marine Aquatic Life
Ammonia			1.0
Arsenic	5.0	0.05	0.036
Barium	100.0	0.001	0.0005
Beryllium			0.1
Cadmium	0.5	0.0002	0.009
Chloride			
Chlorine		0.002	0.007
Chromium	5.0	0.002	0.05
Copper		0.002	0.002
Cyanide (free)	20.0	0.005	0.001
Fluoride	150.0	1.5	0.15
Iron		0.3	0.05
Lead	5.0	$\leq$0.008	0.002
Manganese		100 - 1000	
Mercury	0.1	0.0001	0.0001
Nickel		0.025	0.007
Nitrates, Nitrites	1000.0	(3)	
pH		6.5 - 9.0	6.5 - 8.5
Selenium	1.0	0.001	0.054
Silver	5.0	0.0001	0.0023
Sulphide		0.002(H_2S)	(3)
Zinc		0.03	0.058

(1)   All values in mg/l except pH which is in standard units and conductivity which is in micromhos/cm.

(2)   Concentration in Waste Extract for classification as a Special Waste.

(3)   See Reference.

British Columbia

In contrast to California, British Columbia uses an approach to legislating waste containment systems which is based on an assessment of potential impacts on the receiving water. Source term concentrations derived from analysis of the waste stream liquid effluent or carbonic acid leach test, (short or long term conditions as appropriate), are used together with seepage analyses and background hydrogeological or hydrological

conditions to assess net changes in baseline concentrations in receiving waters (groundwater or surface water). British Columbia, however, is prolific in fish, which are protected under both Provincial and Federal legislation by very strict requirements. Examples of special waste criteria and maximum allowable concentrations for selected constituents in receiving waters are shown on Table 3.

This approach differs from California in that the design of any waste containment facility can be varied to meet the specific receiving water criteria, depending on actual waste characteristics and site specific conditions.

Current Trends

California has traditionally led the way in establishing environmental protection legislation in the United States. Sub-chapter 15 was promulgated into law in 1984. Although some of its limitations are recognised by all, proposed changes have been difficult to formulate due to substantial differences in the perceived limitations between legislators and industry.

Other States have recently introduced specific mining waste disposal legislation which attempts to build on precedents established by California. In some cases, such legislation is needed and appropriate, but in others, attempts to go one better than California are based more on perception than on sound engineering principles. An example is an increase in the required thickness of outer clay liner from 2 feet (0.6 m) to 3 feet (0.9 m). This may increase the time for penetration of an initial wetting front by a few months, but will have no influence on total seepage quantity.

A significant aspect of liner design that is specifically discounted by current prescriptive standard legislation is the potential beneficial effect of attenuation in clay liners and natural geological materials. Attenuation is the process whereby concentrations of constituents in the seepage decrease with passage through the material. Its contribution can be significant and may be appropriate in certain site specific designs as a means towards achieving the overall objectives of non degradation of downstream beneficial water usage.

364

EUROPEAN LEGISLATIVE ENVIRONMENT
Historical Context

Waste Disposal Legislation within Europe has a more complex history than in the United States, having developed from a long history of mining. Metalliferous Mining in Europe has been undertaken for in excess of 2,000 years, the Romans and the Phoenicians, for instance, are known to have traded both tin and copper in Cornwall and Spain. Mining on an industrial scale can be considered to have commenced in Europe, around the 16th Century reaching its peak in the UK during the 19th Century when the Cornish mines for instance were the largest producers of copper in the world. Mining legislation has thus developed in response to industrial activity and to public awareness of safe working practices, rather than to pollution control.

In the UK specific mine waste disposal legislation was first introduced in 1969. Prior to this strict liability was imposed by non specific Common Law and legal precedent (Rylands Fletcher, 1866) on a person/persons keeping potentially dangerous objects or carrying on a dangerous operation on his land, interpreted to include inter alia water supply reservoirs and colliery and mine spoil heaps.

The post war period produced a number of legislative instruments in the UK broadly covering industrial safety and pollution control on a broad front.

- The Mines and Quarries Act (1954) which dealt exclusively with underground safety and with safety of operations on all mines and quarries but not with disposal or its environmental consequences.
- The Clean Air Act of 1954, enacted in response to the estimated 4,000 deaths in London following the severe smog of December 1952; an Act which, was well in advance of environmental legislation throughout the industrialised world and which, despite its shortcomings, was responsible for the reduction of air pollution in the UK.
- The 1951 Prevention of Pollution Act in Rivers, instigated to prevent uncontrolled river discharges and, for the first time, imposing licenses on discharges into rivers.

Each of the above were non specific with general guide-lines rather than global or regional quality levels. Nevertheless significant impact on industrial discharges followed from this legislation.

The Introduction of Mine Dump Legislation in the UK

In 1966, the problem of the stability of mine tips and spoil heaps was brought to public attention with dramatic force by the disaster at Aberfan. This spoil tip failure involved a flow slide of just 140000 yd^3 of mine waste but lead to the loss of 144 lives; 116 of which were children. According to Bishop (Ref 13) "this brought to the attention, not only of the public but also to most professional engineers and geologists, even to those concerned with mining, that it (disposal of mining waste) was a problem to which they had given little, if any, serious attention".

The subsequent government report on the Aberfan disaster recommended extensive and specific controls on the disposal of all mine and quarry wastes. The resulting Mines and Quarry (Tips) Act 1969 was subsequently enacted within the general provisions of the 1954 Act. The Mines and Quarries Regulations specified that all mine and quarry tips were to be controlled with respect to safety, stability and regulation of dumping. The legislation defined levels of responsibility for owners and managers and imposed specific requirements for inspection and design of all tips by so called "competent persons".

World-wide this legislation was seen as probably the most comprehensive enacted at that time, thus in 1970 when the Mufulira disaster occurred on the Zambian copperbelt, resulting in the deaths of 89 people following the inundation of the mine workings by tailings, the United Kingdom Mines and Quarries Act was used virtually word for word as the basis for Zambian mine tips law. Since that time the regulations have been used as the basis for other mine dump and tailings dam safety legislation.

Current UK Pollution Controls

The Mines and Quarries (Tips) Act and the subsequent 1971 Regulations do not impose pollution controls other than those relating to safety i.e. control is imposed on the disposal of solids alone. The pollution element in the UK has historically been controlled by the Town and Country Planning Act and air and water pollution legislation previously mentioned. These were considered to be highly developed planning controls designed to ensure that the environmental and other consequences of individual developments were fully considered before being allowed to proceed. County Planners thus undertook a controlling role to the benefit of the public and of the environment but generally in a non-specific guise. Where appropriate as for mine dumps, advice was taken at the planning stage from relevant statutory consultees such as water authorities who imposed regulations on discharge consent limits. However, the lack of guide-lines for specific environmental elements together with absence of local mining experience often led to a lack of uniformity in discharge controls for new mining ventures.

Since 1970, with increasing public awareness further legislation has been prepared, initially in advance of European Community legislation, but more recently in direct response to Brussels. The 1989 Water Act and the establishment of the independent water regulating authority, the NRA, imposed real controls on industrial discharges. Similarly the Town and Country Planning Regulations 1988 enshrined formal environmental impact assessment in the planning process for all new developments. More recently the 1990 Environmental Protection Act of the United Kingdom has been enacted which with the 1989 Water Act, has established centralised regulatory authorities ie. the Inspectorate of Pollution, the Health and Safety Executive and the National Rivers Authority. The Act is a move towards stricter regulation introduced integrated pollution control rather than EC element specific standards. The introduction of integrated pollution control will enable the UK to move towards a more site specific approach, and alone amongst its European competitors enable consideration, at any site, of the effect of individual elements, from any process on air, water and soil. This system, once fully enacted, will enable rational site specific consent and discharge controls to be imposed on all future mining projects and provide greater emphasis on good engineering practice to the benefit of all.

Present European Controls

The European Community has had an increasing influence on environmental legislation throughout Europe and will continue to do so particularly as a result of the move towards the creation of a genuine single market.

Early directives from Brussels were aimed at specific pollutants the so called "fire fighting" directives which set limits and standards for the control of specific substances. These led to the 1976 EC directive on control of discharges to water limiting such elements as mercury, cadmium and arsenic. At this stage legislation was reactive but by 1977 a preventive or precautionary approach was evident. In 1985 environmental assessments were enshrined in legislation and became a centre piece for all future development plans under EC regulations and in 1987, care of the environment, was enshrined in the Treaty of Rome. An increasing number of environmental controls have been imposed on member states since this time.

The aim of the European legislation has been to provide specific limits for air and water discharges and for land contamination with the aim of establishing common standards across Europe. This approach though laudable is fraught with anomalies, particularly where geographical features conflict with EC standards, natural radon emissions in Cornwall are such an example. The proposed controls therefore make little allowance for individual site conditions and the following example indicates the potential effect on new mine projects.

A new mining venture in the UK required consent to discharge effluent from the tailings dam which would be based under EC guidelines on a consent standard for the receiving water. In the specific locality the receiving water standard to be achieved was the directive for salmonides 78/659/EEC. Quality analysis indicated that the natural river quality upstream of the site was, for both hydrogeological and geomorphological reasons, acidic with a natural heavy metal content resulting from contact with mineral outcrops within the river valley. The natural level of copper in the river was significantly above the requirements of the EC directive. This standard therefore could not logically be applied at this site on tailings dam discharges. For this project a more flexible approach based on site conditions was clearly required and could be employed under new UK integrated pollution control criteria.

A strict but more flexible approach has been taken in Ireland which has a shorter history of mining but has had some notable environmental problems from old tips.

In particular full protection against dust and of groundwater is require (80/68/EEC) to prevent discharges of significant quantities of "certain dangerous" substances, Table 4. Full containment by an impervious liner is not mandatory but it must be demonstrated that the quantity and quality of seepage to groundwater is acceptable.

TABLE 4

EC DIRECTIVE ON PROTECTION OF GROUNDWATER AGAINST POLLUTION CAUSED BY CERTAIN DANGEROUS SUBSTANCES (80/68/EEC)

List I of Families and Groups of Substances

List I contains the individual substances which belong to the families and groups of substances enumerated below, with the exception of those which are considered inappropriate to List I on the basis of a low risk of toxicity, persistance and bioaccumulation.

Such substances which with regard to toxicity, persistance and bioaccumulation are appropriate to List II are to be classed in List II.

1. Organohalogen compounds and substances which may form such compounds in the aquatic environment.
2. Organophosphorus compounds
3. Organotin compounds
4. Substances which possess carcinogenic mutagenic or teratogenic properties in or via the aquatic environment.[1]
5. Mercury and its compounds
6. Cadmium and its compounds
7. Mineral oils and hydrocarbons
8. Cyanides

List II of Families and Groups of Substances

List II contains the individual substances and the categories of substances belonging to the families and groups of substances listed below which could have a harmful effect on groundwater.

1. The following metalloids and metals and their compounds:

1.	Zinc	11.	Tin
2.	Copper	12.	Barium
3.	Nickel	13.	Beryllium
4.	Chrome	14.	Boron
5.	Lead	15.	Uranium
6.	Selenium	16.	Vanadium
7.	Arsenic	17.	Cobalt
8.	Antimony	18.	Thallium
9.	Molybdenum	19.	Tellurium
10.	Titanium	20.	Silver

2. Biocides and their derivatives not appearing in List I.
3. Substances which have a deleterious effect on the taste and/or odour of groundwater, and compounds liable to cause the formation of such substances in such water and to render it unfit for human consumption.
4. Toxic or persistent organic compounds of silicon, and substances which may cause the formation of such compounds in water, excluding those which are biologically harmless or are rapidly converted in water into harmless substances.
5. Inorganic compounds of phosphorus and elemental phosphorus.
6. Fluorides
7. Ammonia and nitrates.

[1] Where certain substances in List II are carcinogenic, mutagenic or teratogenic, they are included in category 4 of this list

Control of seepage must involve best appropriate technology given the site conditions. However, the temptation is to impose full containment by an impervious liner regardless of site conditions. New mines thus face significant costs at the planning stage in order to avoid imposition of inappropriate quality standards by the strict interpretation of EC directives.

SUB-AERIAL DEPOSITION, LINER SYSTEMS AND LONG TERM ABANDONMENT

Recent developments in the design and operation of waste management facilities have focused on systems that utilise the characteristics of the waste materials, site specific climatic conditions and advances in liner design to achieve basic objectives, without reliance on prescriptive standard systems.

Sub-aerial deposition of tailings slurries, in contrast to sub-aqueous deposition into a saturated environment, involves the systematic, rotational deposition of the tailings in thin layers on an exposed tailings beach, with continuous decanting of surface water. Low energy laminar flow of the tailings slurry over a large area results in good liquid solid separation of the slurry, analogous to a lamellar thickener. The discharge area is systematically rotated and the settled solids are left exposed for a period to drain, consolidate and, if climatic conditions permit, achieve the further benefit of air-drying. Underdrainage at the base of the tailings will also assist in consolidation and drainage.

The deposition technique results in a drained, consolidated deposit of laminated tailings which itself provides a control over vertical seepage during operations, and, since the tailings are fully drained and consolidated, allows for relatively simple decommissioning and the elimination of long term environmental concerns.

In controlling leakage from lined storage facilities, soil liners have some distinct characteristics. Seepage will take place over the whole liner area, however, the permeability of the soil will generally decrease with increased loading or confining pressure, reducing the rate of seepage. There may also be a significant reduction in the concentrations of waste constituents in the seepage water as a result of attenuation. Construction of effective soil liners requires a proper application of moisture conditioning, placing and compaction requirements.

Incorrect procedures can render soil liners virtually useless, and a poor understanding of these requirements has contributed to a large extent to documented failures.

Synthetic liners are also commonly used, with high and low density polyethylene currently favoured to provide very effective, low permeability liners. However, even these materials have a finite permeability due to pinhole leaks and poor seaming techniques. In almost all cases where an effective leak detection system has been installed, some finite leakage is detected. Typical values for leakage from HDPE liners for even the best installations range from 45 to 450 litres per day per hectare (30 to 300 gpd per acre).

One characteristic of leakage through a synthetic liner is that it is likely to be confined to certain discrete areas where defects occur. By installing a synthetic material in intimate contact with the top of a clay liner, a synergism exists and the characteristic imperfections of each liner are significantly reduced. Seepage through the synthetic liner will occur at the localised imperfections only, but under a greatly reduced hydraulic gradient due to the underlying clay. Similarly, seepage through the clay will occur directly beneath the imperfections and the total area of seepage will be greatly reduced. A compound liner consisting of a synthetic material (HDPE or VLDPE) in intimate contact with a low permeability clay liner offers the best current practical technology for seepage control. A single compound liner provides for better protection than the California Group "B" double liner prescribed by Sub-chapter 15.

A combination of an appropriately designed liner system, a drainage system above the liner and sub-aerial deposition of the waste material will result in complete protection of groundwater during operations, and a fully drained, consolidated waste pile on completion of operations suitable for long-term decommissioning. Removal of all excess water during operations will eliminate the potential for long-term seepage. Air-drying and consolidation of the tailings will facilitate surface capping and revegetation measures.

CASE HISTORIES

Jamestown Mine, California

Sonora Mining Corporation constructed the Jamestown mine, a 7,000 ton per day open pit gold mine with a froth flotation ore concentrator in 1986. The flotation tailings,

classified as a California Group B mining waste, are pumped to a tailings management facility which is designed as an engineered alternative to the State's prescriptive standards requiring a double liner. The facility is designed to achieve a fully drained, stable mass of tailings suitable for immediate decommissioning on completion of operations, with separate storage of all process liquids and precipitation run-off.

During the initial design stage for the tailings management system (TMS) it became apparent that strict adherence to prescriptive standards would result in a large basin of saturated, low density tailings that would pose a continuous threat to water quality, would be unstable and difficult to construct a final cover on, and would require extensive long-term monitoring and contingency planning. An engineered alternative was proposed which consists of a drained tailings storage facility with a single outer clay liner, and a separate double-lined process water pond. Rotational deposition of tailings is carried out around the perimeter of the storage facility to produce a sub-aerial tailings beach sloping towards the centre of the embankment, where surface water is continuously decanted to the process water pond. Seepage from the base of the tailings resulting from consolidation is also continuously removed by the LCRS overlying the outer clay liner. A schematic section through the facility is shown on Figure 1.

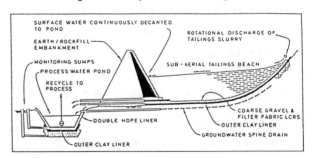

FIG 1   JAMESTOWN MINE TAILINGS MANAGEMENT SYSTEM
- SCHEMATIC CROSS SECTION

The initial stage of tailings management facility was constructed in 1986 with on-going expansions of the liner system and embankment raises. Performance of the TMS has been carefully monitored by piezometers and flowmeters within the facility and water quality testing on process streams and groundwater monitoring wells. The monitoring programs confirm that the system is performing as designed and that the water quality within the facility is very good, and little different to the natural groundwater. Detection monitoring for process reagents has confirmed that there is zero impact by the TMS on local groundwater.

Gold Processing Plant, Greece

Preliminary designs have been prepared for the tailings storage facility for a proposed gold processing plant in Greece. The process involves pressure oxidation of an arseno-pyrite concentrate, and carbon-in-leach gold recovery, with a tailings product consisting of gypsum, calcium carbonate, ferric hydroxide and ferric arsenate.

Waste characterisation testwork was carried out on pilot test samples of the tailings as outlined in Section 2.0, resulting in a classification as non-hazardous based on the EP toxicity test criteria. The carbonic acid leach test indicated leachable constituent concentrations slightly above drinking water standards for a few parameters. The leachate generated in the carbonic acid leach was then passed through the proposed clay liner material to investigate the effects of attenuation, which was indicated to be significant.

The resulting proposed design of the tailings storage facility consisted of a clay liner with overlying drainage blanket over the base of the facility, an impervious confining embankment, and a separate recycle pond for the underdrainage and decant water. The recycle pond was designed with a compound (HDPE over clay) liner. The proposed design would achieve the basic objective of non-degradation of downstream beneficial water usage.

On-going work on this project has been suspended pending resolution of the project location.

Wheal Jane Mine, UK

The Wheal Jane tin mine in Cornwall was developed in an existing mining area where records show underground workings had commenced during the 16th Century and been continuous since that period. The ore body to be developed was adjacent to an extentive group of 19th Century copper mines and therefore in an area where mine workings had affected both river discharges and the environment.

The mine was developed by Consolidated Goldfields in 1970 with a throughput of 1000 t/d and was the first major tailings dam to be designed and constructed under the UK Mine and Quarries Regulations.

The tailings were to be separated by gravitational and flotation methods and deposited behind a dam constructed in an adjacent valley. The tailings dam was designed to incorporate the coarse fraction of the waste within the confining wall. Thus fully drained tailings comprised the dam wall whilst the fine tailings were deposited into the reservoir by controlled beaching to achieve a semi-drained state (Fig. 2). The zoning of the

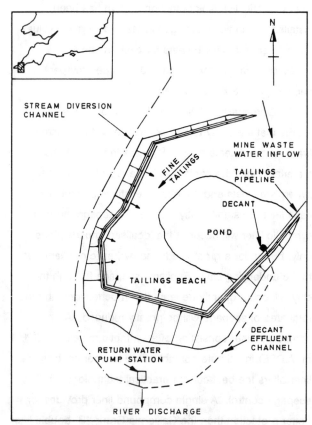

FIG 2   CLEMOWS VALLEY TAILINGS DAM - SCHEMATIC LAYOUT

depository and the deposition method limited the percolation through the tailings into the groundwater system. As a result of the pyrite content of the orebody and the volume of acidic water pumped to surface in order to dewater the mine, the tailings dam was also designed as an industrial water supply reservoir receiving underground water and neutralising it, prior to recycling to the mill or discharging back into the water course.

The development of the mine was undertaken under UK planning procedures in 1968 with a parallel application for consent to discharge to the Cornish River Authority. The environmental implications of the development of the mine were submitted with the planning application. The relevant environmental concerns were as follows:-

- Noise and dust controls imposed by County
planning authorities to the then UK national
standards but with local adaptation as considered
appropriate particularly with respect to possible
dust emissions from the tailings dam and the
potential impact on local agriculture.

- River diversion, control of effluent discharge and
industrial water supply controlled by the local
water authority based on a discharge licence
related to volume of discharge and prime consent
limits for specific elements related to condition of
existing rivers.

- Construction of the containment facility and
disposal of waste undertaken in accordance with
the Mines and Quarries Tips Regulations involving
detailed discussion with the Health and Safety
Executive and the Department of the Environment.

The project received planning consent in 1970 and
under the ownership of Consolidated Goldfields, RTZ and
currently Carnon Consolidated Tin Mines Ltd. tailings
have been deposited behind the Clemows Valley Tailing
dam for in excess of twenty years. The tailings dam is
now some 43 m high and stores some 3.5 million tonnes
of mine waste (Fig. 3). The various regulating bodies
have continued to take an interest in effluent discharges,
stability, etc of the tailings dam. Discussions have been
held with the relevant Inspectorate with respect to each
design change necessary following variations in
production, ore type and mining method (Ref. 14). The
continued interest shown by the Mines and Quarries
Inspectorate has resulted in few changes to the overall
disposal scheme in the valley despite the tightening of
legislation during the production period.

Subsequent to the 1989 Water Act and the
establishment of the independent regulatory authority
(National Rivers Authority) a review of consented
discharge limits was undertaken in 1990. Under the new
licence the quality objectives remain unchanged from
those consented in 1970. Despite the much publicised
"laxity" of UK environmental legislation Wheal Jane is an
example where a conscientious operator has undertaken
the development of a mine and disposal of waste for
some 20 years in an environmentally acceptable manner
consistent with the conditions existing on the site.

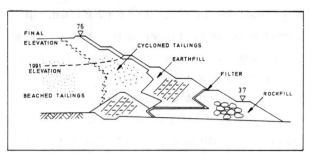

FIG 3   CLEMOWS VALLEY TAILINGS DAM - CROSS SECTION

Neves Corvo Mine, Portugal

Portugal by comparison, where industrial scale mining is
relatively modern, had no specific environmental
guidelines for new mine projects. The development of a
new mine under such conditions often leads to the
adoption of the severest environmental criteria. Such an
approach may restrict development potential or may
reduce design flexibility such that disposal of mine waste
may be neither as efficient nor as practicable as possible.

The Neves Corvo mine in Southern Portugal is a
new 1×10^6 t/yr copper/tin mine which commenced
production in 1988. The mine waste to be generated by
the extraction of the copper and tin ores was to be
deposited in an adjacent valley to the mine site behind
the Cerro do Lobo dam. With no mining background, the
Portuguese authorities took an extreme view and
considered that for reasons of environmental protection,
particularly acid generation potential of the high sulphide
tailings, a zero discharge facility should be constructed.

The Cerro do Lobo dam was therefore designed
as a total containment structure providing both tailings
disposal capacity and water supply storage for the
effluent products from the mill. The Stage I dam was a
28m high clay cored rockfill shouldered dam and was
constructed between 1987 and 1988, with Stage II
construction to raise the dam by 4 m undertaken during
1990 (Ref. 15). To satisfy environmental and mine
planning constraints tailings disposal by sub-aqueous
means was undertaken primarily to prevent acid
generation and any dust from the extensive tailings
beaches which would otherwise be formed. Construction
of the tailings dam was preceded by extensive
environmental baseline studies including, in particular, a
hydrogeological study of the reservoir basin to check the
likely effects of tailings disposal on the existing
groundwater regime. In 1988, tailings disposal was

commenced with distribution via a number of feed pipes located on the dam and on the eastern abutment (Fig. 4). The inundation of the deposited tailings to reduce the acid generation potential of the sulphide waste will be effective in the short term but in the longer term poses significant abandonment problems, due to positive evaporation, a fresh water supply would be required to ensure the waste deposit remained flooded in perpetuity.

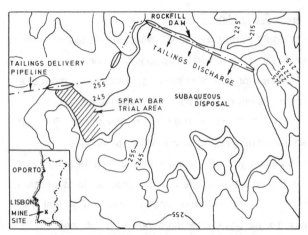

FIG 4   CERRO DO LOBO TAILINGS DAM - SCHEMATIC LAYOUT

Knight Piésold were commissioned to review tailings disposal processes in 1988 and to consider methods of improving the deposition characteristics of the tailings. Initial laboratory studies had indicated that significant density enhancement could be achieved without acid generation in the tailings solids. Sub-aerial deposition trials using spray bars were therefore undertaken within the tailings depository. The spray bar trials were undertaken during the hottest and driest period of 1989, and measurements of pH and moisture content were carried out on the deposited tailings on a frequent basis. Accurate measurements of the dry density of the semi-solid tailings was not undertaken but an indication of density was obtained from the moisture content relationship and showed a significant improvement, by a factor of between 25 and 30%. At the same time, where tailings deposition was controlled, a pH value of above 7 was maintained throughout the trials showing little evidence of acid generation. Where uncontrolled deposition occurred, rapid oxidation and depression of pH occurred, resulting in low pH values of between 2 and 3 locally.

The trials indicated that controlled deposition of the tailings could be undertaken to improve storage capacity, which would not, detrimentally, affect the acid generating potential of the tailings nor depress the pH of the water and, more importantly, would provide the potential for long-term restoration to other than an inundated state. Further trials and consideration of the restoration potential on this dam will be needed prior to the adoption of a final restoration scheme. However, the trials are a positive indication that with a more flexible approach to deposition, appropriate technology can be applied to prevent environmental effects, improve geotechnical characteristics, and therefore long-term restoration potential.

CONCLUSIONS

The experience of mine waste disposal schemes worldwide against a background of increasingly restrictive legislation indicates that economic development is not necessarily served by rigid global standards. A more flexible approach will promote good engineering practice and environmental protection. Future legislation should take account of the need for site specific requirements for new developments and promote technologically appropriate solutions to the improvement of industrial activity and environmental protection.

References

1.   California Administrative Code, Title 23, Chapter 3, Sub-chapter 15 "Discharge of Waste to Land".

2.   Pommen L.W. Approved and Working Criteria for Water Quality. Government of British Columbia, Ministry of Environment and Parks, Water Management Branch, Resource Quality Section. April, 1987.

3.   Guidelines for Freshwater Aquatic Life. Canadian Council of Resources and Environment Ministers (CCREM).

4.   Special Waste Regulation. Waste Management Act, Government of British Columbia, April, 1988.

5.   Knight R.B. and Haile J.P., Sub-Aerial Tailings Deposition with Underdrainage. Seventh Pan American Conference on Soil Mechanics and Foundation Engineering. June 19-24, 1983, Vancouver, B.C.

6.   Haile, J.P. and East, D.R., "Recent Developments in the Design of Drained Tailings Impoundments".

Geotechnical and Geohydrological Aspects of Waste
Management. 1986, Fort Collins, Colorado.

7.   Haile, J.P. and Brouwer, K.J., "Design and
Construction of the Montana Tunnels Tailings Disposal
Facility". 89th Annual General Meeting of the Canadian
Institute of Mining. May 3 - 7, 1987, Toronto, Ontario.

8.   East, D.R., Haile, J.P. and Dew, H.P, "Sub-aerial
Deposition of Phosphatic Clay Wastes". Symposium on
Consolidation and Disposal of Phosphatic and other
Waste Clays. May 14 - 15, 1987, Lakeland, Florida.

9.   Haile, J.P, "Air-entry Permeameter Provides Rapid
Assessment of In-situ Permeability of Clay Liners".
Published in Environmental and Waste Management
World, Volume 2, Number 2, February, 1988.

10.   Haile, J.P. and Kerr, T.F., "Design and Operational
of the Myra Falls Tailings Disposal Facility". Vancouver
Geotechnical Society's 4th Annual Symposium on
Geotechnical Aspects of Tailings Disposal and Acid Mine
Drainage. May 26, 1989, Vancouver, B.C.

11.   N. Fernuik, M.D. Haug, J.P. Haile, "Comparison of
Laboratory and In-situ Field Soil Liner Permeability
Measurements". 42nd Canadian Geotechnical
Conference. 1989, Winnipeg, Manitoba.

12.   Skolasinski, D.Z., Haile, J.P. and Smith, A.C.,
"Design Objectives and Performance of the Tailings
Management System for the Jamestown Mine, California".
Society of Mining, Metallurgy and Exploration, Western
Regional Symposium on Mining and Mineral Processing
Wastes. May 30 -June 1, 1990, Berkeley, California.

13.   Bishop, A.W. The stability of tips and spoil heaps.
Quarterly Journal of Engineering Geology. Vol. 6 Nos. 3
and 4. 1973.

14.   Cambridge, M. and Coulton, R.H. Geotechnical
Aspects of the Construction of Tailings Dams - two
European Studies. Proceedings of the British Dam
Society 6th Conference, The Embankment Dam, Sept.
1990.

15.   Cambridge, M. & Maranha das Neves, E. The Use
of Textured Geomembrane Sheeting for the staged
Raising of the Cerro Do Lobo Dam. Water Power and
Dam Construction. June 1991.

Contaminated soil treatment technologies

Eckart F. Hilmer
Lurgi-Umwelt-Beteiligungsgesellschaft mbH, Frankfurt, Germany

SYNOPSIS

Contaminated Soils are destinated to become a
major problem in densely populated countries.
Land disposal prohibitions will require that
these soils must be treated using best demonstrated available technology.

Various possibilities exist in soil decontamination, organic contaminants can be in-
cinerated, separated by washing or biologically
treated.

Inorganic contaminants can be separated by
washing, extracted or leached by acids and can be
solidified.

Lurgi has broad experiences in all kind of
waste incineration; primarily applied are incineration in the grate incinerator, multiple
hearth furnace, fluidized bed incinerator,
atmospheric circulating fluidized bed incinerator, special combustion chambers (incineration of TNT contaminated soils) as well as
the rotary kiln.

For remediating contaminated soils, Lurgi has
developed a wet mechanical separation process
using only water for cleaning the soil. Plants
are designed as semimobile systems.

Dredging mud, a disposal problem or raw material for the building industry. To solve this
problem Lurgi developed a stepwise process, separating the mud into sand and silt, dewatering
the silt fraction, producing silt pellets and
converting the silt pellets into hard-burnt
ceramic products replacing natural gravel for
concrete production by thermal treatment. Harbour
sediments with high amounts of organic and inorganic contaminants are thus converted to
building material, a contribution to save raw
materials by recycling.

Contaminated soils are one of the most urgent
political and technological challenges facing
Germany today, especially the eastern states.

Protecting the environment form pollution and
remediating contaminated soils are among the most
important objectives for the years to come.

So far, experience with various remedial me-
thodes is available for emergency cases, i. e.
cases where the groundwater is already contaminated with leachates from abandoned industrial
sites or closed-down landfills, such that the
drinking water supply is jeopardized.
Another example is when the construction of industrial facilities, residential areas or railway
lines is brought to a halt because of soil
contamination.

In most of these cases, excavation of the
affected soil alone is no longer sufficient as
it no more than shifts the problem to another
location. Equally unsatisfactory are in-situ
stabilization methods involving the immobili-
zation of the contaminated soil in an monolithic
block or containment by liner systems to prevent
contaminats from being released to the ambient
air, ground and surface water and the surrounding soil. So far, nothing is known about the
long-term behaviour of the different liner systems so that the problem may recur at a later

time.

In contrast, high hopes are placed in genuine remedial methods where the contaminants are converted into harmless substances or separated and solidified, i. e. vitrification.

Lurgi, a process-oriented engineering and contracting company has been engaged in the development and design of environmental protection technology for many decades.

THERMAL TREATMENT

Thermal processes can be used for the decontamination of contaminated soils in two ways; on the one hand, organic compounds can be broken down into their harmless basic components by incineration; on the other hand, methods like vitrification and sintering of the soil can immobilize inorganic contaminants, especially heavy metals, to such an extent that their release to the environment is ruled out even under the most extreme conditions such as salt water, acid rain etc.

Lurgi offers several thermal processes for the treatment of contaminated soils such as thermal treatment by exothermal or endothermal reactions in fluidized bed systems, multiple-hearth incinerators, rotary kilns and other special combustion systems of various designs as well as the travelling grate process for vitrification.

THERMAL TREATMENT IN THE ROTARY KILN

Rotary kilns are the most widely used systems for the thermal treatment of contaminated soils. Because of their great operating flexibility, they are ideally suited to the broad spectrum of contaminants typically encountered in soils and building rubble. Rotary kilns may be directly or indirectly fired, operate under oxidizing or reducing conditions, at atmospheric conditions or under vacuum. Directly fired rotary kiln lend themselves to both co-current and countercurrent operating modes.

The rotary kiln system is proposed for the clean up of the closed-down Marktredwitz chemi-

cal plant. Before the Marktredwitz chemical plant was closed down in 1985 it produced a great variety of mercury products. Negligence and deficient production methods have led to massive mercury contamination in the soil and parts of the buildings. In total, about 80,000 tons of soil and building rubble require remedial treatment.

To clean the material of the mercury and various mercury compounds, Lurgi has proposed a rotary kiln fuelled directly with natural gas (Fig. 1). At operating temperatures of abt. 1000° C, the mercury compounds are vapourized and carried out of the kiln with the flue gas.

CFM   Hg - Decontamination

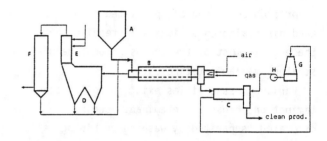

Fig. 1 Directly fired rotary kiln

In a special downstream flue gas cleaning system, the mercury vapours are condensed and the flue gas is consistently cleaned to residual Hg levels of less than 0.05 mg/m^3 STP. The residual Hg concentration of the treated soil is below 10 mg/kg compared to 1000 mg Hg/kg in the raw soil. The plant is designed to process 6 tonnes of soil per hour.

THERMAL TREATMENT IN THE FLUIDIZED BED

Fluidized bed combustion systems are commonly employed for the incineration of oil-containing sludges, industrial effluents and sewage sludge as well as for the thermal treatment of industrial residues. This process has also been successfully employed for treating the contaminated fines fractions of soils and debris.

The use of fluidized bed systems is subject to the following constraints:

- The softening point temperature of the material must be above 950° C.
- Particle decomposition must not interfere with the build-up of a stable fluidized bed.

A distinction is made between the following types of fluidized bed combustion systems:

- Conventional fluidized bed process with a defined bed surface.
- Expanded circulating fluidized bed.

The conventional fluidized bed furnace consists of a vertical cylindrical combustion chamber provided with a refractory lining. The fine-grained soil is decomposed, dried and its organic constituents are burnt in a suspended bed of quartz sand. Uniform air distribution, which is crucial to complete destruction, is achieved through a special configuration of the nozzle grate (Fig. 2). The residue is entrained with the flue gas and collected in downstream dust collectors.

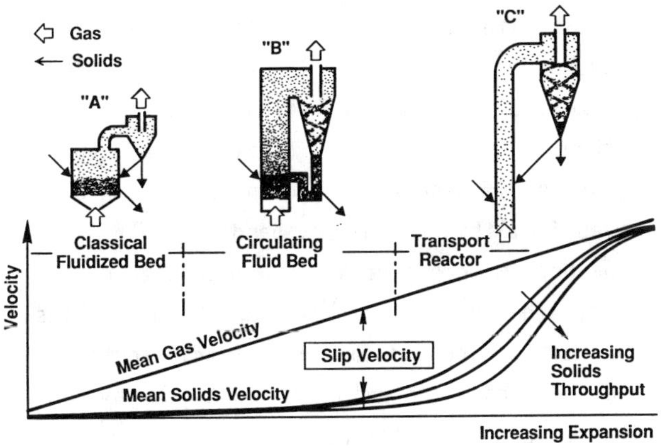

fluidized bed process lies in the long residence times achieved through solids recirculation to the combustion system (Fig. 3).

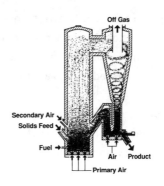

Fig. 3 Circulating fluidized bed process

In some cases, special combustion systems are needed to achieve the required clean up effect. A typical example of such a case is the DAG site in Stadtallendorf/Hesse (Fig. 4). In this remediation projekt, concentrated TNT sludge obtained form an upstream lurgi DECONTERRA soil washing plant is injected into a vertical combustion chamber. Using natural gas as a fuel, the sludge has to be burnt at a high air-to-fuel ratio to achieve complete destruction of the trinitrotoluene.

Fig. 2 Development of Fluidized Bed Technology

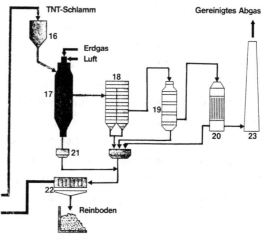

16 Schlammbunker   21 Absetzbehälter
17 Brennkammer   22 Entwässerung
18 Wärmerückgewinnung   23 Abgaskamin
19 Gaswäscher
20 Naßelektrofilter

Fig. 4 Spezial combustion system
Stadtallendorf/ Hesse

Fluidized bed incinerators are characterized by their high heat capacity and good heat and mass transfer characteristics. Circulating fluidized bed combustors use high-velocity air to entrain circulating solids and create a highly turbulent combustion zone for the destruction of toxic compounds. The strength of the circulating

EXTRACTION PROCESS

Apart from thermal processes, extraction methods are increasingly used to clean up contaminated soils. A typical representative of extraction technology is the so-called soil washing process.

An important strategy in operating a soil washing system relies on the fact that the contaminants are primarily sorbed on the clay and silt minerals, the hydrous oxides and organic matter and are therefore concentrated in the fines fraction. The cation exchange capacity of clay minerals permits large amounts of inorganic and organic matter to be sorbed. The contaminants adhering to the coarse material can be washed off by means of either plain water or water with detergent additions.

In principle, the mechanisms involved in the separation of organic and inorganic contaminants from soil are purely physical. The main problem is to achieve a sufficiently high energy input to liberate the contaminants from the soil and transfer them into a fines suspension.

Drawing in its broad knowhow in the wet mechanical processing of minerals, especially ores, salts and sediments from still and running surface waters, Lurgi has developed the so-called DECONTERRA soil washing process. This process uses an attrition scrubber to separate the soil into a clean coarse fraction and a contaminant concentrate. In the attrition drum, soil aggregates are broken up and the coarse materials intimately mixed with the fine material using the coarser material as grinding bodies. Attrition scrubbers are operated at net energy inputs of 4 to 16 kW/t throughput and drum speeds in the range of 50 to 90 % of the crictical speed, i. e.

$$n_{crit} = \frac{42.4}{\sqrt{D}} \ (min^{-1})$$

The suspension dischanged from the attrition scrubber is segregated into coarse and fines fractions which are then routed to further treatment.

Fig. 5 shows a flow diagram of the standard Lurgi DECONTERRA process.

ATTRITION SCRUBBING

In a first major process step, the raw soil is processed through an attrition scrubber using water as a washing agent. Depending on the soil to be treated, specific energy inputs may be as high as 16 kW/t raw soil. This pre supposes that the process is operated and controlled within very narrow constraints. Attrition scrubbing results in the practically complete desintegration of the soil minerals and components down to the micron range.

Another important aspect is that the clay agglomerates are largely broken up, thus increasing the surface area available for adsorption of the contaminants separated from the coarser fractions.

CLASSIFICATION

After attrition scrubbing, the soil is classified applying cut off levels suited to the downstream sorting stages and the specific soil characteristics.

A mesh size of 20 mm is typically employed for the upper screen deck. The 20 mm + fraction is processed through a downstream impact crusher and then returned to the first attrition scrubbing stage. If necessary, wooden debris can be removed from this circuit by hand.

The bottom screen deck has a mesh size of 1 to 2 mm. The screen overflow minus 20 mm to plus 1 to 2 mm is directed to a gravimetric sorter where the highly contaminated light-gravity constituents are separated out.

The reasons for the selection of a bottom screen mesh of 1 mm were as follows:

1) In the presence of fibrous components in the soil, a cut off level of 1 mm is just about the limit that can be handled reliably by industrial screens in a continuous operating mode.

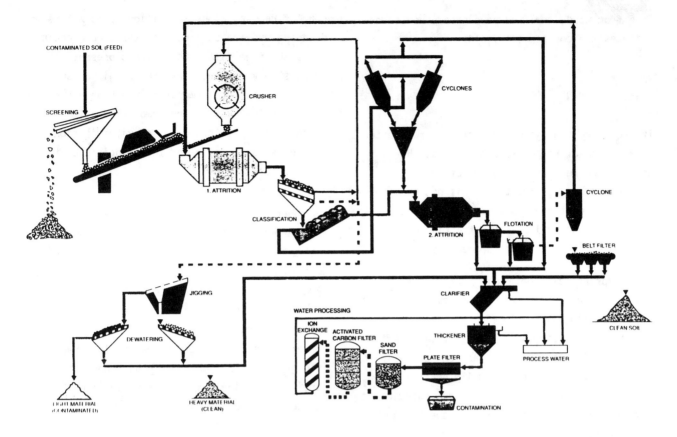

Fig. 5   LURGI Deconterra - Process Flowsheet

2) Paticle sizes of less than 1 mm have an adverse effect on gravimetric sorting.

3) Particles coarser than 1 mm interfere with the subsequent selective flotation process as they tend to be sedimented.

The size fraction minus 1 mm makes the highest demands on the sorting and classification process and hence requires sophisticated process technology. Aprat from the particle size distribution, Lurgi also determines the specific surface area of the solids using the air permeability method to obtain a precise picture on the type of bulk material involved. These measurments yielded specific surface areas of less than 4000 cm^2/g for the fines fraction of the soil. In individual cases, values up to 16,000 cm^2/g were analyzed. It is a known fact that the adsorption capacity rises with decreasing parti-cel size and increasing porosity. However, as the desorption capability decreases to the same extent, this means that desorption of the conta-minants by mechanical methods is practically

impossible. Therefore the fines fraction has to be separated and discharged as residue.

In the DECONTERRA process, fines separation, i.e. desliming, is carried out in hydro-cyclones applying a d 80 mesh of separation. The cyclone overflows with particle sizes of 0.015 to 0.02 mm are descharged as slurry. This fraction is routed to the treatment residue without any further treatment.

The specific surface area of the particles in the cyclone underflow is abt. 1000 cm^2/g.

TREATMENT OF COARSE FRACTION

The coarse fraction is further processed in a jig where the material is lifted up and down by a stream of pulsating water while being slowly moved in horizontal direction. In the process, the heavy material is separated from the light material, each fraction being drawn off sepa-rately. The two fractions can be sorted into one light and one heavy fraction each in another downstream sorter. The heavy fraction which is

free from wood, coal and porous components constitutes the coarse soil yield.

FINES TREATMENT

If required, the cyclone underflows are treated
in a second attrition scrubbing stage. Here,
they are washed in the absence of the highly
contaminated fines fraction. This not only fore-
stalles re-contamination but is also beneficial
to subsequent sorting as the fraction 0.6 mm +
is reduced through the scouring effect.

The principal contaminant sources such as
coal, wood and roots have to be removed from the
fines fraction as completely as possible. This
is best achieved by selective flotation. After
comprehensive testwork, it has been possible to
find a suitable and ecologically neutral collector. The latter attaches to the contaminants
and, after addition of a foaming agent, causes
them to be concentrated in a layer of froth on
the liquid surface. The highly contaminated froth
is then united with the cyclone overflow and
dewatered in a thickener.

The froth flotation unit can be operated with
either aerators or agitators.

RESIDUE TREATMENT

The two fines fractions - cyclone overflows and
flotation froth - are flocculated and thickened.
Subsequently, the thickener underflow is dewatered on a filter press. This normally yields a
filter cake with a residual moisture of < 30 %
wt. As this filter cake contains more than 95 to
99 % of the contaminants originally present in
the raw soil, it has to be disposed of by thermal methods.

AGGLOMERATION AND VITRIFICATION OF HARBOUR SEDIMENTS

The highly contaminated harbour sediments obtained from dredging operations pose a special
challenge to remediation technology. In the harbour of Hamburg alone, 2 million cubic meters of
dredging mud are obtained per year. Here again,
it is the fine silt fraction that presents the
greatest environmental hazard. Lurgi has solved
this problem by agglomerating the highly contaminated silt fraction and immobilizing it in a
ceramic matrix (Fig. 6), i. e. vitrification.

A brief description of the principle of the
process is given below:

The dredging mud is feed from the sand fraction
and the highly contaminated silt fraction dewatered to a residual moisture content below
50 % in a separation and dewatering circuit. The
dewatered silt is then mixed with ground return
fines (so-called soft-fired pellets) from the
travelling grate process.

Next, the silt/return fines mix is formed
into green pellets on continuously operating
Lurgi pelletizing disks before being indurated in
the Lurgi travelling grate.

Return fines are added as a leaning agent to
lower the silt's organic carbon content to a
level which rules out uncontrolled temperature
peaks in the pellets during the subsequent firing process.

In a final process step, the green pellets
are subjected to controlled heat treatment on the
Lurgi travelling grate. After cooling in the
discharge-side section of the travelling grate,
the pellets are discharged followed by separation into product pellets and soft-fired pellets.

The flue gases generated by the firing process
are reacted out in a post-combustion chamber
before being routed to a flue gas cleaning system for desulfurization, dust collection and
mercury removal. The flue gas cleaning system is
equipped with the latest control technology and
ensures consistently low emission levels.

The strengths of this process can be summarized as follows:

Exact control of the temperature profile and a
sufficiently long retention time in the high-temperature range, i.e. at temperatures just below

the softening point of the dry silt, while ensuring ample oxygen supply in the reaction zone.
The end produkt takes the form of ceramic pellets, abt. 10 to 15 mm diameter, featuring good
compressive, shear and temperature cycling
strengths.

Treatment at temperatures above 1070° C reliably destroys organic pollutants and volati-
lizes heavy metals such as mercury, cadmium and
lead. At the same time, this temperature promotes
the agglomeration process, so that non-volatile
non-ferrous metals are incorporated in a ceramic
matrix and chemically immoilized to such an extent that they can no longer be released to the
environment.

Apart from these processes, Lurgi's remediation technology programm includes groundwater re-
habilitation and soil ventilation, in-situ remediation of soils and groundwater contaminated
with volatile organic compounds by vacuum extraction and dual vacuum extraction, the latter process lending itself to the simultaneous remediation of liquids and vapours.

COSTS OF REMEDIAL METHODS

A crucial point for the decision of one or another remedial method is the financial aspect.
Some estimated costs are given below.

Thermal treatment methods range between 350.-
and 1000.- DM/tonne depending on the application.
Costs for Lurgi-DECONTERRA soil washing range
between 150.- and 200.- DM/tonne not including
disposel costs for the residue.

Costs for In-situ vacuum extraction are in
average of 70.- DM/tonne, but can climb up to
200.- DM/tonne, depending upon requirements for
off-gas or waste water treatment.

Recovery of non-ferrous metals from residues of integrated steel works

A. Kaune
Metallgesellschaft AG, Frankfurt, Germany
K.-H. Peters
U. Härter
Thyssen Stahl AG, Duisburg, Germany
M. Hirsch
K. Janssen
Lurgi AG, Frankfurt, Germany

SUMMARY

The finegrained filter-dust fraction of the different steps of steel production in integrated steel mills contains Zn and Pb to an extent, which is prevented inplant recycling without appropriate treatment. A satisfying alternative to dumping does not yet exist. Due to the relative low NF-Metal content and with regard to profitability only energy saving process routes can be taken into consideration. METALLGESELLSCHAFT Group (MG) therefore has developed - in cooperation of Berzelius Umwelt Service (B.U.S) and LURGI with Thyssen Stahl AG - a process to treat these materials in a Circulating Fluidized Bed (CFB) reactor under reducing conditions. Lab-scale tests at LURGI's R&D-center were successful. Zinc and lead were removed and enriched in a secondary dust for further treatment via existing Waelzkiln operations, while the iron content was recovered as an internal recycable product either in oxidic or metallic form.

At present a pilot plant is under construction in cooperation with Thyssen Stahl AG at Thyssen's Duisburg-Hamborn works with an hourly throuput of 4 t, corresponding to an annual capacity of 30 000 t. The start-up is scheduled for autumn 1991. It is expected to furnish design data for large-scale plant and it will also processs sample materials from other smelters. The pilot plant will be integrated into the gas-power system of the existing smelter. The process does not produce any wastes by treating waste material of the steel mill.

In integrated steel mills in the Federal Republic of Germany some 500 kg of residues per ton of raw steel are arising [1]. Most of them are metallurgical slags from pig iron and steel production, followed with a wide margin by dust and sludge from the filter units associated with the blast furnaces and steel smelters. Non-negligible quantities of residues are also produced in the rolling mill, e.g. mill scale and sludge.

Table I illustrates various residue groups and their average percentages in the total waste from an integrated steel mill.

Table I. **Structure and use of wastes in iron and steel production [1]**

Total volume of waste material

About 500 kg/t produced steel

Breakdown of waste obtained

Blast furnace slag	54 %
Steel mill slag	19 %
Other slags	7 %
Works rubbish	6 %
Dust and sludges	8 %
Mill scale and sludge	5 %
Others	1 %
Total	**100 %**

Use of waste

For sale	69 %
In-plant recycling	21 %
Dumping	10 %
Total	**100 %**

Attempts at recycling the residues or upgrading them to marketable products have a long-standing tradition in the steel industry. Slags, for instance, always have been reused almost completely for road construction, as fertilizers or, to a lesser extent, as a lime base for the sinter mix. Oil-free mill scale, consisting entirely of iron Oxide, is another completely recycled residue since it is highly welcome as an iron source for the sinter mix.

Table I also describes in broad terms what the steel mill residues are used for, and these figures make it again clear to what extent this industry today reuses its residues.

381

382

Table II shows in a summary a more detailed picture of the specific amount of wastes obtained in the different steps of iron- and steelmaking processes; the figures are average values of the German works.

The present use of these residues is also marked in this table. It can be seen that an amount of about 20 - 25 kg wastes per ton of steel produced today is dumped although it contains a high amount of iron.

In particular, these materials are

- sludge from the blast furnace (BF) gas scrubbers

- fine dust ore sludge from the basic oxygen furnace (BOF) operation

- electric arc furnace (EAF) dust with zinc contents of less than 10%

- oil contaminated mill scale and sludge with oil contents of more than 0.5%

- from time to time dust from the electrostatic sinter-gas cleaning systems due to its high alkaline content after several recycling circuits.

Common characteristics of these materials are

- The non-ferrous metal and/or alkali content and/or oil contamination do not allow to recycle them in the conventional reprocessing route of integrated steel works, that means via sinter plants and blast furnaces.

- On the other hand, the non-ferrous metal content of NF-bearing residues is of such a low level that an external treatment just on the basis of NF recovery is not feasible economically.

- A relatively high iron content, partially in metallic form.

- Material of extremely fine grain size and sludges.

With respect to the enviromental situation, dumping of these materials cannot be considered as a long-term solution. It also represents a loss of substantial quantities of iron as well as zinc and lead for the national economy.

Since dumping will become more and more restricted, it is only a matter of time until all companies will be forced to look for other possibilities to get rid of these materials.

As the requirements to keep the air and water clean have become more and more stringent in recent years, the quantities of filter dust and sludge have been increasing consistently .

It was therefore mandatory to develop a process to recycle these residues right back into the process and thus reduce the quantities to be landfilled.

Table II. **Yield and Use of Wastes in Iron and Steel Production**

Process step Kind of Wastes	Specific amount in kg/t	Main current use
I. Sinter Plant		
Dust from ESP	2 kg/t Sinter	Direct recycling to sinter feed
II. Hot Metal Production in BF		
BF slag	ca. 300 kg/t HM	Road constr., etc.
Flue dust	8 - 12 kg/t HM	In-plant recycling to sinter feed
Washing tower sludge	4 - 6 kg/t HM	Mainly in-plant dumping
III. Crude Steel Production		
1. BOF		
BOF slag	ca. 100 kg/t St	Road constr., fertilizer, lime for sinter feed
BOF dust, coarse	3 - 4 kg/t St	In-plant recycling to sinter feed
BOF dust, fine	ca. 12 kg/t St	Mainly dumping, minor part disposed of by outside companies
2. Electric Arc Furnace		
EAF slag	ca. 100 kg/t St	Road constr., dumping
EAF dust Grade "A"	10 - 15 kg/t St	>20% Zn content transfer to B.U.S Waelz kiln
Grade "B"	10 - 15 kg/t St	10 - <20% Zn recycled to increase Zn content
Grade "C"	10 - 15 kg/t St	dumped or disposed of by outside companies
IV. Rolled Steel Production		
Mill scale, coarse	ca. 40 kg/t RSt	<0.5% oil content: recycling to sinter feed
Oil-contaminated mill scale sludge	3 - 4 kg/t RSt	Mainly in-plant dumping, rest disposed of by outside companies

PROCESS SELECTION

An essential condition for the selection was that the process should be based on proven elements wherever possible so that the time and money to be spent on the development could be kept to a minimum; the process should also be tolerant to different residue mixes and reliably remove all impurities to ensure that the iron materials can be safely added to the production feed. In addition, the Zn and Pb enrichment of the secondary product should be high enough to make it suitable as non-ferrous smelter feed.

All the steel mill residues to be treated occur in the form of sludge or dust, i.e. of fine particulates. In order to reduce capital expenditure these particulates should be treatable without agglomeration. This at the same time would ensure short heating and reaction periods, and hence high specific plant throughputs, owing to the large reaction surfaces of the particulates. However, since the Fe contents of the residues are not comparable with those of rich ores and since the residues may contain certain impurities, such as sulfur, the Fe product is unsuitable for direct use in the steel mill. No attempt is therefore made to achieve complete reduction to metallic iron, and this in turn leads to savings of energy and reducing agent.

This approach is supported by the possibility of using the fuel content in the mix for process purposes; 20 % carbon in the sludge from the blast furnace gas scrubber, 4 - 10 % oil in the mill sludge, and the metallic iron contained in BOF dust are used for covering a part of the process heat.

Any feasible concept for the treatment of steel mill residues and the recirculation of the products into the industrial cycle has to follow these lines.

This was the starting point for the METALLGESELLSCHAFT Group to initiate the development of an overall concept for the treatment of steel mill residues by Berzelius Umwelt Service (B.U.S) and LURGI in contact with THYSSEN STAHL AG [2]. This concept provides for a Circulating Fluidized Bed (CFB) to recover materials suitable as feed for the ferrous and non-ferrous metallurgy sector. The principle of Circulating Fluidized Beds was developed by LURGI and today is proven technology. CFB reactors are used in numerous industrial processes of the chemical and metallurgical industry. Their applications range from simple physical processes to complex gas/ solids reactions. As evidenced by numerous applications such as calcining of alumina or combustion of coal in power stations, LURGI has for more than 20 years now employed and fully controlled the circulating fluidized bed technology also in large industrial complexes.

A special feature of the circulating fluidized bed is that it can handle fine particulate material of some particle size distribution without any previous conditioning. This ensures that the residues can be treated as a mix even if the particle sizes of the individual components exhibit considerable variations.

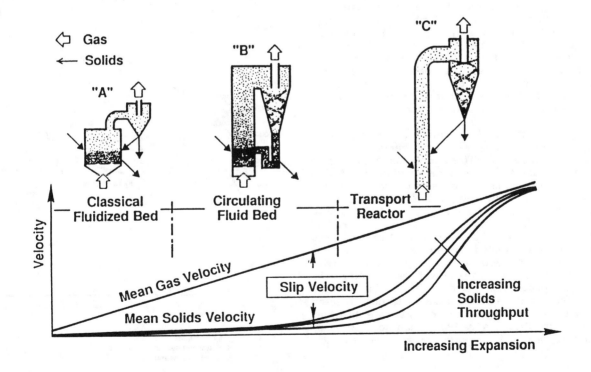

Fig. 1. Fluidized bed systems

PRINCIPLE OF THE CIRCULATING FLUIDIZED BED TECHNOLOGY

The CFB principle will now be explained on the basis of Figure 1 [3,4]. If gas is blown from belowinto a bed of fine-grained solids, i.e. with particle size up to approx. 0.5 mm, this bed will form a stationary, bubbling fluidized system once a minimum gas flow rate is exceeded. Such a stationary fluidized bed is characterized by a defined surface, low solids entrainment, and a marked difference in densities between the bed and the gas space above it (Fig. 1, type A). As the gas flow rate increases, the fluidized bed expands until the solids are more evenly distributed over the total reactor height, and more and more solids are entrained from the reactor at its top together with the off-gas. These solids are precipitated in downstream separators (cyclones) and looped back into the reactor. The recirculation cyclone is directly associated with the reactor and forms an integral part of the CFB-system (Fig. 1, type B).

In addition to this external circulation, an internal recirculation results from the continuously changing formation of clusters of the particles in the reactor itself. This leads to fairly high relative velocities between the solids and the gas, and hence to an extremely intense mixing of the reactants.

PROCESS FOR THE TREATMENT OF STEEL-MILL RESIDUES BY SELECTIVE REDUCTION IN A CIRCULATING FLUIDIZED BED

The reaction conditions of the Waelz process (Fig. 2) are such that zinc and lead can be volatilized only if most of the iron oxides are reduced to metal [5,6]. If the reducing reaction in the material bed of the Waelz kiln is separated from heat generation through combustion in the free kiln space, the resulting Zn partial pressures of 0.1 - 0.3 bar in the reducing bed require conditions which can be reached only in the metallic iron area.

Fig. 3 shows the reduction potential as a function of the temperature at various Zn partial pressures. The Waelz process area lies in field 1 where the endothermic reduction of FeO leads to high energy consumption rates. In the case of dusts containing 6 % Zn, for instance, more than 80 % of the total energy is consumed by FeO reduction. Since, however, the product still contains slag-forming impurities and may also be contaminated by sulfur, its quality is not comparable with that of sponge iron from concentrates and thus does not justify the higher production costs. The fine-grained zinc-free oxidic ferrous residue is best treated in a sinter machine, but in this case the metallic iron "fuel" is rather a handicap in comparison with coke breeze because of its slow combustion.

Selective reduction of zinc and lead at temperatures of 900 - 1000°C without producing metallic iron is possible by lowering the Zn partial pressure. If reduction and heat

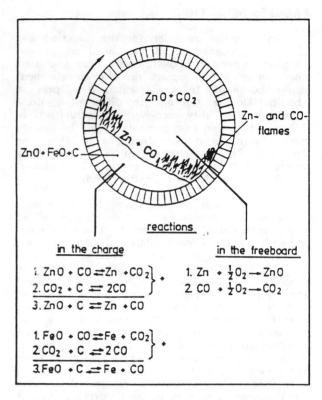

Fig. 2. Zinc volatilization and reduction of iron oxides in the rotary kiln

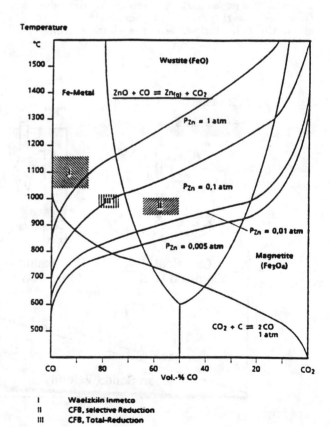

Fig. 3. Dependence of Zn-reduction in CO/CO_2-mixtures and Zn-partial pressure

generation take place simultaneously by partial oxidation of the fuel in a reactor, the system can be adjusted to Zn partial pressures between 10^{-3} and 10^{-2} bar.

The reaction range of a fluidized bed system can therefore be described by fields II and III in <u>Fig. 3</u>. The relatively low Zn partial pressure in this system is defined by the gas throughput and the Zn content in the residue feed. This makes it possible to remove most of the zinc with low energy input already in the wuestite area.

Basic investigations on this subject published by IRSID in 1980 [7] showed that more than 90% Zn volatilization was reached at 1000°C, 6% Zn and low carbon content in the samples with a $CO/(CO + CO_2)$ ratio of 0.3 - 0.5.

By means of orienting tests the advantages applying fluidized bed technology were worked out for the volatilization of zinc by selective reduction. These tests were carried out in a bench-scale fluidized bed plant with 50 mm inner diameter at LURGI's laboratories.

TESTS WITH A BENCH-SCALE FLUIDIZED BED SYSTEM

<u>Fig. 4</u> shows the schematic arrangement of the 50 mm fluidized bed system. <u>Table III</u> illustrates the test parameters and two representative results of these tests.

Table III. **Parameters and results in bench scale tests**

Test parameters

Sample weight	50	g
Particle size	0,5 to 1.0	mm
Bed temperature	1000	°C
CO/CO_2-ratio	1	
Gas flow	430	l/h

Duration of test:		
Heating up to 1000 °C	14	minutes
at nitrogen flow	400	l/h
Reduction gas at 1000 °C	20	minutes

Chemical analysis of samples

component	mixture feed 50.0 g	test 7 discharged 40.2 g	test 8 discharged 42.0 g
	wt-%	wt-%	wt-%
Fe tot	55.1	64.3	64.8
Fe^{++}	14.4	41.8	50.2
Fe met	3.7	.6	.8
Zn	2.1	.06	.21
Pb	.33	< .05	< .05
K	.76	.20	.23
C	4.8	.18	.48
S tot	.38	.38	.42
Cl	.51	.006	.009
Zn-removal		97.6	91.5
Pb-removal		> 87.0	> 87.0
K-removal		77.5	74.3
Fe-bilance	100.0	93.8	98.8

When a residue mix consisting of high-Zn/Pb sludge from blast furnace gas scrubbers, Zn/Pb-laden dust from a top-blown steel mill and dust from the chain conveyors of a sinter plant was treated with weakly reducing gases in a fluidized bed system at 1000°C, it was possible to remove more than 90 % of the zinc and far more than 87 % of the lead. The potassium contents were reduced by more than 70 %. These encouraging results led to further tests with residues from THYSSEN STAHL AG in a LURGI CFB pilot plant between October and December 1989 in order to corroborate these findings.

TESTS IN A CFB PILOT PLANT

1. Plant Description

Essentially, the plant consists of a heat-resistant vertical steel tube of 6 m height and an inside diameter of 200 mm, two downstream cyclones, and one preheater each for primary and secondary gas. The plant can be operated continuously and is designed for a solids throughput of 20 - 25 kg/h.

<u>Fig. 5</u> shows a flow scheme of the plant. Both the reactor and the cyclones are indirectly heated by means of electricity and insulated with fibrous material to compensate heat losses through the surface of the CFB-reactor system. The gases are taken from steel cylinders.

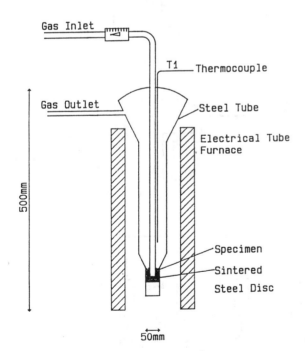

Fig. 4. Fluidized bed apparatus, bench scale

386

The dust is recovered from the waste gas by a heated ceramic filter.

Solids and lignite are introduced by a screw feeder and coke breeze through another one into the CFB-reactor.

The primary product, i.e. the treated residues, is discharged from the CFB reactor bottom by means of a rotary valve and a water-cooled screw.

The residue mix with its zinc, lead and alkaline contents is treated in a slightly reducing atmosphere. The reactor temperature of approx. 1000°C is reached by partial combustion of coke and lignite in the reactor. The coke is also very useful as an anti-sticking agent, and the lignite plays an important role in adjusting the system to the desired CO/CO_2 ratio of approx. 1.2.

An expanded fluidized bed is established in the CFB-reactor. The residue particles react in the gas flow, however iron oxide is not reduced to metallic iron. Most of the solids entrained with the gas are collected in a cyclone and recycled to the reactor. The first cyclone is followed by a second one in which another part of the solids is separated and also looped back into the reactor.

2. Feeds

THYSSEN STAHL AG supplied various dusts and sludges for the tests. LURGI broke these materials down to a particle size of less than 1 mm and predried them to a residual moisture content of approx. 4 % to improve their flow characteristics. The chemical analysis is illustrated by Table IV. No oily rolling mill sludge was added at this stage.

Table IV. Chemical analysis of residue mixture in pilot test

Component	wt-% dry	component	wt-% dry
Fe tot	53.9	Na	.36
Fe^{++}	14.7	P	.05
Fe met	3.4	Mn	1.3
SiO_2	2.4	Pb	.38
Al_2O_3	.77	Zn	1.6
CaO	5.9	C tot	6.7
MgO	.64	S tot	.4
TiO_2	.15	Cl	.58
BaO	< .05	F	.09
K	.98		

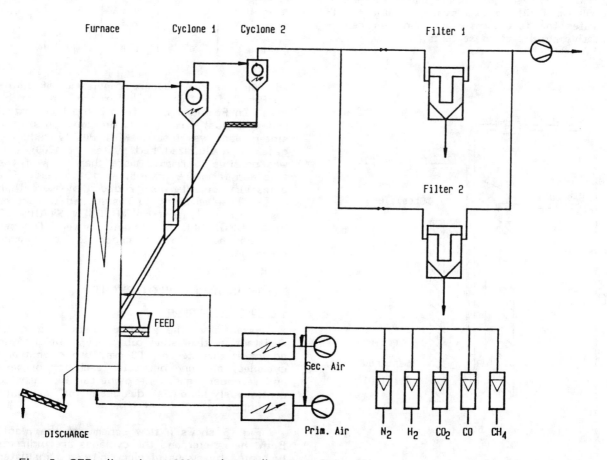

Fig. 5. CFB-pilot plant; 200 mm inner diameter, flow scheme

387

3. Test Results

The pilot plant was operated in several campaigns between October and December 1989. Different process parameters were varied and the influences of the reduction gas analysis and of the temperature, in particular, were investigated. As the campaigns proceeded, the ratio of residue mix to lignite was reduced from 2.5 to 1.0 in order to obtain a CO/CO_2 ratio of more than 1. The furnace temperature was raised to approx. 1000°C. Owing to the coke in the bed, sticking was no problem.

a) Zinc Removal

The increase in furnace temperature to approx. 1000°C and in the CO/CO_2 ratio of the reduction gas to more than 1.0 brought the zinc removal up to more than 80 % at the end of the campaign and reduced the residual zinc content of the carbon-free bed discharge to approx. 0.3 %. It became clear that higher temperatures and a higher CO/CO_2 ratio have a positive effect on zinc recovery.

b) Lead Removal

During all the test campaigns, it was possible to recover with relatively little effort. The lead removal was mostly around 80 % or clearly higher and the residual Pb content in the bed discharge was < 0.05 %.

c) Potassium Removal

The potassium contents were normally reduced by approx. 30 % and some 0.7 % K were still found in the bed discharge. This unsatisfactory potassium recovery is due to condensation in the colder part of the waste gas system, as in the secondary cyclone the temperature already decreased slightly.

Summary of test results with the CFB pilot plant are as follows:

- Zn: approx. 0.3%, equivalent to 80 - 83% collection efficiency
- Pb: < 0.05%, equivalent to > 80% collection efficiency
- K: approx. 0.7%, equivalent to approx. 30% collection efficiency.

The fact that zinc recovery was lower than during the bench-scale tests can be explained by the non-ideal conditions in the reduction zone.

The unsatisfactory potassium removal was due to condensation effects already in the colder secondary cyclone.

Fig. 6 shows the test results from the last three days.

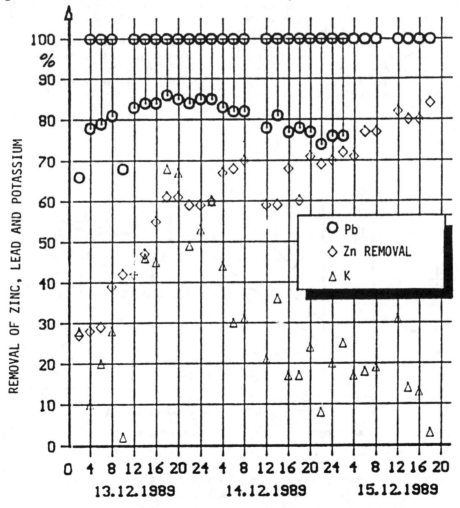

Fig. 6. Removal of Zn, Pb and K from steel mill residues in CFB-pilot plant tests

388

FURTHER DEVELOPMENT

After the preliminary tests on a bench scale and in the pilot plant at LURGI's R & D Centre have proved successfully, the preparations are now under way for tests on an extended pilot scale in order to develop the technical systems, optimize the process and derive the basic data required for a commercial plant.

For this purpose, a pilot plant with a throughput of approx. 4 t/h currently is under construction on the premises of THYSSEN STAHL AG.

A process flow diagram of this pilot plant is shown in Fig. 7.

The granulate is fluidized by means of preheated coke oven gas which is blown into the reactor through the nozzle grate. This primary gas is heated to approx. 600°C in a gas heater. A second gas heater raises the temperature of the combustion air to approx. 750°C before it is blown into the CFB reactor as secondary gas. Oxygen may be added to the combustion air to increase the reactor throughput.

The injection of primary and secondary gas leads to the formation of an expanded fluidized bed in the CFB reactor in which the particles from the gas flow are reacted without reducing the iron oxide to metallic iron. A certain percentage of the solids which are entrained with the waste gas at the reactor top are removed in a recycle cyclone and fed back to the reactor bottom via a pressure seal.

Another part of the solids is collected in a secondary cyclone and led to the primary product discharge system. If necessary this part of solids can be recirculated to the CFB-reactor as well.

The use of hot cyclones ensures that most of the volatilized elements which are unwanted in the steel mill are separated from the solid residue.

As a consequence of evaporation and re-oxidation Zinc, lead and alkaline compounds are concentrated in the waste gas dust load in the form of oxides.

The hot gases leaving the secondary cyclone are cooled directly by injecting water into an evaporation cooler. This cooler is followed by a multi-stage fine dust collection system including a cyclone and a cartridge filter. The secondary dust recovered in this system contains Zn/Pb oxides. It is further cooled, stored in a bin, wetted for dust-free transport, and prepared for shipment to downstream processing units. Some of the coarser fractions of this secondary dust can be recycled to the process, upstream of the granulation stage, in order to obtain higher percentages of zinc and lead in the fine dust as secondary product.

The clean gas is led into the existing blast furnace gas mains so that the steel mill can benefit from its remaining calorific value. This process thus does not produce any residues that have to be landfilled nor does waste water need further treatment.

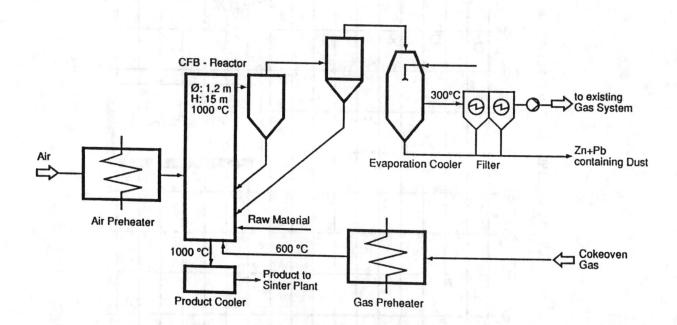

Fig. 7. Process flow diagram of the extended CFB-pilot plant at THYSSEN STAHL AG under construction

OUTLOOK

The selective reduction process was verified by tests with residues from THYSSEN STAHL AG in a modified 0.2 m CFB plant at LURGI's R&D facility. An extended pilot plant to be used for long-term tests and to derive upscaling data is currently being constructed on the premises of THYSSEN STAHL AG. To ensure that all aspects of CFB technology can be covered, this pilot plant will also be equipped to reduce the iron content of the residues to metallic iron.

After some 20,000 t of THYSSEN residues have been treated, the data collected by then will allow us to say whether the concept is technically and economically feasible on an industrial scale.

Further tests will then follow to find out whether the know-how derived in this way can also be applied to steel mill residues from other producers.

3 to 4 regional CFB-plants - integrated in the gas power system of the respective smelter - will be able to treat the whole amount of the said residues of the German steel industry. We believe, that the presented CFB-process will be a contribution to the solution of an increasingly burdensome environmental problem.

REFERENCES

1. J.A. Philipp, R. Gören et al., "Umweltschutz in der Stahlindustrie, Entwicklungsstand - Anforderungen - Grenzen", Stahl und Eisen 107 (1987), No. 11, pp. 507-514.

2. M. Hirsch, A. Kaune and H. Maczek, "Recovery of Zinc and Lead from Steel-Making Dusts, in particular by the Circulating Fluid Bed", 28th Annual Conference of Metallurgists, Halifax, 1989.

3. L. Reh, "The Circulating Fluid Bed Reactor", First International Conference on Circulating Fluidized Beds, Halifax, 1985.

4. M. Hirsch, K. Janssen and H. Serbent, "The Circulating Fluid Bed as Reactor for Chemical and Metallurgical Processes", First International Conference on Circulating Fluidized Beds, Halifax, 1985.

5. F. Johannsen, "Die Technik des Wälzverfahren", Metall und Erz (1984), pp. 235239.

6. H. Maczek and R. Kola, "Recovery of Zinc and Lead from Electric-Furnace Steelmaking Dust at Berzelius", UNEP Industry and Environment, July/August/September 1986, pp. 11.

7. R. Pazoly and J.M. Steiler, "New Treatment Possibilities of BF/BOF Zinc (and Lead) bearing Dusts", First Process Technology Conference, Washington, DC (1980).

This paper is a revised version of that presented to the 50th Ironmaking conference of AIME in Washington, D.C., U.S.A. 14-17 April, 1991.

Simultaneous microbial removal of sulphate and heavy metals from waste water

L. J. Barnes
J. Sherren
F. J. Janssen
Shell Research Ltd, Sittingbourne Research Centre, Sittingbourne, Kent, England
P. J. H. Scheeren
J. H. Versteegh
R. O. Koch
Budelco BV, Budel-Dorplein, The Netherlands

SYNOPSIS

The groundwaters beneath well established metal-refining sites processing metal sulphides are often contaminated with sulphate and heavy metals. With the increased demand in industrial areas for potable water from underlying aquifers, and with the advent of tighter environmental controls, it has become essential to prevent contaminated groundwaters from spreading. Geohydrological control systems may be installed to control such groundwater movement. However, the water extracted by these systems must be purified before it can be released into the public domain or re-used as process water. An anaerobic sulphate-reducing microbial process has now been evaluated for decontaminating this water. The process uses ethanol, both for growth of organisms and as energy source for the reduction of sulphate. When heavy metals occur as contaminants they are precipitated as very insoluble sulphides. This process is therefore able to produce an aqueous effluent acceptable to the authorities.

Following two years of laboratory research with bench scale reactors, a $9m^3$ experimental process demonstration reactor was installed at Budelco's zinc refinery to treat their underlying groundwater. A 6000-fold reactor scale-up was successfully achieved. It has demonstrated that this anaerobic microbial process can remove a wide range of heavy metals and sulphate to the required levels. Extended operation of the demonstration unit has shown that the process is robust, can withstand wide variations in feed composition,

rapidly recovers from operational upsets and is easy to start up and shut down. The metal sulphide sludge produced can be processed in a metal refinery to recover both metals and sulphur.

INTRODUCTION

Historically the metal industry has produced substantial amounts of sludges and solid wastes that contain heavy metals or sulphate or both, and were often stored on production sites. In several instances leaching of the solids by percolating rain and surface waters has caused substantial contamination of the underlying groundwater. Authorities are insisting that contaminated water under such sites should be prevented from spreading. A geohydrological control system can be used to solve the problem but this produces an effluent contaminated with sulphate and heavy metals, see Table 1, which must be treated.

Table 1 Budelco's groundwater

Component	mg.l^{-1}
Sulphate	1300
Zinc	135
Cadmium	1.5
Copper	0.8
Cobalt	0.1
Iron	4
Calcium	320
Ammonium	1

In the mid-eighties Shell Research initiated a programme to examine candidate processes suitable

392

for removing heavy metals from the effluent to levels acceptable to the authorities. The three most suitable processes selected for research were, ion exchange, liquid membrane extraction and anaerobic microbial sulphate reduction. Of these three processes only the last will remove sulphate economically.

This paper concentrates on the experimental work carried out in establishing the feasibility of the Sulphate-Reducing Bacteria (SRB) process. A second paper[1] describes further process development work at Budelco b.v., Budel-Dorplein, The Netherlands, a company owned jointly by Billiton (Shell) and Pasminco; this will culminate in commercialisation of the process in 1992.

The key features of this process are _in situ_ generation of H_2S by microbes in a reactor and concomitant precipitation of heavy metals as highly insoluble sulphides (Table 2).

Table 2 Divalent metal ion solubilities for metal sulphides

Metal	$mg.l^{-1}$
Zinc	2×10^{-7}
Cadmium	1×10^{-9}
Copper	1×10^{-14}
Cobalt	1×10^{-8}
Calcium	110

The development of the SRB water treatment process involved two experimental stages:

(1) Laboratory investigations, which included:

- screening studies to identify suitable mixed bacterial cultures;

- identification of a simple, readily available, economically attractive carbon substrate;

- process studies in continuously operated 1.6 l reactors using a synthetic waste water to ascertain the operating parameters;

(2) Scale up:

- using a $9m^3$ sludge blanket reactor at Budelco's manufacturing site to establish the feasibility of the process with the site's groundwater.

THE SRB PROCESS

Background

Many natural anaerobic aqueous environments, as found in oceans, lakes and sediments[2-14], contain micro-organisms which use organic compounds for growth and as the energy source for reducing sulphate to sulphides. When heavy metals are present, they will precipitate as extremely insoluble sulphides. These organisms play a major role in the formation of certain sedimentary metal sulphide deposits as well as producing the iron sulphides found in coal[15-17]. The overall reaction in such an anaerobic environment can be represented by:

Metal sulphate + Carbon substrate $\longrightarrow$

Metal sulphide + CO_2 + H_2O + cells

SRB normally occur as part of a consortium of interdependent anaerobic organisms. Such a consortium exists because of the presence of a complex mixture of sedimented carbonaceous compounds. Because SRB utilise a limited range of organic substrates they take advantage of products arising from primary organic degrading organisms. Table 3 gives some examples of specific sulphate-reducing organisms together with their maximum growth temperature, preferred substrates and products. In order to sustain active growth of SRB, two important environmental conditions must be satisfied[18], these are :

(1) **Neutral pH**. In nature this is achieved through chemical equilibria involving either precipitation or dissolution of carbonates and sulphides, which are products of organism growth;

(2) **Low redox potential**. This is maintained in nature by the presence of sulphide ions in the aqueous phase and the absence of oxygen.

Table 3 Some sulphate-reducing organisms

Organisms (Max. Temp. °C)	Growth Substrate (Product)
Desulfovibrio vulgaris (44)	La (Ac)
Desulfomonas pigra (45)	yeast (Ac + H_2)
Desulfobulbus propionicus (43)	La,Pr,Et (Ac)
Desulfococcus multivorans (40)	La,Ac,Et,Me (CO_2)
Desulfobacter postgatei (40)	Ac,Et (CO_2)
Desulfosarcina variabilis (38)	La,Ac,Et,Me (CO_2)
Desulfonema magnum (35)	Be (CO_2)
Desulfotomaculum orientis (38)	H_2 + CO_2 (−)

La = Lactate, Ac = Acetate, Pr = Propionate
Be = Benzoate, Me = Methanol, Et = Ethanol

Since these conditions can be attained in an anaerobic bioreactor, the expected concentrations of heavy metals in the aqueous phase of such a reactor will be insignificant. The bioreactor may operate under non-sterile conditions because sulphide is known to be a strong inhibitor for other micro-organisms, therefore preventing microbial contamination.

It should be noted that the quantity of sulphate reduced is proportional to the growth of the SRB, which in turn is related to the energy available in the carbonaceous growth substrate. The sulphate conversion can therefore be controlled by varying the amount of carbon substrate supplied to the growing culture. Acetate, which is often produced by growing SRB, is only slowly degraded by these organisms. However, methanogens, which normally exist in the naturally-occurring organism consortium, rapidly degrade this compound.

EXPERIMENTAL

Laboratory studies

Test feed compositions

Synthetic solutions, based on Budelco groundwater (Table 1), were used as feed media during the studies in order to investigate the effect of various carbon substrates, nutrients (nitrogen and phosphorus), potential microbial inhibitors and heavy metals. General purpose grade chemicals were used to prepare these solutions.

Screening for suitable SRB organisms

Batch screening of the environmental samples was carried out by inoculating 100 ml of non-sterile synthetic medium, in a septum sealed 150 ml bottle, with 5 ml of each sample. The bottles were made anaerobic by addition of sodium sulphide (10 mg) and purging with nitrogen. Resazurin was added as a redox indicator to show when an excess of hydrogen sulphide was produced by active SRB. The synthetic media used for these experiments were doped with a variety of carbonaceous substrates. Lactate was used since it is known[3,4,18-21] to be an excellent growth substrate for most SRB. Both ethanol and methanol were used as they have been reported[2,6-8,19,21-23] to be suitable carbon substrates for some SRB and are cheaper than lactate. Since acetate is known[2,3,21,23,24] to be a metabolite of SRB when grown on lactate or ethanol, acetate analysis was used to monitor complete degradation. The bottles were inspected daily for solution colour changes and for sediment colony growth, which initially showed as small black specks in the grey sediment. Small samples (5 ml) of the aqueous phase were taken weekly, using a hypodermic syringe, and analysed for substrate, acetate, sulphate and various metals.

Environmental samples were obtained from a variety of locations, including an oil refinery, sewage works, effluent treatment plant, river estuary, stagnant pools, water channels, the Budelco zinc refinery and an anaerobic biotreater. The majority of these cultures showed SRB activity when grown on lactate and a large number of the cultures were also able to utilise ethanol. Although methanol was consumed by some cultures practically no sulphate was reduced.

Continuous experiments

In total, eight continuous reactor runs were carried out in the laboratory using three different types of reactors.

1. Stirred tank

Stirred tank reactors are very suitable for studying the effects of various operating parameters on the chemistry and microbiology of the process since they offer consistent, reliable and rapid equilibrium conditions.

Basic data required for the process were obtained from this reactor type. Three runs were carried out in 1.6 litre stirred (200 r.p.m.) reactors with pH and temperature control. Figure 1 is a schematic diagram of the equipment used.

Figure 1 Schematic of laboratory reactor system

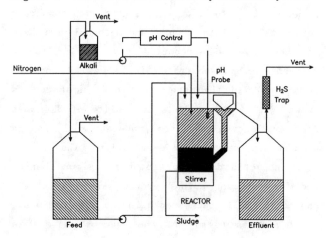

2. Fixed bed

The absence of high shear forces in a fixed-bed reactor avoids dislodging the micro-organisms from the metal sulphide sludge, thus achieving a higher biomass concentration on the particles than in a stirred reactor. This leads to shorter residence times. Very good microbial immobilisation was demonstrated in our laboratory reactor.

One run was performed in a 1.8 litre column reactor which was initially operated at ambient temperature and then controlled at 35°C. Although perfect steady state operation was never achieved in this reactor system some valuable information on residence time and sludge bed buffer capacity was obtained. Complete degradation of ethanol and precipitation of heavy metals were obtained at a residence time of only 4 hours. However, during operation we noticed channelling (possibly owing to compaction).

3. Raked sludge blanket

Biomass can be retained in a properly designed sludge blanket reactor where the sludge blanket is mechanically raked or suspended by liquid recycle to prevent compacting or channelling or both of the high density sludge. Successful operation of laboratory scale raked sludge blanket reactors was demonstrated.

The last four runs were carried out in a 0.75 litre reactor where the reactor temperature was controlled at 35°C. This reactor configuration overcame the sludge compaction and channelling that occurred in the fixed bed reactor.

Process demonstration reactor

Process description

Based on the data collected in the laboratory it was decided, in April 1989, to erect an experimental reactor at the Budelco site to demonstrate scale-up of the SRB process and confirm the laboratory data. A schematic diagram of the $9m^3$ experimental process demonstration reactor is shown in Figure 2, which includes the following:

(a) $9m^3$ raked SRB reactor;

(b) buffer tank for the process feed;

(c) feed tanks for the solutions of nutrients, ethanol, alkali, ferric chloride and zinc sulphate;

(d) in-line mixer for blending the various chemicals with the groundwater feed. The alkali was added to the feed close to the reactor to minimise the chance of blockage by precipitating metal hydroxides;

(e) in-line heat exchanger to test the process at various temperatures;

(f) reactor to remove hydrogen sulphide from the gaseous product stream and soluble sulphide from the aqueous effluent stream;

(g) pH, temperature and feed flow control system.

Figure 2 Schematic of the demonstration unit

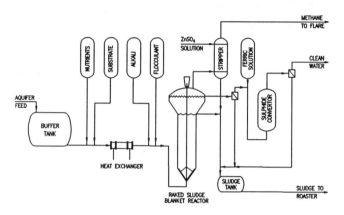

Nutrients (nitrogen and phosphorus) essential for microbial growth, as well as a carbon substrate (energy source) are mixed with the aqueous feed to be treated. The total mixture is supplied to a so-called upflow sludge blanket column reactor. This feed may be pH or temperature controlled or both. In the reactor the desired amount of sulphate is reduced microbially to hydrogen sulphide thus promoting precipitation of the heavy metals present in the feed. The sludge blanket consists of metal sulphides to which the micro-organisms adhere. This results in high biomass concentration within the reactor and achieves a high volumetric reaction rate. Since the reactor produces aqueous, gaseous and solid effluents, all containing hydrogen sulphide, these need to be treated before they can be discharged.

Process operation

Data from the bench scale reactors showed that a 10%v inoculum made up from one of Budelco's environmental samples and an anaerobic biotreater sludge (1:1 volume ratio) could be used to start-up the demonstration reactor. During the early part of these studies it became apparent that the settling behaviour of the sludge particles was different from that observed in the initial laboratory studies. This caused substantial organism washout when residence times shorter than 30 hours were applied. This problem could be avoided by addition of a flocculant. Laboratory studies showed that 'Kemfloc 254' (a flocculant already used at Budelco) was microbially acceptable as it had no deleterious effect on the mixed culture at a concentration (15 mgl^{-1}) five times higher than that required for adequate settling of the sludge particles. A liquid residence time of less than 7 hours could thereafter be achieved without substantial sludge particle carry-over in the reactor's aqueous effluent.

RESULTS AND DISCUSSION

Operating window for the micro-organisms

Organism selection

The data given by the microbial screen show that SRB are widely distributed in the environment. However, although the majority of the microbially active environmental samples consumed methanol, methane was produced in preference to the reduction of sulphate[2,5,6]. This demonstrates that methanol will not be a good carbon source for an envisaged SRB process, even though literature shows[10] that some pure SRB cultures are capable of using methanol to reduce sulphate. The screening experiments also showed that many cultures would grow on either ethanol or lactic acid. However, all cultures formed acetate which was only very slowly degraded by the SRB to carbon dioxide and water. This adversely affects process economics because an additional biodegradation step is needed before the effluent can be disposed.

Organism adaptation

Attempts to use cultures initially grown on ethanol as sole carbon source gave acetate as the main by-product. Even after a 105 day adaptation period the level of effluent acetate was too high unless a long residence time was employed.

Lactate was initially used as sole carbon source to ensure that the largest range of SRB was selected. However, complete degradation of the acetate by-product was never achieved with this substrate; see Table 4.

Table 4 Lactate utilisation at various conditions

	Feed mg.l⁻¹	Effluent mg.l⁻¹					
R/T (h)	–	33	31	5	29	28	5
Temp. (C)	–	22	22	22	31	31	31
Lactate	1140	<1	20*	310	<1	<1*	<1
Acetate	0	640	1900	500	640	2080	680
Sulphate	1590	1050	480	1200	970	150	1210
Zinc	115	<0.01	<0.01	<0.01	<0.01	0.13	0.5
Cadmium	1.8	0.01	<0.01	<0.01	<0.01	<0.01	<0.01
Cupper	1.4	<0.01	0.02	0.03	<0.01	<0.01	<0.01
Cobalt	1.6	0.05	0.03	0.03	0.03	0.01	0.01
Iron	3.5	0.06	0.04	0.03	0.02	0.03	0.04
Manganese	5.9	0.10	<0.01	0.05	0.03	0.03	0.03

R/T = Residence Time * — Feed contained 3380mg.l⁻¹ Lactate

Adaptation of this mixed culture for growth on ethanol was carried out by using methanol as an intermediate substrate, which accelerated the growth of methanogens and hence complete degradation of the acetate formed (Table 5).

Table 5 Effect of changing growth substrate

mmol in Medium		mmol in Effluent	
Lactate	Ethanol	Sulphate Reduced	Acetate Produced
12.6	0	6.4	10.7
4.3	11.8	7.6	3.3
1.3	12.1	7.1	0.6
0	12.5	8.6	<0.1

As the mixed culture of sulphate reducing organisms and methanogens used in this work were obtained from the natural environment, they have not been specifically identified.

Carbon source

Cheap complex natural organic wastes were considered as a potential carbon/energy source for our process[25-27] but such ill-defined materials were rejected on two accounts. Firstly, an additional aerobic biotreater would be required to degrade the unused carbon components and give an acceptable biological oxygen demand[27,28] in the effluent for environmental discharge. Secondly, the questionable reliability and consistency of supply.

Steady state data from experiments using stirred tank reactors show that a mol of sulphate is reduced for each mol of ethanol consumed. This ratio was found to be independent of operating conditions, indicating that sulphate reduction is directly coupled to the growth of the organisms.

Nutrients

The micro-organisms require nitrogen, phosphorus and trace elements for growth. Ammonium salts or urea are ideal nitrogen sources. Nitrate is not acceptable since it is reduced before being assimilated, therefore consuming carbon substrate without reducing sulphate. Phosphates are the most suitable source of phosphorus. All other trace elements are present in the groundwater.

Data suggest that complete degradation of ethanol to carbon dioxide, methane and biomass is achieved if the molar ratios of ethanol/nitrogen, and ethanol/phosphorus are maximally 20 and 500, respectively (ie., 18kg ammonia and 4kg phosphate per tonne of ethanol). Insufficient addition of nutrients leads to reduced organism activity which shows initially as an increase in the reactor acetate concentration. If this condition prevails, it is followed by a lowering of sulphate reduction and appearance of ethanol in the reactor effluent.

pH requirement

The data showed that the optimum pH for sulphate reduction and biomass growth is about 7.5;

however, these micro-organisms will survive pH excursions to values in the range from 5 to 9. The groundwater currently being considered for treatment has a pH of about 4.5. However, it may not be necessary to neutralise the feed before it enters the reactor since it has been shown in the laboratory and in the demonstration sludge blanket reactor that the sludge bed can have substantial buffer capacity (see Table 6). This is due to the presence of carbonates and sulphides, the products of microbial growth and sulphate reduction, in the reactor. However, a neutral pH must be maintained within the reactor.

Table 6 Effect of feed pH on culture performance without pH control

| Feed pH | 7.5 | 5.8 | 3.2 | Concentration in the Feed |
| Reactor pH | 7.8 | 7.4 | 7.1 | |
Concentration in Effluent, mgl^{-1}				mgl^{-1}
Ethanol	<1	<1	<1	572
Sulphate	120	92	27	1250
Zinc	0.032	0.029	<0.004	140
Cadmium	0.010	0.007	0.005	3.71
Cobalt	0.027	0.027	<0.004	2.24
Copper	<0.004	<0.004	<0.004	2.57

Liquid residence time

The minimum residence time required in a sludge blanket reactor to achieve the desired sulphate reduction and complete degradation of carbon substrate and microbial products (in our case ethanol and acetate, respectively) may be limited by either the active microbial concentration or the settling velocity of the bed particles. The latter is influenced by the characteristics of the solid particles. Minimum liquid residence time for a sludge bed reactor was not determined as this required true steady state conditions, which were never achieved in our experimental work because of the prohibitively long time required (months). Laboratory data from the column reactor suggested that the minimum liquid residence would

be about 4 hours, this being determined by the degradation rate of acetate. Although we were never able to operate the laboratory raked sludge bed reactors at a 4 hour residence time, the process demonstration unit studies showed that:

(1) A short residence time (<4 hours) was indeed achievable after addition of flocculant, this gave a higher sludge concentration and better immobilisation of biomass in the reactor.

(2) The minimum residence time that can be applied in the commercial reactor will be dictated by the particle settling rate rather than by sulphate reduction or acetate degradation.

Effect of temperature

Laboratory data showed a maximum operating temperature between 40 and 45°C (Figure 3) as the mixed culture will not survive higher temperatures. Only at temperatures less than 15°C will the ethanol/acetate degradation rate take over from sludge particle settling in determining the minimum achievable liquid residence time in the reactor.

Figure 3 Temperature effect on sulphate reduction

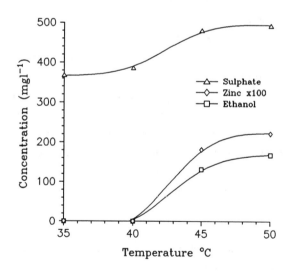

Effect of redox potential

Experimental results show that a low redox potential in the reactor is essential to maintain a stable culture and maximise sulphate reduction.

A maximum redox potential of -100 mV (with respect to a standard calomel electrode) is sufficient for SRB activity[8,18], however, the methanogens require a value nearer to -300 mV for optimum activity. This can be maintained by the presence of soluble sulphide[28]. If the redox potential is too high for methanogen growth then acetate will not be degraded, the reactor pH will drop and the SRB activity will decline. Short excursions to a high redox potential have no detrimental effect on the organisms[18], however, recovery times are long unless the redox potential is lowered by addition of soluble sulphide. In addition, high redox potential will increase the metal concentration in solution until the solubility of metal hydroxide (or carbonate) is reached, the final concentration being highly pH dependent.

Efficiency of heavy metal removal

Efficient removal of heavy metals is possibly the most important attribute of the SRB process. Providing sufficient sulphate is reduced and soluble sulphide is present to maintain a redox potential below -100mV, then heavy metal concentrations of only parts per billion will remain in solution in the reactor's aqueous effluent. Whenever our reactors were operated under conditions of good microbial growth, heavy metal removal was practically complete even for reactor feeds containing up to $1gl^{-1}$ metals, see Table 7. Table 4 and 6 show additional examples of such data. However, even though the metal concentrations in solution are extremely low, the practical efficiency of heavy metal removal is determined by separation of solid metal sulphides from the reactor effluent.

Fate of calcium and magnesium

Because the sludge produced in the SRB reactor will be added to the zinc refinery feed a build up of calcium and magnesium in the sludge should be minimised to prevent calcium precipitation in the zinc refining process and magnesium build-up in the electrowinning circuit. Experiments show that

Table 7 Effect of high metals in feed

Component	Feed mg.l^{-1}	Effluent mg.l^{-1}
Ethanol	810	<1
Sulphate	3000	1300
Zinc	916	<0.05
Cadmium	14.7	<0.01
Copper	5.7	<0.02
Cobalt	0.4	<0.02
Lead	0.09	<0.01
Iron	65	0.15

the majority of both elements end up in the aqueous effluent if excess phosphate in the feed is kept to a minimum. The effect of pH on phosphate solubility in the presence of calcium is shown in Figure 4.

Figure 4 Effect of pH on phosphate solubility in the presence of 300mgl^{-1} calcium

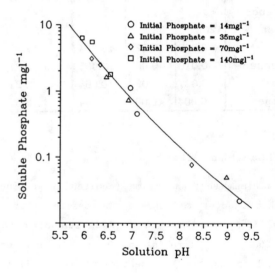

Inhibitory effects of feed components

The inhibitory effect of potential feed components on microbial growth and hence sulphate reduction must be considered. As expected, alkali and alkaline earth cations were found to have no deleterious effect since the organisms used in this process can be isolated from marine environments[4]. Heavy metal cations appear to be non-toxic, presumably because the presence of excess soluble sulphide maintains a very low metal concentration in solution (see Table 2). Certain anions are, in general, known to be powerful microbial inhibitors. However, the following anions were found not to inhibit microbial growth or sulphate reduction at the concentrations tested (mgl^{-1}): molybdate (10), selenate (3.5), arsenate (3.3), fluoride (48) and sulphide (500).

Mass balance over the SRB reactor

The build up of a SRB sludge blanket in the reactor is very slow, and it can only be considered to be in perfect equilibrium with the feed and effluents after a long period of operation (months), providing the feed composition and reactor conditions have remained constant during the whole period. However, a sensible and reasonably accurate overall mass balance can be attained from analysis of the feed, outflow liquid and gas streams assuming the remaining elemental balance is associated with the sludge. Figure 5 shows such a mass balance over a SRB reactor which has been operated under pseudo-steady state conditions.

Figure 5 Mass balance over the SRB reactor (g)

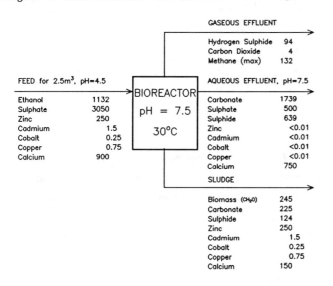

It should be noted that although there is a multitude of equilibria within the reactor, only the elemental ratios of sulphur and carbon compounds in each effluent change drastically.

These depend on the level of sulphate reduced, the heavy metal concentration in the feed, and reactor pH providing optimal microbial conditions exist within the reactor.

CONCLUSIONS

Tighter environmental control within some Western European countries requires that sulphate and heavy metal contaminated groundwaters should be cleaned up to levels demanded by the authorities. This has resulted in the development of a sulphur and heavy metal removal/recovery technology based on the capability of some types of bacteria to reduce sulphate in an anaerobic environment.

From data obtained during batch and continuous experiments in the laboratory and during the demonstration unit operations, we conclude that:

1. Removal of a wide range of heavy metals and sulphate from aqueous streams can be achieved with the anaerobic Sulphate-Reducing Bacteria (SRB) process.

2. SRB utilise ethanol as carbon/energy substrate for reducing sulphate and subsequently form acetate as a major product. The acetate can be successfully degraded to carbon dioxide, methane and biomass by introducing methanogens into the SRB culture.

3. The quantity of sulphate that can be reduced by the SRB process is controlled by the amount of ethanol fed to the reactor. One mol of ethanol is capable of reducing one mol of sulphate.

4. Dissolved heavy metal concentrations in the aqueous effluent from the SRB reactor are extremely low (ppb) provided sufficient sulphate is reduced and the redox potential in solution is maintained below -100 mV.

5. Nitrogen and phosphorus concentrations in the reactor feed must be greater than 5 %mol and 0.2 %mol, respectively, of the ethanol consumed to sustain microbial growth. Excess

phosphate should be minimised to avoid calcium precipitation.

6. Temperature effect on the performance of the SRB reactor is negligible in the range 15°C to 40°C. For the organisms investigated, temperatures higher than 42°C must be avoided to prevent killing the organisms.

7. pH control in the reactor may not be required for treatment of groundwater with a pH greater than 4 owing to the buffer capacity of the system.

8. Addition of flocculant to the SRB reactor feed is essential if short liquid residence times and adequate solid retention are to be obtained within a single reactor.

9. Unwanted sulphide present in the three product streams (i.e. gas, liquid and sludge) from the SRB reactor must be removed before discharge.

10. The anaerobic mixed culture employed is robust, can handle many potentially inhibitory cations and anions, and recovers readily from (substantial) process upsets.

11. Energy consumption in the process is low since it operates near ambient temperature.

REFERENCES

1. Scheeren, P.J.M., Koch, R.O., Buisman, C.J.N., Barnes, L.J., and Versteegh, J.H. A new biological treatment plant for heavy metal contaminated groundwater. This conference.

2. Laanbroek, H.J., Geerligs, H.J., Sijtsma, L. and Veldkamp, H. Competition for sulphate and ethanol among Desulfobacter, Desulfobulbus and Desulfovibrio species isolated from intertidal sediments. Applied and Environmental Microbiology, vol. 47, no. 2, Feb. 1984, p. 329-334.

3. Laanbroek, H.J. and Pfennig, N. Oxidation of short-chain fatty acids by sulphate-reducing bacteria in freshwater and in marine sediments. Archives of Microbiology, vol. 128, 1981, p. 330-335.

4. Bergey's Manual of Systematic Bacteriology. Volume 1. Krieg, N.R. and Holt, J.G. Published by Williams & Wilkins. 1984, p. 663-679.

5. Oremland, R.S. and Polcin, S. Methanogenesis and sulphate reduction: competitive and noncompetitive substrates in estuarine sediments. Applied and Environmental Microbiolology, vol. 44, no 6, Dec 1982, p. 1270-1276.

6. Oremland, R.S., Marsh, L.M. and Polcin, S. Methane production and simultaneous sulphate reduction in anoxic, salt marsh sediments. Nature. vol. 296, 11 Mar. 1982, p. 143-145.

7. Widdel, F. and Pfennig, N. Studies on dissimilatory sulphate-reducing bacteria that decompose fatty acids II. Incomplete oxidation of propionate by Desulfobulbus propionicus gen. nov., sp. nov. Archives of Microbiology. vol. 131, 1982, p. 360-365.

8. Battersby, N.S., Stewart, D.J. and Sharma, A.P. Microbiological problems in the offshore oil and gas industries. Journal of Applied Bacteriology Symposium Supplement. 1985, p. 227S-235S.

9. Widdel, F. and Pfennig, N. A new anaerobic, sporing, acetate-oxidising, sulphate-reducing bacterium, Desulfotomaculum (emend.) acetoxidans. Archives of Microbiology. vol. 112, 1977, p. 119-122.

10. Klemps, R., Cypionka, H., Widdel, F. and Pfennig, N. Growth with hydrogen, and further physiological characteristics of Desulfotomaculum species. Archives of Microbiology. vol. 143, 1985, p. 203-208.

11. Sorensen, J., Christensen, D. and Jorgensen, B.B. Volatile fatty acids and hydrogen as substrate for sulphate-reducing bacteria in anaerobic marine sediment. Applied and Environmental Microbiology. vol. 42, Jul. 1981, p. 5-11.

12. Smith, R.L. and Klug, D.J. Electron donors utilised by sulphate-reducing bacteria in eutrophic lake sediments. Applied and Environmental Microbiology. vol. 42, Jul. 1981, p. 116-121.

13. Widdel, F. and Pfennig, N. Studies on dissimilatory sulphate-reducing bacteria that decompose fatty acids I. Isolation of new sulphate-reducing bacteria enriched with acetate from saline environments. Description of Desulfobacter postgatei gen. nov., sp. nov. Archives of Microbiology. vol. 129, 1981, p.395-400.

14. Stott, J.F.D. and Herbert, B.N. The effect of pressure and temperature on sulphate-reducing bacteria and the action of biocides in oilfield water injection systems. Journal of Applied Bacteriology. vol. 60, 1986, p. 57-66.

15. Biogeochemical Cycling of Mineral-Forming Elements. Trudinger, P.A., Swaine, D.J. Published by Elsevier. 1979, p. 401-430.

16. Microbial Biogeochemistry. James E. Zajic. Published by Academic Press, 1969, p. 87-95.

17. Howarth, R.W. Pyrite: its rapid formation in a salt marsh and its importance in ecosystem metabolism. Science. vol. 203, 5 Jan. 1979, p. 49-51.

18. Brown, D.E., Groves, G.R. and Miller, J.D.A. pH and eH control of cultures of sulphate-reducing bacteria. Journal of Applied Chemical Biotechnology. vol. 23, 1973, p. 141-149.

19. Nanninga, H.J. and Gottschal, J.C. Properties of Desulfovibrio carbinolicus sp. nov. and other sulphate-reducing bacteria isolated from an anaerobic-purification plant. Applied and Environmental Microbiology. vol. 53, no. 4, Apr. 1987, p. 802-809.

20. Le Gall, J. and Postgate, J.R. The physiology of sulphate-reducing bacteria. Advances in Microbiological Physiology. vol. 10, 1973, p. 81-133.

21. Bryant, M.P., Campbell, L.L., Reddy, C.A. and Crabill, M.R. Growth of desulfovibrio in lactate or ethanol media low in sulphate in association with H_2-utilising methanogenic bacteria. Applied and Environmental Microbiology. vol 33, no. 5, May 1977, p. 1162-1169.

22. Braun, M and Stolp, H. Degradation of methanol by a sulphate-reducing bacterium. Archives of Microbiology. vol. 142, 1985, p. 77-80.

23. Isa, Z., Grusenmeyer, S. and Verstraete, W. Sulphate reduction relative to methane production in high-rate anaerobic digestion: technical aspects. Applied and Environmental Microbiology. vol. 51, no. 3, Mar. 1986, p. 572-579.

24. Laanbroek, H.J., Abee, T. and Voogd, I.L. Alcohol conversions by Desulfobulbus propionicus Lindhorst in the presence and absence of sulphate and hydrogen. Archives of Microbiology. vol.133, 1982, p. 178-184.

25. Jenkins, R.L., Scheybeler, B.J., Smith, M.L., Baird, R., Lo, M.P. and Haug, R.T., Metals removal and recovery from municipal sludge. Journal of Water Pollution Control Federation. vol. 53, no. 1, Jan. 1981, p. 25-32.

26. Maree, J.P. and Strybom, W.F. Biological sulphate removal from industrial effluent in an upflow packed bed reactor. Water Research. vol. 21, no. 2, 1987, p. 141-146.

27. Maree, J.P., Geber, A. and Hill, E. An integrated process for biological treatment of sulphate-containing industrial effluents. Journal of Water Pollution Control Federation. vol. 59, no. 12, Dec. 1987, p. 1069-1074.

28. Atlas of Electrochemical Equilibria in Aqueous Solution. Marcel Pourbaix. Published by National Association of Corrosion Engineers, Houston, USA.

New biological treatment plant for heavy metal contaminated groundwater

P. J. H. Scheeren
R. O. Koch
Budelco BV, Budel-Dorplein, The Netherlands
C. J. N. Buisman
Paques BV, Balk, The Netherlands
L. J. Barnes
Shell Research Ltd, Sittingbourne Research Centre, Sittingbourne, Kent, England
J. H. Versteegh
Budelco BV, Budel-Dorplein, The Netherlands

SUMMARY

Soil and ground water underneath the zinc production plant of Budelco (The Netherlands) are contaminated with heavy metals and sulphate. To avoid contamination of nearby drinking water aquifers in the distant future, a Geohydrological Control System (GCS) and a treatment plant for the extracted contaminated ground water have been developed.
For the design of the GCS system ground water analysis data and geological profiles of the site were used to model the existing geohydrological situation. From the model several combinations of water extraction wells were deduced. The model was then improved by introducing additional field work data. Finally the calculated results of the GCS options were compared to the objectives of the system, leading to the choice of a design base. Some 6,000 m³/day ground water will have to be extracted from a combination of 17 shallow and deep wells.
From the extracted water heavy metals and sulphate must be removed before discharging the water to the public domain. Several water treatment options have been studied and piloted. A process based on activity of Sulphate Reducing Bacteria (SRB) combines sulphate removal and heavy metal removal in one process step. The anaerobic bacteria reduce the sulphate to sulphide and consequently metal sulphides will precipitate. A demonstration plant showed promising process results. A Dutch company specialized in anaerobic water treatment was contracted for the design of a commercial scale plant. This design should include effluent treatment for excess sulphide and solids in order to meet the discharge criteria. The proposed design was tested in a pilot plant including all main process units. Pilot plant test results for all process units are presented. The collected information was used to adjust the plant design. Final design for the commercial scale plant is described.

INTRODUCTION

Budelco, a company jointly owned by Billiton (The Netherlands) and Pasminco (Australia), is a zinc refinery in the Netherlands producing over 200,000 tpa of zinc. The soil and ground water underneath the plant have been contaminated with heavy metals and sulphate due to 100 years of industrial activity. Ground water analysis data, linked with computer based aquifer simulation models indicate that the contamination will eventually surpass the site perimeter. In order to avoid contamination of nearby drinking water aquifers in the distant future, a Geohydrological Control System (GCS) has been developed. This GCS is designed to contain the contaminated ground water within the site borders by pumping up ground water from strategically located wells. The GCS will produce water contaminated with sulphate and zinc as well as traces of other heavy metals.

Before using the water in the zinc production process or discharging the water to the public domain, heavy metals and sulphate have to be removed.
Different water treatment processes have been studied. Three options were selected for further investigations :

- Ion Exchange (IX)
- Liquid Membrane Permeation (LMP)
- Sulphate Reducing Bacteria (SRB).

IX

The IX process was studied by Billiton Research Arnhem (1). The process is based on an exchange of heavy metal ions with hydrogen ions. In practice the water is led through a column packed with a cation exchange resin. Heavy metals are extracted from the water to the resin. The resin has to be regenerated by a stripping liquor, in this case an acid solution. Thus the process is semi-continuous : one or more columns are in use for heavy metal extraction producing clean water while another column is stripped producing an acid metal solution.
In summary it was concluded that :

- the process would remove heavy metals to acceptable levels
- the process would not remove sulphate
- the process could be considered to be standard technology and thus upscaling to a commercial plant would be relatively easy

- for the regeneration of the resins a relatively large volume of stripping liquor would be needed, resulting in a low metal concentration bleed stream to be handled by Budelco's production process
- depending on the cation exchange resin the process would only be effective for a certain (group of) metal(s)
- operation costs of the process were estimated to be relatively high.

LMP

The LMP process is a development of the University of Graz (Austria) (2). The process is a solvent extraction: metal ions in the water are exchanged with hydrogen ions of the extractant. Intensive surface contact is achieved by mixing the extractant with the water in a column in counter-current mode. Again metals have to be stripped from the extractant. In the LMP option the stripping liquor is emulsified in the extractant before entering the extraction column. By breaking up the emulsion after the extraction, the strip liquor, containing the metals, is separated from the extractant. The extractant is then ready for reuse. Thus LMP can be designed as a continuously metal removing process.
The process was piloted at Budelco's site in 1980, first in a one step and later on in a two step pilot plant. (3)
In short, conclusions of the test work were :

- LMP would remove heavy metals to levels just acceptable by the authorities
- the process would not remove sulphate
- extractant efficiency could be affected by iron hydroxide in the feed water
- removal was selective for some metals
- the stripping liquor was an acid solution with high metal concentration and thus easy to handle by Budelco's zinc production process.

A 200 m³/h scale process design was made at the beginning of 1989. From the design it appeared that capital costs would be relatively high.

SRB

The role of Sulphate Reducing Bacteria in the environment is well known. Possible application of SRB's in a process for waste water treatment has been studied by Shell Sittingbourne Research Centre (SRC) since 1987. The bacteria reduce sulphate to sulphide. Consequently heavy metals in the water will precipitate as metal sulphides because of the extremely low solubility products of these components. In principle the process is capable of removing metals as well as sulphate.

Several electron donors for the process have been tested. First indications suggested that operation costs for the process, based on a lactate carbon source, would be too high. However, by the end of 1988 SRC reported that ethanol was very efficient as a carbon source for the SRB's. Budelco supported a labscale study by SRC to apply the SRB process for treatment of Budelco's GCS water.

Based on the positive results achieved (4), it was decided to run a process demonstration test at Budelco's site. A 9 m³ demonstration plant was developed with support from SRC. This plant was operated for 10 months to prove the process on a large scale and to establish operational ranges for process parameters.
Results of lab tests and demonstration runs are presented in a separate paper (5).

Because of the proven capability of the process to remove sulphate and heavy metals simultaneously at acceptable operation costs, Budelco decided to start the development of a full scale water treatment plant based on the SRB process. Paques, a Dutch company with experience in anaerobic water treatment systems, was contracted for Basic Design. Although the SRB process was proven, the subsequent treatment of the process effluent to remove excess sulphide and remaining solids still had to be developed. In addition Paques needed more information on design parameters. An integrated pilot plant was therefore installed, containing all basic components needed for a complete water treatment. The plant was operated for 8 months to prove the efficiency of all components and to fix design parameter values.

This paper describes the design of the GCS, results of pilot plant tests and the design of a commercial scale SRB water treatment plant.

GEOHYDROLOGICAL CONTROL SYSTEM

Background

In 1892 the Kempensche Zinkmaatschappy was founded in Brabant (The Netherlands). The process used was a thermal reduction of oxidic zinc ores in horizontal retorts using coke as the reductant. Later the oxidic ores were replaced by roasted sulphidic ores. Annual production was 50,000 tpa zinc. The residues of the process, the so-called zinc ashes, were used as landfill to level the swampy site areas and make them suitable for plant extensions. The zinc ashes, containing up to 15 % of heavy metals were at that time considered to be inert material. As we now know, heavy metals are leached by percolating ground and rain water. Zinc and cadmium levels up to 200 mg/l and 20 mg/l respectively are found at a depth of ten to twenty meters underneath the zinc ashes.
In 1973 an electrolytic zinc plant was erected to replace the thermal process.

Zinc is recovered with 98 % efficiency from sulphidic ores by roasting and subsequent hot acid leaching. Annual design capacity was 150,000 tonnes of special high grade zinc, capacity today is 210,000 tpa. The residue of the electrolytic zinc process is an iron-ammonium- sulphate compound called jarosite. Jarosite, containing traces of heavy metals, is considered to be chemical waste and is stored in HDPE lined ponds. Monitoring of the zinc concentration in the drainage systems indicated that one pond is leaking. The leakage contributes to ground water contamination.

In the seventies a geophysical electrical resistivity investigation of soil and ground water underneath the zinc plant was conducted. The resistivity measurements proved to be unsuited to quantify the extent of the contamination.

After 1983 sampling probes were installed to investigate the soil and ground water contamination caused by the leaking jarosite pond. Results indicated a relatively small contribution to the overall contamination although locally high concentrations of heavy metals were found.

Two further studies on contamination by leaching zinc ashes showed severe contamination of ground water to a depth of 30 meters. Contamination is however confined within the industrial site.

It was concluded to design a geohydrological system to contain the contamination within the site perimeter (6).

Design of Geohydrological Control System (GCS)

For the design of the GCS two companies were contracted : Delft Geotechnics for model development and calculations, Tauw Infra Consult for field work, design and engineering.

The design program of the GCS involved 6 phases :

1) modelling of the existing situation
2) modelling of proposed control system options
3) additional field investigations
4) reworking of the model with phase 3) data
5) selection of final option
6) detailed engineering and definitive cost estimate.

Phase 1

Budelco provided information collected from 50 drill holes including geological profiles, lysimetric data (capillary gauge) and concentrations of zinc, cadmium and sulphate. Water balance data were used in a regional and local geohydrological model as well as general information on soil permeability and retardation known from previous investigations. The model assumptions were adjusted until a good fit was achieved between predicted ground water movement and actual field measurements.

Phase 2

In this phase about 10 combinations of deep and shallow extraction wells were developed. The influence on ground water movement was calculated from the model for each combination. Most promising options were selected. Additional field information required for model improvement was identified.

Phase 3

The additional field work, mainly consisting of drilling new holes, was executed. Because of new uncertainties arising from interpreting the data, the scope of the field work was extended and the number of wells increased to 100.

Phase 4 and 5

Reliability of the model was improved by introducing the additional field information in the model. The adjusted model was then applied to calculate the effect of proposed control options. Combining the results with the objective to contain the contamination within site perimeters led to the selection of the control system.

Phase 6

The Basic Design of the selected GCS was made by Tauw. The design includes pumps, control valves and pipeline manifolds. A PLC based flow control system has been developed.

Results

Figure 1 shows the regional soil profile with Budelco located at the centre. The region forms part of the "Central Slenk" of the Brabant massive and is marked by NW-SE parallel faults. The top layer has medium permeability. Underneath this layer a water carrying formation with good permeation is found, referred to as first aquifer. The clay layer at increased depth is of poor water permeability. Underneath the clay layer again a water carrying formation is found referred to as the second aquifer. Most water winning for domestic and industrial use is from the second aquifer.

Figure 2 gives the profile of the contamination. The shaded area displays a ground water contamination exceeding the Dutch "C" levels of 800 mg/l zinc and 10 mg/l cadmium. Simulations with the geohydrological model show that it will take several hundred years for the contamination front to reach any water winning area. One result of the calculations is shown in Figure 3. The contours represent the 10 ug/l cadmium front after the number of years indicated.

Figure 4 is an isohypse contour diagram representing the contours of equal freatic height. It is concluded that ground water movement is from SW to NE.

The "Zuid-Willemsvaart" is a canal which provides a constant input to the freatic body.

Vertical velocity of rainwater penetrated in the top layer is estimated to be 0.2 m/year. Horizontal velocity in the sand layer of the first aquifer is 10 m/year.

Options for the control system ranged from only top layer drainage to deep well extraction. Top layer drainage requires a large number of holes. The contamination is then pulled back into the top layer. Deep drainage requires fewer holes but contamination is pulled downwards.

406

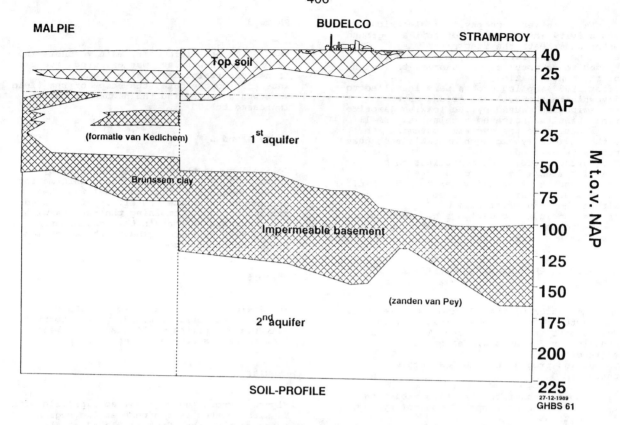

Figure 1   Soil profile underneath Budelco and surroundings

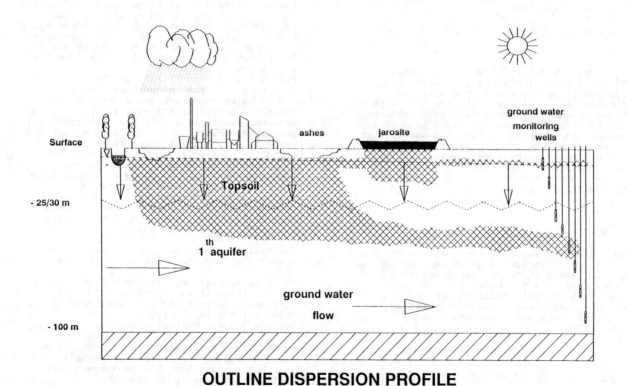

OUTLINE DISPERSION PROFILE

Figure 2   Contamination profiles (marked areas). At the right
side of the refinery the leaking jarosite pond.

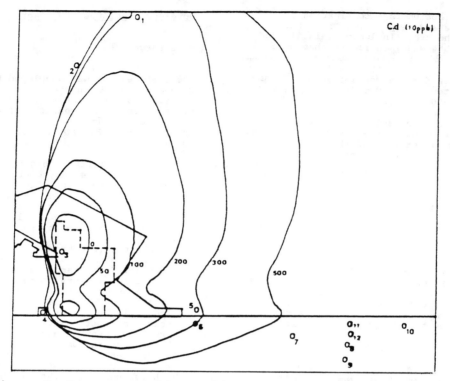

Figure 3 Isoconcentration lines for future cadmium contamination. (Refinery at left side "-----")
The numbers at the lines denote the number of years to pass before the profile is reached.

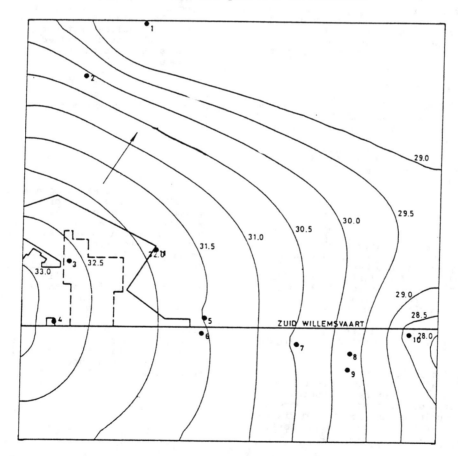

Figure 4 Isohypse profiles. (Refinery at left side "-----")
Numbers at the lines denote the height above sea level of the freatic surface.

The selected option is a combination of
shallow wells and deep wells. The water
balance of the area (including rainfall and
influx from the west) indicates that some
6,000 m³/day will have to be extracted through
17 wells.
Best estimates for the contamination of the
extracted water are : 50 mg/l zinc, 0.1 mg/l
cadmium and 500 mg/l sulphate.

The heavy metals and sulphate in the
extracted ground water must be removed before
the water is either used in the zinc
production process or discharged to the
public domain.

SRB PILOT PLANT

Background

A demonstration run of the SRB process,
reported in a separate paper (5), showed the
process capable of simultaneous removal of
heavy metals and sulphate. It was decided to
develop a commercial scale water treatment
plant based on the SRB process to treat
7,000 m³/day.
Objectives of the development were :

- Design of a commercial scale anaerobic
  reactor for the SRB process including
  treatment of the gas produced. Normal
  anaerobic treatment systems for organic
  contaminations produce methane and carbon
  dioxide. In the SRB process the gas will
  also contain hydrogen sulphide.
- Development of a removal step for the
  excess sulphide present in the SRB process
  effluent. After reduction of most of the
  sulphate and precipitation of metal
  sulphides, the excess sulphide is present
  as HS⁻/H₂S in the effluent. Because of the
  toxicity and odour of this component, the
  sulphide has to be removed almost
  completely.
- Design of a removal step for solids flushed
  with the SRB aqueous effluent. The solids
  mainly consist of metal sulphides, which
  would be converted to soluble metal
  sulphates after discharge of the water.
  Therefore, removal of the solids determines
  the final metal concentration in the
  discharged water.

Budelco contracted Paques for the Basic
Design of the plant. Main plant units
proposed were :

- an Upflow Anaerobic Sludge Blanket (UASB)
  reactor for the SRB process including a gas
  catching system and gas burner
- a Submerged Fixed Film (SFF) reactor for
  the aerobic conversion of sulphide in the
  SRB effluent to solid sulphur
- a Tilted Plate Settler for solids removal
  (TPS),
- a continuous sand bed filter (DynaSand) as
  a solids polishing step before discharge of
  the water.

Budelco rented a pilot plant from Paques
which included all main units. A test program
was executed during 8 months at Budelco's
site. Objectives of the test program were to
determine the units efficiency and to fix the
design parameters for the commercial plant.

Development of the SRB plant

UASB reactor

The UASB reactor was developed by the
University of Wageningen, The Netherlands,
(7) for anaerobic treatment of waste water
contaminated with organic compounds. A Paques
version of the reactor has been patented
under the name BIOPAQ (Figure 5). The reactor
consists of :

- An influent system designed to spread the
  incoming water evenly over the active
  biological bed. Hydraulic capacity of the
  influent system is such that the sludge bed
  is gently mixed to avoid the channelling
  problems of a compacted sludge. Mixing of
  the bed is improved by the produced gas
  bubbles.
  The ethanol (electron donor) is mixed with
  the influent. The SRB convert sulphate to
  sulphide. The ethanol is partly converted
  to acetate. In turn acetate is converted to
  methane and carbon dioxide by the so-called
  methanogens. Bacteria must attach to the
  sludge blanket in order to achieve low
  residence times.
- A reaction chamber normally designed for
  3 - 6 hours residence time.
- A three phase separator at the top. The
  separator consists of 3 layers of inverted
  V-shaped channels. The gas produced is
  caught by these "hoods". The separator is
  located below the water surface and after
  building up sufficient pressure the gas is
  released to a gas buffer. Subsequently, the
  gas is periodically burned in a flare.
  Because of gas removal, the top water layer
  is relatively undisturbed and facilitates
  settling. Solids in the top water layer
  will settle at the tilted plates of the gas
  hood system and subsequently sink to the
  sludge bed.

The pilot plant had a PLC controlled 12 m³
UASB reactor preceded by a 10 m³ mixing tank
for neutralizing and dosing of substrate and
nutrients.

SFF reactor

Biological sulphide oxidation is developed by
the University of Wageningen (8).
The SFF reactor consist of a water inlet
system at the bottom, a Pall ring packed
reaction chamber and an air inlet system.
Aerobic bacteria convert sulphide to
elemental sulphur by oxidation. The bacteria
and sulphur adhere to the rings until sheer
forces caused by stirrer action of the forced
air stream detach the solids. Oxygen for the
reaction is supplied by the air whereas the
carbon source (carbon dioxide) and the
nutrients (N, P) are present in the water. At
high redox values and high oxygen supply,
sulphide can be oxidized further to sulphate.
To avoid an increased sulphate concentration
in the water this reaction is minimized by
appropriate oxygen supply. The pilot plant
comprised a 1.5 m³ SFF reactor.

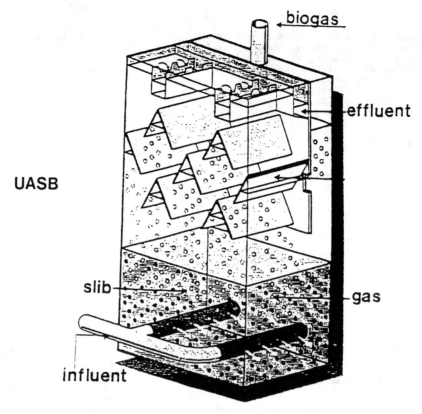

Figure 5 Schematic view of UASB reactor.

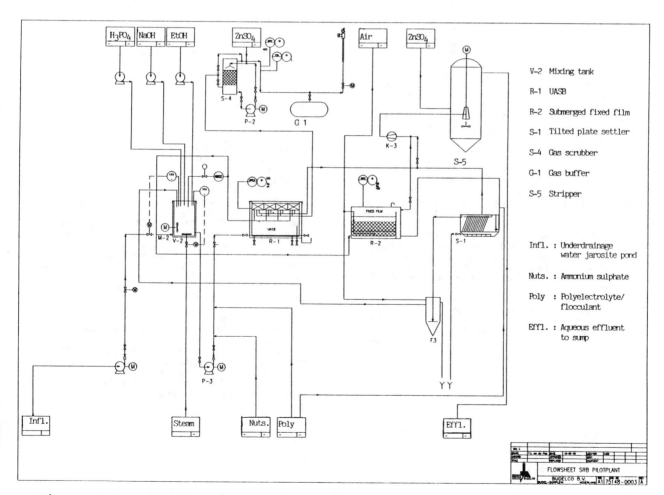

Figure 6 Process Flow Diagram of SRB pilot plant.

TPS

The Tilted Plate Settler is a well known
settling unit for solids. Settling is
improved by shortening the settling distance
through parallel tilted plates. The settler
removes solids consisting of sulphur produced
in the SFF, metal sulphides and biomass
flushed with SRB effluent. The pilot plant
settler had a volume of 12 m³.

DynaSand filter

The filter is a continuously refreshed sand
bed. The water flows up through the bed
leaving the solids at the bottom of the bed.
Air is pumped through a small diameter pipe
at the centre of the filter. This airlift
transports the sand particles with attached
solids to the top of the filter. During
transport the sand is separated from the
solids by sheer forces. At the top of the
filter the sand is separated from the wash
liquor by gravity and the solids pass out as
a slurry. The pilot plant contained a
DynaSand filter with a bed height of
approximately 1.2 m and a surface of 0.6 m².

A schematic view of the plant is shown in
Figure 6.

Operation conditions

As a result of the demonstration tests of the
SRB process the operational ranges of process
parameters were known :

pH   :   6 - 8
T   :   20 - 38 °C
residence time   :   2 hours for
                     conversion, in
                     practice determined
                     by solids settling
substrate   :   1 mole ethanol per
                     mole sulphate

Inoculation of the UASB reactor was achieved
by transferring biosludge from the
demonstration reactor.
The GCS ground water was simulated with
underdrainage water from a jarosite pond.
This water normally is contaminated with more
than 200 mg/l zinc and 1 g/l sulphate.
A flocculant (Synthofloc) was added to the
UASB influent. Different kinds of flocculant
were also applied to improve settling in the
TPS.
For design parameter determination the off-
gas of the UASB reactor was monitored. The
gas catching system allowed for in-line gas
analysis of methane and carbon dioxide.
The first part of the program was aimed at
the objectives mentioned previously. The
latter part of the test period was used to
investigate variances of output
concentrations during stable operation of the
pilot plant. Finally several start-up trials
were performed to determine an optimal start-
up procedure.

Analysis

During the tests the following process flows
were analyzed :

- feed water   :   heavy metals
                       and sulphate,
                       redox
- UASB effluent   :   heavy metals,
                       sulphate,
                       sulphide,
                       acetate,
                       phosphate,
                       redox
- UASB off-gas   :   methane, carbon
                       dioxide,
                       hydrogen
                       sulphide
                       (partly)
- SFF effluent   :   heavy metals,
                       sulphate,
                       sulphide,
                       phosphate,
                       redox,
- TPS effluent   :   heavy metals,
                       sulphate
- DynaSand effluent : heavy metals,
                       sulphate,
                       sulphide,
                       phosphate,
                       redox.

Heavy metals are measured on clear solutions
to check the conversion efficiency. Analysis
on heavy metals of the total solution
(including solids) show the efficiency of
solid removal steps. Under normal operating
conditions solids content is too low to be
measured directly.

Pilot plant test results

Although pilot plant trials were affected by
feed concentration variations, the test
objectives were achieved.

UASB results

After start-up with non-growing biomass SRB
activity was restored within 1 week. However,
build-up of methanogen activity, necessary to
convert acetate to methane, was slower. The
resulting high acetic acid concentration can
stop the process because of low Ph values in
the reactor. Specific start-up conditions are
described later.
The liquid residence time in the reactor is
dependent on biomass activity and flush out
of solids. The flush out is the determining
factor and thus the solid settling. A
residence time of 4 hrs could be achieved by
appropriate flocculant dosing.
Table 1 shows average results under the
operating conditions mentioned.

Synthofloc must be dosed in a range of 0.5 to
1.0 mg/l. If less flocculant is dosed, solids
will flush out with the effluent. Higher
dosing leads to gas inclusion and big flocks
start to flotate with an eventual danger of
blockage. For easy operation of the
commercial scale plant a lower ionic strength
flocculant might be better.

Table 1 : Average UASB influent and
effluent concentrations. Effluent
analysis on <u>clear</u> solution.

Component	Unit	Concentration		Reactor conditions
		Influent	Effluent	
Zn	mg/l	250	< 0.05	**Influent flow:**
Cd	mg/l	1.10	< 0.001	3000 l/h
Fe	mg/l	33	0.03	**Residence time:**
Pb	mg/l	13	< 0.02	4 h
Cu	mg/l	2.2	< 0.02	**Temperature:**
Co	mg/l	0.2	< 0.02	20 °C
Ni	mg/l	0.3	< 0.01	**Ph:**
Mn	mg/l	35	20	7
Mg	mg/l	40	35	**Ethanol : Zn**
Ca	mg/l	300	290	0.7 : 1
Na	mg/l	60	370	**Redox potential:**
As	mg/l	0.010	< 0.005	-390 mV
SO_4^{2-}	g/l	1.4	0.120	**Sludge load:**
NH_4^+	mg/l	12	19	100 g/l

The hydraulic operation range is 1.5 to 3 m^3/h (residence time of 4 - 8 hrs). Higher flow velocities cause solids flush-out. At lower flow rates the sludge bed mixing is insufficient and the sludge solidifies causing channelling and inactivity.

The gas production was 7 m^3/day at 2 m^3/h feed flow rate. The gas contained 70 - 80 % methane, 5 - 10 % carbon dioxide and 5 - 10 % hydrogen sulphide. Gas production is strongly dependent on activity of the methanogens for acetate conversion.

SFF results

Table 2 shows a mass balance of sulphur components over the UASB and SFF reactors.

Sulphide removed by precipitation of metal sulphides was 50 mg/l. Sulphide removal via off-gas was 140 g/m^3 which can be converted to 22 mg/l S. Sulphide stripped in SFF was 3 g/m^3 to be converted to 40 mg/l S.

Efficiency of the SFF reactor is high : up to 300 mg/l sulphide in the SFF influent are decreased to less than 1 mg/l in the effluent under aerobic conditions. Incidentally the sulphate concentration in the effluent increases because of excess oxidation occurring during underload conditions but increase never exceeds 50 mg/l SO_4.

The conversion is strongly dependent on the air supply. (Table 3)

The redox and dissolved oxygen concentration in the effluent were measured to check if both parameters can serve as control parameters for the conversion.

The oxygen measurement was troublesome because of calibration problems. The redox value shows a sudden increase at low sulphide levels which can be expected from theory. The sulphide concentration is the best control parameter.
During the tests a blockage of the reactor occurred caused by a flocculant overdose. For final reactor design cleaning facilities are important.

TPS and DynaSand results

The efficiency of TPS and DynaSand filter is demonstrated in Table 4.

The TPS is very effective for sulphur removal. In different tests at which different flocculants were applied the efficiency was always better than 89 %.

The DynaSand filter is an effective polishing step in the process. It is a rugged filter that needs only a minimum maintenance. However, an overdose of flocculant can cause poor separation of sand and solids. The result is a that sand is flushed out with the washing liquor.

At stable operation solids concentration in the UASB effluent is less than 100 mg/l. This will result in less than 1 mg/l solids in the DynaSand filter effluent and the corresponding total zinc discharge is less than 0.3 mg/l.

412

Table 2 :   Sulphur mass balance UASB and
            SFF.
            Influent flow: 1.5 m³/h, UASB off-
            gas flow 0.24 m³/h, air flow 20
            m³/h.
            All components expressed as mg/l
            S.

component	influent UASB	influent SFF	effluent SFF
sulphate	450	220	* 150
sulphide	0	245	5
sulphur	0	0	290
total	450	465	445

* Sulphate concentration is decreased in the
  reactor because SRB are still active under
  these conditions (9).

Table 3 :   Influence of air supply on
            sulphide conversion in SFF.

air flow m³/h	sulphide influent mg/l	sulphide effluent mg/l
30	156	< 1
28	136	8
26	156	13

Table 4 :   Solids removal in TPS and
            DynaSand filter in mg/l during a
            typical run.

component	influent TPS	effluent TPS	effluent DynaSand
solids mg/l	2400 *	385 (86 %)	37 (90 %)
zinc (total) mg/l	114	19 (83 %)	1.3 (93 %)
sulphur mg/l	336	15 (96 %)	-

* Solids value in the TPS influent was high
  because of inappropriate flocculant dosing.

Table 5 :   Variance test of SRB process for
            zinc and cadmium.

component	feed concentration	effluent concentration	effluent conc. stand. dev.
Zn (mg/l)	180	0.3	0.25
Cd (ug/l)	650	7.6	2.5

Variance test

A test was executed to determine the stability of the process during normal operation. Although the test was disturbed by feed water variations (flow and concentrations) and mechanical failures in the pilot plant, stable operation could be achieved over a 3 day test period, which is 18 times the residence time of the UASB reactor. Results are presented in Figure 7 for total zinc and cadmium concentrations in the pilot plant effluent. Table 5 shows the essential results of the test.

The variations mainly result from variations in the hydraulic load of the UASB reactor. If a feed flow variation occurs the solids concentration in the UASB overflow varies strongly if flocculant dosing is not adjusted. Because TPS and DynaSand filter operation is not changed, the absolute solids discharge varies, resulting in variations in total metal discharge. It is therefore concluded that hydraulic load of the UASB should be fixed. Flocculant dosing should be adjusted to total solids load in the UASB reactor.

Start-up improvements

Last trials performed with the Paques pilot plant were start-up trials. Preceding tests showed problems with restarting the SRB process after a period of inactivity caused by inadequate process parameter start values. In order to anticipate start-up difficulties for the final plant a correct start-up procedure was established.
Important parameters at start-up are:

- pH in UASB reactor
- redox in UASB reactor
- influent flow
- temperature of influent
- sulphate concentration in influent
- phosphoric acid in UASB effluent.

Additional critical parameters during start-up are:

- acetic acid concentration in UASB effluent
- solids settling behaviour in UASB.

Under accurate initial process conditions the SRB become active within 3 days. The bacteria then produce acetate which will lower the pH in the reactor. At a pH lower than 6 SRB become inactive. Because of the relatively slow growth rate of methanogens and thus slow decomposition of acetate, the influent and recycle flow should be low. A high influent flow will flush out the methanogens. A high recycle flow results in a build-up of the acetate in the reactor. Analysis of acetate in the UASB effluent indicates the activity of the methanogens. Another indication is the methane produced, which is 70 - 80 % of the total gas production. Thus a low gas production volume means low methanogen activity.
The sulphate concentration should be high enough to produce excess sulphide after metal sulphide precipitation in order to keep the redox value in the reactor low.

The phosphoric acid dosed as a nutrient will precipitate as metal phosphate at pH values above 7. For this reason at start-up phosphoric acid should be overdosed and checked at the UASB outlet.
As soon as methanogen activity is detected the pH can be lowered. As a result precipitated phosphate will dissolute and phosphate dosage must be decreased strongly to avoid new pH problems.

Total start-up (no acetate in effluent) under these conditions will take 2 to 3 weeks although sulphate reduction and metal removal are at acceptable level after 3 days.

Conclusions

The main units of the proposed design for the commercial scale SRB plant have been tested on pilot plant scale. Results show that the plant design meets the targets for heavy metal and sulphate removal.
A 4 hours residence time is required for the SRB process in the Upflow Anaerobic Sludge Blanket (UASB) reactor. Flocculant dosage in the UASB influent is essential for plant performance. Hydraulic load of the reactor should be fixed during normal operation, which can be attained by a recirculation system. The hydraulic load value chosen is dependent on the influent pipeline system. For the 12 m^3 pilot reactor the influent flow range is 1.5 to 3 m^3/h.
Start-up of the SRB process in the UASB reactor is strongly dependent on operation conditions; operation parameter values should be adjusted to changing SRB and methanogen activity. Required start-up time for metal/sulphate removal and acetate removal is 3 days and 2 weeks respectively.
The UASB off-gas contains 70 - 80 % methane, 5 - 10 % carbon dioxide and 5 - 10 % hydrogen sulphide. The gas can be easily burned in a flare. However, for environmental reasons the hydrogen sulphide has to be scrubbed from the gas first to avoid sulphur dioxide emission from the flare (90 tons per year for the commercial scale plant)
Sulphide removal from the UASB effluent in the Submerged Fixed Film (SFF) reactor is very efficient. Concentrations up to 300 mg/l sulphide are decreased to less than 1 mg/l without considerable sulphate production. Residence time needed is less than 0.5 hour. The best control parameter for the reaction is the sulphide concentration in the SFF effluent.
The Tilted Plate Settler (TPS) and the DynaSand filter remove solids with an efficiency of 90 % or more. Because these systems are complementary an overall efficiency of solids removal of more than 99 % can be achieved. Both systems require only minor attention and maintenance. Flocculant addition improves solids removal but is not essential in this stage of the process.

Total ZINC in DynaSand filter effluent

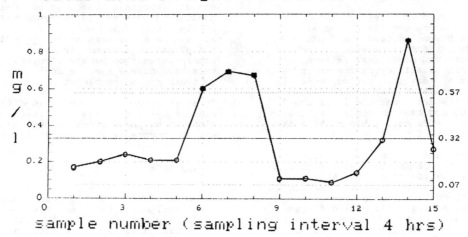

sample number (sampling interval 4 hrs)

Total CADMIUM in Dynasand filter effl.

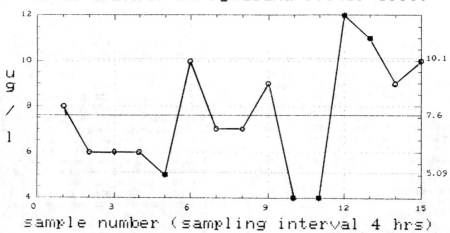

sample number (sampling interval 4 hrs)

Figure 7   SRB time series total zinc and cadmium in pilot plant effluent, including averages and upper and lower limits at a standard deviation of "one sigma".

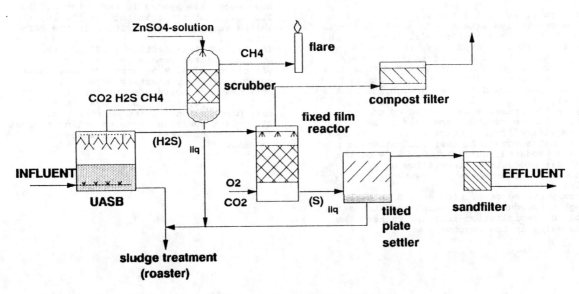

Figure 8   Process block diagram SRB plant.

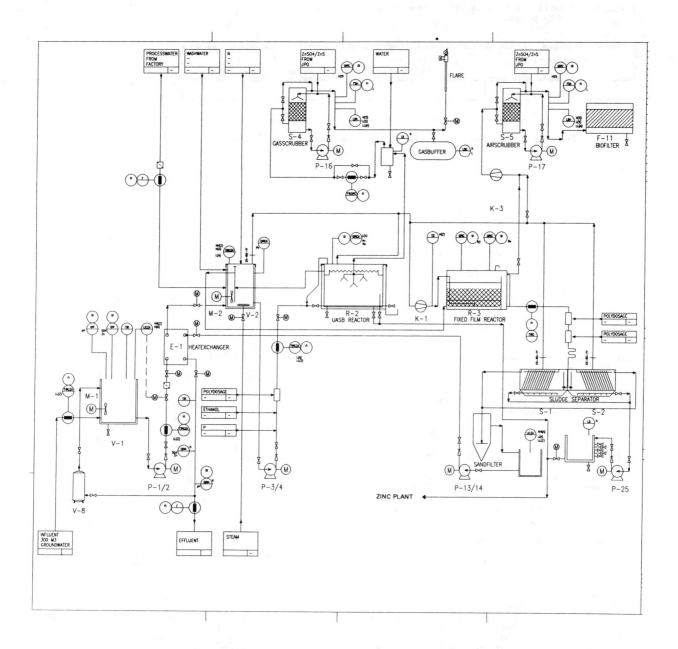

Figure 9   Schematic overall process flow diagram SRB plant.

Table 6:   Design criteria for commercial
            SRB plant.

component	influent concentration	effluent concentration
Zn (mg/l)	100	< 0.3
Cd (mg/l)	1	< 0.01
SO_4^{2-} (g/l)	1	< 0.2

Final Design

Design of the commercial plant is based on
the line-up tested in the pilot plant :

- buffer tank to buffer variations of
  influent flow and concentrations
- mixing tank for addition of substrate and
  nutrients
- anaerobic UASB reactor for sulphate
  reduction
- aerobic SFF reactor for sulphide oxidation
- Tilted Plate Settler for solids removal
- Dynasand filters as a final polishing step
- Gas Handling System for sulphide scrubbing
  and methane burning.

The concentration design criteria chosen are
briefly described in Table 6.

Because the technology involved in the
process is new other design parameters have
been set conservatively :

- influent flow   : 300 m³/h
- temperature   : 25 °C
- residence time UASB : 6   hrs
- residence time SFF   : 0.55 hrs
- residence time TPS   : 0.4 hrs
- filter surface   : 30 m² (10 m³/m².h).

Special attention has been paid to safety and
environmental aspects.
Gas scrubbing will be performed by
precipitation of the sulphide with a
concentrated zinc sulphate bleed stream from
the zinc production process. A preliminary
design of the scrubber is currently being
tested.
To avoid odour problems all relevant plant
units work at a slight underpressure. The
evacuated air will be scrubbed in a separate
scrubber and subsequently fed to a biofilter.
Parts of the plant must be considered as
explosive zones because of the gas produced
and equipment is designed accordingly.
The plant is designed for fully automated
operation. On-line analysis equipment for
measurement of main process parameters is
connected to a central process computer for
plant control. Software controlled actions
will be taken at major deviations from
parameter setpoints to avoid safety problems
and to assure fast recovery of the operation.

A process block diagram and a schematic
overall process flow diagram are presented in
Figure 8 and 9 respectively.

REFERENCES

1) A.J.A. Konings, Private communication.

2) J. Draxler, R.Marr, Emulsion liquid
    membranes, part I, Chem. Eng. Proc. 20
    (1986), 319.

    D. Lorbach, R. Marr, Emulsion liquid
    membranes, Part II, Chem. Eng. Proc. 21
    (1987), 83.

3) J.P. van 't Sant, LMP pilot plant
    tests, private communication.

4) L.J. Barnes, private communication.

5) L.J. Barnes, F.J. Janssen, J. Sherren,
    J.H. Versteegh, R.O. Koch,
    P.J.H. Scheeren, Simultaneous microbial
    removal of sulphate and heavy metals
    from waste water. (Submitted for
    publication and presentation at the EMC
    Conference, Brussels, 1991)

6) J.H. Versteegh, The Design of a
    Geohydrological Control System,
    Proceedings Symposium : Dealing with
    the environmental problem in the mining
    and petroleum industry, T.U. Delft, The
    Netherlands, 1990.

7) G. Lettinga et al, Use of the Upflow
    Anaerobic Sludge Blanket (UASB) reactor
    concept for biological waste water
    treatment, especially for anaerobic
    treatment, Biotech. & Bioeng, 22.

8) C.J.N. Buisman et al., Biotechnological
    Process for Sulphide removal with
    sulphur reclamation, Acta Biotechnol.,
    9, (1989), 255

9) C.J.N. Buisman et al., Sulphur and
    Sulphate reduction with acetate and
    propionate in aerobic process for
    sulphide removal, Appl. Microbiol.
    Biotechnol., 32, (1989), 363

Silver plating from thiosulphate baths

A. Hubin
G. Marissens
J. Vereecken
Department of Metallurgy, Electrochemistry and Materials Science, Vrije Universiteit Brussel, Brussels, Belgium

SYNOPSIS

Electroplating of silver is a process of major industrial significance. In the present work, silver plating from baths containing silver thiosulphate complexes is investigated. Thereto, a combination of electrochemical (linear sweep voltammetry on a rotating disc electrode) and surface analytical (Auger electron spectroscopy and scanning electron microscopy) techniques is used.

The reaction follows a mechanism involving mass and charge transfer and chemical reaction steps. The rate equation, describing the kinetics of the plating process, is determined.

The deposits are shown to be composed of polycrystalline silver, without incorporation of thiosulphate ions. In a wide range of plating conditions, smooth deposits are obtained.

Since smooth and pure silver deposits can be realized at high plating rates, the development of thiosulphate baths to replace the commonly used, but toxic, cyanide baths for silver plating on an industrial scale is very promising.

INTRODUCTION

Silver is plated for its outstanding electrical, thermal and mechanical properties and its corrosion resistance towards various acids and organic agents[1,2].

For silver plating, mainly silver cyanide solutions are used, and this is extensively treated in the literature[1-5]. Yet, in view of the growing concern for environmental problems, they will have to be replaced because of their toxic properties.

Deposition of silver from solutions in which silver is present as free ions, such as $AgNO_3$ or $AgClO_4$ offers no alternative. It proceeds at high rate and results in deposits with a dendritic structure[6,7], what is unacceptable in plating applications. The presence of a complexing agent is indispensable to slow down the reduction reaction.

In the present work, the appropriateness of thiosulphate ($S_2O_3{}^{2-}$), a complexing agent for silver of common use in photographic applications, as an alternative for cyanide in plating baths is investigated.

In the case of deposition of a metal in the presence of a complexing agent, the metal ion is generally not undergoing the charge transfer step directly[8]. Fig.1 shows a schematic representation of the steps that can possibly be involved.

For the investigation of silver plating from silver thiosulphate baths, three aspects of the process are considered. The kinetic characteristics of the reduction are determined, since they fix at which rate the plating can proceed. Prior to the kinetic study, requiring measurements in solutions of varying composition, the thermodynamic data on complex formation between Ag^+ and $S_2O_3{}^{2-}$ ions are sorted out. And finally, the composition and the morphology of the silver deposits are examined as a function of the plating conditions. Those properties are of major importance because they are strongly related to the physical and mechanical behaviour of electroplated pieces.

For this investigation, a combination of electrochemical and surface analytical techniques is used.

EXPERIMENTAL

The solutions were prepared with twice demineralized water and the following chemicals : $AgNO_3$ (p.a. Agfa Gevaert), $Na_2S_2O_3$ and $NaNO_3$ (p.a. Merck). They were deaerated by nitrogen bubbling.

Measurements were carried out on a rotating silver (99.9% Johnson, Matthey) disc electrode with a radius of 2 mm. This type of electrode was selected, to be able to characterize the mass transfer step, represented in Fig.1. The theory of rotating disc electrodes is treated in[9,10]. A platinum (Johnson, Matthey) counter electrode with large area and a calomel electrode with saturated KCl solution (Tacussel) were used. It was possible to work under thermostatic conditions ($25.0 \pm 0.1°C$) by making use of a double walled electrolytic cell. Before each experiment, the electrode was polished with diamond spray (Buehler Metadi, compound 5 and 1 μm) and rinsed ultrasonically successively in twice demineralized water and chloroform.

Linear sweep voltammetry experiments were performed at a scan rate of 1 mV s^{-1} using a potentiostat/galvanostat with built-in generator (Princeton Applied Research, PAR model 273) controlled by an IBM/PS2 personal computer.

Characterization of the silver deposits was done with three surface analytical techniques : Auger electron spectroscopy, AES, (PHI 545 with cylindrical mirror analyser), scanning electron microscopy, SEM, (Jeol JSM 50-A) and electron diffraction (Seifert). Auger spectra were recorded in the E[dN(E)/dE] derivative mode under constant conditions[10].

RESULTS AND DISCUSSION

Thermodynamic Aspects of Solution Composition

In AgNO$_3$ - Na$_2$S$_2$O$_3$ - NaNO$_3$ solutions, all silver is, in good approximation , present as Ag(S$_2$O$_3$)$_2^{3-}$ and Ag(S$_2$O$_3$)$_3^{5-}$ [10-12]. The amount of uncomplexed Ag$^+$ is negligible, but its activity is nevertheless an important thermodynamic quantity. Between S$_2$O$_3^{2-}$ and Na$^+$ the complex NaS$_2$O$_3^-$ is formed[13]. Beside the complexes, uncomplexed S$_2$O$_3^{2-}$ and Na$^+$ are present.

For the calculation of the concentrations, activities and activity coefficients (hereinafter called concentration parameters) of all components in the solution, a set of 12 equations with 16 unknowns is formulated[10,12]. This means that 4 of the unknowns must be measured or calculated independently.

The activity of the uncomplexed silver ions $(a_{Ag^+})_f$ is deduced from the measurement of the equilibrium potential E_0 of the solution versus a silver electrode[14] :

$$E_{\circ Ag/Ag^+}/NHE = 0.799 + 0.059 \log (a_{Ag^+})_f \qquad (1)$$

The activity coefficients of uncomplexed S$_2$O$_3^{2-}$ and Na$^+$, $(y_{S_2O_3^{2-}})_f$ and $(y_{Na^+})_f$ and of the complex NaS$_2$O$_3^-$, $y_{NaS_2O_3^-}$, are calculated with the ion interaction model, using tabulated values of specific parameters[10,15]. This model is chosen because of the high ionic strength of the solutions ($I > 0.1$ mol kg^{-1}).

Details concerning the combined measurement/calculation routine are found in[10,13]. The concentration parameters of the AgNO$_3$ - Na$_2$S$_2$O$_3$ - NaNO$_3$ solutions are given in Table I as a function of the total silver and thiosulphate concentrations.

Determination of the Mechanism and the Kinetic Parameters

In order to determine which of the elementary steps, represented in Fig.1, are actually occurring and at which rate,

n°	$(c_{Ag^+})_t$	$(c_L)_t$	$(a_{Ag^+})_f$	$(c_L)_f$	$(y_L)_f$	$(c_{NaL^-})_f$	$(y_{NaL^-})_f$	$c_{AgL_2^{3-}}$	$y_{AgL_2^{3-}}$	$c_{AgL_3^{5-}}$	$y_{AgL_3^{5-}}$
1	$1.2 \cdot 10^{-3}$	$2.7 \cdot 10^{-2}$	$1.0 \cdot 10^{-13}$	$2.3 \cdot 10^{-2}$	0.65	$1.6 \cdot 10^{-3}$	0.87	$1.2 \cdot 10^{-3}$	0.51	-	-
2	$4.9 \cdot 10^{-3}$	$6.0 \cdot 10^{-2}$	$1.0 \cdot 10^{-13}$	$4.3 \cdot 10^{-2}$	0.58	$7.2 \cdot 10^{-3}$	0.85	$4.9 \cdot 10^{-3}$	0.39	-	-
3	$1.1 \cdot 10^{-2}$	$9.7 \cdot 10^{-2}$	$1.2 \cdot 10^{-13}$	$6.2 \cdot 10^{-2}$	0.53	$1.3 \cdot 10^{-2}$	0.85	$1.1 \cdot 10^{-2}$	0.35	-	-
4	$2.0 \cdot 10^{-2}$	$1.4 \cdot 10^{-1}$	$1.3 \cdot 10^{-13}$	$8.2 \cdot 10^{-2}$	0.49	$1.9 \cdot 10^{-2}$	0.91	$2.0 \cdot 10^{-2}$	0.31	-	-
5	$4.0 \cdot 10^{-2}$	$5.8 \cdot 10^{-1}$	$1.6 \cdot 10^{-14}$	$2.9 \cdot 10^{-1}$	0.39	$2.0 \cdot 10^{-1}$	0.98	$2.4 \cdot 10^{-2}$	0.24	$1.4 \cdot 10^{-2}$	0.09
6	$1.0 \cdot 10^{-1}$	$7.2 \cdot 10^{-1}$	$4.2 \cdot 10^{-14}$	$3.0 \cdot 10^{-1}$	0.32	$1.8 \cdot 10^{-1}$	0.89	$6.3 \cdot 10^{-2}$	0.18	$3.8 \cdot 10^{-2}$	0.06
7	$7.0 \cdot 10^{-2}$	$9.2 \cdot 10^{-1}$	$1.0 \cdot 10^{-14}$	$4.0 \cdot 10^{-1}$	0.35	$3.7 \cdot 10^{-1}$	1.04	$3.4 \cdot 10^{-2}$	0.18	$3.0 \cdot 10^{-2}$	0.06
8	$1.1 \cdot 10^{-1}$	1.0	$1.2 \cdot 10^{-14}$	$4.1 \cdot 10^{-1}$	0.31	$3.4 \cdot 10^{-1}$	0.96	$5.2 \cdot 10^{-2}$	0.11	$5.7 \cdot 10^{-2}$	0.03
9	$1.4 \cdot 10^{-1}$	1.1	$2.1 \cdot 10^{-14}$	$4.2 \cdot 10^{-1}$	0.30	$3.3 \cdot 10^{-1}$	0.95	$7.3 \cdot 10^{-2}$	0.13	$7.0 \cdot 10^{-2}$	0.03
10	$1.1 \cdot 10^{-1}$	1.3	$0.9 \cdot 10^{-14}$	$4.6 \cdot 10^{-1}$	0.32	$5.1 \cdot 10^{-1}$	1.04	$4.8 \cdot 10^{-2}$	0.11	$6.1 \cdot 10^{-2}$	0.03
11	$2.0 \cdot 10^{-1}$	1.4	$1.5 \cdot 10^{-14}$	$4.6 \cdot 10^{-1}$	0.27	$4.7 \cdot 10^{-1}$	0.97	$9.0 \cdot 10^{-2}$	0.08	$1.2 \cdot 10^{-1}$	0.01
12	$1.3 \cdot 10^{-3}$	$2.5 \cdot 10^{-1}$	$1.7 \cdot 10^{-15}$	$1.7 \cdot 10^{-1}$	0.50	$8.0 \cdot 10^{-2}$	0.98	$9.9 \cdot 10^{-4}$	0.36	$3.3 \cdot 10^{-4}$	0.18
13	$5.5 \cdot 10^{-3}$	$5.1 \cdot 10^{-1}$	$1.9 \cdot 10^{-15}$	$2.8 \cdot 10^{-1}$	0.46	$2.2 \cdot 10^{-1}$	1.08	$3.4 \cdot 10^{-3}$	0.27	$2.0 \cdot 10^{-3}$	0.11
14	$1.3 \cdot 10^{-2}$	$7.8 \cdot 10^{-1}$	$1.9 \cdot 10^{-15}$	$3.6 \cdot 10^{-1}$	0.43	$3.8 \cdot 10^{-1}$	1.15	$6.8 \cdot 10^{-3}$	0.20	$6.1 \cdot 10^{-3}$	0.07
15	$2.5 \cdot 10^{-2}$	1.1	$1.5 \cdot 10^{-15}$	$4.1 \cdot 10^{-1}$	0.41	$5.9 \cdot 10^{-1}$	1.21	$1.1 \cdot 10^{-2}$	0.12	$1.5 \cdot 10^{-2}$	0.03

Table I : Concentration parameters at T = 298 K for silver thiosulphate complexes. L = S$_2$O$_3^{2-}$; concentrations (c) and activities (a) are in mol l^{-1}; activity coefficients (y) are dimensionless. Indexes t and f mean respectively total and free (= uncomplexed).

the reduction rate is measured as a function of a number of parameters : the potential (to study the influence of the charge transfer step), the rotation speed of the electrode (for the characterization of mass transfer) and the concentrations of $Ag(S_2O_3)_2^{3-}$ and $Ag(S_2O_3)_3^{5-}$ and free $S_2O_3^{2-}$. Hereto, linear sweep voltammograms, giving the current density or reduction rate as a function of potential, on a rotating disc electrode are recorded for the solutions of Table I. Figs.2-3 give in illustration a few typical examples of voltammograms, respectively as a function of solution concentration and of rotation speed .

In[10,16] the voltammograms are analysed in detail, giving the following result. Mass and charge transfer determine the rate of the reduction. Free silver ions and different complexes are present in the solution, in equilibrium with each other. In the bulk, $Ag(S_2O_3)_2^{3-}$ and $Ag(S_2O_3)_3^{5-}$ are the predominant species. Mass transfer to the electrode is determined by their diffusion, each with its own diffusion coefficient. At the electrode Ag^+ and $Ag(S_2O_3)^-$ are reduced at different rates, with a ratio inversely proportional to the $S_2O_3^{2-}$ concentration, respectively to Ag and $[Ag(S_2O_3)^-]^-$ in equilibrium with each other. The electrode reaction is schematically represented in Fig.4.

The kinetics comply to the following rate equation[16] :

$$\frac{1}{j} = \frac{1}{j_c} + \frac{1}{j_m} \qquad (2)$$

$$j_c = - F(k_{0,0} + k_{1,0} \beta_{11} a_{S_2O_3^{2-}}) \exp\left(\frac{-\alpha FE}{RT}\right) a_{Ag^+} \qquad (2a)$$

$$j_m = - F\, 0.62\, \nu^{-1/6}\, \omega^{1/2} (D_{1,2}^{2/3} c_{1,2} + D_{1,3}^{2/3} c_{1,3}) \qquad (2b)$$

with :

$k_{0,0}$	= $1.1\ 10^5$	m s^{-1}
$k_{1,0}$	= $2.9\ 10^{-2}$	m s^{-1}
α	= 0.77	
β_{11}	= $6.6\ 10^5$	mol^{-1} m^3
$D_{1,2}$	= $6.8\ 10^{-10} - 6.4\ 10^{-11}$ I	m^2 s^{-1} (I mol l^{-1})
$D_{1,3}$	= $7.2\ 10^{-10} - 1.3\ 10^{-10}$ I	m^2 s^{-1} (I mol l^{-1})
T	= 298 K	

In the equations are :

. j, j_c and j_m respectively the total current density, and the contributions of charge and mass transfer to it;

. α and $k_{i,0}$ the kinetic parameters of charge transfer, respectively the transfer coefficient and the potential independent rate constant of $Ag(S_2O_3)_i^{1-2i}$, with i = 0 or 1;

. $D_{1,i}$ the diffusion coefficient of $Ag(S_2O_3)_i^{1-2i}$, with i = 2 or 3 as a function of the ionic strength I of the solution, characterizing mass transfer;

. ν the kinematic viscosity of the solution, tabulated as a function of composition in[16];

. $a_{S_2O_3^{2-}}$, a_{Ag^+} and $c_{1,i}$ respectively the activity of uncomplexed $S_2O_3^{2-}$ and Ag^+ and the concentration of $Ag(S_2O_3)_i^{1-2i}$;

. E, the potential, and ω, the rotation speed of the electrode, the driving parameters for respectively charge and mass transfer;

. R the universal gas constant (= 8.3145 J mol^{-1} K^{-1}) and F Faraday's constant (= 96486 A s mol^{-1}).

Composition and Morphology of the Silver Deposits

The composition and the morphology of the silver deposits, electrochemically formed according to the reaction scheme of Fig.4, are important criteria in the decision whether $S_2O_3^{2-}$ is acceptable as complexing agent in plating baths or not.

The reaction scheme predicts the formation of a deposit of Ag and $[Ag(S_2O_3)^-]^-$. Their ratio depends on the $S_2O_3^{2-}$ concentration, but also on the value of β_{11}^- which cannot be deduced from the kinetic measurements. Therefore, the composition of the silver deposits is determined as a function of the $S_2O_3^{2-}$ concentration, mainly by means of Auger electron spectroscopy (AES).

The morphology is characterized as a function of the plating conditions with scanning electron microscopy (SEM).

Composition

The Auger analysis gives a wide variety of informations about the surface layer of specimen of about 10 Å thickness: not only component identification[17-19], but also chemical state identification[20-26] and structural information[27]. Component identification is straightforward and goes by the determination of the energy of the Auger peaks. Chemical and structural state identification are based on the line shape analysis of the Auger peaks, and this requires the comparison of the spectra of the unknown sample with those of standard samples of known composition.

The complete analysis is described in detail in[10,16] (silver deposit and standard sample preparation, line shape analysis of Ag, S and O peaks on deposits and standards, identification by comparison), but only the results are reported here.

A typical example of an Auger spectrum of the surface of an electrodeposited silver layer is given in Fig.5, showing the presence of S, Ag, O and Na. O and Na are not present all over the surface, but always together. S is detected everywhere, but the intensity of its peak, relative to the intensity of the Ag peak, is varying.

From the line shape analysis in[10,16] is known that Ag is not bounded to an oxygeneous anion, such as $S_2O_3^{2-}$. S is thought to be present as an S^o adsorption layer, in accordance with the results of[28]. Remnants of the supporting electrolyte $NaNO_3$ on the surface are responsible for the presence of O.

Depth profiles indicate S, O and Na only to be present in the top layers as contaminants. The bulk of the deposits is composed of silver, and is polycrystalline, since the diffractograms completely correspond to those of pure polycrystalline silver[10,16,29].

Morphology

For the morphological characterization of the deposits, SEM pictures are taken.

Silver is deposited galvanostatically (imposed current) under varying conditions. The parameters are : magnitude of the current density relative to the limiting current density (or deposition rate), deposition time (or thickness of the deposit), concentration of silver and complexing agent.

An extensive report of the investigation is found in[30]. The results are summarized in Fig.6, where three predominant types of morphologies are shown.
At the limiting current density (maximum plating rate) the deposits show a dendritic structure (Fig.6c). At lower current densities, smooth deposits are formed (Fig.6a). Thick deposits tend to roughen (Fig.6b). The concentration of $S_2O_3^{2-}$ has little influence.

Current density, total silver concentration and deposition time determine interactively the working conditions for the formation of smooth deposits, but as long as the current density is smaller than the limiting current density, the adjustment is not critical.

CONCLUSION

Using a combination of electrochemical and surface analytical techniques, the possibility to perform silver plating from thiosulphate instead of cyanide containing baths is examined.

The kinetic investigation reveals two important facts. The kinetic parameters are such that current densities of the same order of magnitude as in the case of cyanide baths (100 A m^{-2} [1]) can be achieved. So, from a kinetic point of view, the behaviour of the thiosulphate complexes is very promising.
The mechanism predicts the formation of a deposit composed of Ag and $[Ag(S_2O_3)]^-$, entailing that $S_2O_3^{2-}$ is incorporated in it. This is undesirable, but it is shown that polycrystalline silver is deposited, independent of the thiosulphate concentration.

Also for the morphological aspects, the results are good. In a wide range of plating conditions, smooth deposits are obtained. A compromise must be reached between levelness of the deposit and plating rate.

Based on the conclusions of this investigation, it is worth while to examine the engineering aspects of silver plating from thiosulphate baths.

REFERENCES

1. LUCE B.M., FOULKE D.G.
"Modern Electroplating", ed. F.A. Lowenheim, 3rd Ed., J. Wiley and Sons.

2. TOURNIER R.
Galvano-Organo Traitements de surface, 503 (1980) 123.

3. VIELSTICH W., GERISCHER H.
Z. physik. Chem., N.F., 4 (1955) 10.

4. BALTRUSCHAT H., VIELSTICH W.
J. Electroanal. Chem., 154 (1983) 141.

5. FLEISCHMANN M., SUNDHOLM G., TIAN Z.Q.
Electrochimica Acta, 31-8 (1986) 907.

6. TAJIMA S., KOMATSU S., BABA N.
Electrochimica Acta, 19 (1974) 921.

7. VEREECKEN J., WINAND R.
Electrochimica Acta, 22 (1977) 401.

8. DESPIC A.R.
"ComprehensiveTreatise of Electrochemistry, Vol.7", ed. B.E. Conway, J.O'M. Bockris, E. Yeager, S.U.M. Khan, R.E. White, Plenum Press, New York and London, 1983.

9. LEVICH V.G.
"Physicochemical Hydrodynamics"
Prentice-Hall, Inc. Englewood Cliffs, New York, 1962.

10. HUBIN A.
"Bijdrage tot de Studie van de Elektrochemische Reductie van Zilverthiosulfaat- en Zilverthiocyanaatcomplexen", Ph.D. Thesis, Vrije Universiteit Brussel, Brussels, 1989.

11. "Gmelins Handbuch der anorganischen Chemie"
Springer Verlag, Berlin-Heidelberg-New York, achte Auflage, Ag[B3], 1975.

12. HUBIN A., VEREECKEN J.
J. Electroanal. Interfacial Electrochem., paper submitted.

13. GIMBLETT F.G.R., MONK C.B.
Trans. Farad. Soc., 51 (1955) 793.

14. WEAST R.C., ASTLE M.J., BEYER W.H.
"CRC Handbook of Chemistry and Physics", CRC Press Inc., 64th Edition, 1983.

15. PYTKOWICZ R.M.
"Activity Coefficients in Electrolyte Solutions" Vol.1 CRC Press Inc., 1979.

16. HUBIN A., VEREECKEN J.
J. Electroanal. Interfacial Electrochem., paper submitted.

17. BRIGGS D., SEAH M.P.
"Practical Surface Analysis by Auger and X-ray Photoelectron Spectroscopy", J. Wiley & Sons, New York, 1983.

18. THOMPSON M., BAKER M.D., CHRISTIE A., TYSON J.F.
"Auger Electron Spectroscopy", J. Wiley & Sons, New York, 1985.

19. CHUNG M.F., JENKINS L.H.
Surf. Sci., 22 (1970) 479.

20. MADDEN H.H.
J. Vac. Sci. Technol., 18-3 (1981) 677.

21. CARRIERE B., DEVILLE J.P., HUMBERT P.
J. Microsc. Spectrosc. Electron., 10 (1985) 29.

22. QUINTO D.T., ROBERTSON W.D.
Surf. Sci., 27 (1971) 645.

23. CHATTARJI D.
"The Theory of Auger Transitions", Academic Press, 1976.

24. RAO C.N.R., SARMA D.D., HEGDE M.S.
Proc. R. Soc. Lond., A370 (1980) 269

25. WEISSMANN R.
Solid State Communications, 31 (1979) 347.

26. YASHONATH S., HEGDE M.S.
Proc. Indian Acad. Sci. (Chem. Sci.) 89-5 (1980) 489.

27. TERRYN H., LAUDET A., VEREECKEN J.
"Proc. Fifth Int. Conf. Quantitative Surface Analysis" (1988).

28. PERDEREAU M.M.
C.R.Acad. Sc. Paris, 274 (1972) 448.

29. "Joint Committee on Powder Diffraction Standards" (1974)

30. MARISSENS G.
"Invloed van de experimentele omstandigheden op de structuur van Ag-afzettingen bereid door de reductie van Ag-S$_2$O$_3^{2-}$ complexen", Final work, Vrije Universiteit Brussel, Brussels, 1990.

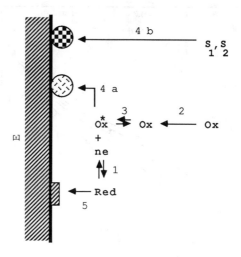

Fig.1 : Schematic representation of the elementary steps for the electrochemical formation of a metal deposit:

E = electrode
Ox = initial state of the oxidant
Ox* = electroactive state of the oxidant
Red = reductant, deposited on the electrode
S$_1$,S$_2$ = electroinactive components
step 1 = charge transfer
step 2 = mass transfer
step 3 = homogeneous chemical reaction
step 4 = heterogeous chemical reaction (4a and 4b are adsorptions of respectively Ox* and S$_1$)
step 5 = electrocrystallization

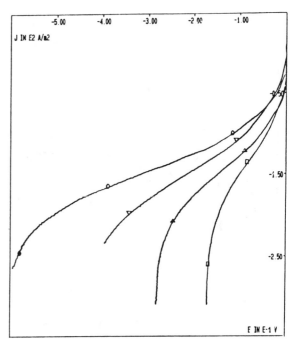

Fig. 2 : Voltammograms as a function of solution concentration :

$(c_{Ag^+})_t$ = 0.04 ($\square$), 0.07 (Δ), 0.10 (∇), 0.14 (o) mol l^{-1};

scan rate = 1 mV s^{-1}; rotation speed = 1375 rpm;
Ohmic drop corrected.

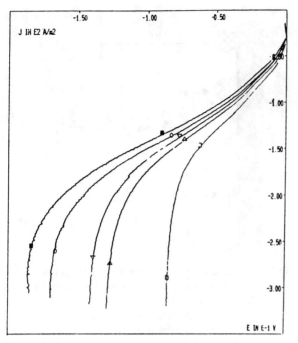

Fig. 3 : Voltammograms as a function of rotation speed of the electrode :

ω = 344 (□), 773 (△), 954 (▽), 1375 (o), 1611 (■) rpm;
$(c_{Ag^+})_t$ = 0.04 mol l^{-1}; scan rate =1 mV s^{-1};
Ohmic drop corrected.

$$Ag(S_2O_3)^- + e \xrightarrow{k_1} [Ag(S_2O_3)^-]^-$$

$$\beta_{11} \downarrow\uparrow \qquad\qquad \downarrow\uparrow \beta_{11}^-$$

$$Ag^+ + S_2O_3^- + e \rightarrow Ag + S_2O_3^-$$

$$k_o$$

Fig. 4 : Mechanism of the reduction of silver thiosulphate complexes. k and β are respectively the rate and the equilibrium constant.

a)

b)

c)

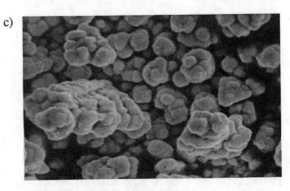

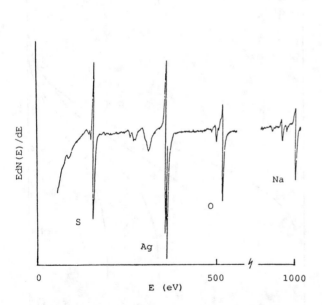

Fig. 5 : Auger spectrum of the silver deposit.

Fig. 6 : SEM pictures of the silver deposit :
(a) at low current density,
(b) for a thick deposit,
(c) at limiting current density.
Magnification : 3000 x.

Use of peroxygens in treating cyanide effluents from gold processing

E. N. Wilton
P. J. Wyborn
J. A. Reeve
Interox Research and Development, Widnes, England

SYNOPSIS

Cyanide has been in use in the extraction of gold for the last 100 years, and so cyanide containing effluents have had to be disposed of in some way for this period of time.

With the current expansion of the gold mining industry and the generally increased environmental awareness, cyanide detoxification is now becoming of paramount concern to mine operators.

Laboratory tests have been carried out to destroy both free and complexed cyanides using peroxygens. The results have shown that although hydrogen peroxide can detoxify both free and zinc and copper cyanides, in all cases Caro's acid, (a mixture containing peroxymonosulphuric acid) was found to be significantly faster.

A plant trial was carried out using Caro's acid demonstrating that cyanides could be detoxified on a large scale down to the required level <0.5 mg/l CN⁻. The trial also brought to light the fact that Caro's acid will destroy thiocyanates.

INTRODUCTION

Source of Cyanide

Cyanide is used extensively in the extraction of precious metals such as gold and silver. In the case of gold, the cyanide is used to dissolve out the metal from the crushed ore by formation of the soluble complex represented by the following reaction:-

$$4 \text{ Au} + 2H_2O + 8 \text{ CN}^- + O_2 \xrightarrow{pH \sim 10.5} 4 \text{ Au (CN)}_2^- + 4OH^-$$

The ore also contains other metal containing minerals such as copper, zinc and iron, which can be leached out. Once the gold is leached and recovered, a solution remains containing free and complex cyanides such as copper and zinc cyanides, and other metals that have been extracted. Before discharge, these free and complex cyanides have to be removed and/or treated so as to render the solution acceptable to the environment.

Cyanide Destruction

"Natural degradation" utilising large tailings ponds remains the most common method of cyanide removal. The main mechanisms by which cyanides are naturally degraded are:- volatilisation of HCN, biodegradation, photo- decomposition, chemical precipitation, and hydrolysis. The rate of removal of cyanide is thus heavily dependant upon several factors; pH, cyanide concentration and type of cyanide species, temperature, amount of uv light (sunlight), and bacterial activity. Due to a general tightening of regulations around the world on the discharge of toxic effluents into the environment and the possible impact on wildlife, this method is becoming much less acceptable.

Therefore, commercial processes are now being adopted for the removal of the cyanide. Several methods are now employed throughout the world including:

   i) H_2O_2/Cu^{2+} [1]
  ii)   Inco SO_2/air [2]
 iii)   Alkaline chlorination [3]
   v)   Acidification, volatilisation, and recovery (AVR)[4]

The final method is not a destructive process, a low strength cyanide effluent remains, requiring treatment after cyanide recovery is complete.

The two main detoxification processes are the Inco SO_2/air and H_2O_2 processes. The alkaline chlorination process is more of an emergency treatment for gold mine effluents.

With regard to the H_2O_2 process, previous studies[5,6,7] have shown that peroxygens can be used to oxidise certain cyanides to cyanates, namely free cyanide and the more labile complexes e.g. $Cu(CN)_4^{3-}$, $Cu(CN)_3^{2-}$, $Zn(CN)_4^{2-}$.

Cyanides react with peroxygens in the following manner:-

Hydrogen Peroxide

$$CN^- + H_2O_2 \longrightarrow CNO^- + H_2O$$

Caro's Acid

$$CN^- + SO_5^{2-} \longrightarrow CNO^- + SO_4^{2-}$$

In the H_2O_2 detoxification reaction, a catalyst such as copper is often required to allow a faster reaction[8]. This catalyst may have to be added, if it is not already present in the effluent. The cyanate produced from the detoxification reaction is then hydrolysed by alkali to give carbonate and ammonia,

$$CNO^- + H_2O + OH^- \longrightarrow CO_3^{2-} + NH_3$$

or by acid to give carbon dioxide and ammonium,

$$CNO^- + 2H^+ + H_2O \longrightarrow CO_2 + NH_4^+$$

The hydrolysis products are of much lower toxicity than cyanide. Depending on discharge regulations, cyanide can be measured as:- Free cyanide (CN^-_{free}); weakly acid dissociable cyanide (CN^-_{WAD}) or total cyanide (CN^-_{TOT}). Of these, CN^-_{WAD} is the most commonly controlled as this contains the free and labile cyanide complexes which form the toxic part of the CN^-_{TOT}. The discharge limits, which vary according to local regulations and the type of analytical method used for control, determine which cyanides need to be treated.

Peroxygens and Their Advantages

Hydrogen Peroxide

Hydrogen peroxide is a colourless, virtually odourless liquid which can be used in various oxidation applications. If stored correctly it will remain stable for up to 1 year, with a loss of only 1% of its available oxygen. It is particularly suitable for environmental applications as its decomposition
products are only water and oxygen.

Caro's Acid

Caro's acid is an equilibrium mixture containing hydrogen peroxide, (H_2O_2), permonosulphuric acid, (H_2SO_5), sulphuric acid, (H_2SO_4), and water. It has a higher oxidation potential than H_2O_2, and is thus even more effective at cyanide detoxification. It will remain stable for several months if stored at a temperature of <5°C, but is normally made on-site.

Interox have developed a cheap, portable on-site unit to produce Caro's acid.

Caro's acid is particularly suited to environmental applications as the decomposition products are only sulphate, oxygen and water.

LABORATORY STUDIES

Experimental Procedures

i) Generation of Caro's Acid

Caro's acid was prepared in the laboratory by the controlled addition of 70% H_2O_2 into a cooled, stirred flask containing 98% H_2SO_4.[9] These concentrations were chosen as the most convenient to handle safely in the laboratory. The reaction is exothermic and its temperature was kept below 10°C, to prevent decomposition.

The reaction can be represented as:-

$$H_2O_2 + H_2SO_4 \quad H_2SO_5 + H_2O$$

By varying the mole ratio of the reactants, (the ratio of H_2SO_4:H_2O_2 allowed to react) different equilibrium mixtures (e.g. high H_2SO_5/low H_2O_2 or vice versa) can be produced. For cyanide detoxification applications because the pH of the reaction mixture must be kept at >10, a low H_2SO_4:H_2O_2 mole ratio is preferred to minimise neutralisation requirements, and hence costs.

ii) Hydrogen Peroxide

The H_2O_2 used in the detoxification experiments was commercial 35% w/w grade available from Interox.

iii) Preparation of Cyanide and Cyanide Complexes

Three different cyanide complexes as well as KCN were used in the detoxification experiments:
$$K_3[Cu(CN)_4]$$
$$K_2[Cu(CN)_3]$$
and $$K_2[Zn(CN)_4]$$
which are often found in the effluent emanating from gold plants. These complexes were prepared in the following manner[5]:-

$K_3[Cu(CN)_4]$

Evaporation of an aqueous solution containing the correct ratios of KCN and CuCN, gives pale yellow crystals of $K_3[Cu(CN)_4]$.

$K_3[Cu(CN)_3]$

Prepared as above, but adjusting the ratios of CuCN and KCN to give the required product. Cream coloured crystals are produced of $K_3[Cu(CN)_3]$.

$K_2[Zn(CN)_4]$

ZnO was added to 4.5M NaOH solution and the

mixture boiled for 30 minutes. To the hot solution was added the stoichiometric amount of KCN, and the solution stirred for a further 10 minutes until most of the cloudiness had disappeared. The hot solution was then filtered to remove unreacted ZnO, and this solution allowed to cool and evaporate, leaving white crystals.

iv) Analysis of Cyanide Solutions

Two methods were used to determine cyanide in solution:-

a) Classical Wet Method

This method involves an acidification, reflux distillation, and titration of the resultant solution[10]. An aliquot of the solution to be analysed is placed in a distillation flask, excess acid is added, and the apparatus resealed. The reaction mixture is then refluxed for 1 hour, while HCN gas is driven from the solution, (using a N_2 gas purge), which is readsorped in a solution of NaOH, forming NaCN. This solution is then titrated using silver nitrate solution and a suitable indicator, such as Rhodamine.

For the above method, we favour a variation on the indicator. This is known as "Denige's" modification, and involves the use of ammonia and potassium iodide as a turbidometric indicator. This method has been found to be easier to use and is more accurate than the conventional indicator.

b) Cyanide Analyser

For the analyses reported, a cyanide analyser from Yugoslavia, (ISKRA MA5400) has been used. This is basically an automated version of the classical wet method. However, instead of a final titration, the NaCN solution is passed over a silver wire electrode, and the cyanide measured potentiometrically.

The advantage of this analyser is that very small samples are required (5ml), and the analysis is completed in ~5 mins (c.f. ~1.5 hours for the classical wet method).

v) Detoxification Experiments

The initial work was carried out using simple stirred beaker reactions containing 1g/l equivalent CN⁻ of synthetic cyanide effluent. To this was added known amounts (in one portion) of either H_2O_2 or Caro's acid, to give a pre-determined reaction stoichiometry, initially 1:1 H_2O_2:CN⁻. At pre-set time intervals, aliquots were removed from the reaction mixture, and the residual cyanide measured. A reaction profile was produced, and the time taken for the cyanide to be reduced to below detectable levels, determined. The reactions were carried out at 25°C.

This procedure was repeated at several stoichiometries of both H_2O_2 and Caro's

acid.

vi) Control of pH during detoxification with Caro's Acid

Three methods were used to neutralise the acidity of Caro's acid.

Method a) - Preneutralisation of Effluent

In the experiments using Caro's acid, the pH of the cyanide solution was increased to ~12.5 with NaOH. This was to ensure that as the Caro's acid was added, the contained acidity did not drop below 10.

Method b) - Preneutralisation of Caro's Acid

A second method was also investigated. This involved analysing the Caro's acid prior to use, and adding enough NaOH to the Caro's acid to just neutralise the acidity prior to addition to the effluent. However this caused unacceptable decomposition of the Caro's acid, thereby precluding adequate detoxification.

Method c) - Use of pH Stat

This method used a pH meter coupled to an automatic titrator to maintain a pre-determined pH in the reaction beaker.

Experiments were carried out statted at pH 9, 10 and 11. At pH9, the stat had trouble in maintaining the pH accurately, in this area the Caro's acid caused the pH to drop to 8, at which point the last drop of alkali caused a rapid pH rise up to ~10.5.

RESULTS OF LABORATORY WORK

In the reaction between peroxygens and free cyanide, at 1:1 peroxygen: cyanide reaction stoichiometry, (Fig 1a), hydrogen peroxide reduced the cyanide level to <0.1 mg/l CN⁻(wad) in approximately 24 hours. This should be compared with 12 minutes which is the time required to reach the same residual cyanide level using Caro's acid. Figs 1b. 1c and 1d show the results of the reactions between peroxygens and the three cyanide complexes studied (again at reaction stoichiometries of 1:1). Fig 1b shows the results obtained for the $K_3[Cu(CN)_4]$ complex, hydrogen peroxide reduced the cyanide level to <0.1 mg/l CN⁻(wad) in 4 hours whereas Caro's acid took 10 minutes. Similar results were obtained with the other two complexes ie for $K_2[Cu(CN)_3]$, (fig 1c), hydrogen peroxide reduced the cyanide level to <0.1 mg/l CN⁻(wad) in 2.5 hours whereas Caro's acid again took 10 minutes and for $K_2[Zn(CN)_4]$, (fig 1d), hydrogen peroxide achieved <0.1 mg/l CN⁻(wad) in 4 hours but Caro's acid took only ~1 minute.

The faster reaction time of hydrogen peroxide with the $K_2[Cu(CN)_3]$ complex compared to the $K_3[Cu(CN)_4]$ complex can be rationalised in terms of their relevant stability constants (for $K_2[Cu(CN)_3]$

$Bn=29M^{-n}$ as opposed to $Bn=31M^{-n}$ for $K_3[Cu(CN)_4]$. The mechanism of oxidation of cyanide complexes by hydrogen peroxide is believed to proceed via breakdown of the complex to liberate free cyanide and hence the lower the stability constant of a complex the faster its reaction with hydrogen peroxide should be.

The $K_2[Zn(CN)_4]$ complex has a lower stability constant ($Bn=19M^{-n}$) and therefore would be expected to have a shorter reaction time. However the reaction time measured was similar to that found for the $K_3[Cu(CN)_4]$ complex. A possible explanation for this observation is that the oxidation of copper complexes introduces copper into the solution, which is able to catalyse the oxidation.[5] Zinc compounds do not catalyse the reaction between hydrogen peroxide and cyanide and hence although $K_2[Zn(CN)_4]$ has a lower stability constant than $K_2[Cu(CN)_3]$ the reaction is not catalysed and hence requires a longer reaction time to achieve the same residual cyanide level.

Fig 1e shows that varying the pH between 9 and 11 does not have any significant effect on the rate of oxidation of free cyanide using Caro's acid.

CONCLUSION

All the laboratory work carried out to date has shown that Caro's acid can be used with great effectiveness, in terms of its speed of reaction and its ability to detoxify many cyanide complexes, for the treatment of cyanide containing effluents.

FIG 1a

COMPARISON OF H2O2 WITH H2SO5 AT SAME STOICH

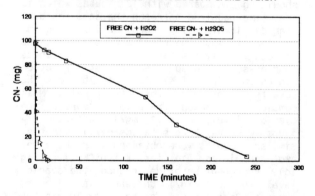

FIG 1b

COMPARISON OF H2O2 WITH H2SO5 AT SAME STOICH

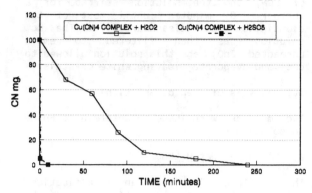

FIG 1c

COMPARISON OF H2O2 WITH H2SO5 AT SAME STOICH

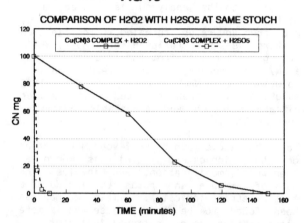

FIG 1d

COMPARISON OF H2O2 WITH H2SO5 AT SAME STOICH

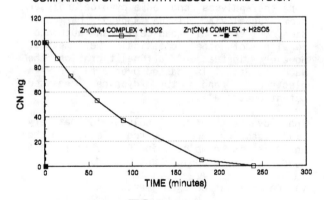

FIG 1e

Reactions of Cyanide and Caro's Acid
at Various pH

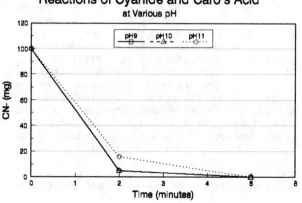

MINE TRIAL

The objective of running a trial on a mine site was to demonstrate that Caro's acid could detoxify actual effluent down to a required limit, and also to test the effectiveness and reliability of a prototype Adiabatic Caro's Acid unit, (ACU).

i)   Trial Procedure

A gold production plant has been extracting gold by the treatment of old and current tailings. The slurry leaving the plant is treated with iron to reduce the level of CN^- (wad) prior to settling. The majority of the clear settled liquor is detoxified both for use in non-cyanide processes and the hydraulic mining of the tails and for disposal.

The total plant required the treatment of $80m^3$/hour of effluent, containing a maximum level of 160 mg/L CN (wad) to give a discharge of <0.5 mg/L CN (wad). This provided an ideal opportunity to test the practicality of adiabatically generated Caro's acid in the field and the effectiveness of Caro's acid.

Due to the size of the prototype unit batches of ~2000 l of effluent were treated during each detoxification run. A schematic diagram of the Interox process is shown in fig 2

The initial level of cyanide in the effluent was determined from tank 1. To this tank was then added the required amount of 30% w/w NaOH solution, to maintain the pH >10.5 after addition of Caro's acid. The effluent flowed by gravity from tank 1, to tank 2, the main detoxification reactor, where the Caro's acid (of varying reaction stoichiometries) from the ACU was continuously added at a fixed rate. The pH of the solution was monitored in accordance with safe working practices, and the detoxification tank thoroughly agitated. The treated effluent was then passed to a final tank, tank 3, where analyses for residual cyanide CN^- (wad) and pH were undertaken prior to discharge.
In all 8 separate detoxification runs were completed. Throughout all of the detoxification runs, the mole ratio of $H_2SO_4:H_2O_2$ in the Caro's acid was maintained between 1:1 and 2:1 so as to keep neutralisation requirements as low as possible.

TRIAL RESULTS

The main results of the trial are presented in Table 1.

The first two detoxification runs 1 and 2 were carried out with the Caro's acid being added to the bulk of the solution (position A, Fig 2) and the stirrer located centrally. The initial results were not as expected from the laboratory work ie. 8 mg/l CN^-(wad) cf <0.5 mg/l CN^-(wad).

In order to optimise the use of Caro's acid a number of modifications to the system were made:

Caro's acid dosing point

The point of addition of the Caro's acid was varied by repositioning it to a much more turbulent one (position B, fig 2). A direct comparison of the results from these two different addition points (cf. Run 1 vs Run 3) shows how critical this factor is and the reduction in residual cyanide which can be achieved by using the optimum dosing point. [8 mg/l CN^-(wad)].

Mole Ratio of $H_2SO_5:CN^-$

Having optimised the Caro's acid dosing point the effect of the $H_2SO_5:CN^-$ mole ratio was investigated. Runs 4-8 were carried out using the same basic conditions except for the initial CN^- concentration which could not be controlled and the $H_2SO_5:CN$ ratio. As the ratio was increased (3.5:1-6.5:1) the residual cyanide level dropped and the optimum ratio was found to be 5:1. Using a higher mole ratio 6.5:1 did not give any added benefit (Run 7 & Run 8). A comparison of the results from Run 3 and Run 7 indicates that the effectiveness of using a 5:1 mole ratio is not dependent on the initial cyanide concentration and that the ratio arbitrarily used in Run 3 was actually the optimum. The results of the other detox runs merely confirm the above findings.

Table 1: Results of Plant Scale Cyanide Detoxification runs with Caro's Acid

Run No.	1	2	3	4	5	6	7	8
Initial CN^- conc mg/l	130	129	94	115	160	115	145	102
CN^- conc after treatment mg/l	8.4	7.9	0.4	10	3.8	2.0	0.2	0.5
Caro's Acid Mole Ratio $H_2SO_4:H_2O_2$	2.3:1	1.2:1	1.7:1	1.8:1	1.8:1	1.4	1.1	1.9
$H_2SO_5:CN$ Mole ratio	5.8:1	5.2:1	5.3:1	3.5:1	4.0:1	4.0:1	5.0:1	6.5:1

428

INTEROX MINE TRIAL PROCESS

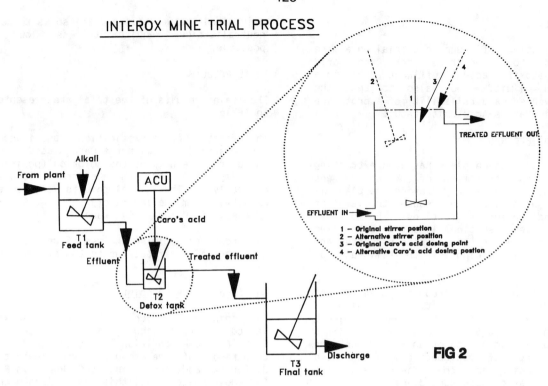

FIG 2

1 - Original stirrer postion
2 - Alternative stirrer position
3 - Original Caro's acid dosing point
4 - Alternative Caro's acid dosing postion

DISCUSSION OF MINE TRIAL

The laboratory tests on synthetic solutions of cyanide and its complexes showed that only slightly above the theoretical requirement of Caro's acid was required to destroy the cyanide. During the trial it was found that in order to achieve an acceptable residual cyanide level (<0.5 mg/l) in the effluent the quantity of Caro's acid required was much higher, i.e. 5.0:1, $H_2SO_5 : CN^-$ mole ratio.

Analysis of the effluent found that it contained high levels of thiocyanate which was also being detoxified by the Caro's acid, according to the reaction:

$$4SO_5^{2-} + SCN^- + H_2O \longrightarrow 5SO_4^{2-} + CNO^- + 2H^+$$

Therefore the theoretical amount of Caro's acid required to treat the thiocyanate and cyanide combined gave a mole ratio of slightly over 4.0:1 ($H_2SO_5 : CN^- + SCN^-$), only slightly in excess of the theoretical requirement.

Test work carried out in the laboratory before the trial showed that although hydrogen peroxide could be used to detoxify the cyanide component of the effluent, the mole ratio of the $H_2O_2 : CN^-$ required was considerably higher than the hydrogen peroxide equivalent contained in the Caro's acid. The effluent contained copper at a level of ~100 mg/l and so no copper was needed to catalyse the test work using hydrogen peroxide. Hydrogen peroxide under normal conditions does not react with SCN^-.

The solution after detoxification using hydrogen peroxide was fawn coloured and opaque, whereas that treated with Caro's acid contained a precipitate which coagulated very rapidly to give a clear solution and compact residue.

The optimum addition point of Caro's acid was such that it indicated that the detox reactor could have been scaled down in size ie. the mixing volume near the addition point/weir was small.

CONCLUSIONS OF TRIAL

The laboratory work and the trial work indicated that Caro's acid can be effective in the destruction of cyanide. The trial showed that the adiabatic generator tested on site can efficiently generate Caro's acid.

Caro's acid not only destroys cyanide but also thiocyanate. As yet, there are are no stringent regulations on the discharge of thiocyanates, however as the legislation moves towards a "total toxicity" as a criterion for discharge legislation, thiocyanate levels will play a much more important role.

REFERENCES

1. Knorre, H., Automatic Detoxification of Effluent Containing Cyanide Using Hydrogen Peroxide. Brennst-Waerme - Kraft, 1988, 40(3) M48-50.

2. Zaidi A., Conard B., Devuyst E., Schmidt J. and Whittle L. Performance of a Full Scale Effluent Treatment System using the SO_2/Air Process for Removing Free and Complexed Cyanide. Proceedings of the Industrial Wastes Symposia, 59th Annual WPCF Conference, Los Angeles,

California,   October   5-9,   1986,
p.284-297.

3. The Chemistry, Analysis, Toxicity and
    Treatment of Cyanidation Wastewaters,
    Society of Mining Engineers Conference,
    Las Vegas, February 26th 1989, Section
    6, p.41.

4. McNamara. V.M.   The AVR Process for
    Cyanide Recovery, and Cyanogen Control
    for Barren Recycle and Barren Bleed.
    Proceedings of the GOld Mining Effluent
    Treatment   Seminars,   Mississauga,
    Ontario, March 22-23, 1989.

5. Leahy C.D.   The Oxidation by Peroxides
    of   Cyanides , Cyanide Complexes and
    Related Species.   PhD Thesis, Imperial
    College, London 1990.

6. Teixeira L.A.   and Tavares L.Y. The
    Treatment   of   Cyanide   Containing
    Effluents   Using   Hydrogen   Peroxide.
    Paper from 14th Encontro Nacional de
    Tratamento   de   Minerois   E
    Hidrometalurgia,   9-12   September 1990,
    Salvador, p. 925-934.

7. Ingles J., Scott J.S., State-of-The-Art
    of Processes for the Treatment of Gold
    Mill   Effluents.   Mining, Mineral and
    Metallurgical   Processes   Division,
    Industrial Programs Branch, Unpublished
    Report of the Environmental Protection
    Service, Environment Canada, March 1987.

8. Polyblank   G.R.,   Copper   Catalysed
    Oxidation   of   Cyanide   by   Hydrogen
    Peroxide.   B.Sc Thesis, University of
    Sydney, 1985.

9. Caro's   Acid Laboratory Procedure for
    Hydrometallurgy.   Interox   Published
    Brochure, AO.5.5.

10. Standard Methods For the Examination of
    Water and Wastewaters.   15th Edition,
    APHA-AWWA-WPCF, 1980 Page 319.

11. A   Textbook of Quantitative Inorganic
    Analysis.   Vogel A.I. 3rd ed. London:
    Longmans, 1961.

Mechanisms of uptake of metal complexes and organics on carbon and resins: their modelling for the design of multi-component adsorption columns

J. S. J. van Deventer, F. W. Petersen
Department of Metallurgical Engineering, University of Stellenbosch, Stellenbosch, South Africa
M. A. Reuter
Department of Metallurgical Engineering, University of Stellenbosch, Stellenbosch, South Africa
(at present, Institut für Metallhüttenwesen und Elektrometallurgie, RW-TH Aachen, Aachen, Germany)

ABSTRACT

The effect of organics on the adsorption of metal cyanides on activated carbon and ion-exchange resins in a packed column is described. At low loadings of organics on either the resin or carbon, the adsorption of metal cyanides is influenced in a kinetic way, which can be ascribed to pore blocking. At higher loadings of organics, both a kinetic as well as an equilibrium influence are observed. This can be attributed to two effects, viz.: (1) permanent pore blocking, where the pores of the adsorbent are totally sealled off by the organic species, and (2) competitive adsorption for active sites between the metal cyanide and organic foulant.

Factors such as pretreatment of the adsorbent and simultaneous adsorption of metal and organic species are considered as most important in the design of adsorption columns. A multi-component Freundlich isotherm is found to be adequate in describing the mechanisms of such simultaneous adsorption. It is also illustrated that KBS modelling is a valuable tool in the effective design of multi-component adsorption columns.

INTRODUCTION

In view of the increasingly environmental awareness it becomes imperative to remove organics and metals from effluent streams to acceptable standards. The literature on the design of columns used for the treatment of waste-water is quite substantial. Many of the principles associated with the design of such columns can be derived from the design of adsorption columns for metal complexes, such as those used in the gold mining industry. In the recovery of metals from cyanide leached pulps, the pulps usually contain a large variety of organic and inorganic substances which can cause a significant effect on the adsorption behaviour of activated carbon or ion-exchange resins. These substances may be derived from recycled treated sewage effluent, organics associated with the solid raw materials derived from reclaimed sand dumps and slimes dams, and organic process reagents. For the recovery of gold from plant effluents and slimes dam return waters, the carbon-in-column process is widely practised [1]. Central to the design of such columns are accurate forecasts of the different operating conditions and performance characteristics of the adsorbents. Both activated carbon and ion-exchange resins could be used as adsorbents for removing organics and dissolved metals from plant-water.

Many physical and chemical interactions occur during the simultaneous adsorption of multiple species of metal complexes and organics, which complicates the design significantly. Also, the identities of many of the organic and inorganic species in industrial effluent streams are not known. A number of approaches have been published to describe the equilibrium of multi-solute organic mixtures on activated carbon [2,3]. Van Deventer [4] investigated the competitive effect of one metal cyanide on the adsorption capacity of other metal cyanides. No research, however, has been published on any aspect regarding the simultaneous adsorption of a metal complex and

431

an organic foulant. Unless these effects are properly accounted for in models used for the design of adsorption columns, such models will not be reliable to predict performance over the complete range of operation.

EXPERIMENTAL

Materials

Potassium aurocyanide, potassium tetracyanocuprate and potassium argentocyanide were used as adsorbates. The latter was dissolved in a potassium cyanide solution at a pH of 8 and temperature of $90°C$. The reacting solutions used in adsorption tests contained almost no free cyanide, so that the effect of CN^- was not considered in the present work. All reagents were of analytical grade and distilled deionised water was used throughout. A wide range of organics were used as depicted in Table I. The 1.0 to 2.0 mm size fraction of a commercial coconut shell activated carbon, Le Carbone G210 AS, with apparent density of $835 kg/m^3$ was used. Before adsorption tests, the carbon was soaked in water and then oven dried at $120°C$ for 4 hours. Duolite A161, a strong base anion exchange resin with a polystyrene matrix, manufactured by Duolite International, was also used. The effective particle size of this resin was 0.9 mm, the value of the wet-settled density was $700 kg/m^3$, and the voidage ϵ of the wet-settled resin occupied by resin particles was 86 %.

Column experiments

These experiments were conducted in a glass column with an internal diameter of 31 mm and a

height of 150 mm. The packed column contained 12,32 g of dry carbon and had a void fraction of 0,38, while in the case of resin it contained a volume of 0.075 L of wet-settled resin with a void fraction of 0.86. A sintered glass disk at the bottom of the column facilitated an even distribution of solution. A metering pump was used to provide a constant flow rate of $1.4 \times 10^{-6} m^3/s$ from a continuously stirred 600 liter stainless steel tank. The flow rate had to be monitored continuously to compensate for the decrease in the level of the solution in the feed tank. A pH of 9 was maintained throughout the experiment.

For modelling purposes, the equilibrium loading relationship between the silver, copper or gold solution and the adsorbent, had to be determined. These tests were carried out in stirred vessels for a period of 3 weeks. Simultaneous or competitive isotherms involving gold or silver in the presence of humic acid, were conducted in a manner similar to that for the single component isotherm.

Analytical methods

A Varian Techtron AA-1275 atomic absorption spectrophotometer (AA) was used with an air-acetylene flame and an absorbance peak at 328.1 nm for the analysis of silver cyanide in the solution, an absorbance peak of 324.7 nm for copper cyanide in solution, and an absorbance peak of 242.8 nm in the case of gold cyanide. In the presence of organics in solution the AA readings had to be corrected as explained by Petersen and Van Deventer[5]. The organic concentrations were determined by measuring the

TABLE I DIFFERENT ORGANIC FOULANTS

Organic	Loading (mg/g)	Molecular mass (g/mol)	$Dx10^{12}$ for Au (m^2/s) carbon	$Dx10^{12}$ for Au (m^2/s) resin	Solvent
no organic	–	–	3.80	3.65	–
methanol (MT)	47	32	2.65	–	water
phenol (PN)	48	89	3.32	3.42	ethanol
heptane (HT)	51	100	2.86	–	ethanol
potassium ethyl-xanthate(PEX)	52	134	2.70	3.32	water
humic acid (HA)	51	1154	2.42	–	ammonia

ultraviolet absorbance on a LKB Biochrom 4050 ultraviolet spectrophotometer. The pH of all solutions was monitored by using a Beckman combination pH-electrode coupled to a Beckman Φ 71 pH-meter.

MATHEMATICAL MODEL

A dual resistance model involving both external film diffusion and intraparticle surface diffusion was applied to the profiles for the uptake of gold, copper and silver cyanide by activated carbon and ion-exchange resins in a packed column in order to estimate kinetic parameters. This model is similar to the model used by Van Deventer [6] for the adsorption of gold cyanide.

The following main assumptions have been made in the development of the model:

(a)   The carbon and resin particles can be treated as equivalent spheres for modelling purposes,

(b)   The radial transport of silver or gold cyanide, and the organics into the pores can be described by a surface diffusion mechanism. Pore diffusion is assumed to be negligible,

(c)   Accumulation of metal cyanide or organics in the liquid phase within the pores of the carbon or resin is negligible,

(e)   It is assumed that the adsorption reaction on the carbon, and the ion-exchange reaction on the resin occur instantaneously, so that equilibrium exists at the solid-liquid interface,

(f)   In the packed column the liquid is in plug flow, and the carbon or resin is assumed to consist of a number of completely mixed reactors.

The liquid phase material balance for a packed column is:

$$\epsilon A \frac{\partial c}{\partial t} + v \frac{\partial c}{\partial x} + \frac{6k_f(1-\epsilon)A}{d_p}(C-C_s) = 0 \quad (1)$$

The mass balance for surface diffusion inside the pores is:

$$\frac{\partial q}{\partial t} = \frac{D}{r^2} \frac{\partial}{\partial r} \left[r^2 \frac{\partial q}{\partial r} \right] \quad (2)$$

If no accumulation occurs at the external surface of the carbon, the boundary condition becomes:

$$k_f(C-C_s) = D\sigma \frac{\partial q}{\partial r} \bigg|_{r=R} \quad (3)$$

For the resin $\sigma = 1/\Phi$

Local equilibrium at the particle surface:

$$q_s = ac_s^n \qquad \text{[for carbon]} \quad (4)$$

$$q_s = \frac{k_1 c_s}{k_2 + c_s} \qquad \text{[for resin]} \quad (5)$$

Figures 1 and 2 show the isotherm results for gold, silver and copper cyanide and the fitted equilibrium expressions for activated carbon and resin respectively.

When equations (3) and (4) or (5) are applied to a packed column, the value of q refers to a specific height in the column. The value of c_s is guessed at each step and the calculation of equations (2) and (3) repeated with a fourth order Runge-Kutta routine until equation (4) or (5) is satisfied. Equation (1) is discretized using a backwards difference method and solved together with the other equations.

This study also involved the simultaneous adsorption of two components on activated carbon. Although simultaneous asdorption was also performed with ion-exchange resins as adsorbents, the latter did not reveal any equilibrium influence, and will therefore only be considered to changes in the kinetic parameters. The equilibrium model equations are therefore developed on the basis of two solutes as described by Sheindorf et al. [3]:

$$q_{s,1} = a_1 c_{s,1}(c_{s,1} + B_{12}c_{s,2})^{n1-1} \quad (6)$$

$$q_{s,2} = a_2 c_{s,2}(c_{s,2} + B_{21}c_{s,1})^{n2-1} \quad (7)$$

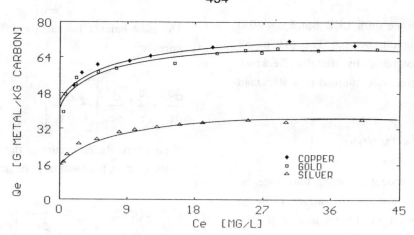

FIGURE 1: Equilibrium adsorption of copper, gold and silver cyanide on activated carbon.

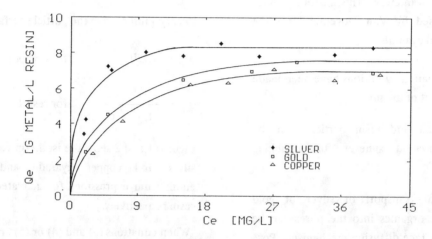

FIGURE 2: Equilibrium loading of silver, gold and copper cyanide on DU A161.

Equations (6) and (7) can be written in the form:

$$f(C_{s,1}) = \left[\frac{a_1 C_{s,1}}{q_{s,1}} \right]^{\frac{1}{1-n1}} - C_{s,1}$$

$$= B_{12} C_{s,2} \qquad (8)$$

$$f(C_{s,2}) = \left[\frac{a_2 C_{s,2}}{q_{s,2}} \right]^{\frac{1}{1-n2}} - C_{s,2}$$

$$= B_{21} C_{s,1} \qquad (9)$$

With $f(C_{s,1})$ and $f(C_{s,2})$ plotted against $C_{s,2}$ and $C_{s,1}$ respectively, the competition coefficients B_{12} and B_{21} can be obtained from the slopes of the curves.

Equations (6) and (7) are based on the assumption that species 1 and 2 both follow the Freundlich isotherm in a single solute solution. By replacing equation (4) or (5) with equations (6) and (7) and by duplicating the remaining equations for each adsorbate, the above mentioned model can be applied successfully to a two component system using packed column.

A computer programme was developed for the numerical solution of these simulation models. This programme, written in Turbo Pascal 4, can be used on an IBM compatible personal computer equipped with a 8087 mathematical coprocessor. The value of the film transfer coefficient k_f can be estimated from short-column experiments [7], while the surface diffusivity D can be estimated by using the model in a least squares routine.

RESULTS AND DISCUSSION

Van der Merwe and Van Deventer [8] showed experimentally that the reversibility of the adsorption isotherm of a metal cyanide $M(CN)^{2-}$ (where M = Au, Cu or Ag) depends on the ratio of $M(CN)^{2-}$ to MCN on the carbon. This ratio was determined by eluting the loaded carbon in a cyanide free solution, and thereby determining the reversibility of adsorption of the metal cyanide. The carbon used in the work presented here revealed no MCN, so that adsorption of gold, silver or copper cyanide was fully reversible. This suggests that these metal cyanides adsorbed on the carbon as ion-pairs [9].

Sensitivity Analysis

To test the flexibility of the model to different operating conditions a sensitivity analysis was performed as depicted in Figures 3 to 7. Except where other values are specifically mentioned, the following parameters were used in the sensitivity analysis:

volumetric flowrate: $1.4 \times 10^{-6} \, m^3/s$

length of packed bed: 0.15 m

initial concentration: 20 mg adsorbate/L

incremental time-step: 30 seconds

film transfer coefficient: $3.61 \times 10^{-5} \, m/s$

macropore diffusivity: $3.80 \times 10^{-12} \, m^2/s$

constant in isotherm: 46.3

exponent in isotherm: 0.105

void fraction in column: 0.38

Figure 3 shows that variations in the value of the Freundlich coefficient "a" have a significant influence on the later stages of the breakthrough curve, while variations in the value of the intraparticle diffusivity D affect more the early period of adsorption, as indicated in Figure 4. This is a clear indication that a kinetic influence will be observed at the lower part of the breakthrough curve, while an equilibrium influence will affect the later stages of adsorption. These observations

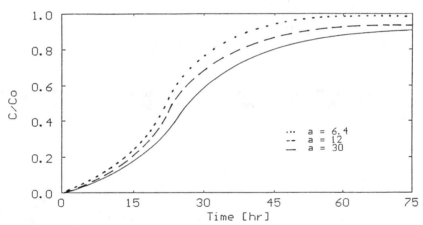

FIGURE 3: Sensitivity of the packed column to variations in the Freundlich parameter a.

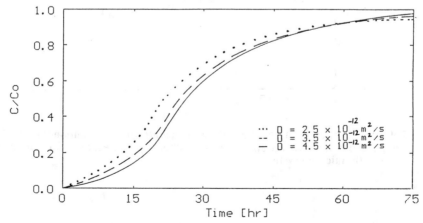

FIGURE 4: Sensitivity of the packed column to variations in the surface diffusivity D.

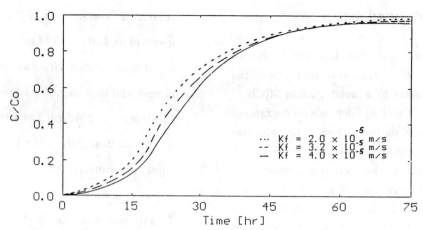

FIGURE 5: Sensitivity of the packed column to variations in the film transfer coefficient k_f.

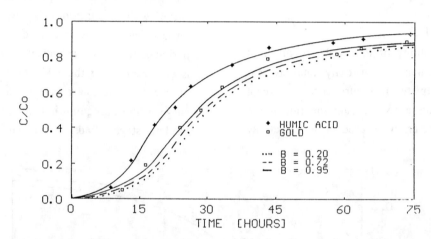

FIGURE 6: Sensitivity of the effluent concentration of component 1 (gold) from the packed column to variations in the competition B_{12} for the gold/humic acid system.

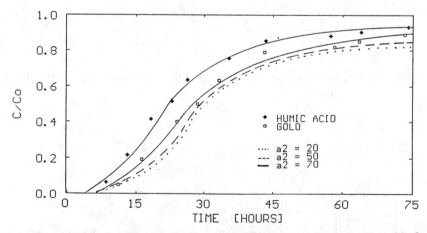

FIGURE 7: Sensitivity of the effluent concentration of component 1 (gold) from the packed column to variations in the isotherm parameter a_2 (for humic acid) for the gold/humic acid system.

obtained here, will be used in the next section to interpret the experimental data.

Variations in the size as well as the external roughness of the carbon particles can influence the effective thickness of the liquid film surrounding the carbon or resin particle. Variations in the resistance to bulk phase transport are depicted in Figure 5 by varying the film transfer coefficient k_f. It is evident that the model is more sensitive to changes in k_f during the early adsorption period than during the subsequent slower period of adsorption when intraparticle diffusion controls.

Variations in the competition factor B_{12} and the equilibrium parameter a_2 in the two-component adsorption system are shown in Figures 6 and 7 respectively. It appears as if variations in a_2 have a more pronounced effect on the adsorption of component 1 than do variations in B_{12}. Also, it is significant that the adsorptivity of component 2 affects the adsorption behaviour of component 1 (gold) already in the early stages of a column run. Variations in B_{12} in Figure 6 did not affect the adsorption of component 2 (humic acid), and seemed to be the same for the three different values of B_{12}. However, variations in a_2 as depicted in Figure 7, revealed an influence in the adsorptivity of component 2, which shown a downward shift as the value of a_2 increases (The curve for component 2 as indicated in Figure 7, presents only the case where $a_2 = 50$).

Equations 6 and 7 show clearly that, if B_{12} is relatively low, the effect of a_2 on the adsorption behaviour of component 1 will be less significant. The nature of the interaction between different adsorbates during the adsorption process is poorly understood in the literature, and deserves much more attention.

Pre-treatment prior to adsorption

The resin and carbon in Figures 8 to 11 were pretreated in solutions of organics, rinsed with water, and then contacted with a clear solution of silver, gold or copper cyanide. The organic loadings adsorbed onto the resin or carbon during the pre-treatment period varied from 46 mg organic/g adsorbent to 520 mg organic/g adsorbent. UV-spectrophotometry confirmed a loss of only 5 to 7 % of the organic loading from the resin or the carbon during the adsorption of metal cyanide.

Figures 8 to 10 clearly show that low loadings of organics on either activated carbon or ion-exchange resins, inhibited the rate of uptake of gold, silver or copper cyanide, but did not affect the equilibrium attainment, and confirms earlier results by Petersen and Van Deventer [10]. It has to be stressed that the organic loadings recorded on the carbon or resin used in these experiments varied between 46 and 51 mg organic/g carbon or

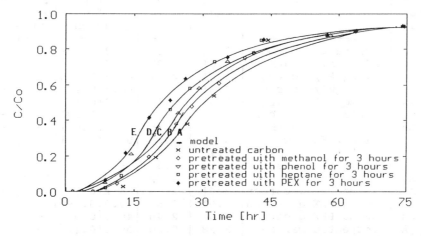

FIGURE 8: Inhibited mass transfer (kinetic influence) of gold cyanide to activated carbon owing to fouling of organics in a packed column. $C_0 = 20$ mg Au/L; $l = 15$ cm; $v = 1.4 \times 10^{-6}$ m^3/s; M = 12.32 g.

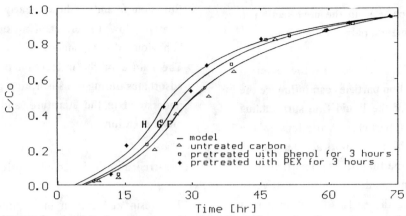

FIGURE 9: Inhibited mass transfer (kinetic influence) of copper cyanide to activated carbon owing to fouling of organics in a packed column. $C_O = 20$ mg Cu/L; $l = 15$ cm; $v = 1.4 \times 10^{-6}$ m^3/s; M = 12.32 g.

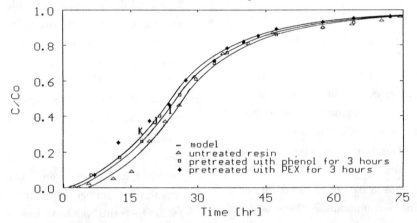

FIGURE 10: Inhibited mass transfer (kinetic influence) of silver cyanide to DU A161 owing to fouling of organics in a packed column. $C_O = 20$ mg Ag/L; $l = 15$ cm; $v = 1.4 \times 10^{-6}$ m^3/s; V = 0.075 L.

TABLE II SENSITIVITY OF KINETIC AND EQUILIBRIUM PARAMETERS TO ORGANIC FOULING

No.	Ad-sor-bate	Conditions	$k_f \times 10^5$ [m/s]	$D \times 10^{12}$ [m^2/s]	A	n	k_1	k_2
					carbon		resin	
A	Au	untreated carbon	3.61	3.80	46.3	0.105	–	–
B	Au	48 mg MT/g carbon	3.61	3.32	46.3	0.105	–	–
C	Au	51 mg PN/g carbon	3.61	2.71	46.3	0.105	–	–
D	Au	47 mg HT/g carbon	3.61	2.65	46.3	0.105	–	–
E	Au	52 mg PEX/g carbon	3.61	2.60	46.3	0.105	–	–
F	Cu	untreated carbon	3.75	4.10	52.5	0.091	–	–
G	Cu	50 mg PN/g carbon	3.75	3.85	52.5	0.091	–	–
H	Cu	51 mg PEX/g carbon	3.75	3.68	52.5	0.091	–	–
I	Ag	untreated resin	3.10	3.65	–	–	11.4	3.29
J	Ag	42 mg PN/L resin	3.10	3.42	–	–	11.4	3.29
K	Ag	45 mg PEX/L resin	3.10	3.32	–	–	11.4	3.29
L	Au	150 mg PEX/g carbon	3.61	2.55	44.2	0.102	–	–
M	Au	290 mg PEX/g carbon	3.61	2.50	42.0	0.099	–	–
N	Au	462 mg PEX/g carbon	3.61	2.40	40.8	0.089	–	–
O	HA	untreated carbon+	3.28	5.20	50.3	0.111	–	–
P	Au	untreated carbon+	3.61	3.70	46.3	0.105	–	–

+ simultaneous adsorption of gold and humic acid (HA)

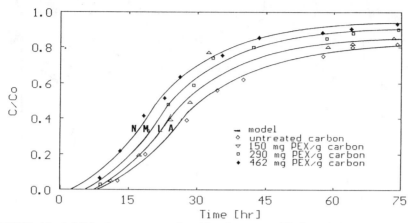

FIGURE 11: Inhibited mass transfer (kinetic + equilibrium influence) of gold cyanide to activated carbon owing to fouling of organics in a packed column. C_o = 20 mg Au/L; l = 15 cm; v = 1.4 x 10^{-6} m^3/s; M = 12.32 g.

22 and 36 mg organic/L resin. The results obtained from these Figures (8 to 10) indicated a constant equilibrium value, so that a mechanism of competitive adsorption between the metal cyanide and organic species, is irrelevant. Furthermore, it appeared that fouling of the resin or carbon had no effect on the value of the film transfer coefficient k_f, but had a marked effect on the surface diffusivity D. This implies a mechanism of pore blocking, whereby the organic foulants that had diffused into the pores of the adsorbent, partially blocked the pores and thereby hindered the passage of metal cyanides to such an extent that diffusion remained possible, but at a reduced rate. This is supported by the kinetic parameters in Table II estimated for each of the sets of data A to K in Figures 8 to 10. When a mechanism such as pore blocking by organic foulants is operative during the adsorption of metal cyanides, the relationship between the size and shape of the foulant and the pore distribution of the adsorbent becomes important. Basically, the pore wall reduces the freedom of movement of larger organic molecules, resulting in retarded diffusion. This is confirmed by the results depicted in Table I which indicates that the value of the surface diffusivity decreased with an increase in the molecular mass of the organic foulant.

At higher loadings of organics (more than 120 mg organic/g adsorbent) both the rate of adsorption as well as the equilibrium loadings are affected, as depicted in Figure 11. Although relatively high loadings of organics were recorded on the resin as well, the latter did not show any changes in the equilibrium values. At these high loadings of organics on the carbon, it was impossible to obtain acceptable fits for the model by changing only the values of D. It was therefore necessary to use different values of A and n in the Freundlich isotherm as summarized in Table II. The values of A and n were determined from equilibrium experiments on the pre-treated carbon. The mechanism of adsorption onto activated carbon is extremely complex and largely unknown. It is possible that the competitive adsorption between organics and metal cyanides is due to competition for active sites, but it is also possible that the organics merely block certain regions in the carbon and thereby rendering them inaccessible to metal cyanides. It was impossible to obtain acceptable fits of the data in Figure 11 by using only different A and n values. The surface diffusivity D also decreased, which means that a further inhibition of intraparticle transport occurred when the loading of organic increased.

Simultaneous adsorption

In most studies on the simultaneous adsorption of organics onto activated carbon, the Ideal Adsorbed Solution (IAS) theory was used to correlate results [4]. In the work presented here, the Freundlich-type multicomponent isotherm was used to describe the simultaneous adsorption of gold or silver cyanide and the organic specie onto activated carbon, simply because the IAS theory

440

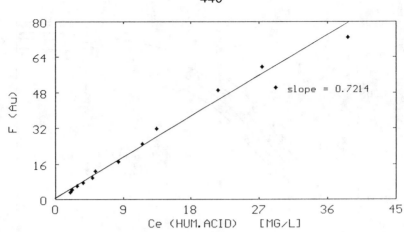

FIGURE 12: Estimation of the competition factor B_{12} for the competitive adsorption of gold cyanide and humic acid on activated carbon.

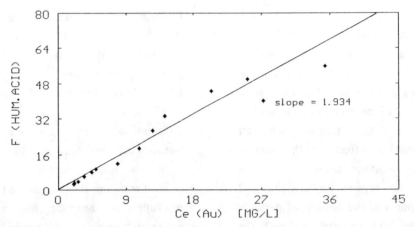

FIGURE 13: Estimation of the competition factor B_{21} for the competitive adsorption of gold cyanide and humic acid on activated carbon.

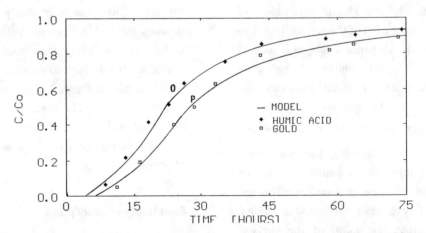

FIGURE 14: Rate curves for the simultaneous adsorption of gold cyanide and humic acid on activated carbon in a packed column. $C_0(Au) = 20$ mg/L; C_0(humic acid) $= 70$ mg/L; $l = 15$ cm; $v = 1.4 \times 10^{-6}$ m^3/s; M = 12.32 g.

does not hold for partially dissociated solutes. Figures 12 and 13 present the results of the bi-solute isotherm of gold cyanide and humic acid in linear form. From the linear relationship in Figure 12, the slope of the line, which represents the competition that humic acid offers to the adsorption of gold cyanide, was determined by least squares regression to be 0.7214. The equilibrium gold loading is then given by:

$$q_{e,Au}=46C_{e,Au}(C_{e,Au}+0.72\,C_{e,hum.acid})^{0.10-1}$$

By using the slope of 1.934 from Figure 13, the equilibrium humic acid loading may be given as:

$$q_{e,ha}=50C_{e,ha}(C_{e,ha}+1.93\,C_{e,Au})^{0.11-1}$$

As explained in the theory by Sheindorf et.al [3], the one competition factor should be the reciprocal of the other in a bi-solute system. The fact that this is not the case here, shows that these equations are merely convenient empirical expressions.

Figure 14 shows the simultaneous adsorption of gold cyanide and humic acid on activated carbon in a packed column. If the results obtained in Figure 14 are compared to that obtained in Figure 8, which involved both gold cyanide, the surface diffusivity D is the only parameter that did not remain constant. The decrease in the intrinsic diffusivity can be explained through the fact that two mechanisms were operative at the same time namely, the competitive adsorption for active sites on the carbon, as well as that of pore blocking.

APPLICATION OF KNOWLEDGE-BASED SYSTEMS

Nature of KBS systems

In most industrial columns used for the removal of metals and organics from plant effluent, complex multicomponent solutions are encountered. Existing phenomenological models are usually incapable of considering the effects of pH, ionic strength, carbon deactivation, and all the complex interactions between the many species in solution. The more conventional fundamental models discussed and applied in the previous sections of this paper are useful to demonstrate the sensitivity of column operation to well-defined variables, but cannot be applied conveniently when the available knowledge on adsorbates is incomplete. Moreover, it is frequently necessary to include qualitative information and order-of-magnitude estimations in the design of columns.

This type of ill-defined problem is not unique to the minerals industry, and has received attention in the literature. Recently, knowledge-based systems (KBS) have found increasing application for treating ill-defined problems, especially in the chemical industry. Most KBS in the chemical industry have been formulated for fault diagnostics, steady-state circuit design and simulation, process selection, operator training, data management and scenario analysis by means of "what if?" questions. However, the dynamic simulation of ill-defined processes by use of KBS is a new concept, and as such has received little attention in the literature. Reuter and Van Deventer [11,13] have explained the methodology involved in detail in previous papers.

An *expert system* is a computer program that behaves like an expert in a usually narrow field of application. One of the features of an expert system is the separation of knowledge from the techniques that are used to think about this knowledge. In view of the fact that expert systems require expert knowledge in some form, they are also called KBS. An expert system should be able to explain its behaviour and decisions to the user. In addition, a KBS should be capable of dealing with uncertainty and incompleteness of information. An expert system consists usually of a *knowledge base*, an *inference engine* and a *user interface*. Although different architectures are used to define a KBS, most KBS use both *qualitative* (experiential and heuristic) as well as *quantitative* (equation orientated) knowledge, also termed *shallow* and *deep* knowledge respectively.

Treatment of adsorption data

The basis of the adsorption model is batch concentration-time data, and the corresponding loading-time curve that covers the adsorption profile up to the equilibrium loading. From such a batch adsorption curve, the "rate variable" k[q(t)], which is defined here to be a function of only the loading q(t), may be determined. This "rate variable" is subsequently used to predict the change in concentration and carbon loading at any position in the column at any moment in time *at the prevailing chemical process conditions.*

It must be noted here that such KBS models do not attempt to suggest a mechanism for the reaction under consideration. What they do imply, however, is that a kinetic curve which describes a process up to equilibrium is a *finger-print* of that reaction and therefore of the reaction mechanism. Consequently, the KBS model can be used as a basis to describe the reaction under all possible conditions, given that the mechanism does not change. The database will obviously cover all possible mechanisms for a particular system. This basic finger-print curve and its associated shallow and deep level knowledge, defined as *objects,* are termed the *pivot-data,* or the *standard condition.* These *pivot-data* serve as a reference with which other curves are compared, and which can subsequently be used for fault-diagnosis or process identification.

Any condition that differs in whatever way from this standard condition or the pivot-data, is considered to be *non-standard.* Consequently, any change in the empirical parameters or the profiles of loading should be considered as the combined effect of different deviations from the standard condition, and therefore as taking all interactions into account. Such changes in the adsorption behaviour may be caused by changes in the chemistry of the feed solution, deactivation of the adsorbent, or changes in the nature of competition between the known and unknown adsorbates. It is convenient to express deviations from the *pivot-data* or standard parameters as percentages, which are estimated through experience, directly from the plant or from experimental data. This KBS approach of modelling developed at the University of Stellenbosch has been applied successfully to the simulation of small-scale columns under various modes of operation, including that of continuous countercurrent flow.

Overall structure of the KBS

Figure 15 illustrates the overall structure of the KBS, which consists mainly of: (1) A data-base for entering/editing of all relevant experimental and heuristic information; (2) An inference procedure for system diagnostics and the simulation of different reactor and column systems; (3) A working memory which contains the current input and status of the specific adsorption column or other reactor being investigated. The different modules of the system communicate via files in order to increase the working memory; (4) A user-friendly menu-driven interface used to communicate with the above three components.

As expected, the KBS is heavily dependent on the range, accuracy and population density of the available data.

CONCLUSIONS

Both activated carbon and ion-exchange resins used in adsorption columns for the removal of metals from plant effluent are fouled by organics usually present in industrial effluent streams. At low loadings of organics on either the resin or carbon the adsorption of metal cyanides is influenced in a kinetic way, and can be ascribed to pore blocking. At higher loadings of organics, both a kinetic as well as an equilibrium influence were observed. This can be attributed to two effects, viz.: (1) permanent pore blocking, where the pores of the adsorbent are totally sealled off by the organic species, and (2) competitive adsorption for active sites between the metal cyanide and organic foulant. It has also been illustrated that KBS modelling is a valuable tool in effective design of multi-component adsorption columns.

NOMENCLATURE

a   parameter in Freundlich isotherm

A   flow area of the column [m^2]

B   equilibrium competition factor

C   solution phase concentration [g/m^3]

D   surface diffusion coefficient [m^2/s]

d_p particle diameter [m]

f   function defined in Eqs. (8) and (9)

k_f external film transfer coefficient [m/s]

k_1 parameter in Langmuir isotherm expression

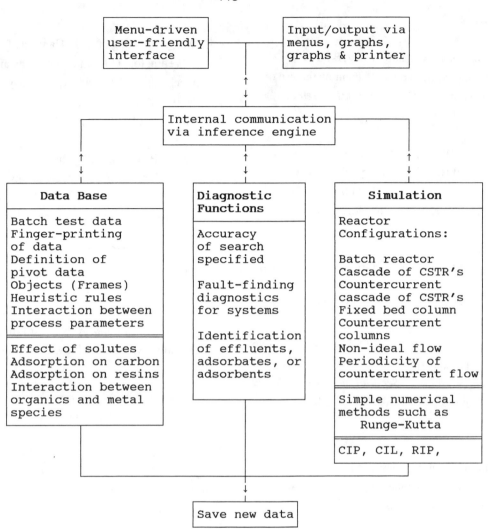

Figure 15 Overall structure of KBS.

k_2	parameter in Langmuir isotherm expression
M	mass of carbon [kg]
n	exponent in Freundlich isotherm expression
N	rotational speed of impeller [r.p.m.]
q	loading of adsorbent on carbon [g/kg] or resin [g/m^3 wet-settled resin]
r	radial variable [m]
R	radius of adsorbate particle [m]
t	time variable
v	volumetric flow rate [m^3/s]
V	volume of liquid in reactor [m^3]
x	distance variable along column length [m]

Greek Letters

ϵ	void fraction in bed of carbon
σ	apparent density of the carbon [kg/m^3]
Φ	volumetric fraction of wet-settled resin occupied by resin [m^3 of resin/m^3 of wet-settled resin]

Subscripts

e	equilibrium condition
o	initial condition
s	carbon surface

444

REFERENCES:

1. Davidson, R.J., and Strong, B.,"The recovery of gold from plant effluent by the use of activated carbon", *J.S. Afr. Inst. Min. Metall.*, vol.83, no.8, 1983, pp.181-188.

2. Fritz, W. and Schlünder, E.U. "Competitive adsorption of two dissolved organics onto activated carbon-I. Adsorption Equilibria., *Chemical Engineering Science*, vol. 36, 1981, pp. 721-730.

3. Sheindorf, C.H., Rebhum, M. and Sheintuch, M., "A Freundlich-type multicomponent isotherm", *Journal of Colloid and Interface Science*, vol. 79, no. 1, 1981, pp.136-142.

4. Van Deventer, J.S.J., "Competitive equilibrium adsorption of metal cyanides on activated carbon", *Separation Science and Technology*, vol. 21, no. 10, 1986, pp. 1025-1037.

5. Petersen, F.W. and Van Deventer, J.S.J., "The influence of organics on the determination of gold, silver and copper by atomic absorption spectrophotometry", *Minerals Engineering*, vol. 3, no. 5, 1990, pp.415-420.

6. Van Deventer, J.S.J., "Kinetic models for the adsorption of gold onto activated carbon", *MINTEK-50 Int. Conf. Mineral Science & Technology,* MINTEK, Randburg, South Africa, L.F. Haughton (ed.), vol. 2, 1985, pp. 487-494.

7. Weber, W.B. and Liu, K.T., "Determination of mass transport parameters for fixed-bed adsorbers", *Chem. Eng. Commun.*, vol. 6, 1980, pp. 49-59.

8. Van der Merwe, P.F. and Van Deventer, J.S.J., "Studies on the interaction between metal cyanides, oxygen and activated carbon", *Proceedings of the Perth International Gold Conference,* Randol International Ltd, 1988, pp. 258-260.

9. McDougall, G.J., Hancock, R.D., Nicol, M.J., Wellington, O.L. and Copperthwaite, R.G., "The mechanism of the adsorption of gold cyanide on activated carbon", *J.S. Afr. Inst. Min. Metall.* vol. 80, no. 9, 1980, pp. 344-356.

10. Petersen, F.W. and Van Deventer, J.S.J., "The inhibition of mass transfer to porous adsorbents by fine particles and organics", *Chemical Engineering Communications,* in press, 1991.

11. Reuter, M.A. and Van Deventer, J.S.J., "A knowledge-based system for the simulation of batch and continuous carbon-in-pulp systems", *Extraction Metallurgy '89*, Institution of Mining & Metallurgy, London, 1989, pp. 419-442.

12. Van der Merwe, I.W., Van Deventer, J.S.J. and Reuter, M.A., "Knowledge-based computer simulation of gold leaching in batch and continuous systems", *14th Congress of the Council of Mining and Metallurgical Institutions*, Institution of Mining & Metallurgy, London, 1990, pp. 147-160.

13. Reuter, M.A. and Van Deventer, J.S.J., "A knowledge-based system for the simulation of batch and continuous carbon-in-leach systems", *Proceedings of APCOM '90,* TUB-Dokumentation, Technical University of Berlin, vol. 51, no. 1, 1990, pp. 343-356.

Light Metals

Status and challenges for modern aluminium reduction technology

Harald A. Øye
Institute of Inorganic Chemistry, The Norwegian Institute of Technology, Trondheim, Norway
Reidar Huglen
Hydro Aluminium AS, Karmøy Fabrikker, Håvik, Norway

SYNOPSIS

A short history of the development of the Hall-Heroult technology for aluminium electrolysis is presented. Then the characteristics of modern cell technology and operational philosophy are discussed. Important characteristics are size, magnetics, off-gas cleaning, feeder systems, process control, mathematical modelling, bath chemistry, choice and handling of materials, start-up, operational and managerial philosophy. No new processes are expected in the first decade, but further improvements are expected in environmental conditions, choice of materials, control strategy and mathematical modelling. Important research challenges are the development of inert or semi-inert sidewalls, cathodes and anodes.

HISTORICAL IMPROVEMENTS OF THE HALL-HEROULT PROCESS

Since the turn of the century, large improvements have been obtained for the operational results of the Hall-Heroult process.[1-4] Current efficiency has been improved from 70 to 95 %, energy consumption is lowered from 40 to less than 13 kWh/kg Al and gross carbon consumption is reduced from 900 to close to 500 kg C/t Al. An obvious driving force in the improvements of the Hall-Heroult process has been the previous low metal yields and the big difference in actual and theoretical minimum energy consumption which is 6,34 kWh/kg Al. Also the competition from other materials has favoured an improvement of operational results to reduce production cost. Likewise, the large growth of consumer markets (airplanes, cars, buildings, cans) has triggered the increase of cell size (amperage) and the reduction of specific investments to compensate for the inherent low productivity of the aluminum electrolysis process compared to other electrometallurgical processes.

Fig. 1 illustrates the historical development of current efficiency and potline amperage. A plateau of about 90 % current effciency and a cell size for the largest newly built pots of about 150 kA was reached in the 1960's. In the 1970's basic knowledge was put into computer models of cell design, magnetics, heat balances, process control and bath chemistry. This resulted in an improvement of current efficiency to the current level of 95 %, a large increase in pot size to 300 kA and 400 kA cells being under trial.

The underlying driving force for the application of high amperage cells to be seen in the years to come, is cost competitiveness. Another important cost issue to be addressed both by existing and green-field smelters, is optimation of internal plant logistics. This will influence the choice and development of auxilliary equipment and systems, as well as the layout of the smelter complex.

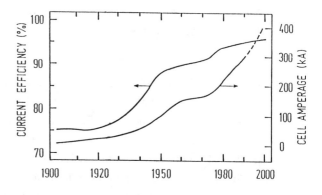

Figure 1: The progress in current efficiency (%) and cell amperage (kA) of newly developed cells in this century.

CHARACTERISTICS OF MODERN CELL TECHNOLOGY

Old potlines with fully paid-back investment costs and possibly also linked to a hydro-electric power station built long ago may be cost effective. Nevertheless such plants are steadily phased out due to increased direct production cost, but especially due to stricter requirements to the aluminum industry´s effect on the inner and outer environment. It is also found that some plants with relatively simple technology show good technical results close to the more advanced designs, due to workers´ dedication and good management. But a steady modernization is occurring in the aluminum industry either in form of more or less deep-going retrofit programs or by building new plants.

1. **Size.** Small cells are usually more labour intensive and less energy efficient. Cells with amperage less than 150 kA are hardly considered modern to-day.

2. **Magnetic Compensation.** A direct consequence of the higher amperage is the need for magnetic compensation to reduce metal and bath velocity and counteract the tendency to a non-horizontal metal surface. New cells are usually arranged side by side with distributed quarter rizers for more efficient compensation of the magnetic fields.

3. **Off-gas Cleaning.** Dry-scrubbing systems are applied thus reducing the emission to less than 0.5 kg fluorides per ton aluminum produced. The off-gas fluorides and hydrocarbon vapors are absorbed on the alumina constituting the ore feedstock to the cells.

4. **Inner Environment.** Modern plants emphasize the protection of workers from PAH and fluoride vapors and dust. Sometimes it is accomplished by personal protection equipment, but the long-time solution is an improved pot-room atmosphere determined by cell-emissions, hooding efficiency and suction capacity.

5. **Point feeders.** Introduction of point feeders not only for Al_2O_3 but also by some smelters for AlF_3, has made it possible to add alumina at known amounts and thereby monitor alumina and bath composition more closely. Point feeders also require specific properties of the alumina which is different from previous experience with side- and centerbreak cells.

6. **Process control.** The point feeder and close monitoring of alumina concentration is the basis for the modern cell control. Presently, the only continuously measured control parameters are the cell voltage and potline current which also enable calculation of the cell resistance. This resistance depends on the alumina content in the cell, and is the basis of the control system which contains rather sophisticated logics with additional inputs and safeguards. It is possible to control the alumina concentration, keep a nearly constant anode-cathode distance and also to spot irregular cell behaviour at an early stage which in turn stabilizes bath composition and operational temperature. The improved alumina concentration control will also reduce the anode effect frequency, usually to about 0.1 a day.

7. **Mathematical Models.** Mathematical models based on knowledge of physico-chemical properties of the materials and the processes coupled with accurate experimental data on cell behaviour are the corner-stones of modern cell technology. The models will be the basis for designs, choice of materials, effect of mechanical forces, process control, and will describe voltage drops, magnetics, fluid dynamics, heat balances, time dependent changes, and disturbances, evaporation and transport of impurities.

8. **Bath Chemistry.** The bath composition has been optimized with respect to current efficiency and energy consumption and to-day´s electrolytes contain 7 - 13 % excess AlF_3 added to cryolite. In some cells LiF and MgF_2 are added. With point feeding of alumina a smaller difference between the liquidus and solidus of the electrolyte can be allowed.

9. Cathode Materials. There is a tendency to use more graphitic cathode block materials to reduce expansion due to sodium intercalation and reduce cathode voltage drop. Redesign of collector bars contributes also to reduced cathode drop. The thickness of refractory and insulation materials are tailored to each design, and it is often used three kinds of materials to obtain the optimum positions of the isotherms.

10. Steel Shell. It is found that a strong and elastic steel shell prolongs life for the high amperage cells.

11. Handling of Materials. An aluminum plant may have the appearance of a busy storage building with trucks, cars, and conveyor belts going in all directions. Shortened flow lines of raw-materials, products and production equipment are emphasized in modern plants. Alumina is transported automatically directly to the different cells.

12. Pot-Tending Machines. Pot-tending machines are multipurpose and substitute the work that previously was carried out by specialized trucks or by operators.

The self-baking continuous Søderberg anode is currently not considered to be modern technology. The inherent advantages of the continuous anode have been more than offset by:
- Less efficient fume collection
- Higher anode voltage drops
- Stud pulling is a difficult and often an unpleasant work operation
- Point feeders cannot be placed in the middle of the cell
- A lower practical limit to the cell size
- Inferior anode quality compared to prebake due to lower baking temperature
- More prone to dusting.

There are, however, efforts going on to modernize the existing Søderberg technology. Several companies have installed point feeders on a limited number of cells. Point feeders are a major step forward in order to achieve improved feed control and higher current efficiencies. This way of feeding also creates new options for improved fume collection through reduced crust-breaking and design modifications for the gas skirts. To overcome the main Søderberg drawbacks it is, however, also necessary to have a major reduction of the anode fumes and get a stable and improved anode quality with less dusting and anode problems. The ability to do so is likely to be the only way to a viable Søderberg technology for the future.

CURRENT PLANT OPERATION

Competitive advantages are also created by the ways the potlines are operated. During the last decade, the dominating aluminum reduction technology licensee has been Aluminium Pechiney with its 175/180 kA cells. Even though the companies that have acquired AP technology have received the same specifications and instructions, there are significant differences in operational results.[5] For the period 1982 - 1988 differences in obtained current efficiencies of more than two percent have been observed. The corresponding differences in energy consumptions are about 0,3 kWh/kg Al. Generally, potlines with the highest current efficiency also have obtained the lowest energy consumption. Notably, one potline has both achieved the highest current efficiency (>95 %) and the lowest energy consumption (12,7 kWh/kg Al).

For a potline of 120.000 tons/year, two percent current efficiency amounts to 2.400 tons of aluminum, the related production cost being that of alumina, solely. Conversely 0,3 kWh/kg Al amounts to 37 GWh of electrical energy saved annually. The operational performance data illustrates a general trend of learning by experience in that practically all potlines have improved their current efficiency by one percent during the time of operation.[5]

As pointed out in two previous papers[6,7] when the amperage of the reduction cells increases, an increase of productivity is obtained by a reduction in manhours/ton Al. Likewise, as the total production of a smelter complex increases, an increase in overall productivity should be expected.

Figures 2 and 3 illustrate the productivity of a random sample of smelters measured over a wide range against cell amperage (cell size) and overall tonnage (smelter size). At first glimpse, the illustrations show no correlation between productivity and size, neither cell nor smelter size. The extremes in productivity are found for small and large size, respectively, but in-between a large variety is observed.

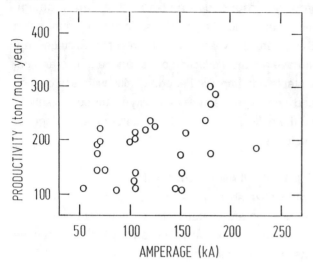

Figure 2: Productivity (tons per worker-year) vs. cell amperage (kA) for a random sample of smelters.

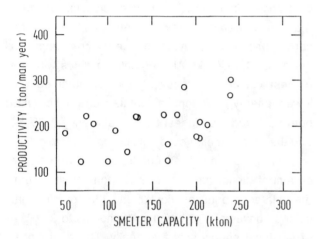

Figure 3: Productivity (tons per worker-year) vs. smelter size (annual aluminum production) for a random sample of smelters.

For larger smelters (>200,000 t/yr.), productivity ranges from 170 to 300 tons per worker-year, while for small smelters (50,000-100,000 t/yr.), the productivity span is from 120 to 220 tons per worker-year. For newly built smelters with a capacity of more than 200,000 t/yr., a productivity of 250-300 tons per worker-year is typical. For very large, single-technology smelters a productivity in the range 400 tons per worker-year may be reached. It appears that potlines with cell amperages more than 175 kA combined with large scale are required to achieve productivity of more than 250 tons per worker-year.

There may be some apparent reasons that explain the observed differences:

- The plant may consist of one or two technologies (i.e. prebake and Søderberg, or two prebakes of different amperage).
- Different working hours due to different shift arrangements.
- Whether the plant has carbon facilities or not.
- Differences in casthouse complexity.
- Differences in process automation and mechanisation due to age.
- Degree of subcontracted services.
- Differences due to plant management attitudes.

The variations in productivity depicted in Figs. 2 and 3 may be ascribed to one or a combination of several of these factors. The need for high productivity is especially important in high cost countries.

In particular, relatively high productivity at low smelter capacity is connected with high cell amperage. Conversely, relatively low productivity at high smelter capacity may be connected with several potlines and low amperage. Normally, prebake smelters with capacity significantly above 100.000 t/year, have their own carbon plants.

However, underlying all combinations of smelter complexity and size, including cell amperage, is the fundamental effect of management attitudes. In this respect tendencies towards relative positions of high and low productivity is strongly influenced by management capability.

DIFFERENCES IN OPERATIONAL PHILOSOPHY

The major mission of all businesses is to comply with the demands of customers and society, justifying the use of resources in the long term which also means being profitable.

The location of aluminum smelters during the last century has been very strongly connected to the location of readily available (abundant) energy sources, and where the energy production potential and capacity has been greater than the local communities could absorb themselves. Also, the availability and cost of capital and labour have influenced both location and operational philosophy. The aluminum smelting industry is both energy and capital intensive.

These factors have been determinant for the historical difference between the operational philosophy of the North-American and the European aluminum smelting industries. Capital and labour have been traditionally expensive in the USA, while energy so far has been readily available. This has favoured a tendency towards high current densities and high current efficiencies in the USA at the expense of high energy consumption. In continental Europe, with high population density, electrical energy have been an in-demand resource as well as not very abundant, which in turn has generated a tendency towards lower energy consumption in European aluminum smelters. Typically, the difference in energy consumption between the two continents has amounted to 1 - 2 kWh/kg Al.

Table 1 summarizes operational data (from 1989 and 1990) from eight different types of prebake reduction technologies. Types II and IV data are an average for several potlines with similar technology. Types I and II are pointfed cells, Type IV have centerbreakers, while data for Type III both represent pointfed and centerbreak cells. Types III and IV are more than fifteen years old.

Bath chemistry and average operational temperatures seem to be fairly similar for the technologies I - III. Also the current densities for these technologies are not too different. Current efficiencies

and energy consumptions are significantly better for technologies I and II due to this being the latest released technologies where the impact of computer modelling and up-to-date process control have brought the state of the art forward.

For normally well operated end-to-end prebake cells with modified magnetics and thermal balance including up-dated process control, current efficiencies of more than 93 % and energy consumption down below 14 kWh/kg Al can be achieved.

Carbon consumption numbers are clearly different between point-fed and centerbreak cells as seen between the groups including Types I and II, Type III, and Type IV, respectively. The difference is most clearly displayed in the frequency of anode effects.

The reduction technology represented by the operational data in column I, illustrates an extreme in several aspects of aluminum reduction technology. It is an old low- amperage technology that has been boosted to a current density of 1.33 A/cm^2 by employing low melting lithium-magnesium containing bath and SiC as cathode side lining to dissipate

Table I Operational data from different prebake reduction technologies

	I	II	III	IV	V	VI	VII	VIII
Amperage (kA)	65	155	177	180	214	220	235	300
Current Density (A/cm^2)	1.33	0.79	0.83	0.77	0.79	0.80	0.81	0.80
Cell Voltage *(V)	4.75	4.70	4.76	4.2	4.42	4.29	4.58	4.42
Current Efficiency (%)	90.8	92.5	93.9	94.6	92.0	94.5	92.4	92.3
Energy Consumpt. (kWh/kg Al)	15.6	15.1	15.1	13.2	14.3	13.5	14.8	14.3
Iron Content (wt %)	0.15	0.15	0.13	0.11	-	0.13	0.12	0.12
Silicon Content (wt %)	0.05	0.05	0.04	0.04	-	0.03	0.04	0.03
Gross Carbon Consumpt.(kg/t Al)	672	600/540	565	568	598	535	579	560
Net Carbon Consumpt.(kg/t Al)	485	470/430	415	413	417	425	411	430
AlF$_3$ Consumption (kg /t Al)	NA	20.0	22.0	19.9	-	20.0	18.8	21.0
Cathode Drop (mV)	350	360	-	405	433	360	354	-
Excess AlF$_3$ (wt %)	1.5	10.2	12.0	11.0	12.0	10.5	12.0	12.0
CaF$_2$ (wt %)	**	5.4	6.6	4.9	5.4	5.0	6.4	5.5
Bath Temperature (°C)	954	958	957	963	951	964	959	951
Feed System	CB	CB/PF	PF	PF	PF	PF	PF	PF
Anode Effects (per day)	1.0	1.0/0.1	1.4	0.15	-	0.2	0.86	1.6

* including cell to cell

** 4.4 wt % CaF$_2$ + 2.0 wt % MgF$_2$ + 3.0 wt % LiF

PF: Point Feed, CB: Center Break

excess heat. Current efficiency and energy consumption are in line with the very best operated Søderberg potlines, but the technology has to pay a high carbon consumption due to centerbreakers and high current density. This technology is probably an example of an old potline where for a limited period of time, there will be available inexpensive electrical energy locally, and the metal output is maximized. The technology also illustrates the robustness and flexibility of old cells when using up-dated knowledge to improve cell characteristics. It would be unthinkable with modern technologies to do such drastic changes to the original cell design.

The cells represented by column II - VIII are all high amperage and magnetically compensated, some built a few years ago, some up to 28 years old. Some common features are apparent. The current density varies within a narrow range being between 0.77 and 0.83 A/cm^2. Current efficiency is above 92 %, the highest being 94.6 %. The content of Si + Fe is between 0.15 and 0.20 wt %. The AlF$_3$ consumption is within the narrow range 18.8 - 22 kg/t Al. The bath chemistry is similar, CaF$_2$ being between 4.9 and 6.6 wt % and excess AlF$_3$ between 10.2 and 12.0 wt %. This is reflected in a narrow operational temperature range between 951 and 963 °C.

The energy consumption varies between 13.2 and 15.1, a 14 % difference. It is typical that the oldest designs II, III and VII have the highest energy consumption. Again it is demonstrated that high current efficiency often means high energy consumption. The difference between centerbreak and point feeders is illustrated by cell-type II. Centerbreak gives a considerable higher carbon consumption and a higher frequency of anode effects. For the other cells the carbon consumption is not clearly correlated to cell parameters, and the 5 % difference in net consumption probably reflects differences in raw materials, efficiency of the carbon plant and operational procedures.

In many areas there is general consensus how to design and operate a successful modern aluminum cell. There are, however, some notable differences in opinion:

Shell Design. It is generally agreed that high-amperage cells need strong pot-shell, but opinions vary whether the shell should be deep or shallow. A shallow pot-shell with a shallow metal pad (10 cm) has a smaller heat-loss area, uses less materials and may have a more stable side ledge. On the other hand the refractory layer is thinner with greater chances for complete penetration of the bath. This design may also prevent the use of graphitic cathode blocks due to their higher thermal conductivity.

Insulation. The insulation design may show large variations illustrated by the following ranges in heat loss:

Top: 35 - 70 %
Side: 20 - 45 %
Bottom: 5 - 10 %

This is a consequence of the initial design and the need to keep a proper heat balance. Changes may also be related to the wish to boost the amperage when the aluminum price is high. The heat loss from the top is the only easily adjustable parameter by varying the ore cover.

Automation and Mechanization. Although all smelters see the benefit of point-feeders, the complexity of pot-tending machines varies greatly. The cost aspect, skill and fine-tuning of present operation are important factors in deciding the degree of automation. It was for instance at some time built pots with individually adjustable anodes. Presently it is found that satisfactory results are obtained without this extra option, and individual anode regulation may even disturb the interpretation of cell resistance measurements.

Preheating. The most common preheating methods are resistive heating with a coke bed and gas burner heating. Coke bed preheating has the advantage that no special equipment is needed as the line current provides the power source. The success of the method is, however, dependent on an even current distribution to avoid local overheating. This problem is reduced by some smelters with the use of shunting avoiding full line current at the start, but at the expense of a longer heat-up period, which in turn is favorable for minimum thermal stresses. Gas heating gives a more uniform heat-up rate and improved temperature distribution in the cathode introducing minimum thermal stresses. Although gas heating appears to be a simpler method, difficulties are experienced, the most notable being: Potential air-burn damage, more complex cut in, preheating time being about the double of resistor pre-heat. A more complete discussion is given.[8]

Start-up. Most smelters use an acid electrolyte (7 - 13 % excess AlF_3), for normal operation, but the start-up bath may either be close to the operating composition (acidic) or basic (excess NaF). It is well known that sodium penetration into the cathode is faster from a basic bath. If a slow sodium penetration is wanted, an acidic start-up bath would be preferential. The smelters advocating basic start-up claim that their procedure impregnates the lining with high-melting materials and prolongs the cell life. The smelters with a successful acidic start-up have higher early productivity and detrimental effects on pot-life are not reported even though no comparison has appeared. It may be concluded that a proper start-up without overheating the cell, both at start and later on, is extremely important for cell life, but the operative mechanisms are not fully understood.

LiF Additions. Addition of LiF and sometimes also MgF_2 give increased productivity, increased current efficiency and decreased emissions. But there is no consensus whether it is cost effective or not.

Al_2O_3 - Content. The average levels are generally between 2 - 2.5 wt % or 3 - 3.5 wt %. The lower level seems to be preferred for high amperage cells. They have also the highest current efficiency.

Al_2O_3 Quality. Only pot-lines without dry scrubbing have to-day a choice between low-surface floury alumina or high surface sandy alumina. In dry scrubbers sandy alumina with a minimum surface area of 45 m^2/g is a must.

Control Algorithm. All modern control systems use resistance curves as parameters. The algorithms are however different. The success of a particular system is dependent on the logics used to describe the process and the logics ability to spot and correct irregularities.

TO-DAYS CHALLENGES
There is no doubt that the technological development of the Hall-Heroult process is the backbone of aluminum's advance to remain as a competitive metal to serve growing customer markets. Aluminum has inherent competitive properties of weight, ductility, strength, surface, energy content and recyclability to make it a desired material for use in modern society.

However, for the day-to-day operations of smelters and competition in the metal markets, other factors may be just as important as the technical level of the reduction technology. It has been shown that different results have been achieved with the same technology and that great improvements of operational results have been obtained from year to year. Basically, these differences are due to different management attitudes and commitment for human oriented relations. To search these potentials, a too rigid technological thinking may be a barrier. However, if the human relations program thinking is right, a good technological understanding may create better results. There is an increased awareness that teaching of underlying principles, training, motivation and quality of manual work are of paramount importance.

Outsiders seeing an aluminum smelter may not associate the operation with high technology. But there are real challenges in operating a process successfully around 950 °C for more than 6 years with a liquid that can oxidize, reduce or dissolve nearly all known materials. Knowledge, persistence, awareness of details and sometimes also imagination and intuition are important factors for a successful result. A small error or failure to see how different factors are interrelated may give catastrophic results.

Size and plant logistics are important factors for economic competitiveness. With respect to size it appears that total plant size may be even more important than the cell size.

The technology must also be adapted to the technological and economic environment. Robustness has been the hallmark of the previous work-horse cells with low amperage and high energy consumption. The new high performance cells perform presently more like race-horses with the inherent sensitiveness to disturbances.

Another challenge is to have cells suitable for electrical load levelling on a short- and medium term basis. The cost of peak power is high in many areas, and the aluminum industry will strengthen its position in the public and improve its power base when participating in load-levelling.

It is also to be remembered that the products of the smelters and the final measure of smelter success are the casthouse products. Product quality, customer orientation and service may be just as important for the smelters viability as the current state of the reduction part of the smelter.

There will be a continuous pressure for internal and external environmental protection. Internally, the allowed limits for workers exposure to dust, fluorides, PAH-volatiles and heat stress will be discussed and questioned. Answers to which cell design and operational philosophy to choose will have to be adjusted. Hooding efficiency and suction capacity of cells may be questioned since it is the two factors mostly affecting internal and external emissions. Also, the collection of SO_2 may generally be demanded giving in-land smelters a cost-handicap compared to sea-shore smelters where less expensive sea-water scrubbing can be applied.

In the long run, it will be aluminum´s total ecological position that will give the guidelines for future development both for reduction technology and markets. All man use of fossil fuel and the production of metals from metal oxides are connected with the release of CO_2. Given the alternatives, it may look like that aluminum has a favourable position as a material due to the possibilities of weight reduction of structures, energy saving in transportation and particularly the convenience and low energy consumption for recycling and remelting of the metal. It must be remembered that the total energy demand of society is made up of the product of population, standard of living and overall energy efficiency.

FUTURE DEVELOPMENTS

It might be concluded that no new process appears to be in sight in the first decade, with the exception that chloride electrolysis may again be subject to serious considerations. There is also another factor today that limits development of new processes. Due to the rightly greater emphasis on safety and environmental standards, combined with generally shorter working hours, the cost of process development has risen dramatically. An estimate is that it may take 3 times as many man-years to obtain the same results. In addition these man-years are associated with very skilled persons of which there is a shortage. With these conclusions in mind it is appropriate to look into further improvements of the current Hall-Heroult process.

In the future, further improvements are expected in the control strategy and in mathematical modelling of the design and the process. An important challenge is to develop a control system that reacts better towards dynamic changes such as start-up, tapping, anode changes, alumina and aluminum fluoride concentrations, anode effects, temperature changes and occasional anode malfunctions. Better control systems and tight process specifications are prerequisites for increased automation of the electrolysis, a feature that is necessary if the cost of aluminum shall remain competitive. Additional sensors are also highly desirable.

Although the problem of gas emission to the surroundings and the general pot-room atmosphere are at a satisfactory level in modern smelters, there are still specific operations where improvement is desired from an environmental point of view. A satisfactory and economic recycling of spent pot-lining is also a challenge that needs to be addressed.

The size increase up to 300 kA has generally been accomplished by making the cell longer, still having a rather large surface to volume ratio. A major break-through in energy consumption will be obtained if this surface to volume ratio could be reduced, but several difficult operation and design problems will then have to be solved. On the other hand, the increased magnetic field with amperage may limit the cell size to less than 500 kA.

Another important break-through would be the development of a semi inert sidewall. This will eliminate the heat loss associated with maintaining a frozen ledge. This could also give flexibility in operation by using heat exchangers in the side.

The ultimate goals are to develop an inert cathode and an inert anode. They will have to be developed separately as no single material is expected to withstand the severe conditions of electrolyzing aluminum from a fluoride bath. With inert cathodes and anodes developed one could finally look into the possibility of obtaining an enclosed, multipolar cell.

Nevertheless the chance of developing a truly inert anode seems slim. On a long range scale possibly a workable semi-inert anode can be developed, i.e. an anode that may have a life of months instead of weeks. This will then eliminate the release of CO_2. On the other hand the reversible cell voltage will increase one volt due to the lack of carbons depolarizing effect. This voltage may, however, be partly or fully offset by reduced ohmic resistance and anode overpotential.

ACKNOWLEDGEMENT

This contribution is to a large extent based on the paper: Harald A. Øye and Reidar Huglen: "Managing Aluminium Reduction Technology. Extracting the Most from Hall-Héroult", which appeared in JOM **42**, Nov. 1990, p. 23-28.

References

1. Haupin,W. in Hall-Heroult Centennial, 1886-1986. "History of Electrical Energy Consumption by Hall-Heroult Cells", The Metallurgical Society, Warrendale, PA, (1986), pp. 106 -113.

2. Richards, N.E. in Hall-Heroult Centennial, 1886-1986.."Evolution of Electrolytes for Hall-Heroult Cells", The Metallurgical Society, Warrendale, PA, (1986), pp. 114-119.

3. Welch, B.J. in Hall-Heroult Centennial, 1886-1986. "Gaining That Extra 2 Percent Current Efficiency", The Metallurgical Society, Warrendale, PA,, (1986), pp. 120-129.

4. Belitskus, D. in Hall-Heroult Centennial, 1886-1986. "Carbon Electrodes in Hall-Heroult Cell: A Century of Progress", The Metallurgical Society, Warrendale, PA, (1986), pp. 130-143.

5. Langon, D. and Peyneau, J.M. "Current Efficiency in Modern Point Feeding Industrial Potlines", Light Metals. (1990), pp. 267-274.

6. Welch, B.J. "Aluminium Reduction Technology Entering the Second Century", J. Metals. 40, (1988) No. 11, pp. 19-25.

7. Grjotheim, K. and Welch, B.J. "Aluminium Smelting as the Industry Enters the 21st Century, J. Metals. 41, (1989) No. 11, pp. 12-16.

8. Sørlie, M. and Øye, H.A. Cathodes in Aluminium Electrolysis. Chap.II, Aluminium-Verlag, Düsseldorf (1989).

New approach to acid methods for special alumina production

Rumen D. Kanev
Kurillo Metal SPE, Polymet Co., Sofia, Bulgaria

SYNOPSIS

A new method for producing high-purity alumina has been developed. Actually, that is an acid technology created on the basis of a new approach - the chemical transformation of the preliminary obtained and purified ammonium aluminium sulfate salt by means of neutralization with ammonium bicarbonate to the basic ammonium aluminium carbonate salt, with the following thermal decomposition and calcination of the same at a comparatively low temperature, and with full utilization of the gaseous phase containing less corrosive ammonia, carbon dioxide and water vapour.

In the present paper some theoretical aspects of this newly developed method concerning mainly the new approach to the basic ammonium aliminium carbonate salt formation and its decomposition to a high-purity alumina are considered. On the basis of the complete process flowsheet some results of the pilot plant investigation for determination of optimal process parameters are given.

The main conclusion of the paper is that this new approach to the acid methods can be used very effectively for high-purity alumina production.

INTRODUCTION

In the field of the special aluminas production many different methods have been developed [1-3]. For some types of these products with normal soda and silica content, the modifications of conventional Bayer process can be used [4]. For the types with low or very low soda and silica content, however, this process is not suitable, and the acid methods have to be used.

Nowadays, the most widely used acid methods [5] for producing high-purity aluminas are those, based on the thethermal decomposition of the preliminary purified to the given extend ammonium aluminium sulfate [6]. The technological shortcomings of these methods however, are connected with the comparatively high temperature used in this process and the difficulties arising during the separation and utilization of a high-corrosive gaseous phase containing SO_2, SO_3 and water vapour [7].

In the pilot plant scale a new method for producing high purity alumina has been developed [8]. Actually, that is an acid technology created on the basis of a new approach - the chemical transformation of the preliminary obtained and purified ammonium aluminium sulfate by means of neutralization with ammonium bicarbonate to the basic ammonium aluminium carbonate salt, with the following thermal decomposition and calcination of the same at a comparatively low temperature with a full utilization of the gaseous phase, containing less corrosive ammonia and carbon dioxide, and water vapour, back in the process.

This newly developed method has the following productional steps:

- percolation leaching of aluminium wastes with diluted sulfuric acid to obtaining pure aluminium sulfate;

- ammonium aluminium sulfate salt formation by mixing aluminium and ammonium sulfates followed by purification of this semi-product by means of counter-current precrystalization, using complex-forming agents;

- neutralization of the purified to the given extend ammonium aluminium sulfate with ammonium bicarbonate to formation of basic ammonium aluminium carbonate salt and separation into the solution ammonium sulfate, used back in the process;

- thermal decomposition of the basic ammonium aluminium carbonate salt to the ammourphous alumina and its calcination at elevated temperatures to a final product - high-purity alumina with full utilization of the separated gaseous phase as a mixture of ammonium carbonate and ammonium bicarbonate and its return back to the previous productional step.

This method combines the following advantages of few well known in the practice for acid alumina recovery processes:

- high speed of aluminium wastes acid leaching by means of percolation with preliminary liquid aluminium sulfate purification using a reduction and cementation phenomena;

- high degree of ammonium aluminium sulfate salt formation;

- high degree of ammonium aluminium sulfate salt purification by means of counter-current precrystalization with complex-forming agents;

- high degree of basic ammonium aluminium carbonate formation by means of neutralization of ammonium aluminium sulfate into the ammonium bicarbonate solution;

- low temperature of ammourphous alumina formation by means of basic ammonium aluminium carbonate thermal decomposition with full utilization of the separated gaseous phase, and

- direct ammourphous alumina calcination to high-purity alpha-alumina at lower temperature.

In the present paper some theoretical aspects of this method, concerning all productional steps and mainly the new approach to the basic ammonium aluminium carbonate salt formation and its decomposition to a high purity alumina are concidered, and the results of the pilot plant investigation for determination of optimal process parameters with monitoring the impurities content into the semi-products and final product are given.

EXPERIMENTAL

Some theoretical aspects

The main chemical interactions which take place in this method can be summarized as follows:

- Aluminium wastes acid leaching:

$$2Al + 3H_2SO_4 + nH_2O = Al_2(SO_4)_3.nH_2O + 3H_2O \quad (1)$$

- Aluminium and Ammonium sulfates mixing:

$$Al_2(SO_4)_3.nH_2O + (NH_4)_2SO_4 =$$

$$(NH_4)2Al_2(SO_4)_4.24H_2O + (n-24)H_2O \quad (2)$$

- Ammonium aluminium sulfate purification:

$$(NH_4)Al_2(SO_4)_4.24H_2O.Impur. + Compl.Agent =$$

$$= (NH_4)2Al_2(SO_4)_4.24H_2O + Impur.Compl.Agent \quad (3)$$

- Ammonium aluminium sulfate neutralization:

$$(NH_4)2Al_2(SO_4)_4.24H_2O + 8NH_4HCO_3 + nH_2O =$$

$$= NH_4Al_2CO_3(OH)_2 + 4(NH_4)_2SO_4 + (n+24)H_2O +$$

$$+ 7CO_2\uparrow + NH_3\uparrow \quad (4)$$

- Basic ammonium aluminium carbonate salt thermal decomposition and calcination:

$$NH_4Al_2CO_3(OH)_2 \xrightarrow{400^oC} am\text{-}Al_2O_3 + NH_3\uparrow + CO_2\uparrow +$$

$$+ H_2O\uparrow \xrightarrow{1150^oC} alpha\text{-}ALUMINA \quad (5)$$

The leaching of aluminium wastes with sulfuric acid is a very well known process for producing aluminium sulfate.Usually, the wastes are treated with sulfuric acid diluted to some chosen extend and metallic aluminium reacts according to the above shown reaction (1). This reaction is exothermic and proceeds with comparatively high reaction speed and heat and hydrogene evolution at the beginning. However, as the reaction proceeds and the amount of free acid in the system decreases, the reaction speed and heat evolution decrease too. Therefore , to achieve a moderate final reaction speed additional external heating is used. From the other hand, into the aluminium wastes coexist as impurities some metallic and non-metallic inclusions. Due to their comparatively high chemical affinity towards sulfuric acid they also react and pass quantitatively into solution and coexist in the final product as corresponding metal sulfates.

These basic problems of that well known process are overcomed to a great extend in this productional step by percolation leaching of the wastes using a specially designed percolation column and creating conditions for impurities separation by attaining maximum possible effect from cementation and reduction phenomena [9].

The direct mixing of liquid aluminium sulfate and crystaline ammonium sulfate is also a well known process for producing ammonium aluminium sulfate. Usually this process takes place within the temperature range from 65 to 95°C and after a complete dissolving of crystaline phase the solution is cooled and obtained according to reaction (2) crystals of ammonium aluminium sulfate are separated by filtration, leaving into the liquid phase the surplus of ammonium sulfate. This process has the following basic problems: a complete aluminium transformation from aluminium sulfate to the ammonium aluminium sulfate, and impurities separation from the crystal semi-product. In the newly developed method the above mantioned problems of this productional step are overcomed by using into the stage of mixing around 120 - 130 pct stoichiometric surplus of ammonium sulfate and so, creating conditions for complete aluminium utilization, and for impurities separation, a step-by-step precrystalization of the crystals is used. For attaining a biger purification effect however, in any precrystalization step complex-forming agents are added. (see reaction 3).

The amount of these reagents is calculated upon the base of the impurities content as 120 - 150 pct stoichiometric surplus. In this manner for reaching the requisite level of impurities content under 10 ppm in the finally purified crystals only three stages of counter-current precrystalization are nessesary (against six or seven without usage of complex-forming agents). In the first two stages the salicilic acid prove to be the best complex-formiming agent, and for the last one, EDTA II (ethylene-diamine tetraacetic acid) can be used successfully. In many methods developed for producing high-purity alumina the purified ammonium aluminium sulfate is the final semi-product from which the high-purity alumina can be produced. From the technological and technical point of view however, its thermal decomposition is a very difficult process connected with handling and utilization of the separated within the temperature range from 100 to nearly 750 °C highly corrosive gaseous phase with low sulphur oxides and ammonia content.

In view of the above, in the newly developed method for production of high-purity alumina an additional productional step is involved. That is the neutralization of the purified ammonium aluminium sulfate with ammonium bicarbonate in the presence of a small amount ammonium sulfate, according to reaction (4). This reaction however, is highly endothermic and has to be carried out very carefully into the frameworks of the previously determined conditions, because, there exist some possibilities as a final product, instead of basic ammonium aluminium carbonate salt, the aluminium hydroxide to be obtained. Because of its colloidal nature and pure sedimentation and filtration behaviours this semi-product is highly undesirable for further processing.

By contrast with aluminium hydroxide, the basic ammonium aluminium carbonate salts has a very well-formed small crystals which settle down in the suspension very quickly and possesses highly desirable sedimentation and filtration behaviours. From the other hand, its thermal decomposition takes place within the temperature range from 90 to 265 °C (according to DTA-curves) with separation of a less corrosive gaseous phase containing ammonia, carbon dioxide and water vapour. The full utilization of that off gases can be accomplished in a very sophisticated way by means of absorbtion and the solution obtained can be used back into the process for dissolving new amounts of ammonium bicarbonate.

The technological manner developed for successful neutralization of ammonium aluminium sulfate with ammonium bicarbonate has a few stages [10]. First, into the solution, made by mixing of wash waters from the washing of final semi-product and deionizide water, and absorbtion water, containing certain amounts of ammonium ions,the ammonium bicarbonate is dissolved by means of permanent stirring within the temperature range from 35 to 60°C. The total amount of ammonium

ions in the resulting solution has to be at least 125 pct of that nesessary for basic ammonium aluminium carbonate salt formation. Second, into the prepared solution, as it was described above, the crystaline ammonium aluminium sulfate is added with preliminary determined rate, but not more than 0,3 kg/dm³.min-1. After finishing the addition of the crystals the resulting suspension is leaving with stirring for ageing 15-30 min. Third, after finishing the ageing the crystals of basic ammonium aluminium carbonate are separated from the suspention by means of simple settling and suction of the overflow, which actually contains very pure ammonium sulfate and can be used for washing of the obtained in the previous productional step ammonium aluminium sulfate crystals. The separated crystals of basic ammonium aluminium carbonate undergoes three steps of counter-current washing and the outstream is used for dissolving of ammonium bicarbonate at the beginning of this productional step. The final crystals has 36-37 pct alumina content and can be used directly into the following productional step.

The final productional step of this method is the thermal decomposition and calcination of the obtained basic ammonium alluminium carbonate. Usually, it can be accomplished in the static conditions using a corundum crusibles within two temperature ranges from 100 to 550°C and 550 to 1150°C. Within the first temperature range the decomposition of that salt with separation of ammonia, carbon dioxide and water vapour takes place. The very important fact here is, that the time for decomosition has to be sufficient for complete separation of the volative components from the salt, otherwise there exists a possibility for the secondary permutation among the reaction product (see reaction 5) with ammonium aluminate formation - NH_4AlO_2. This semi-product has a higher point of thermal decomposition and actually can affect the intermediate products of calcination. Within the second temperature range only polymorphous transformation of alumina takes place - from ammourphous phase through gamma-modification to the final alpha-modification. The product obtained possesses all chemical and physical properties of high-purity alumina.

In fig. 1 the flowsheet of the method is given .

Pilot plant investigation

The investigation on development of this method in a close-circuit productional steps have been carried out in a specially designed and constructed pilot plant unit with capacity of 4 tpy. The main equipment and apparatus assemblied in the different productional sections of this plant were as follows:

In the section for percolation leaching two percolation columns (dia 300 mm and height 8 m), assembled with glass cylindrical sections, heaters with internal serpantine worm-pipes, lids and valves and two PVDF percolation

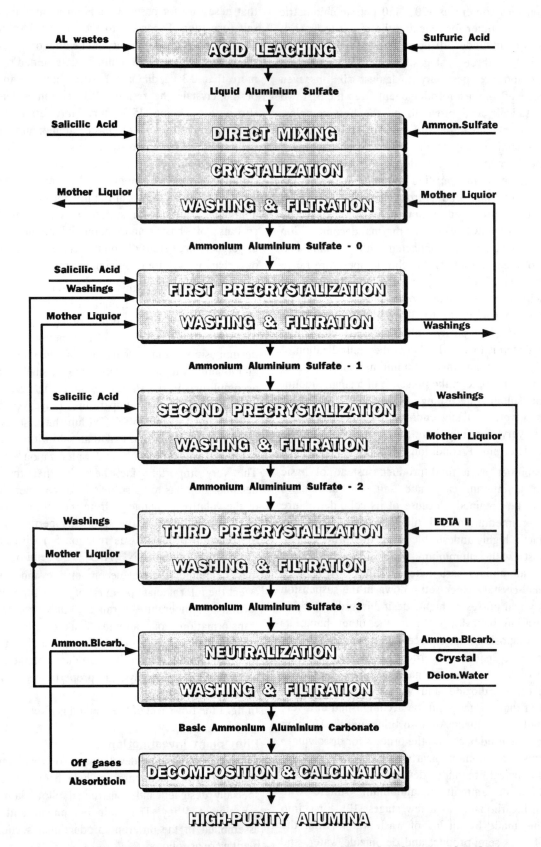

Fig.1. Flowsheet of the method for high purity alumina production

pumps (MDR 105V2 JOHNSON PUMN,Sweden) were fitted. There were also a horizontal vacuum-filter and four glass made vessels with total capacity of 0.8 m³ for filtration and transitional storaging of aluminium sulfate obtained.

The total productional rate of one run was 0.4 m³ per column.

In the section for ammonium aluminum sulfate formation a cylindrical glass vessel with capacity of 0.15 m³ for aluminium sulfate dosing , internaly epoxinated stainless steel reactor with steam-jacket for mixing of aluminium and ammonium sulfate (0.2 m³) , two heated with steam-jacket horizontal vacuum-filters, two internaly epoxinated stainless stell reactors for crystalization of the filtered hot solution (0.4 m³ each) and finaly one glass made horizontal vacuum-filter for filtration of the suspension containing ammonium aluminium sulfate crystals and mother liquior were fitted.

In the section for purification of ammonium aluminium sulfate salt three sets of apparatus for the first, second and third precrystalization were fitted.In any set there were glass made reactor with internal serpantine worm-pipe heating (0.1 m³) for dissolution of the crystals; two glass made reactors with external cooling-jacket (0.05 m³ each) for crystalization of the hot solution and one glass made horizontal vacuum-filter for filtration of the suspension containing purified crystals and mother liquior.

In the section for neutralization of the purified ammonium aluminium crystals there were four glass made reactors with external steam-jacket (0.05 m³) for neutralization and six glass made reactors with external steam-jacket (0.05 m³) for washing of basic ammonium aluminium salt obtained. The suspention was filtered on three glass made vacuum-filters.

All equipment and apparatus of the above mantioned sections were connected in accordance with technological flowsheet by glass and PVC made pipes, tubes and pumps.

The main equipment in the last technological section for thermal decomposition and calcination of the basic ammonium aluminium carbonate salt was an electrical furnace NABERTHERM W1000 with moving loading-car (1360 ºC -1 m³). This equipment was fitted with externaly water-cooled pipe-line and glass made (dia 300 mm and height 5 m) absorbtion column for off gases utilization.

For storaging and dosing of different mother liquiors and washing waters twenty four glass made vassels (0.15 m³ and 0,2 m³) were used, and for deionized water production there were fitted five glass made apparatus with total capacity of 0.24 m³/h (PCL-60 and PCL-30,KAVALIER,Chehoslovakia).In the unit there were also a sulfuric acid storage with dosing system, vassels for deionized water with dosing system and different ventilation systems and aspirations.

All glass made vessels, apparatus , reactors, pumps and pipes were disigned as a standart equipment by KAVALIER and GLASSEXPORT from Chehoslovakia.

In the pilot plant investigations the following raw material and chemical reagents were used:
- technical grade sulfuric acid (92 pct);
- deionized water;
- technical grade ammonium sulfate;
- chemicaly pure ammonium bicarbonate;
- complex-forming agent Salicilic Acid and EDTA II;
- flocculants TIOFLOC B31; MAGNAFLOC 342 and DRIMAX (produced by ALLIED COLLOIDS LTD,Bradford,England)

As a main raw material the aluminium containing cuttings (Si - 10.11 pct; Fe - 0.18 pct; Mg - 0.31 pct, Cu - 0.01 pct; Pb - 0.01 pct, Zn - 0.05 pct; Mn - 0.01 pct and Ti - 0.001 pct) were used.

All chemical analysis for the impurities content into the semi-products, mother liquiors, washing waters and final product were made using Perkin Elmer AAS PC 5100.

The pilot plant investigation have had as a main object determination of the impurities separation and their actual destribution within the different productional steps at the preliminary determined process parameters. Therefore, the results are commented only for the particular impurity level and its degree of separation in any productional step.

RESULTS AND DISCUSSIONS

Aluminium wastes acid leaching

In this productional step the following optimal process parameters were used:
- leaching temperature (max) - 85 ºC ;
- leacing time (total) - 260 min;
- acid concentration - inlet 220 g/dm³; outlet 7 g/dm³;
- alumina concentration in the semi-product - 73.9 g/dm³.

In table 1 the results for impurities content at the inlet and outlet of the process and degree of their separation in connection with created conditions for cementation and reduction phenomena are given.

The results shown in table I prove the statement that the percolation leaching of aluminium wastes with sulfuric acid within the limits of the optimal process parametes is a very effective way for impurities separation by means of cementation and reduction phenomena, especially for most of the heavy metals cations. The degree of K and Na ions separation however is extremely low in this process conditions and that have to explained with their higher acidic actifity than that of aluminium.

Aluminium and ammonium sulfates mixing

In this productional step the following process parameters were used :
- mixing temperature - max - 85ºC ;
- mixing time - max - 75 min;
- complex-forming agent - salicilis acid 125 pct to the impurity content ;

Table I. Impurities content and degree of separation

| Impurity | Content(ppm) | | Degree of |
	Inlet	Outlet	separation (pct)
- Si -	4808	96.16	98
- Fe -	87	26.10	70
- Ca -	0.0381	0.008	79
- Mg -	148.88	13.40	91
- K -	0.0053	0.004	24
- Na -	0.0052	0.004	26
- Cu -	0.0125	0.001	92
- Pb -	0.4667	0.014	97
- Zn -	24.75	0.99	96
- Cd -	0.02	0.001	96
- Mn -	0.02	0.001	95
- Ni -	0.0238	0.005	79
- Co -	0.0318	0.007	78
- Cr -	0.0028	0.001	65
Total :	5069.2562	136.35	96

- rate of crystalization - 1,45 to 1,55 °C/min;
- rate of crystal filtration - 0,5 to 0,3 m³.m²/sec;
- washing temperature - avarage 45°C;
- washing time - avarage 30 min;
- washings - mother liquior with avarage ammonium sulfate content 20 pct;
- solid to liquid ratio - from 1/2,5 to 1/3,5 ;
- ammonium aluminium sulfate salt obtained:
    avarage alumina content -   11,07 pct;
    avarage ammonium ions content - 3,7 pct.

In table 2 the results for impurities content at the inlet and outlet of the process and degree of their separation in connection with created conditions for ammonium aluminium sulfate salt formation and a complex- forming agent are given.

The results shown in table II indicate that within the limits of the optimal process conditions for ammonium aluminium sulfate salt formation the implementation of salicilic acid as a complex-forming agent leads to an effective separation of most of the impurities presented into the inlet semi-products. As a matter of fact however, at this stage of the process the salicilic acid does not has a noticeable effect on K and Na cations separation.

Ammonium aluminium sulfate salt purification

In this productional step the following optimal process parameters (for 1st, 2nd and 3rd precrystalization) were used:
- temperature for direct dissolvatation - max 85°C;
- time for direct dissolvatation - max 30 min;
- complex-forming agents:

Table II. Impurities content and degree of separation

| Impurity | Content(ppm) | | Degree of |
	Inlet	Outlet	separation(pct)
- Si -	173,16	32,90	81
- Fe -	47,05	9,41	80
- Ca -	0,1351	0,05	63
- Mg -	24,15	4,83	80
- K -	0.0051	0,004	21
- Na -	0,0051	0,004	20
- Cu -	0,16	0,04	75
- Pb -	0,163	0,05	69
- Zn -	4,5	0,99	78
- Cd -	0,1	0,01	90
- Mn -	0,417	0,1	76
- Ni -	0,192	0,05	74
- Co -	0,0921	0,07	14
- Cr -	0,3334	0,07	79
- Ti -	0,2857	0,1	65
Total	250,7485	48,678	80,6

1st and 2nd precrystalization - salicilic acid 120 pct to the impurities content;
3rd precrystalization - EDTA 125 pct to the impurities content.
- rate of crystalization - 1,35 to 1,45 °C/min;
- rate of crystals filtration - 0,4 to 0,25 m³.m²/sec
- washings - mother liquiors with avarage ammonium sulfate content 20 pct;
- washing temperature - avarage 40°C;
- washing time - avarage 30 min;
- ammonium aluminium salts obatained:
  after 1st precrystalization -
      avarage alumina content -   10,85 pct;
      avarage ammonium ions content - 3,73 pct
  after 2nd precrystalization -
      avarage alumina content -   10,66 pct;
      avarage ammonium ions content - 3,44 pct;
  after 3rd precrystalization -
      avarage alumina content -   10,73 pct;
      avarage ammonium ions content - 3,63 pc

The results shown in table III indicate that within the limits of the optimal process conditions for ammonium aluminium salt purification the type of complex-forming agent has a great influence over the degree of impurities separation. Salicilic acid prove to be a very effective complex-forming agent for Fe, Ca, Mg, Cu, Pb, Zn, Cd, Mn, Ni, Co, Cr and Ti components. It does not has noticeable effect however, over the degree of K, Na and Si

Table III. Impurities content and degree of separation in precrystalization steps

		1st Precrystal.		2nd Precrystal.		3rd Precrystal		
Impurity	Inlet (ppm)	Outlet (ppm)	Degree of Sep.(pct)	Outlet (ppm)	Degree of Sep.(pct)	Outlet (ppm)	Degree of Sep.(pct)	Total
- Si	32,90	6.91	79.0	3.11	60,0	2,11	13,6	93,6
- Fe	9,41	1,50	85,8	0,04	99,5	0,04	0,0	
- Ca	0,05	0,02	84,3	0,02	0,0	0,02	0,0	
- Mg	4,83	0,003	99,9	0,003	0,0	0,003	0,0	
- K	0,004	8,0	+170	2,89	67,9	1,11	61,6	86,1
- Na	0,005	10,50	+182	2,41	79,6	1,04	56,9	90,1
- Cu	0,04	0,025	44,3	0,025	0,0	0,025	0,0	
- Pb	0,05	5,0	+891	0,06	98,9	0,06	0,0	
- Zn	0,99	0,95	10,0	0,008	98,3	0,008	0,0	
- Cd	0,01	0,009	10,0	0,009	0,0	0,009	0,0	
- Mn	0,1	0,02	82,2	0,02	0,0	0,02	0,0	
- Ni	0,05	0,04	28,7	0,04	0,0	0,04	0,0	
- Co	0,07	0,05	36,3	0,05	0,0	0,05	0,0	
- Cr	0,07	0,05	36,3	0,05	0,0	0,05	0,0	
- Ti	0,1	0,01	91,1	0,01	0,0	0,01	0,0	

Total: 48,369 34,08 8,74 4,587

Avarage degree of total impuritie separation in all precrystalizations = 90,52 pct

components separation and for that purpose EDTA II has to be used.

The reached after third step of precrystalization impurities level in the purified ammonium aluminium sulfate salt of less than 5 ppm is more than it is required to further processing of this semi-product in the following productional steps.

Ammonium aluminium sulfate salt neutralization

In this productional step the following optimal process parameters were used:
- temperature for neutralization - max 45°C;
- time for neutralization - max 100 min;
- liquid to solid ratio - from 2,5/1 to 3/1 ;
- time for suspension ageing - max 30 min;
- temperature for basic salt washing and sedimentation - max 60°C ;
- time for basic salt washing - max 15 min ;
- time for basic salt sedimentation :

    with MAGNAFLOC 342   - 15 min;

    without MAGNAFLOC 342   -125 min;
- solid to liquid ratio - from 1/2,5 to 1/3,5 ;
- way of washing - counter-current, with deionized water at the last step and by intermediate steps for sedimentation and washing with wasing waters and final vacuum filtration;

- basic ammonium aluminium carbonate salt obtained:

    avarage moisture -   56,3 pct;

    avarage alumina content -   37,6 pct;

    avarage ammonium ions content - 14,9 pct.

    avarage particle size (D50) -   1,0 micron.

In table 4 the results for impurities content at the inlet and outlet of this productional step is given.

The results shown in table IV indicate that impurities content in the basic ammonium aluminium carbonate salt obtained by neutralization of ammonium aluminium sulfate salt with ammonium bicarbonate are less than that in the washings separated from that process. That significant fact is one of the advantages of this semi-product,which have to be connected with its very fine, but very cristaline nature. Actually, this means that any type of equipment for washing and sedementation of the basic ammonium aluminium carbonate salt can be used in a very sophisticated way.

Thermal decomposition and calcination

In this productional step for thertmal decomposition and calcination of basic ammonium aluminium carbonate salt to a high-purity alumina the following optimal process parameters were used:

Table IV. Impurities content at the neutralization step

Impurity (ppm)	Total at inlet	In the basic salt	In the washings
- Si -	7,39	3,11	4,28
- Fe -	0,14	0,04	0,10
- Ca -	0,07	0,02	0,05
- Mg -	0,01	0,001	0,009
- K -	3,89	1,22	2,67
- Na -	3,65	3,05	0,50
- Cu -	0,088	0,025	0,063
- Pb -	0,21	0,06	0,15
- Zn -	0,028	0,008	0,02
- Cd -	0,031	0,009	0,022
- Mn -	0,07	0,02	0,05
- Ni -	0,14	0,04	0,10
- Co -	0,17	0,05	0,12
- Cr -	0,17	0,05	0,12
- Ti -	0,035	0,01	0,025
Total	16,092	7,713	8,379

- temperature and time for decomposition :
  100 C ---------- 120 min'
  300 C ---------- 120 min;
  500 C ---------- 120 min; sub-total time - 360 min.
- temperature and time for calcination :
  900 C ---------- 15 min;
  1000 C ---------- 30 min;
  1150 C ---------- 120 min; sub-total time - 165 min.
- total time for thermal treatment - 525 min;

The chemcal composition of the separated during the step for thermal decomposition gaseous phase, directed toward absorbtion column, indicate that the separation of the physicaly and chemicaly combined into a basic ammonium aluminium carbonate salt water starts at 95 C and finish at around 150 C, and that of carbon dioxide starts at 90 C and finish at around 170 C. The separation of ammonia however, starts at around 110 C and finish at the temperatures over 450 C. Having in mind the chemical composition of this salt it can be suggested that during thermal decomposition some secondary phase permutation are implemented and a new metastable intermediate phase are formed. Most probably that may be the case of the secondary ammonium aluminate formation [NH_4AlO_2].

The chemical composition and some physical properties of the final product produced under statet above optimal process conditions are as follows:
- alpha alumina content - more than 90 pct;
- moisture (L.O.I.) - less than 0,5 pct;
- single crystals size destribution : D25 - 0,8 microns; D50 - 1,1 microns; D100 - 1,6 microns.
- Impurities content, less than (in ppm):
  - Si - 8,28   - Na - 8,11   - Mn - 0,05
  - Fe - 0,11   - Cu - 0,07   - Ni - 0,1
  - Ca - 0,05   - Pb - 0,15   - Co - 0,1
  - Mg -0,005   - Zn - 0,02   - Cr - 0,1
  - K - 3,5   - Cd - 0,02   - Ti - 0,1
- Total impurities content , less than - 25 ppm;
- Total alumina content (dry basis), more than - 99,99 pct.

The product by this newly developed acid method possesses all required properties to be qualificated as a high-purity alumina and can be used successfuly in all fields of its applications.

CONCLUSION

1. This newly developed acid method for producing high-purity alumina by a new approach for impurities separation and semi-products obtaining has a remarcable priorities over the known untill now acid methods used for that purpose.The impurities separation starts already in the first productional step - percolation leaching of aluminium wastes with sulfuric acid, using the created process conditions for implementation of cementation and reduction phenomena and proceeds in the following step for ammonium aluminium sulfate salt formation and purification by means of precrystalization with complex-forming agents.

2. The results obtained during the pilot plant investigation show that:
- by percolation leaching more than 80 pct of the impurities presented into the inlet acid solution and dissoluable aluminum can be separated; and, in the followed three stages counter-current precrystalization of the ammonium aluminium sulfate salt, using as a complex-forming agents Salicilic acid (for 1st and 2nd stage) and EDTA II (for 3rd stage), more than 85 pct of the inlet impurities can be separated too, and so, to obtain semi-product with final total impurities content less than 5 ppm;
- the neutralization of the purified ammonium aluminium sulfate salt with ammonium bicarbonate for obtaining a basic ammonium aluminium carbonate salt has a notable effect on the method as whole, because its higher alumina content (aroun 37 pct) than that of ammonium aluminium sulfate salt (around 10,9 pct) and lower

temperature for thermal decomposition and calcination, is a fact leading to a great energy savings and sophisticated off gases utilization.

3.The physical properties and impurities content of the final product prove that this newly developed method can be used successfuly for producing a high-purity alumina.

ACKNOWLEDGEMENT

The results of the present paper are based on the prolonged investigations carried out in the pilot plant unit designed and constructed in KURILLO METALS, on streat from the end of 1990 and now in experimental operation.

The author wants to extend his sincere gratitude to the scientists, engineers and technicians of Light Metals Div.,of KURILLO METALS for the help in the experimental work and to Dr,Z.Sartowski and Dr.J.P.Bublic from Allied Colloids Ltd for their kind and helpful assistance in the experiments on sedimentation and filtration of the semi-products suspensions using ACL MAGNAFLOC and TIOFLOC flocullants.

References
1.Bes de Berg O., Rev.L,Aluminium, 9, 1, 1975,397.
2. Obzornaya informacia dlya polucenia spec glinozemov., Obz.Inf.MINF,Moskva.,1986.
3.Tanev P., Diis.Doc.Deg. Inst.Chem.Techn.,Burgas,1984
4. Moriwake T., Jap.Light Metals Assos.,14,2,1984,67
5. Kirbi R.M., US Pat.Off., 4,216,010, 1979.
6. Kanev R.D. et all., Proc.Inst.Non-Ferr.Met.,Plovdiv,1987.
7. Kanev R.D., Diss.Doc.Deg., Techn.Univ.,Sofia,1986,66.
8. Kanev R.D.,et all., Metallurgia,Sofia,5,1990,6.
9. Kanev R.D., Proc.IMM Conf.,Birmingham,5,1990,111.
10.Kanev R.D. et.all., Bulg.Pat.Off.,88224,1989.

Hazelett twin-belt aluminium strip-casting process: caster design and current product programme of aluminium alloy sheet

Peter C. Regan
Wojtek Szczypiorski
Hazelett Strip-Casting Corporation, Colchester, Vermont, USA

SYNOPSIS

The Hazelett caster is extensively used in the European nonferrous metals industries producing zinc sheet, copper wire rod and copper anode. Its application in the aluminum industry, however, has been in North America and Japan and, more recently, Venezuela, where its proven capability of a high product volume at a relatively low investment cost has been exploited for converting relatively large volumes of scrap or primary smelter metal into hot rolled strip.

Recent improvements in the caster process technology have expanded the alloy production capability and increased the surface quality of the strip. This will allow the implementation of the process in markets such as Europe, where the high volume sheet consumption is concentrated in the more demanding applications. This comes at a time when the capital cost of the conventional DC hot rolling system becomes increasingly difficult to justify.

We will review the history and operation of the Hazelett casting machine, as well as the metallurgical aspects and the downstream processing of the strip.

HISTORY OF DEVELOPMENT

The history of the Hazelett twin-belt caster dates back to 1919. Clarence W. Hazelett, one of the true pioneers of continuous casting, worked with a variety of continuous casting techniques including a single roll, two rolls, and a roll and a ring. He eventually concluded that parallel, endless, moving mold surfaces were desirable. His twin-belt concept first came into commercial production in the 1960s and the work has been continued by his sons. Today, the Hazelett Strip-Casting Corporation remains a family owned, private company employing 110 people dedicated to the design and manufacture of the twin-belt caster.

CASTER DESCRIPTION

A fully assembled 1320mm wide Hazelett Model 23 caster is shown in Fig. 1. The mold consists of an upper and lower belt, with an endless chain of edge dam blocks forming each side. (Fig. 2)

Fig. 1 Hazelett Model 23 Caster

The steel belts are wrapped around grooved pulleys. The upstream pulley contains both belt preheating and water cooling functions. Preheating is employed to pre-expand the belt before entering the hot mold region, thus reducing the thermal distortion in the casting zone, wherein the water cooling is immediately applied. The downstream pulley tensions and drives the belt to attain optimal flatness and steers the belt via axial rotation to prevent it from running sideways. The upper carriage can be hydraulically raised to permit mold maintenance and belt changing.

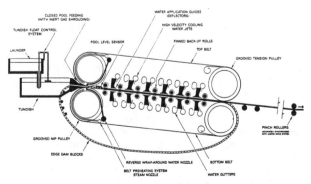

Fig. 2 Hazelett Caster Schematic

Thickness of the cast strip is established by interchangeable carriage spacers and can vary between 12mm and 75mm. Most aluminum strip is cast at a thickness of 19mm.

Casting width is governed by the lateral spacing of the side dam blocks and can vary up to the maximum design capacity of any particular machine. Current designs allow a maximum casting width of up to 2 metres.

For aluminum, the casting angle is set at 6°; a mold length of 1900mm is employed.

BELT COOLING SYSTEM

Mold heat extraction is by fast film water cooling on the back side of the belts, repeated at intervals (Figs. 2 and 3). The unique, high efficiency cooling maintains belt temperatures below 110°C on the casting side and 80°C on the water side.

Belts are supported by grooved back-up rolls which allow full coverage by the cooling water over the full mold area. These can be adjusted to shape the mold.

Exiting water is entrapped and returned to the recirculation pit while an exhaust system controls the humidity in the mold area prior to the casting zone, to prevent condensation on the belts.

Casting belts are low carbon, cold rolled steel strip 1.2mm thick. Side dams are low carbon steel. Belts and side dams are easily replaceable and can be changed in 20 minutes, using specially designed fixtures.

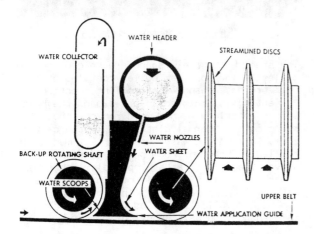

Fig. 3 Belt Cooling & Support

The belts are coated with a ceramic/metallic matrix which is applied by flame spraying. The ceramic and metal compounds used in the Matrix coating are selected for the particular aluminum alloy being cast to provide the desired heat transfer and surface parting characteristics. The coating is permanent and lasts as long as the belt is in use. Casting belts are normally used for one or two weeks, depending on the operator's production program.

METAL FEEDING

In operation, molten metal is "injected" into the mold cavity by a tundish and nozzle (Fig. 4). Inert gas shrouding is applied in the slight gap between the nozzle and the belt. This acts both to preclude oxygen and control heat transfer at the interface between the solidifying metal and the belt.

A starter bar is placed across the mold, and as the pool reaches the desired height, the caster drive is started so that mold speeds match metal entry rate. Casting runs of 24 hours are common; maximum runs exceed 70 hours.

During operation, the operator uses the PLI (Pool Level Indication System) to monitor the level of molten metal inside the mold. Belt temperatures are recorded (Fig.4), showing the operator the metal pool location on the INTERACQ display screen (Fig.6). Normally, the level is kept up close to the tip of the nozzle.

Shutdown of the caster is simple and changeovers of cast width can be accomplished by changing the tundish.

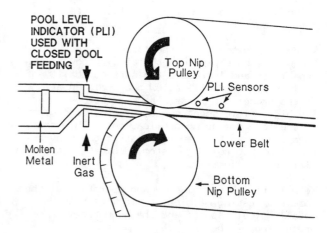

Fig. 4 Metal Feeding System

One or more holding furnaces receive metal either from melting furnaces or, preferably, from ladles directly transported from a smelter. In-line degassing and filtering is accomplished just prior to the tundish (Fig. 5).

CASTING AND ROLLING LINE

A pinch roll isolates the caster from any downstream disturbances in the in-line rolling mills. The speed of the pinch roll must be accurately synchronized with the casting machine, with allowance made for the shrinkage of the cooling strip. Speed differential and torque is monitored and displayed (Fig. 7).

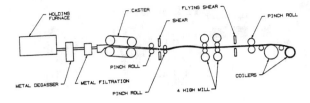

Fig. 5 Casting & Rolling Line

The 19mm thick cast strip is hot rolled immediately in line. Reductions of up to 70% per stand are made in the rolling mill, which can include 1, 2 or even 3 stands in tandem (Fig. 5). This allows the line to produce a hot band or reroll gauge as low as 1mm, thus drastically reducing the amount of comparatively expensive cold rolling and possible thermal treatments. Table I lists the current Hazelett aluminum casting installations and various rolling mill combinations.

TABLE I

HAZELETT ALUMINUM CASTING AND ROLLING INSTALLATIONS

Date	Company	Location	Width (mm)	Rolling Mills
1963	Alcan	Canada	660	1 Std - 2 High
1970	Nihon	Japan	300	2 Std - 2 High
1979	Barmet	U.S.A.	711	2 Std - 4 High
1983	Barmet	U.S.A.	356	3 Std - 2 High
1984	Nihon	Japan	450	2 Std - 2 High
1985	Advanced	U.S.A.	762	2 Std - 4 High
1986	Barmet	U.S.A.	1320	3 Std - 4 High
1987	Vulcan	U.S.A.	1320	1 Std - 4 High
1988	Pivensa	Venezuela	1040	2 Std - 2 High
(1991)	Nichols	U.S.A.	1320	3 Std - 4 High

One or two coilers are generally employed with the rolling mill to allow continuous production of strip. Casting production rates are relatively high -- 20 kilograms per hour per millimetre of cast width -- a linear casting speed of up to 10 metres/minute. Pure aluminum casts somewhat faster, while complex alloys, slower.

Because of the relatively low entry speed, the hot rolling mills can be much simpler than today's high speed mills. Frequently, refurbished rolling stands previously used in antiquated steel or nonferrous strip rolling operations, can be employed with excellent results. All installations in the U.S. employ such mills.

PRODUCTION

The economics of twin-belt casting and rolling of aluminum strip are favorable. The annual production rate of a 1 metre wide installation can be 100,000 TPY, or more. By employing refurbished rolling equipment, the capital cost of such an installation is quite low, when compared to a similar production facility employing either numerous 2-roll casters or a conventional reversing hot mill with finishing train.

Operating cost is likewise low. Normally a total of four operators for the caster and rolling mill. The plant can be economically operated on only one shift. As production requirements increase, additional shifts can be added without any additional capital cost.

Owing to the 1900mm mold length, alloys with long freezing ranges such as 3% or more magnesium can be cast. Table II lists products currently produced and/or under development. However, these complex aluminum alloys require very stable heat transfer conditions in the mold, and careful control of the process parameters using extensive data acquisition and process control techniques.

TABLE II

ALUMINUM ALLOYS CAST ON HAZELETT TWIN-BELT CASTER

Alloy	End Products Cast Commercially
1050	REFRIGERATION FIN STOCK, IMPACT EXTRUSION SLUGS
1060	IMPACT EXTRUSION SLUGS
1070	IMPACT EXTRUSION SLUGS
1100	LICENSE PLATES, HEAVY FOIL, ROLL BOND FOR REFRIGERATORS
1145	
1170	HEAVY FOIL
1200	THIN FOIL
1350	ELECTRICAL CONDUCTOR APPLICATIONS
3003	SIDING, UTENSILS, HEAVY FOIL
3004	FURNITURE TUBING, FLEXIBLE ELECTRICAL CONDUIT, BRIGHT SHEET
3105	RAIN GUTTER, WINDOW SASH, FLEXIBLE CONDUIT, LICENSE PLATES, BRIGHT SHEET

Alloy	End Products Cast Commercially
3015*	TOOLING PLATE
3107	UTILITY SHEET
3207	
5042	BEVERAGE CAN TABS
5052	UTILITY SHEET
5082	BEVERAGE CAN TABS
5454	BEVERAGE CAN TABS
7072	RADIATOR FIN STOCK

	Cast In Test Series
2024	REDRAW ROD (FASTENERS)
3004	BEVERAGE CAN BODIES
5056	REDRAW ROD (SCREENING)
5182	BEVERAGE CAN LIDS
6061	AUTOMOBILE WHEELS
7072	TOOLING PLATE

* MODIFIED WITH 4.3% ZINC

OPERATING VARIABLES AND PROCESS CONTROL

Process variables which must be monitored and controlled include molten metal level and temperature in the tundish, cooling water temperature and pressure, belt temperature and profile, exit slab temperature and shape. Furthermore, the speed relationship between the caster and pinch roll, or the "draw," compensating for longitudinal shrinkage of the moving slab is quite critical. To maintain the neutral point of the slab properly within the caster mold, the slab must be kept slightly in compression by monitoring the torque on the drive components. These functions are monitored by the INTERACQ system which is the operators Interface/Data Acquisition System using personal computers, networks and high powered software.

Real time data display and storage is accomplished with the INTERACQ system using approximately 20 to 30 channels of data. Screen displays are custom generated, with their number in the range of hundreds. (Figs. 6 and 7) Functions such as belt temperature and flatness information and belt steering system data can be displayed to the operator in a fashion that enables rapid interpretation. This information is very important for maintaining uniform heat extraction across the belt width which, in turn, will result in a good, uniform liquation -free surface. The acquired data may be played back, off line, to allow study and analysis of patterns and correlation among the various signals.

Hardware employed to gather the data include displacement probes, which are electronic

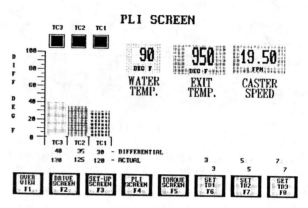

Fig. 7 INTERACQ Display
Metal Pool Control

METALLURGICAL ASPECTS

Hazelett aluminum casters are virtually always installed as part of a continuous melting, casting and rolling line. In addition to the furnaces, caster and rolling mills, we have in-line degassers, filtration and grain refiner equipment. The quality of the product will depend on the quality of operation in each of these areas.

The Hazelett process can be included in the group of high solidification rate processes. From the solidification point of view, it is similar to that of the Alusuisse Caster II (block caster), faster than the conventional DC (direct chill) process and slower than the twin roll TYPE caster. The specific heat removal rate determines metallurgical and mechanical characteristics of alloys cast on the Hazelett caster. Solidification rates can be increased or decreased to a certain extent by using a combination of Matrix^{TM} coatings and inert gas shrouding.

When comparing Hazelett-cast and DC-cast aluminum alloys, the differences will become quite obvious.

A. Grain Structure

Higher solidification rates achieved on the Hazelett caster will result in smaller grain size, finer dendritic cells, smaller size of secondary phase particles and their different morphology as well as higher retention of certain elements in solid solution. This will also ensure the necessary surface quality for very demanding applications such as can stock. However, in comparison with twin roll cast material, the Hazelett-cast strip has bigger grains, coarser dendritic cells, larger size secondary particles and less retention of elements in solid solution.

The high solidification rates ensure fine, equiaxed and randomly textured grains throughout

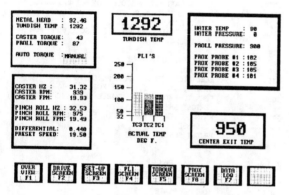

Fig. 6 INTERACQ Display
Caster Parameters

devices capable of detecting belt movements or displacement of other machine components. Sensitivity is 0.25mm with a range of 4mm. The probes do not contact the belt and do not seriously interfere with cooling water.

Belt temperature is sensed by thermocouple-based units which sense the temperature of the water side of the casting belts. They are arranged at various locations, in particular, close to the point where the molten metal first contacts the belt, as shown in Fig. 4.

The caster/pinch roll relationship is monitored by both the actual rotational speed of both components, as well as with industrial quality rotating transformer-coupled strain gauge transducers installed between the caster and pinch roll drive motors and their respective gear systems to provide information on both relative speed and torque.

the whole slab section. To further improve grain structures, grain refiners such as Al-Ti-B (5:1 TiB) alloy in the shape of a rod, is fed continuously to the melt at the location guaranteeing its highest efficiency.

Table III represents some typical grain size measurements for common aluminum alloys cast on the Hazelett caster.

TABLE III

ALLOY	GRAIN SIZE (μm)	
	Surface	Center
AA 1100	150	300
AA 3003	100	180
AA 3004	80	150
AA 3105	130	250
AA 5052	120	200

The cast strip has a grain size gradient between the surface and the center. This is inherent in the process and the size will largely depend on solidification rate and slab thickness.

Heat extraction can be altered by using different combinations of Matrix coatings and inert gas shrouding. A combination of coatings, shrouding and certain aspects of the caster's mechanical features gives quite broad flexibility in adjusting the solidification rates. This is very important when casting different alloys. Some other factors include belt topography, metal head in the tundish and metal temperature.

In general, properties of Hazelett cast strip improve with increasing cooling rates which result in finer and more homogeneous structure, especially for higher alloys.

B. Solid Solution

During the fast cooling process of the Hazelett caster, certain elements such as manganese and magnesium will be retained in solid solution and remain there during hot rolling and, as a result, substantial strengthening is achieved. The DC cast ingot, on the other hand, is homogenized before hot rolling, thereby precipitating Mn as dispersoid.

The solute promotes strain hardening during deformation and has a strong influence on final mechanical properties of the sheet.

The two chemistries in Table IV will result in quite different mechanical properties after following identical hot/cold rolling and annealing practices. The only difference is in chemistry, specifically the Mn content.

TABLE IV

Chemistry of Hot Rolled Strip

CHEMISTRY	Mn	Mg	Cu	Fe	Si	Zn	Cr	Ti	Ni	Pb
I	.31	.26	.13	.49	.17	.030	.008	.013	.005	.007
II	.66	.25	.15	.59	.23	.042	.015	.021	.006	.007

The work hardening curves for the samples of both chemistries are shown on Fig. 8.

As can be seen, Mn content has a significant effect on mechanical properties of Hazelett cast AA3105.

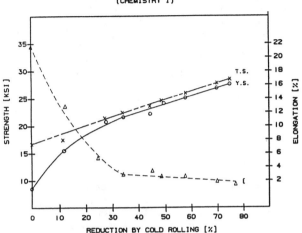

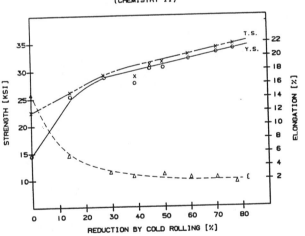

Fig. 8 Work Hardening Curves

C. Hot and Cold Rolling

A heavy reduction of 60 to 70% is usually taken in the first in-line mill stand, with reductions of at least 50% taken in subsequent stands. The 3-stand configuration can reduce strip thickness from 19mm to 1.5mm or less in certain cases.

The exit temperatures in the mill depend largely on casting speeds, mill configuration, cooling techniques, etc. In general, this temperature is not higher than 315°F. Higher temperatures can sometimes be achieved but the strip remains well below the recrystallization temperature. This factor, along with the presence of solute, as mentioned above, means that the Hazelett hot rolled strip will have higher mechanical properties than that rolled from DC ingot of the same chemistry and thickness. For example, in many cases a lower level

of Mn and Mg will be sufficient to match the mechanical properties of the DC material with higher Mn and/or Mg content.

Most products require additional cold rolling. In some cases, the Hazelett cast and hot rolled strip is cold rolled directly to the final gauge; in others, intermediate annealing is introduced. The cold rolled structure is more homogeneous than hot rolled, with finer constituents more evenly distributed. If intermediate annealing is introduced, it is the first high temperature treatment in the whole process. It activates precipitation of manganese dispersoid from the solid solution.

When compared to DC annealing temperatures, the Hazelett strip will require higher temperatures, owing to the differences in the metallurgical structure.

SUMMARY

The Hazelett twin-belt caster creates an opportunity for high capacity production of aluminum sheet at a relatively low capital cost. Recent advances in process control and mold technology have expanded the capability of the process to make it a viable production tool for the expanding European aluminum industries.

Surface engineering of aluminium and its alloys

H. Terryn
J. Vereecken
Department of Metallurgy, Electrochemistry and Materials Science, Vrije Universiteit Brussel, Brussels, Belgium

SYNOPSIS

The increase of the applications of aluminium and its alloys is accompanied by the possibilities of using surface treatments to create specific surface properties. These surface treatments are used to increase the corrosion or wear resistance, or to create special electrical, optical, hydrophilic or adhesion properties

The specific surface characteristics are related to the morphology, structure, composition and even charge of the modified Al surface.

The success of a specific treatment is strongly dependent on the precise understanding of the surface behaviour of the used Al alloy during the treatment.

In this article three types of treatment are considered : AC electrolytic graining, porous anodizing and sealing.

INTRODUCTION

Anodizing and allied processes have been studied widely because of the commercial importance of the formed anodic oxide film, which can give the Al-surface specific properties, for example : a high corrosion resistance, a decorative aspect in architectural applications, adhesive, electrical or optical properties.

Anodizing is defined as an electrochemical treatment in which stable oxide films are formed on the surface of metals[1]. Anodic coatings can be formed on aluminium employing a variety of electrolytes using either AC or DC current or a combination of both.

Anodic coatings are classified according to the dissolving action of the particular electrolyte on the anodic oxide produced in the reaction. Anodic oxide films produced in sulphuric, phosphoric of chromic acid electrolytes are of the porous type[1].

Anodic oxide films produced in electrolytes such as ammonium tartrate, boric acid, ammonium citrate, etc., have little or no capacity to dissolve the oxide. The films are relatively thin (less than 1 μm) and essentially non-porous[1].

Anodizing is usually combined with pre and post treatments in order to obtain the required Al-surface. The type of pre and post treatments is dependent on the application of the anodized Al-surface. It is very important to understand the interaction between the various treatments that are imposed on the Al-surface.

The variety of pre and post treatments that can be imposed is rather large due to the high number of applications. The pretreatments can be classified as mechanical, chemical and electrochemical, commonly used to grain, etch, polish, ..., the surface prior to anodizing[1].

After anodic oxidation, the porous oxide layer is usually sealed to improve the corrosion resistance. When the porous films are heated in water, steam or a salt solution, the pores are closed due to the reaction between the oxide and the sealing medium.

The success of the treatment is related to the precise understanding of the surface behaviour of the selected Al alloy during the treatment, which requires the understanding of interaction between the material input (type of Al-alloy, mechanical and thermal pretreatment) and the parameters of the treatment.

In this article three important surface treatments are discussed, namely AC electrolytic graining of Al which is used as a pretreatments to roughen the Al-substrate, the porous anodizing process itself and its sealing.

The complementary information deduced from electrochemical and surface analytical techniques was used to characterize these treatments in detail[1].

Table I gives an overview of the used methods with their respective characteristics. The obtained results are discussed more in details in a number of publications indicated in table I.

473

Method	Information	References
Electrochemical method	Kinetic parameters of the electrochemical reactions	1,4-17,20
Electrochemical impedance	Thickness of the anodic coating The increase in specific surface area Hydration of the oxide layer Corrosion resistance	7,8,11,13,15,16,21
Roughness measurements	Roughness	1,9,17
Adsorption measurements	Specific surface area porosity	1,9
Transmission Electron Microscopy	Morphology Structure	1,4-6,9,10,20 12,14,17-19
Auger Electron Spectroscopy	Composition of the surface Composition of the oxide coating	2,3
Spectroscopic Ellipsometry	Roughness Composition Oxide layer thickness Hydration	18

Table I

STUDY OF AC ELECTROLYTIC GRAINING OF Al
1,12,17,18.

AC electrolytic graining of Al involves the application of an alternating voltage or current between aluminium electrodes in a suitable electrolyte, e.g. hydrochloric acid or nitric acid. The object of this industrial treatment is to develop a uniformly pitted and convoluted surface topography, which improves the water retentive properties of aluminium offset plates for lithography. It is used as a pretreatment prior to anodizing.

The depth and distribution of the pits are dependent upon particular electrical conditions (applied current density, graining frequency, form of the AC signal) and the electrolyte parameters (composition of the electrolyte, pH, temperature, ...).

In order to understand how the morphology of the grained surface is formed and how the treatment parameters act, it was necessary to determine the AC electrolytic graining mechanism.

The obtained morphology after AC electrolytic graining at 50 Hz in a hydrochloric acid solution is shown in the SEM micrograph of Fig.1. The surface roughness is created by the presence of a large density of hemispherical pits, having no characteristic dimension. Beside the formation of these pits, a thick friable etch film is present, represented in Fig.2, which has to be removed prior to successive treatments.

Through variation of the electrograining frequency in a region from 0.1 to 1000 Hz and by variation of the anodic and cathodic charges of the AC cycle, it was evident that the hemispherical pits propagate through the development of a high population of cubic pits, formed by the (100) faces of the FCC lattice. These pits are formed by the anodic charge. An example of these pits is shown in the TEM micrograph of Fig.3. The size of the cubic pits can be related to four important grain parameters :
1) the number of pits, initiated each time the positive cycle is passed;
2) the amplitude of the sinoidal current;
3) its frequency;
4) the anodic current efficiency in the process.

It is especially on the first point that the electrode pretreatment (selected Al-alloy, thermal and mechanical pretreatment) acts.

It was also proved that during the negative half-cycle the non-uniform etch film is developed over the aluminium surface (Fig.2). By examination of the etch film development as a function of the AC wave form and frequency, it was obvious that the film develops during the negative half cycle only when sufficient alkanity occurs. This can be explained by the hydrogen evolution at negative polarization, causing a significant increase of the pH.

From the determination of the mechanism of the AC electrolytic graining process it is now possible to understand the impact of following parameters :

. current density of the applied signal;

. frequency of the applied signal;

. form of the applied signal;

. electrolyte condition;

. pH of the electrolyte;

. temperature of the electrolyte;

. influence of the Al preconditioning.

Study of the Porous Anodic Oxidation of Al.

(TEM) Transmission electrode microscopy studies of the growth of porous layers in sulphuric and phosphoric acid permitted description of the morphology. A more detailed reviewing can be found in [1,14,19]. In these electrolytes porous oxide films are developed, having a barrier layer in contact with the metal and a porous film above it, usually thicker than the barrier layer. The pores appear as cappilaries through the anodic film. A schematic representation of this porous anodic film on rolled Al is given in Fig.4.

Porous anodic films may be formed under constant current density (or constant voltage) and the voltage (current) time behaviour is characteristic of the way the oxide film grows. Such a voltage time transient is given in Fig.5.

Oxidation under constant current density shows an approximately linear increase in voltage from the beginning.

The growth of a porous oxide layer on aluminium involves a competition between oxide formation and dissolution. Initially a compact barrier layer is formed by ionic migration. Fig.6 shows a cross-sectional TEM view of such a barrier layer. It is believed that Al^{3+} and O^{2-}/OH^- ions are both mobile in the field. The extend of solid film formation at the electrolyte interface depends critically upon the anodizing current density, pH and electrolyte composition. A non porous oxide continues to grow for as long as ionic current continues to flow. The thickness of the barrier layer depends upon the voltage drop across the oxide and is usually characterized by the anodizing ration, expressed in nm/V.

After a time the linear relation ceases and the value increases slowly to a maximum. This behaviour reflects the onset of non-uniform film growth and pore development. After reaching a maximum the voltage declines to a steady-state accompanying the formation of a regular porous anodic film which thickens according to Faraday's law.

Non uniform growth appears in aggressive electrolytes, due to dissolution effects, caused by field assisted dissolution and Joule heating, which results in the formation of pores. Thereafter regular porous anodic film growth occurs and the oxide film thickness in proportion to anodic oxidation time while the barrier layer is maintained at a constant thickness beneath each pore. This regular porous oxide film is shown in the TEM view of Fig.6. The oxide layer formed on flat electropolished surfaces, has a very regular structure. Cell, pore and barrier layer dimensions depend on the anodizing conditions and have different anodizing ratios (nm/V).

The shape of the voltage-time transient described is retained over a wide range of anodizing conditions. The rate of the voltage rise and the steady-state voltage are influenced by the applied current density and decrease according to the relative aggressiveness of the acid used.

The composition and roughness of the aluminium substrate may influence the anodic oxide film formation. A significant part of the literature deals with growth of porous layers upon electropolished surfaces[1], and little information is available about the role of the composition and topography of the aluminium surface. However this is of industrial importance where the oxide film is often formed on extruded profiles and on rolled or even grained surfaces[1].

The growth of porous oxide layers on relatively rough surfaces, such as rolled and AC electrograined aluminium has been studied.

Therefore electrochemical measurements [1,4,5,6,10,14,18, 21] have been recorded during the growth of the duplex oxide layer and combined with high voltage electron microscopy. The study leads to following conclusions. The growth of the porous layer on rolled and AC electrolytic grained aluminium is similar to that of the porous oxide films formed on electropolished flat surfaces. Steady state porous cell dimensions, barrier layer thickness, pore and cell diameter are proportional to he anodizing voltage. Fig.8 shows a cross-sectional view of the porous layer formed on the AC grained surface. Although the steady-state porous cells were observed near the metal-film interface, complete distribution of irregularly shaped pores was observed on the outer surface of the oxide film. This is due to the fact that each pore tends to orientate perpendicularly to the surface. Crossing of the pores and irregularities will particularly occur as the curvature of the aluminium surface increases.

From the study of the porous oxide film formation it is now possible to understand the influence of the following parameters :

. current density;

. potential of the anodic oxidations process;

. composition of the electrolyte;

. temperature of the electrolyte;

. influence of the Al roughness.

Study of the Sealing Process of Porous Oxide Films

Sealing of the porous anodic films is usually carried out to improve the corrosion resistance of the layer. During the sealing process the pores are filled due to the reaction of Al_2O_3 in boiling water, which results in the formation of boehmite[4,6,7,11,18,19]. The structure and composition of the porous layer are significantly changed, especially in the upper layer of the Al_2O_3 film. Beside the closing of the pores, hydrated aluminium products are also deposited on top of the oxide surface. Fig.9 shows a cross section of a thoroughly sealed porous film formed in phosphoric acid. The closing of the pores clearly improves the corrosion resistance. Yet, the presence of the sealing products on the oxide surface hinder the adhesive properties of the Al_2O_3 layer and makes it more unstable in time, again due to the formed boehmite (Fig.10).

Characterization of surface treatments by impedance spectroscopy and scanning ellipsometry

The characterization of surface treatment is important to develop surface analytical techniques, which are able to determine the surface characteristics in a fast, non-destructive way.

Indeed, techniques as TEM, Auger, SIMS, XPS, require high vacuum and often a difficult sample preparation. Impedance and ellipsometry measurements are used to analyse the surface where respectively from the electrical and optical characteristics of the surface information is gained. The attention is focussed on the characterization of oxide thickness (barrier and porous layer thickness) hydration of the oxide layer, roughness, and detection of the corrosion resistance.

Impedance Spectroscopy (EIS)[7,8,11,13,15,16,19,21].

Fig.11 gives an example of a measured impedance spectrum in the form of Bode-diagrams of a porous oxide layer as a function of different sealing times.

As could be seen from TEM research, sealing of the pores in hot water results in the formation of boehmite especially on top of the porous oxide layer, where the pores are closed.

As can be determined from Fig.11, from the impedance amplitude as well as from the phase angle diagram an extra relaxation occurs due to the sealing. From the equivalent circuit a resistance and a capacity can be accorded to this sealing. So it can be concluded that recording of the resistance and capacity as a function of sealing time leads to a quantitative characterization of the sealing. This is one example of the use of impedance method to characterize Al-surface treatments.

Spectroscopy ellipsometry[18].

In the present investigation, spectroscopic ellipsometry is introduced. This technique determines the reflecting properties of the sample towards polarized light at different curve lengths. The benefit of the method lies in the optical nature of the measurement, which is non-destructive and applicable in situ.

It also presents the advantage of being extremely sensitive to minute surface effects, like thin film formation or morphological changes (e.g. roughening). From the measured reflective properties, a characterization of the physical and chemical state of the surface can be deduced by means of simulation and regression procedure. In this procedure, the ellipsometric response of a theoretical multilayer model of the surface structure is fitted to the measured response by a least square regression analysis. The model which allows an optimal accordance is retained as the characterization of the surface structure, provided it is physically realistic. As an example Fig. 12 shows the ellipsometric data recorded on a porous oxide layer, from which the thickness can be easily determined.

CONCLUSION

Using the complementary information, issued from electrochemical and surface analytical techniques it is possible to characterize some important Al-surface treatments and to understand the influence of most important parameters. The results allow to tailor the Al-surface properties in a known way and possible undesirable side-effect can be avoided. The knowledge and experience obtained this study can be extrapolated to other surface treatments.

REFERENCES

1. TERRYN H.
   "Electrochemical Investigation of AC Electrograining of Al and Its Porous Anodic Oxidation", Ph.D. Thesis, Fac. TW., VUB, 1987.

2. TERRYN H., VEREECKEN J.
   "Auger Electron Spectrometry of Porous Layers of Anodized Aluminium", Passivity of Metals and Semiconductors, Ed. Froment, Elsevier, 1983, pp. 747-752.

3. TERRYN H., VEREECKEN J.
   "Auger Electron Spectrometry of Porous Layers on Anodized Aluminium", Proceedings 5th International Symposium on Passivity, 30/5-3/6/83, Bombannes, France, 1983, pp.307-309.

4. TERRYN H., VEREECKEN J., VANHELLEMONT J., VAN LANDUYT J.
   "HVEM Investigation of Al/Al_2O_3 Structures for Offset Purposes", Extended Abstracts 169th Meeting Electrochemical Society, 4-8/5/85, Boston, USA, Vol.86, 1, 1985, p.48.

5. TERRYN H., VEREECKEN J., VANHELLEMONT J., VAN LANDUYT J.
"HVEM Investigation of Al/Al$_2$O$_3$ Structures for Offset Purposes", Proceedings Symposium on Aluminum Surface Treatment Technology, Ed.Alwitt, Electrochemical Society, Vol.86, 11, 1986, pp.291-301.

6. VAN HELLEMONT J., TERRYN H., VAN LANDUYT J., VEREECKEN J.
"HVEM Study of Porous Al$_2$O$_3$ Layers Formed by Anodic Oxidation in Sulphuric and Phosphoric Acid", Proceedings XIth Int. Congres on Electron Microscopy, 9/ 9/86, Kyoto, Japan, 1986, p.1069.

7.VAN DER LINDEN B., TERRYN H., VEREECKEN J.
"The Use of Impedance Measurements to Characterize Various Aluminium Pretreatments", Preprints, Eurocorr'87, 6-10/4/ 87, Karlsruhe, Germany, 1987, pp.579-761

8. HUBRECHT J., TERRYN H., VAN DER LINDEN B., VEREECKEN J.
"Caractérisation des traîtements de surface à l'aide des mesures d'impédance", 2ème Forum sur les Impédances Electrochimiques, 28-29/10/87, Montrouge, France, 1987, pp.43-52.

9. TERRYN, H., VEREECKEN, J., THOMPSON G.
"Investigation of the AC Graining Process of Al", Proceedings of the Anual Technical Conference and Exhibition, 27-29/4/88, Stratford upon Avon, United Kingdom, pp.113-134.

10. TERRYN H., VEREECKEN J., VANHELLEMONT J., VAN LANDUYT J.
"Investigation of the Growth of Porous Oxide Films on Different Pretreated Al Surfaces", Extended Abstracts of the 174 Meeting of the Electrochemical Society, Vol.88-2, 9-14/10/88, Chicago, USA, p. 180.

11.VEREECKEN J., TERRYN H., VAN DER LINDEN B.
"Etude du Colmatage d'Aluminium Anodisé par Mesure d'Impédance", 3ème Forum sur les Impédances Electrochmiques, 24/11/ 88, Montrouge, France, pp. 185-193.

12.TERRYN H., VEREECKEN J., THOMPSON G.
"Investigation of the AC Graining Process of Al", Journal of the Institute of Metal Finishing, Vol.66, 1988, p.116.

13.HUBRECHT J., TERRYN H., VAN DER LINDEN B. VEREECKEN J.
"Characterization of Protective Coatings with Electrochemical Impedance Spectroscopy", Abstracts International Symposium Corrosion Science and Engineering, in honour of M. Pourbaix's 85th birthday, Brussels, 12-15/3/89, pp. 72-73.

14.TERRYN H., VAN HELLEMONT J., VAN LANDUYT J., VEREECKEN J.
"Influence of the Aluminium Pretreatment on the Growth of Porous Oxide Films", The Insitution of Metal Finishing "Surface Finishing", 11-14/04/89, Brighton, England, pp. 135-153.

15.TERRYN H., VEREECKEN J.
"Impedance Analysis of Porous Oxide Films Formed on Different Pretreated Al Surfaces", Extended Abstracts of the 1st International Symposium on Electrochemical Impedance Spectroscopy, 22-26.05.89, Bombannes, France, p. C4.8.

16.VAN DER LINDEN B., TERRYN H., VEREECKEN J.
"Fractal Approach of the Impedance of Al$_2$O$_3$/Al Layers", Extended Abstracts of the 1st International Symposium on Electrochemical Impedance Spectroscopy, 22-26.05.1989, Bombannes, France, p. C2.15.

17.TERRYN H., VEREECKEN, J., THOMPSON G.
"Electrograining of Aluminium in Hydrochloric Acid" part 1 and 2, Accepted in Corrosion Science

18.DE LAET J.
"Studie van oppervlaktebehandelingen met behulp van oppervlakteanalysetechnieken", thesis, Fac.TW, VUB, 1989.

19.VAN DER LINDEN B.
"Karaketerisatie van elektrochemisch voorbehandeling aluminium", thesis, Fac. TW, VUB, 1987.

20.TERRYN H., VEREECKEN J.
"Invloed van de voorbehandeling van aluminium op de opbouw van de poreuze anodisatielaag", A.T.B. Metallurgie XXIX 1-2, 1989.

21.VAN DER LINDEN B., TERRYN H., VEREECKEN J.
"Investigation of anodic aluminium oxide layers by electrochemical impedance spectroscopy", Journal of Applied Electrochemistry, 20, 1990, pp. 798-803.

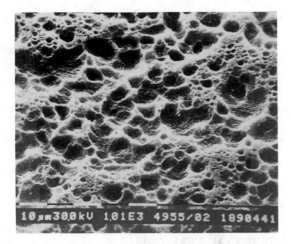

Fig.1 : SEM micrograph of an AC electrolytic grained Al-surface.

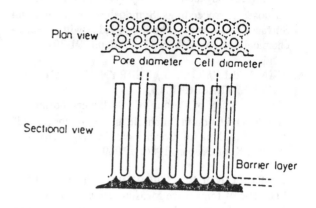

Fig.4 : Schematic representation of the porous layer.

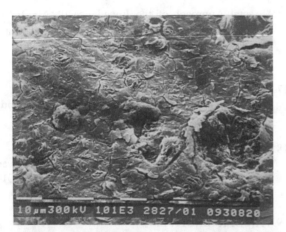

Fig.2 : SEM micrograph of an AC electrolytic grained Al-surface covered with each products.

Fig.3 : TEM cross sectional view of the AC electrolytic grained Al-surface.

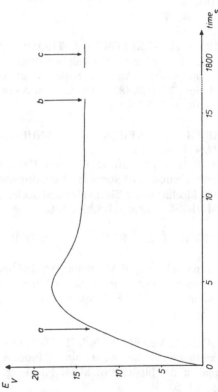

Fig.5 : Voltage-time transient of the porous oxide film formation recorded under constant current density.

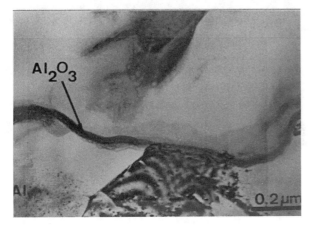

Fig.6 : TEM cross sectional view of a barrier layer.

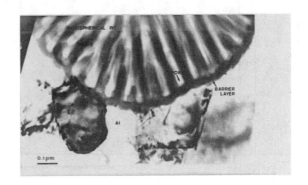

Fig.8b : at a higher magnification in one hemispherical pit.

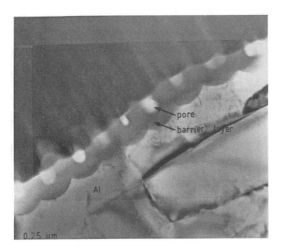

Fig.7 : TEM cross sectional view of a porous layer.

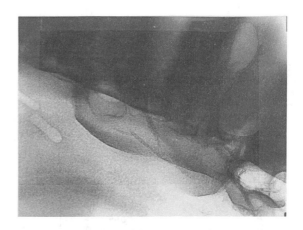

Fig.9 : TEM cross sectional view of a sealed porous oxide layer.

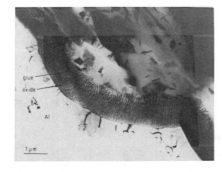

Fig.8a : TEM cross sectional view of a porous oxide layer formed on an AC electrolytic grained Al-surface.

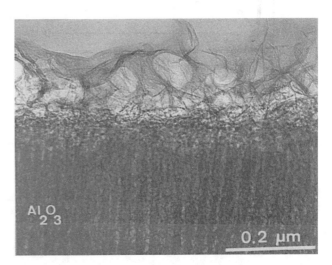

Fig.10 : TEM cross sectional view of the deposition of sealing producs on tom of the Al_2O_3 surface.

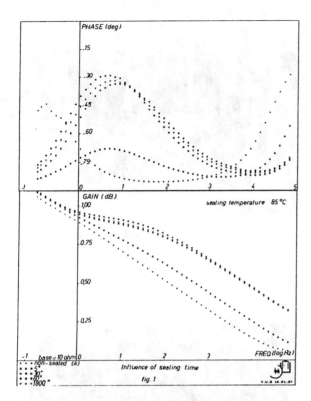

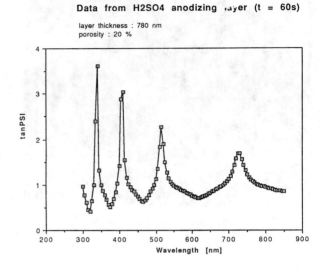

Data from H2SO4 anodizing layer (t = 60s)

layer thickness : 780 nm
porosity : 20 %

Fig.11 : a) Bode diagrammes (amplitude and phaze angle as a function of the frequency) of a porous anodic oxide layer as a function of the sealing time (sealing at 85°C in water).

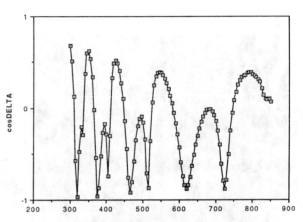

Fig.12 : Example of an ellipsometric spectrum (tan psi and cos delta as a function of the wavelength).

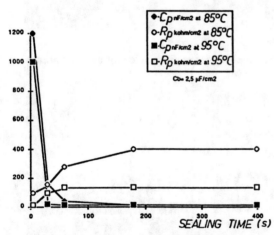

b) Determination of the parameters of the equivalent circuit simulating the obtained data.

Modelling

Time-dependent simulation of Czochralski growth: application to the production of germanium crystals

N. Van den Bogaert
Unité de Mécanique Appliquée, Université Catholique de Louvain, Louvain-la-Neuve, Belgium
J. Braet
MHO—a Division of ACEC–Union Minière, Hoboken, Belgium

SYNOPSIS

We present a numerical method for the simulation of Czochralski growth. Our model is both global, considering conductive and radiative exchanges in the overall furnace, and time-dependent, including the heat capacity of all constituents, the motion of the solid/liquid interface, the varying interface radius and all geometrical changes induced by the lengthening of the crystal. The input parameters are the pulling rate and the heater power. However, calculations can be performed in an inverse way as well and yield the heater power as a response to a prescribed crystal shape. This latter possibility has been applied to the pulling of the head of a germanium crystal. The results show how to control the radius by modulating the heater power in order to avoid the generation of undesirable thermal stresses.

INTRODUCTION

Most semi-conductor crystals are grown in Czochralski furnaces. One of the main difficulties in obtaining high quality crystals in such furnaces is the control of crystal shape, particularly during the shouldering process, i.e., during the transition from conical to cylindrical growth. This control is essential in order to grow a single crystal. The aim is to reproduce a given crystal shape while avoiding the generation of undesirable thermal stresses. To control the radius, one may either regulate the pulling rate or the heater power. Regulating the pulling rate yields a rapid radius response but may cause the generation of thermal stresses due to accelerations of the solidification front. Regulating the heater power while maintaining a constant pulling rate would be optimal but is difficult to achieve in practice due to the high associated time constants.

The numerical model we have developed is particularly well suited for solving such industrial problem. It takes into account two essential aspects of the process. First, in order to accurately model the system when important geometrical changes occur, time-dependent calculations are performed. Second, in order to secure a correct representation of the melt/crystal environment, global calculations are performed, taking into account heat transfer throughout the furnace.

In the next section, we briefly describe our finite element method. We highlight the difficulties of performing time-dependent calculations on a moving geometry.

Results of a numerical simulation of the shouldering process in a typical germanium furnace are presented and discussed in the last section. A given crystal shape is imposed, the pulling rate is maintained at a constant level and the heater power is calculated.

NUMERICAL METHOD

Our numerical method is both global and time-dependent. Different heat transfer modes in the different parts of the furnace are considered: conduction in solid and liquid bodies (crystal, melt, crucible, heater and insulator) and radiation between surfaces. Convection in the melt is not included at this stage, as the complete resolution of Navier-Stokes equations would be far too expensive. However, a simplified model is being developed and will soon be coupled to our thermal model. Calculations are fully time-dependent considering heat accumulation in all solid bodies and taking the continuously changing geometry of the furnace into account. The model computes the temperature field throughout the furnace, the successive solid/liquid interfaces and the shape of the crystal or the heater power as functions of time. Calculations can be performed in a direct or inverse way. In a direct simulation the heater power is imposed as a function of time and the shape of the crystal is calculated. This shape is an immediate result of the time variation of the radius at the tri-junction line (intersection of the melt, crystal and ambient gas). Such direct simulation is a transposition of the real process. In an inverse simulation the real input is considered as an unknown and the real output is imposed: the crystal shape is imposed and the corresponding heater power is calculated. The pulling rate is imposed in both direct and inverse simulations.

The coupling of the different heat transfer modes in the furnace is based on the method used in the quasi-steady state model developed by Dupret et al.[1]. The technique consists in separately analysing the heat transfer within each constituent and imposing continuity of temperature and heat flux on their interfaces. However, in a time-dependent model, heat transfer has to be considered on a moving geometry and thermal inertia of the solid bodies and of the solid/liquid interface is to be taken into account. As the calculation of the solid/liquid interface position and of the radius is decoupled from the global thermal calculation, the final solution for a given instant is obtained by means of an iterative procedure. The shape of the crystal is unknown and is determined by the shape of the meniscus at the surface of the melt. This is done in a

483

similar way as Derby et al.[2]. Flexible meshes are used in order to represent the time dependence of the geometry.

Three types of constituents corresponding to different heat transfer modes are considered. Each type is characterised by a typical equation which is discretised in space and in time if necessary and leads to a local system of equations. These local systems are then coupled in order to obtain a global solution. In what follows, we present each type of constituent.

Radiative enclosures

Radiative exchanges are computed by discretising the radiative integral equation[1], which is based on a diffuse radiation hypothesis. As these exchanges are dependent upon the viewed and hidden parts of the furnace, it is obvious that the continual change of configuration requires the evaluation of such exchanges for every time step. Moreover, a difficulty arises when solving a problem with an unknown radius, as the configuration of the main enclosure, determined by the value of the tri-junction radius, changes during the iterative procedure inside a given time step. Computing view factors for each new value of the radius would lead to prohibitive computation time. We have developed an original method for performing this evaluation at low cost[3]. After space discretisation using one-dimensional finite elements, we obtain a system of equations relating nodal heat fluxes to the fourth power of nodal temperatures. As radiation is transmitted at the speed of light, no explicit time derivatives appear in the system.

Solid bodies

Diffusive heat exchanges in the solid and liquid bodies are governed by the heat equation:

$$\rho c \frac{\partial T}{\partial t} + \rho c \underline{v}(\underline{x}, t) \cdot \underline{\nabla} T - \underline{\nabla} \cdot [k(\underline{x}) \underline{\nabla} T] = r(\underline{x}, t)$$

where T stands for the temperature field, ρ for the specific mass, c for the heat capacity, $\underline{v}$ for the velocity, k for the conductivity, r for a volumetric heat source, $\underline{x}$ for the position with respect to a fixed point of the furnace and t for the time. The first term in the equation accounts for the thermal inertia of the solid body. This equation is discretized in space by a Galerkin/finite element technique and yields a system of ordinary differential equations including time-derivatives of nodal temperatures. The set of equations is further discretized in time by an implicit Euler method which may be written as:

$$\dot{T}^{n+1} = \frac{T^{n+1} - T^n}{\Delta t^{n+1}} \ ,$$

where T^n and $\dot{T}^n$ are the temperature and the time derivative of the temperature at instant t^n and $\Delta t^{n+1} = t^{n+1} - t^n$. A condensation technique allows one to eliminate internal temperatures and to obtain a linear system relating nodal temperatures and nodal heat fluxes on the boundary of the element.

Crystal-melt element

This element may be considered as a solid body. However, as the crystal solidifies, heat is released along the solidification front. This solidification heat includes a quasi-steady state contribution proportional to the pulling rate and a time-dependent contribution dependent upon the motion of the interface with respect to the melt level, due to the inertia of the interface. The latter contribution may reach high values in rapidly changing geometries.

When allowing the tri-junction radius to be free, we determine the shape of the crystal on the basis of a coupling between the Laplace-Young equation governing the meniscus shape with the thermodynamic condition of contact angle (ϕ_0) constancy which relates the meniscus angle ϕ to the growth angle ψ (see figure 1):

$$\psi = \phi - \phi_0,$$

and with the geometric condition which relates the time derivative of the radius at the tri-junction $\dot{r}$ to the growth angle ψ through the growth velocity v:

$$\dot{r} = v \tan\psi.$$

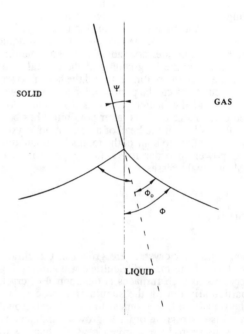

Figure 1. Enlarged view of the tri-junction. Contact angle ϕ_0, meniscus angle ϕ and growth angle ψ.

Global solution

Global computation is performed at every time step by assembling the local equations relating nodal temperatures and heat fluxes on the boundary of each element. After elimination of the heat fluxes, one obtains a non-linear system which is solved by means of a Newton method. A complete description of the method is given in Dupret et al[1].

The overall numerical procedure may be summarized as follows:

- computation of an initial quasi-steady state solution: this solution is obtained by solving the equations with all time-derivatives set to zero;

- for each discrete instant:

i) automatic generation of a new geometry: this includes mesh deformations and generation of the corresponding radiative enclosures;

ii) determination of the local linear systems characterising the solid bodies; this stage requires the knowledge of the temperature field at the preceding discrete instant;

iii) computation of the radiative exchanges on the enclosures whose geometries have been modified;

iv) determination of the non linear local system characterising the melt/crystal element: computation of the geometrical unknowns (solid/liquid interface, radius and meniscus shape);

v) global coupling of all local systems and back to step iv if convergence is not reached;

vi) computation of the internal temperature field in each solid body; this temperature field is necessary for step ii of next discrete instant.

A detailed description of our time-dependent numerical method may be found in Van den Bogaert et al.[3].

RESULTS AND DISCUSSION

As an example, we have modelled the early stages of growth of a germanium crystal. The axisymmetric geometry of the furnace, represented in figure 2 varies in time and is fully defined by the height of the crystal, which is determined by integration of the pulling rate over time. The crucible moves slowly upwards in order to keep the melt level at a constant position with respect to the heater. The position of the bottom of the melt is determined by mass conservation. The germanium melting temperature is 1211 K; its latent heat of fusion is $2.59\ 10^9$ J/m^3. The material data of the different constituents of the furnace are listed in table I. A constant pulling rate of 2cm/h is imposed. The contact angle is 14°. The shape of the crystal is imposed (imposed radius at the tri-junction for every time step) and the heater power is calculated. We thus solve an inverse problem. The initial solution for the time-dependent simulation was obtained by using the quasi-steady state model. The growth angle of the initial crystal is 60°, its radius is 1 cm and its height is 3 mm. The simulation includes three stages: during the first stage, the crystal is imposed to grow at the initial growth angle of 60° until it reaches a radius close to 5 cm; during the second stage the growth angle is imposed to decrease until it reaches 0° : this is the shouldering; during the last stage, a constant diameter is imposed. At the end of the simulation the crystal height is 10 cm. The time step used for this simulation was 200 seconds and corresponds to a growth of 1 mm.

Figure 3 shows the meshes corresponding to different stages of growth. One may observe the way these meshes are deformed in order to conform to the shape of the crystal. Figure 4 shows the corresponding temperature fields in the melt and in the crystal. The configuration of the isotherms in the melt during conical and during cylindrical growth is quite different. The heater power is

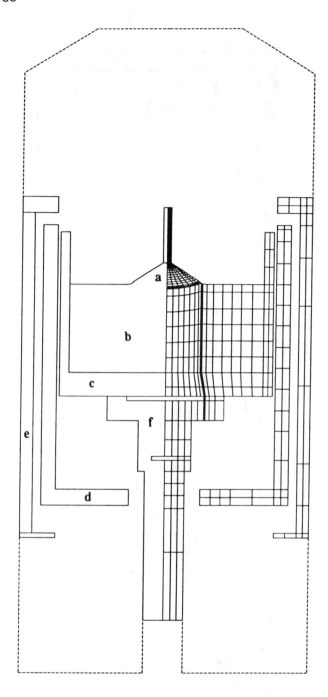

Figure 2. Sketch of the furnace and mesh used for the calculations : (a) crystal, (b) melt, (c) crucible, (d) heater, (e) insulator, (f) pedestal.

material	conductivity W/mK	emissivity	heat capacity J/Km3
solid germanium	25	0.2	$2.3\ 10^6$
liquid germanium	71	0.5	$2.3\ 10^6$
insulating material	1	0.3	$3.3\ 10^5$
graphite (crucible,heater,pedestal)	42	0.7	$3.3\ 10^6$

Table I. Thermal properties of the different constituents of the furnace.

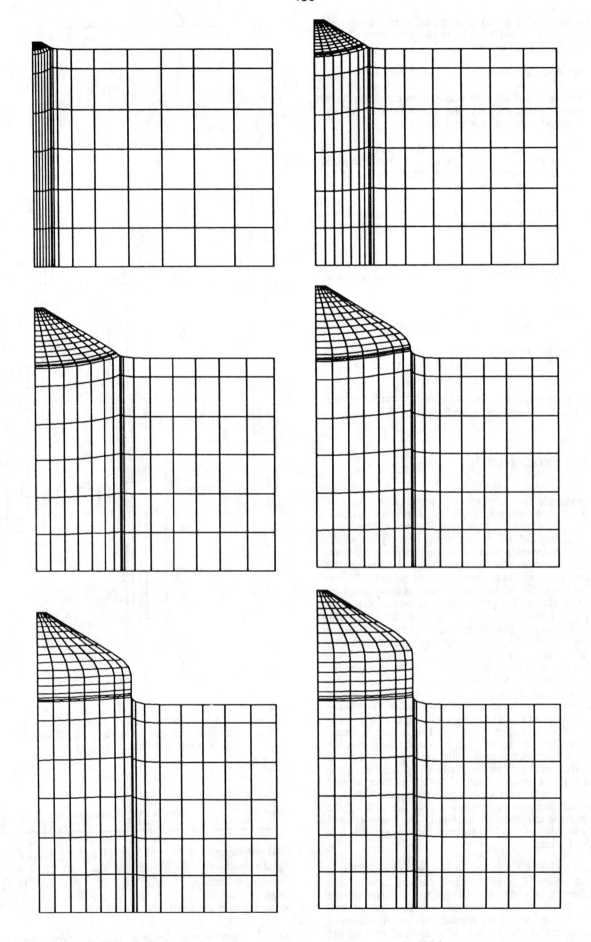

Figure 3.   Crystal/melt meshes for various stages of growth.

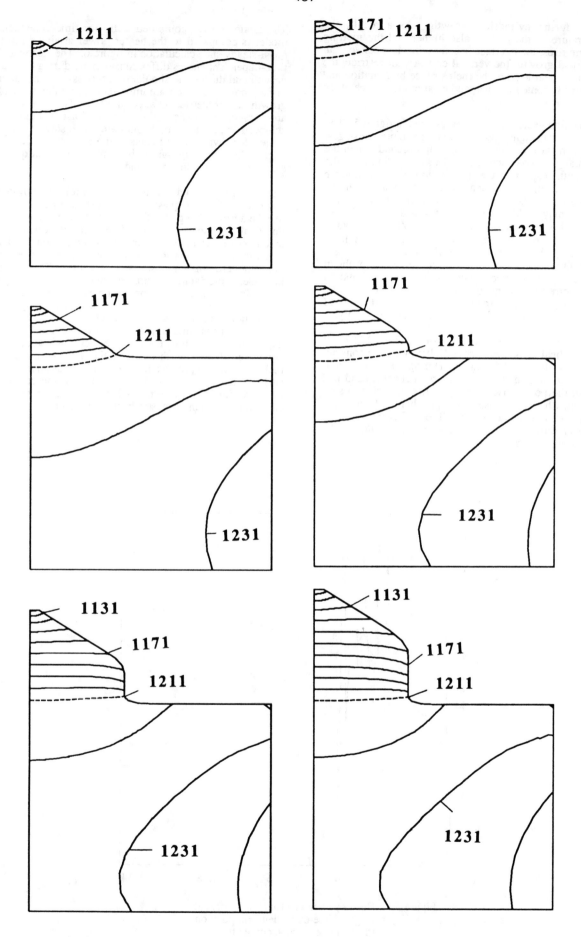

Figure 4. Isotherms in the melt and in the crystal, separated by 10 K for various stages of growth.

higher during cylindrical growth. As a result, the temperature in the melt is also higher and the melting isotherm rises: the interface becomes thus less convex. In cylindrical growth, the vertical crystal wall restricts the amount of heat leaving the melt surface by radiation and isotherms become more perpendicular to the surface of the melt.

The resulting heater power is represented in figure 5. The heater power is highly discontinuous at two precise stages: first, when starting the simulation, and second, when the growth angle starts decreasing. This discontinuity of the result with respect to the data is typical of an inverse problem: a slight modification of the situation in the environment of the tri-junction - induced by the varying radius - may yield an important response of the heater power. This problem is addessed in N. Van den Bogaert et al.[3]. In order to obtain a legible result we have smoothed the heater power time variation curve. To assess the validity of the smoothing, we have solved the problem in a direct way: the smoothed power variation was imposed and the corresponding radius was calculated. The resulting crystal shape was extremely close to the initial imposed shape: the maximum difference was about 0.3 mm only.

We may observe that the heater power is about 20.572 W during conical growth and about 21.195 W during constant radius growth, thus 3 % higher. During the shouldering, i.e., during the transition between conical and cylindrical growth, the power rises rapidly at first in order to stop radial growth and decreases subsequently in order to avoid that the radius starts decreasing. The slight time variation of power during constant radius growth is due to the changes of geometry.

The heater power computed with our time-dependent model is compared to the heater power arising from a quasi-steady state calculation in figure 6. The steady and time-dependent curves differ considerably during conical growth and during the shouldering process. As a matter of fact, transient effects are important during this stage of growth, as geometrical changes are fast. The two curves are very close to each other once a constant radius is achieved as geometrical changes are much slower at that stage of growth. This means that transient effects are negligible during constant radius growth and that quasi-steady state simulations are valid.

During conical growth, the time-dependent heater power is much lower than the quasi-steady state heater power. This may be attributed in part to the release of solidification heat due to the motion of the solid/liquid interface which is taken into account in the time-dependent model. The increase of power during the shouldering proces occurs once the growth angle starts decreasing. The power increases more rapidly in time-dependent calculations as some delay is necessary for the changes arising in the heater to reach the tri-junction line. In quasi-steady state calculations one assumes that the whole power variation effect is felt instantaneously at the tri-junction. In time-dependent calculations the major effect is felt about one hour later, due to the thermal inertia of the crucible and melt. In order to produce the same radius variation in the early instants following the perturbation, the required power variation must be much higher in time-dependent calculations than in quasi-steady state calculations.

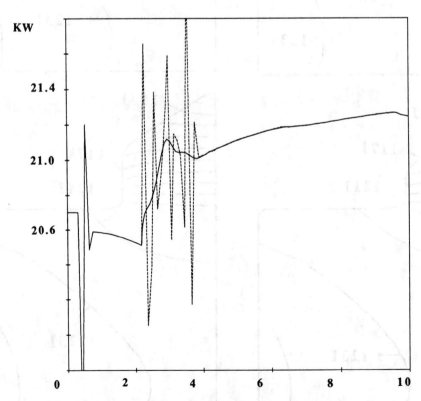

Figure 5.   Heater power as a function of crystal length. Time-dependent computation (dashed line). Same curve after smoothing (solid line).

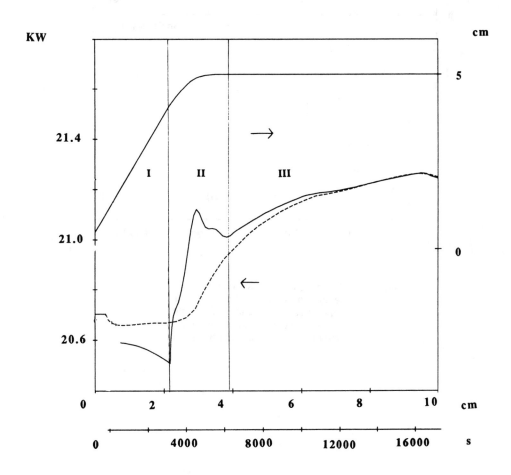

Figure 6.   Heater power as a function of crystal
            length and as a function of time. Time-
            dependent calculations (solid line) and
            quasi-steady calculations (dashed line).
            Imposed radius as a function of crystal
            length.

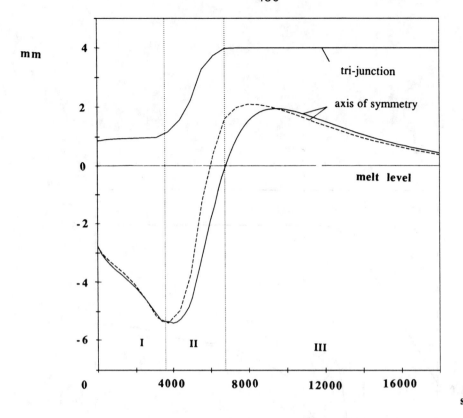

Figure 7. Interface position with respect to the melt level as a function of time: at the tri-junction and on the axis of symmetry. time-dependent calculations (solid line) and quasi-steady calculations (dashed line).

Figure 7 shows the evolution of the position of the solid/liquid interface with respect to the melt level. Two points of the interface are represented: the point on the axis of symmetry and the tri-junction. The interface thermal inertia considered in the time-dependent simulation causes a delay to appear in the interface position with respect to the steady simulation. The position of the interface on the axis of symmetry decreases during conical growth, increases once the growth angle starts decreasing, reaches a peak value during the shouldering, and decreases towards a constant value corresponding to constant radius growth. The interface remains convex with respect to the crystal during the whole process. The difference between the vertical position of the tri-junction line and the interface axial position, also called deflection, is about 5 mm during conical growth, decreases down to 2mm during the shouldering and reaches 4 mm for constant radius growth. The tri-junction height (or meniscus height) is an immediate result of the imposed radius. The initial meniscus height is 0.9 mm corresponding to a growth angle of 60°. It rises rapidly when the growth angle starts decreasing and reaches an almost constant value of 4 mm during cylindrical growth.

Figure 8 represents the heat released at the interface. The quasi-steady state contribution, due to the pulling rate, is proportional to the crystal section. As a result, such contribution varies as the square of the diameter during conical growth and remains constant during cylindrical growth. The contribution due to the motion of the interface is high during the shouldering as the interface rises rapidly. It reaches up to 40% of the quasi-steady state contribution and is of opposite sign. The quasi-steady state contribution corresponds to heat release and the time-dependent contribution corresponds to heat absorption. This latter contribution induces the delay of the interface position which has been observed in figure 7. Once cylindrical growth is reached, contribution due to interface velocity almost vanishes as the solidification front remains close to the same position.

Figure 9 shows the history of the successive interface positions. An interface is represented for every 600 seconds period. One may observe that the interfaces are close to each other during the shouldering.

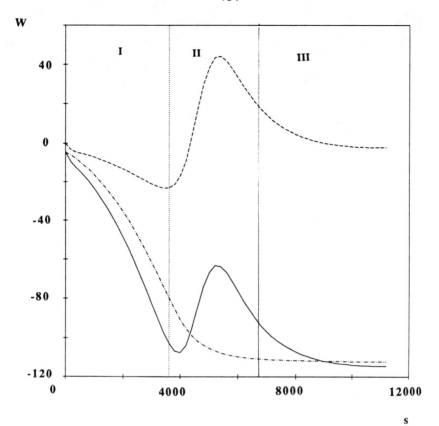

Figure 8.   Solidification heat at the interface as a
function of time: quasi-steady state
contribution, due to the pulling rate (dash-
dotted line); time-dependent contribution,
due to the change of interface position
(dashed line) and total amount (solid line).

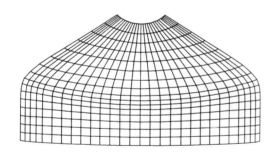

Figure 9.   Successive positions of the solidification
front. An interface is represented for
every 600 seconds period.

CONCLUSION

We have briefly described our numerical method for the
time-dependent simulation of crystal growth.   The
particularities of such a model with respect to a quasi-
steady state model have been outlined. As an application
of our method, we have computed the heater power
variation required to grow the head of a crystal with a
given shape. The comparison of the power variation in
time-dependent and quasi-steady state calculations has
revealed the presence of important transient effects
induced by geometrical changes during the shouldering
process.

References
[1] Dupret F., Nicodème P., Ryckmans Y., Wouters P., Crochet M.J.
    1990 Int. J. Heat and Mass Transfer, 33, 1849
[2] Derby J. J., Brown R. A. 1987 J. Crystal Growth 83, 137
[3] Van den Bogaert N., Crochet M.J., Dupret F. 1991, in preparation.

Steady-state self-regulating simulation model applied to design of a new hydrometallurgical process for recovery of non-ferrous metals

A. Q. Novais
A. P. Barbosa
E. Trincão
Unidade de Simulação e Engenharia de Processos, Laboratório Nacional de Engenharia e Tecnologia Industrial (LNETI), Lisbon, Portugal

SYNOPSIS

An equation-oriented simulation model is developed and presented for a new hydrometallurgical process for recovery of non-ferrous metals from Iberian complex sulphide ore concentrates, by means of concentrated ammonium chloride aqueous solution.

The modelling approach employed for the elementary unit operations is analyzed and discussed together with the strategy adopted for representing the diagram structure, with the objective of conferring on the model a high degree of responsiveness in terms which are described as self-regulating steady-state characteristics.

The reference conditions for processing 20000 kg/h of Portuguese ore concentrate are presented and simulation runs are applied to reassess the validity of some design parameters namely, the extent of the copper reduction reaction, the pulp density and the by-pass stream flow which feeds the silver cementator. The effects are measured in terms of changes in the rate of consumption of by-feeds, process water and oxygen as well as on the solubility and recovery yields of the metal values. On the basis of these results design alternatives are analysed and discussed.

Process simulation is a well established technique in process design which can in addition be highly effective in effort reviewing and planning when experimental development work is also involved.

This is particularly relevant for hydrometallurgical processes whose complexity is usually great both as a result of the composition and constitution of the raw materials which contain a vast number of components and reactional mechanisms not fully understood and diagram configurations based on intricate networks of operations and streams required for satisfactory process performance.

Greater advantage would be made of the simulation model if it were available at the early stages of design in order to provide guidance as to the interplay of the operating variables and as a basis for quantitative evaluation. In practice the development and complexity of a model evolves iteratively by means of on-going validation and integration of the information gathered experimentally with the practical and theoretical background knowledge available.

While for process design it is required of a model that it should simulate adequately, on the basis of mass and heat balances and within an acceptable numerical approximation, the process steady-state conditions, this same objective can be met with a variable degree of elaboration. In particular in the case of highly concentrated electrolyte solutions and complex patterns of interconnecting streams to be found in many hydrometallurgical processes, a simulation model should ideally also enable the estimation of operating parameters and variables which need to be preset in order to meet process specifications.

The development of models which meet these goals while presenting the necessary flexibility and ease of use required for repeated prospective changes and examination of alternative diagram configurations, without excessive programming effort, can be satisfactorily achieved on the basis of the so-called equation-oriented simulation packages of which SPEEDUP is a well known example.

In this study a simulation model with the above characteristics is presented which was set up in the course of the experimental development, under an EEC cofinanced project, of a new hydrometallurgical process for recovery of non-ferrous metals from sulphide concentrates of Iberian ores by direct attack with a highly concentrated ammonium chloride leaching solution.

The process involves two major leaching operations and a number of metal recovery operations which include the solvent extraction of copper and zinc.

The simulation results obtained for the assumed

493

reference conditions are presented for the Portuguese ore concentrate and the discussion is centred on the handling of some by-pass streams and the recycle of the zinc solvent extraction raffinate which controls the dilution required to prevent the precipitation of zinc and/or copper(II) diamines in the crystallization of lead chloride.

SPEEDUP SYSTEM

Speedup is an "equation-oriented flowsheet simulation and optimization system for process engineering systems"[1], typically those with a modular structure.

It is beyond the scope of this paper to discuss the many features of the package but reference is next made to those which influenced directly the modelling approach.

(i) While the calculation of the units within the flowsheet is not order-dependent, the modular structure of the process diagram is maintained both for problem input and data output - physical units are represented by models which can be developed independently; the output can be inspected on an unit basis.

(ii) Provided that the problem is correctly specified in terms of equations and set variables and/or parameters, the unknowns to be calculated may be input or output streams and/or parameters, or combinations of these - by-feeds, make-up streams, process water requirements can thus be calculated together with output streams.

(iii) The models consist essentially of equations which need not to be variable assignments, whose order is arbitrary and which are solved simultaneously within blocks, (matrices of lower order), and of procedures which gather the calculation routines which are structured sequentially or which are trivial and would increase unnecessarily the total number of equations - the non-linear algebraic equations associated with the chemical equilibria are an example of the former while the empirical models employed for leaching and the conversion of physical units (e.g., kmol/h to kg/h) are examples of the latter.

(iv) The same problem input file is used for simulation, optimization or parameter estimation - while the steady-state operating conditions are calculated by simulation, some of the equilibrium constants were estimated from experimental data prior to simulation and one of the recycles requires an optimization procedure. The switching between running modes requires only minor input file modifications.

PROCESS DIAGRAM

Modelling approach

Figure 1 shows the information diagram for the global process whose detailed description can be found elsewhere[2]. The former is a simplification of the latter used as a basis for the computer representation.

It should be noted that batteries of reactors are depicted by single models (R1 to R8) and the solvent extraction operations (E1 and E2) by models of the black-box type. The reasons for this approach are three-fold:

(i) Most reactor models are empirical and based on preset chemical reaction sequences and maximum reaction extents, with the exception of the lead chloride crystallizer (R5) and the sulphates removal unit (R6) which are based on multiple reaction equilibria and the silver and the lead cementators (R4 and R7) which use solubility curves fit to experimental data over a wide range of temperatures and compositions. Thus, while these models respond correctly to changes in flow composition and also to some extent to temperature variations (through the equilibrium constants and solubility curves) they remain insensitive to residence times.
Consequently there are no gains with regard to the calculation of the steady-state operating conditions to extend the number of reactors of each type unless there were intermediate feeds or changes in the operating temperatures, which is not the case.

(ii) The solvent extraction operations have in common a fairly large number of counter-current stages and closed-loop streams which, if incorporated into the global process model, would add considerably to its complexity. This was considered undesirable and the detailed treatment of these two operations was carried out separately and falls outside the scope of this paper.
To illustrate this point in table.I a comparison is drawn between these three simulation systems and it can be noted the considerable magnitude of the solvent extraction systems, in particular the zinc extraction which is the most complex because of its upstream position in the process diagram.

(iii) The main focus of the process is on the interplay of the leaching operations (including copper reduction) and its effect on metal values solubilization and recovery, on the process self-sufficiency in leaching agent, the distribution of the pair NH_3/NH_4+, the inhibition of copper and/or zinc precipitates, the consumption of by-feeds and so on.
These global effects require that a considerable emphasis should be placed on the correct representation of the diagram structure as expressed by the pattern of interconnecting, by-pass and recycle streams and positioning of the make-up units and inlet points

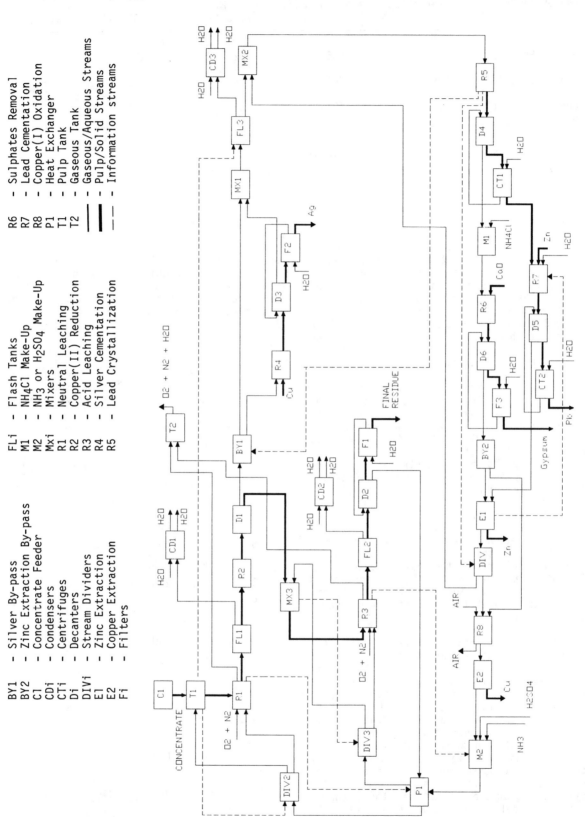

Figure 1 - Information Diagram for the Global Process

Legend:

BY1	– Silver By-pass		FLi	– Flash Tanks	R6	– Sulphates Removal
BY2	– Zinc Extraction By-pass		M1	– NH4Cl Make-Up	R7	– Lead Cementation
C1	– Concentrate Feeder		M2	– NH3 or H2SO4 Make-Up	R8	– Copper(I) Oxidation
CDi	– Condensers		Mxi	– Mixers	P1	– Heat Exchanger
CTi	– Centrifuges		R1	– Neutral Leaching	T1	– Pulp Tank
Di	– Decanters		R2	– Copper(II) Reduction	T2	– Gaseous Tank
DIVi	– Stream Dividers		R3	– Acid Leaching	——	– Gaseous/Aqueous Streams
E1	– Zinc Extraction		R4	– Silver Cementation	**——**	– Pulp/Solid Streams
E2	– Copper Extraction		R5	– Lead Crystallization	– – –	– Information streams
Fi	– Filters					

Table I - Simulation Systems

Designation	Total Number of variables	No.Equat.= = No.Unknowns	Speedup Input File (Kbytes)	Average Validation time (CPU s)	Average Run time (CPU s)
Global Process	6534	5560	402	189	101
Zinc Extraction	2390	2066	106	71	20
Copper Extraction	1151	1053	63	29	10

Hardware - MicroVAXII in Cluster (DEC-Digital)

Software - SPEEDUP (Imperial College+Prosys Technology LTD)

for by-feeds and process water.

Another major simplification which was required for modelling is related to the complex nature of the reacting medium, which is a highly concentrated electrolyte, rich in ammonia and metal ions, kept at approximately 6 molal in chlorides throughout the entire process. Under these conditions the number of species which coexist in soluble form is very vast and include copper and zinc chlorocomplexes as well as mono-, di-, tri and tetramines.

Thus and in order to reduce the complexity and cost of the computer implementation, the composition of the aqueous stream is described merely in terms of the elementary ions (see table.II under the "liquid" heading) and only at critical points of the process the full equilibrium composition of the stream is restored by means of an appropriate property simulation model based on multiple reaction equilibria analogous in structure to the R5 and R6 reactor models. In table.II is also shown the stream format adopted for the solids, air streams and single by-feeds. Also and in order to reduce the sparsity of the computer implementation, as handled by Speedup, the pulps depicted by single bold lines in figure 1, are represented by one "solid" stream and one "liquid" stream, the latter to account for the entrained aqueous phase.

The transfer of information between units not connected by physical streams or, more generally, the transfer of information not contained in such streams is carried out by means of information streams or control signals designated "connections" in the Speedup input language. These are not control signals in the dynamic sense, but rather constraints to be met at steady-state. The network formed by these connections, shown in figure 1, adds complexity to the model response, ensuring what can be appropriately described as self-regulating steady-state characteristics to the process simulation model. Table.III describes the information streams which were implemented. The one designated R5/DIV involves an

optimization procedure which is discussed later.

For the sake of illustration, in figure.2 is shown the leaching and copper reduction section of the process diagram with an indication of the correspondence between physical units and simulation models.

Table II - Streams Formats

Component No.	Stream Type			
	"Liquid"	"Solid"	"Air"	"Element"
1	H_2O	Inerts	O_2	Cu
2	NH_3	$PbSO_4$	N_2	(Zn
3	Cu^{2+}	PbS	H_2O	H_2O
4	Cu^+	Ag_2S		CaO
5	Zn^{2+}	$CuFeS_2$		NH_3)
6	Pb^{2+}	FeS_2		
7	Ag^+	FeAsS		
8	$SO_4^=$	$Zn_{0.9}Fe_{0.1}S$		
9	Cl^-	S		
10	NH_4^+	FeOOH		
11	H^+	CaO		
12	Ca^{2+}	$CaSO_4.2H_2O$		
13	-	Ag		
14	-	Zn		
15	-	Cu		
16	-	$PbCl_2$		
17	-	$Fe_3(SO_4)_2NH_4(OH)_6$		
18	-	Pb		
19	-	$PbCl_{0.5}(OH)_{1.5}$		
20	-	$CuCl_{0.5}(OH)_{1.5}$		

Table III - Information streams

Designation	Information Transmitted	Function
T1/FL3	Water flow required in T1 to prepare concentrate feed pulp	To estimate the water flow to be removed as low pressure steam in FL3.
T1/DIV2	Water flow required in T1 to prepare concentrate feed pulp	To estimate aqueous phase to be fed to T1
R1/P1	Estimated temperature of inlet aqueous phase in R1 for adiabatic operation	To control the extent of heat exchange in P1 and indirectly the head load in R3
R3/M2	H^+ requirements in R3 for leaching the hydroxi-chlorides	To estimate make-up of NH_3 or H_2SO_4 in M2
R5/BY1	(Cl^-) free in solution	To estimate the (Ag^+) at saturation in R5 and either act on the by-pass to allow a main stream flow with the equivalent amount of Ag^+ or apply a 10% split ratio, whichever in less restrictive
R5/DIV	R5 inlet and outlet concentration of $Zn(NH_3)_2^{2+}$, $Cu(NH_3)_2^{2+}$ and free (Cl^-)	To estimate the minimum recycle from DIV to MX2 to avoid diamines precipitation (optimization procedure)
E1/R7	$ZnCl_2$ flow required in zinc extraction for the scrubbing of copper	To estimate the process water feed in R7
E1/BY2	$ZnCl_2$ aqueous solution composition	To estimate the by-pass flow to R8, assuming total Zinc extraction in E1 and (Zn^{2+})=5 g/1000g H_2O

Process description

Leach and Copper Reduction

The concentrate is fed from C1 to a pulp tank (T1) where it is mixed to a pulp with part of the aqueous phase recycle from acid leaching (R3 through FL2, D2, P1 and D1V2), before being fed to the first of a battery of three neutral leaching reactors (R1) which operate at 105 deg.C and 1.5 atm of pressure in an atmosphere of injected oxygen. The first reactor receives in addition the remainder of the aqueous phase recycle from acid leaching.

The reacted pulp is subsequently cooled down to 95 deg.C and concentrated in a flash tank (FL1), before entering the copper reduction reactor (R2). The aqueous phase is then decanted from the residue (D1) and proceeds to silver cementation. The residue is mixed to a pulp (MX3) with part of the raffinate from copper extraction and fed to the first of a series of two acid leaching reactors (R3) where the remainder of the copper raffinate solution is equally fed. The operating conditions of these reactors as well as of the flash tank (FL2), where the reacted pulp undergoes cooling and concentration, are identical to those for neutral leaching.

Silver Cementation

In By1 a fraction of the aqueous phase stream from D1 forms a by-pass where silver is precipitated as a cement by means of metallic copper in a series of three stirred reactors (R4) and the cement undergoes separation and washing in a decanter and press filter assembly (D3 and F2) with internal filtrate recycle, before leaving the process as a final product. The aqueous phase is mixed (M1) with the main stream from By1 and this stream proceeds to lead crystallization.

Lead Crystallization

The aqueous phase stream is cooled down to 50 deg.C and concentrated in a flash cooler (FL3) before being admitted to the crystallizer (R5).

A split fraction of the zinc extraction final raffinate might be added at this point of the process (MX2) to inhibit the precipitation of copper and/or zinc diamines in R5. This split is controlled by an optimization procedure as already mentioned.

The lead chloride crystals are subsequently separated from the mother liquor (DA1) and washed in a centrifuge (CT1) before proceeding to lead cementation. The diluted mother liquor is readmitted to the crystallizer forming a recycle and thus furthering the crystal nucleation. The aqueous phase stream proceeds to sulphates removal.

Lead Cementation

498

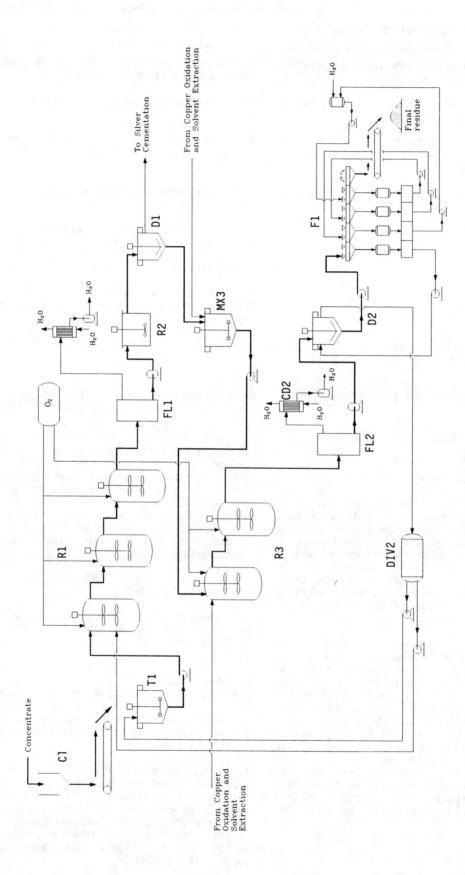

Figure 2 – Process diagram – Leach and Copper Reduction Section

The lead chloride and metallic zinc dust are admitted, together with make-up process water and aqueous zinc chloride recycle solution (this recycle is not shown in figure 1), to a heated stirred reactor (R7) where cementation takes place. The resulting pulp undergoes separation and washing in a decanter and centrifuge assembly (D5 and CT2) with internal recycle. The metallic lead leaves the centrifuge as final product and the aqueous phase stream which leaves the decanter D5 is split between the recycle fed to the reactor and the stream admitted to zinc solvent extraction.

Sulphate Removal

The aqueous phase stream received from lead crystallization is firstly corrected for its chlorides composition (M1) and then admitted to the first of a series of three stirred reactors (R6) where CaO is also fed and the precipitation of calcium sulphate takes place. The resulting pulp is separated in a decanter and belt type filter assembly (D6 and F3) with washing and internal recycling. The gypsum with some entrainment forms a final residue and the aqueous phase stream which leaves the decanter D6 is split in BY2 between zinc solvent extraction (E1) and copper oxidation (R8).

Zinc Solvent Extraction

The aqueous phase stream from sulphates removal is assumed simply to contact the extractant with all the zinc being removed and all the formed acid converted to ammonium ion. On the basis of preset phase ratios for the extraction, scrubbing, washing and stripping, the requirements in extractant, process water, zinc spent electrolyte and zinc chloride are estimated. The aqueous raffinate may undergo splitting in DIV and originate a recycle stream to MX2.

Copper Oxidation and Solvent Extraction

The main stream from zinc extraction enters copper oxidation together with the aqueous stream from sulphates removal. The reaction takes place at 50 deg.C and at atmospheric pressure in the presence of dry air assumed to be 50 percent reactive. The conversion of Cu(I) to Cu(II) is practically total and is controlled by the final concentration of Cu(I) in solution which is set at 1g/1000g H_2O.

The simplified calculation adopted for the copper extraction operation is similar to the one described for zinc extraction. The acidity of the copper raffinate stream is corrected at the make-up unit M2 before entering the heat-exchanger P1 as cold stream and thus closing the process loop.

Feed Composition

The elemental mass composition of the Portuguese concentrate employed in the simulation work

is shown in table.IV. It differs from the actual composition in that traces of Sb and Bi were eliminated. From this elemental composition and taking into account the constituents of the ore and inerts, the mineralogical mass composition shown in table.V is derived by means of a least squares procedure.

Table IV - Concentrate Elemental Mass Composition

Element	(%)
Zn	19.5
Cu	8.3
Ag	0.011
Pb	6.1
Fe	22.3
As	0.3
S	36.3

Table V - Concentrate Mineralogical Mass Composition

Component	(%)
Inerts	5.905
$PbSO_4$	6.137
PbS	2.265
$Zn_{0.9}Fe_{0.1}S$	32.294
$CuFeS_2$	23.900
FeS_2	29.358
FeAsS	0.129
Ag_2S	0.012

Chemical Reactions

Tables VI, VII and VIII show the reactional sequences and maximum conversions employed in the empirical models respectively for the neutral leaching, acid leaching and copper reduction reactors, together with the stoichiometric ratios of the main reactants involved in each chemical reaction.

The models which are based in multiple reaction equilibria involve a very large number of chemical equilibria which originate complex systems of non-linear algebraic equations. In table.IX the equilibria required to describe accurately the experimental results are shown for the elimination of sulphates as gypsum. With proper

Table VI - Neutral Leaching - Reactional Sequence Maximum Conversions and Reactants Stoichiometric Ratios

Component	Sequence Order	Conversion (%)	O_2	NH_4^+
$PbSO_4$	1	100	-	-
PbS	2	95	3.5/4	6/4
$CuFeS_2$	3	90	4/2	2/2
$Zn_{0.9}Fe_{0.1}S$	4	88	3.6/4	5.2/4
Ag_2S	5	90	3.5/4	6/4
FeS_2	6	28	3.75	-
$PbCl_{0.5}(OH)_{1.5}$	7	10*	-	3/2
$CuCl_{0.5}(OH)_{1.5}$	8	0*	-	3/2

*Percent conversion of the solubilized PbS and Cu(II) into the hydroxi-chloride form (in these two reactions the NH_4^+ acts as a product).

adaptations the remaining models are similarly structured.

Table VII - Acid Leaching - Reactional Sequence Maximum Conversions and Reactants Stoichiometric Ratios

Component	Sequence Order	Conversion (%)	O_2	H^+
NH_3	1	100	-	1
$PbCl_{0.5}(OH)_{1.5}$	2	100	-	3/2
$CuCl_{0.5}(OH)_{1.5}$	3	100	-	3/2
PbS	4	70	0.5	2
$CuFeS_2$	5	80	1.25	2
$Zn_{0.9}Fe_{0.1}S$	6	73	2.1/4	7.2/4
Ag_2S	7	70	0.5	2

Table VIII - Copper Reduction - Reactional Sequence Maximum Conversions and Reactants Stoichiometric Ratios

Component	Sequence Order	Conversion (%)	Cu(II)	NH_3
$PbSO_4$	1	100	-	-
PbS	2	85	14/4	8/4
$CuFeS_2$	3	85	14/2	14/2
$Zn_{0.9}Fe_{0.1}S$	4	85	14.4/4	9.2/4
Ag_2S	5	85	14/4	8/4

Optimization procedure

The Speedup equation-oriented approach is highly adequate to handle the models based on multiple reaction equilibria such as sulphates removal shown in table.IX, which generate a vast number of non-linear algebraic equations. But this only applies when all the equilibria are simultaneous. In table.IX equation.27 which corresponds to the precipitation of gypsum is associated to a solubility product which is always met since the operating conditions are such that gypsum is formed.

Table IX - Reactional mechanism for the sulphates removal operation

1 - $Zn^{2+} + NH_3 = Zn(NH_3)^{2+}$

2 - $Zn^{2+} + 2NH_3 = Zn(NH_3)_2^{2+}$

3 - $Zn^{2+} + 3NH_3 = Zn(NH_3)_3^{2+}$

4 - $Zn^{2+} + 4NH_3 = Zn(NH_3)_4^{2+}$

5 - $Zn^{2+} + Cl^- = ZnCl^+$

6 - $Zn^{2+} + 2Cl^- = ZnCl_2$

7 - $Zn^{2+} + 3Cl^- = ZnCl_3$

8 - $Zn^{2+} + 4Cl^- = ZnCl_4^{2-}$

9 - $Zn^{2+} + 3Cl^- + NH_3 = ZnCl_3(NH_3)^-$

10 - $NH_3 + H^+ = NH_4^+$

11 - $Cu^{2+} + NH_3 = Cu(NH_3)^{2+}$

12 - $Cu^{2+} + 2NH_3 = Cu(NH_3)_2^{2+}$

13 - $Cu^{2+} + 3NH_3 = Cu(NH_3)_3^{2+}$

14 - $Cu^{2+} + 4NH_3 = Cu(NH_3)_4^{2+}$

15 - $Cu^{2+} + Cl^- = CuCl^+$

16 - $Cu^{2+} + 2Cl^- = CuCl_2$

17 - $Cu^{2+} + 3Cl^- = CuCl_3^-$

18 - $Cu^{2+} + 4Cl^- = CuCl_4^{2-}$

19 - $Cu^{2+} + NH_3 + 3Cl^- = CuCl_3(NH_3)^-$

20 - $H_2O = H^+ + OH^-$

21 - $Zn^{2+} + SO_4^{2-} = ZnSO_4$

22 - $Cu^{2+} + SO_4^{2-} = CuSO_4$

23 - $HSO_4^- = SO_4^{2-} + H^+$

24 - $Ca^{2+} + SO_4^{2-} = CaSO_4$

25 - $Pb^{2+} + SO_4^{2-} = PbSO_4$

26 - $Pb^{2+} + 2Cl^- = PbCl_2$

27 - $Ca^{2+} + SO_4^{2-} + 2H_2O = CaSO_4.2H_2O$ (s)

28 - $CaO + 2NH_4^+ = Ca^{2+} + 2NH_3 + H_2O$

The recycle stream between DIV and MIX2 is null except when there is an indication that the copper and/or zinc diamines might precipitate in R5. To detect this possibility a model structurally analogous to the sulphates removal is employed but the equations which regulate the solubility of these two species are handled outside the model in a separate section associated with the objective function to be optimized which in this case reduces to minimization of the recycle flow. This is so because the solubility products are only met in the assumed design limiting situation of aqueous phase saturation in diamines. Ordinarily the concentration products of the species involved fall below the corresponding solubility products which therefore act as inequality constraints and can not be handled similarly to equation.27 in the sulphates removal model.

The optimization procedure can thus be summed up as follows

$$\text{Minimize (Recycle flow)}_{DIV-MX2}$$

subject to

$$((\, Zn \, (NH_3)_2^{2+} \,) \, (\, Cl^- \,)^2 \leq \, Kdz)_{1,2}$$

$$((\, Zn \, (NH_3)_2^{2+} \,) \, (\, Cl^- \,)^2 \leq \, Kdc)_{1,2}$$

where Kdz and Kdc are the solubility constants of the zinc and copper diamines and subscripts 1 and 2 refer to the inlet and outlet conditions of model R5, since, as a precaution, a check for diamines precipitation is made before and after lead crystallization.

Simulation work

While the final goal of the modelling and simulation work was to calculate consistent steady-state operating conditions for process design and economic evaluation, it also enabled the development of considerable insight into some of the design assumptions and parameters.

The present work illustrates the contribution of the process simulation model to assess, with regard to these reference operating conditions, the effect of some major parameters by means of simulation tests which are next described.

(i) Extent of the copper reduction reaction

This parameter (R2.Cr) was set at 0.80 for design but it is thought that this value might be relaxed without a detrimental effect on the diamines precipitation and on the process self-sufficiency in leaching agent. Thus an indication of the associated flexibility was sought.

(ii) Density of the pulp

This parameter (T1.dens_pulp) influences directly the volumetric flow of the aqueous phase which is kept in closed-loop, of the make-up flows and of the impregnation losses in final residues. Its design value was set at 100 g/1000g H_2O and also by design the chlorides concentration in the aqueous phase is kept at 6 molal.
It must be noted that the density of the pulp has practical implications on the operation of pumps and other equipment as well as on equipment capacity and hence on the plant economic evaluation,

but in this work only an objective similar to (i) is sought.

(iii) Silver cementation by-pass stream

The design strategy adopted for this stream is indicated in table.III. The reference value of 10 percent is thought to be a conservative estimate which affects directly the silver recovery yield and hence the process economics. An indication of the flexibility associated with this parameter was sought taking into account the likely fluctuation in the concentrate composition and the need to prevent the formation of silver precipitates.

A number of simulation runs was carried out by varying the copper reduction extent (0 to 0.80), the density of the pulp (50 to 150g/1000g H_2O) and for the silver cementation the by-pass stream was reduced (10 to 0 percent) and the silver content in the concentrate was increased by a factor of ten (at the expense of inerts) to investigate whether these drastic conditions could bring about silver chloride formation.

The effects of these tests were measured in terms of changes in the rate of consumption of by-feeds, process water and oxygen as well as on the solubility and recovery yields of the metal values.

Simulation results

Tables X and XI show the results obtained respectively for the process mass balance based on 20 000kg/h of concentrate, under the reference conditions, and the solubilization and recovery yields of the non-ferrous metals.

It can be noted that the assumed operating strategy regarding make-ups, reflected in the position of units M1 and M2 in the diagram, points to the consumption of both NH_4Cl (to adjust to 6 molal the aqueous phase composition in chlorides) and NH_3 (to neutralize the excess of acid required by acid leaching), thus ensuring the balance in leaching agent without the need for H_2SO_4, which was considered as an alternative.　•

The consumption of metallic copper and zinc are related to the cementation, respectively, of silver (in R4, together with the reduction of Cu(II)) and of lead (in R7) and contribute to the observed recovery yields of these two metals.

The process water requirements are related to the assumed phase separation conditions which are based on preset impregnation coefficients of aqueous phase in the pulps (50, 30 and 9 percent wet base respectively for decanters, filters and centrifuge) and the water/impregnation washing ratios for final residues and products (which vary from 1.5 to 3 with the type of filter and equals 9 for the centrifuge). The values reported in table.X correspond to a design compromise between process water requirements for reduction of metal losses by entrainment and excess of water which needs

502

to be removed by evaporation.

Table X - Process mass balance for 20 000kg/h of concentrate

Component	Kmol/h	Kg/h
Main Feed:		
Zn	6.025 E+1	3.939 E+3
Cu	2.605 E+1	1.655 E+3
Pb	5.941	1.231 E+3
Ag	1.937 E-2	2.089
Other Feeds:		
O_2 (80% reactive)	1.569 E+2	5.021 E+3
Air (Dry)		
O_2	5.504 E+1	1.761 E+3
N_2	2.071 E+2	5.802 E+3
CaO (85% reactive)	6.946 E+1	3.895 E+3
NH_4Cl	2.503 E+1	1.339 E+3
H_2SO_4	0	0
NH_3	3.176	5.399 E+1
Cu	4.750 E-1	3.018 E+1
Zn	6.398	4.183 E+2
Water		
Process	1.469 E+3	2.643 E+4
Cooling	3.482 E+4	6.267 E+5
Products:		
Zn	6.335 E+1	4.142 E+3
Cu	2.611 E+1	1.659 E+3
Pb	5.817	1.205 E+3
Ag	1.582 E-2	1.706
Losses:		
Zn	3.297	2.155 E+2
Cu	4.130 E-1	2.626 E+1
Pb	1.240 E-1	2.589 E+1
Ag		3.830 E-1
Evaporated Water	9.888 E+2	1.780 E+4
Final Residue	-	1.259 E+4
Gypsum	5.897 E+1	1.015 E+4

Table XI - Non-ferrous metals solubilization and recovery yields

Metal	Solubilization (%)	Recover (%)
Zn	97.14	95.05
Cu	99.72	98.44
Pb	99.93	97.90
Ag	97.00	81.66

In figures 3 and 4 are shown respectively the copper and zinc diamines concentration products as a function of R2.Cr, with the operating and saturation lines being also indicated.

It is found for these conditions that only the copper diamine may precipitate, for R2.Cr below 10 percent, since the zinc diamine operating line is well below saturation over the entire range of R2.Cr. In figure 3 the optimal values obtained for the recycle, which is required to prevent precipitation, expressed as a fraction of the main stream from DIV, are also indicated.

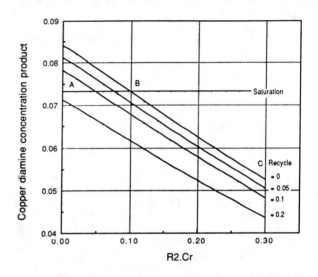

Figure 3 - Copper diamine concentration product versus copper(II) reduction conversion (the operating line <> ABC)

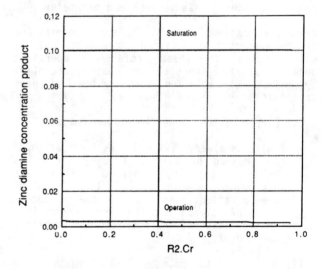

Figure 4 - Zinc diamine concentration product versus copper(II) reduction conversion

Figures 5 to 10 show some major effects which can be observed as a result of variation in R2.Cr. Since reactor R2 is a net contributor to the leaching of the chalcopyrite (tables VIII and V), as R2.Cr decreases the solubilization and recovery of copper also decrease, more markedly in the region below 60 percent, while the oxygen consumption increases as a result of the unreacted sulphides

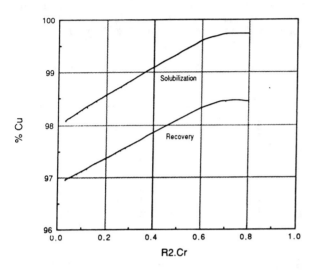

Figure 5 - Copper solubilization and recovery yields versus copper(II) reduction conversion

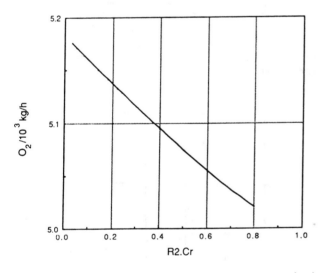

Figure 6 - Oxygen consumption versus copper(II) reduction conversion

in R2 being diverted to R3 where they undergo partial conversion (table.VII). Lower conversions of Cu(II) are also associated to greater consumption of metallic copper since the reduction of Cu(II) to Cu(I) takes precedence over the cementation of silver. An increasing solubilization of copper with R2.Cr also requires a greater consumption of CaO (figure 8) in order to eliminate the sulphates which as a result are being formed in copper reduction, together with those from neutral leaching.

The trends observed for NH_3 and process water consumption (figures 9 and 10) are more complex.

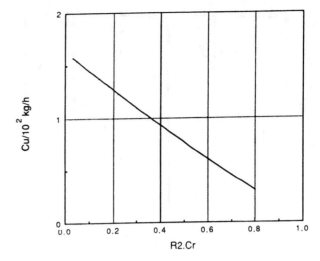

Figure 7 - Copper consumption versus copper(II) reduction conversion

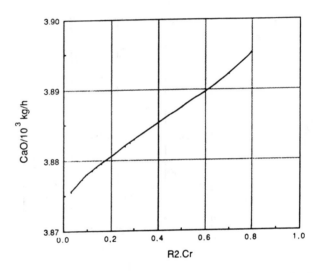

Figure 8 - CaO consumption versus copper(II) reduction conversion

For the former the make-up flow required to adjust the acidity in R3 competes with the NH_3 admitted to this reactor from R2 through the impregnated pulp. The changes in slope reflect the differences in the stoichiometric ratio of NH_3 to sulphides (table.VIII). For the latter the upward trends observed at both ends of the interval relate to an increase in the solid output stream flows. At the lower end a net increase in unreacted concentrate can be expected while for higher R2.Cr a noticeable increase in gypsum formation is found.

When T1.dens_pulp is varied about the design value of 100 g concentrate/1000 g H_2O, it is found that the concentrations of diamines in R5, while increasing with the pulp density, are well below saturation. Figure 11 illustrates the results obtained for copper. The effect of this parameter on the metal solubilization yields is negligible but it is very marked on recovery. A decrease in the recovery yield

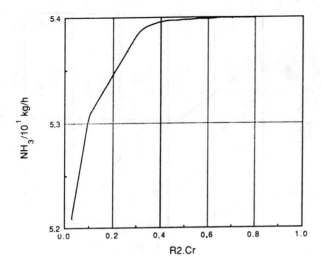

Figure 9 - NH$_3$ consumption versus copper(II) reduction conversion

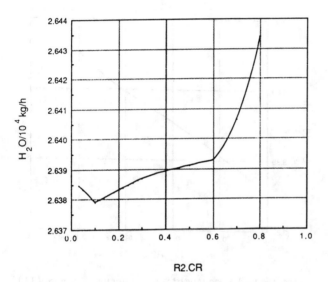

Figure 10 - Process water versus copper(II) reduction conversion

of copper, zinc and silver (figure 12) is observed, while for lead a reverse trend is found (figure 13). A denser pulp implies higher concentrations of metals in the aqueous phase and hence greater losses by impregnation.

The separation of lead by precipitation in the form of lead chloride crystals explains the trend reversal observed for this metal since phase saturation can be attained with less residual lead in solution. In the case of silver a similar effect could be expected but, as already mentioned, by design the by-pass for silver cementation is set at 10 percent of the main stream, which implies a considerable inventory of silver in the process which offsets any gains similar to those found for lead, due to greater losses by impregnation.

Other significant effects are found which are illustrated in figures 14 to 16. In figure

14 an increase in metallic zinc consumption is reported as a result of an higher recovery of lead. Figures 15 and 16 show that less make-up feeds in both NH$_4$Cl and NH$_3$ are required to adjust the composition of the aqueous phase which acts as the leaching agent. The observed decrease in NH$_4$Cl is directly related to the lower volumetric flow of aqueous phase while for the decrease in NH$_3$ a number of factors contribute to explain the greater requirements of acid in R3 namely, the lesser input of NH$_4^+$ from DIV3 due to a reduced recovery of copper and an increase in the NH$_3$ input from MX3 due to impregnation.

Finally figure 17 shows the effect of the

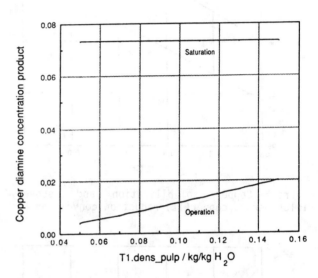

Figure 11 - Copper diamine concentration product versus pulp density

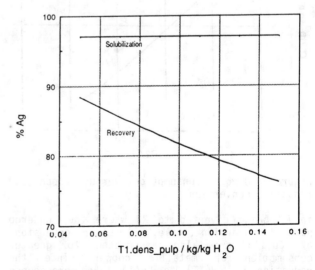

Figure 12 - Silver solubilization and recovery yields versus pulp density

variation over the range 0 to 0.10 of the fraction of the main stream at BY1, which forms the by-pass stream for silver cementation. It is found that even under conditions of non recovery the operating line is well below

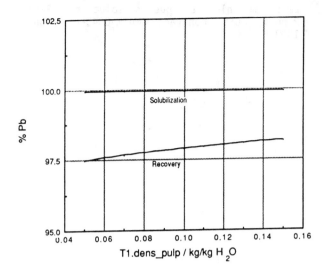

Figure 13 - Lead solubilization and recovery yields versus pulp density

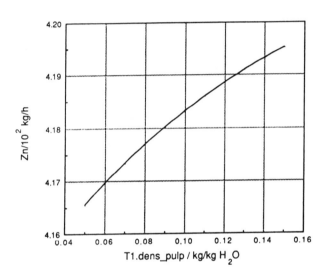

Figure 14 - Zinc consumption versus pulp density

saturation. When, at the design value of 0.10, the concentrate composition in silver is increased by a factor of ten, the value obtained for the silver concentration in R5 is 0.0071 kmol/1000 g H_2O which is still within a satisfactory operating region.

CONCLUSIONS

In this work a simulation model is presented which was set up in the course of the experimental development of a new hydrometallurgical process for the recovery of non-ferrous metals from Iberian sulphide ore concentrates, under an EEC research contract.

The model is based on and takes advantage of the features of the Speedup equation-oriented simulation package and thus ensures the flexibility required for the gradual building-up and exploration of the model. This is an

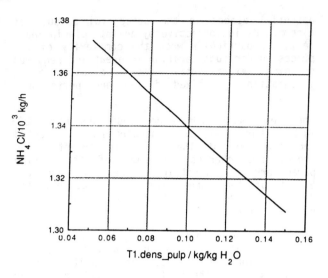

Figure 15 - NH_4Cl consumption versus pulp density

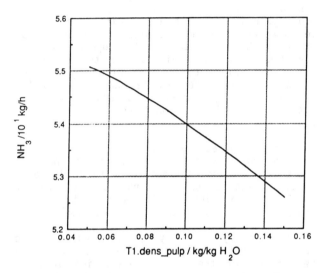

Figure 16 - NH_3 consumption versus pulp density

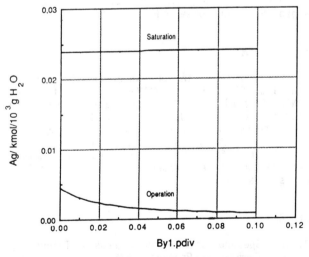

Figure 17 - Silver concentration in lead crystallization versus silver cementation by-pass stream expressed as a fraction of the main stream

essential aspect in the successful pursuit of a common design objective by an interdisciplinary team, in particular when the complexity of the process under consideration is great and requires on-going experimental data validation and integration, in mixed design and performance modes.

The steady-state reference conditions for 20000 kg/h of Portuguese ore concentrate are presented and the use of the model in investigating prospective changes in some of the design parameters is illustrated. These include the extent of the copper reduction reaction, the density of the pulp and the silver cementation by-pass stream.

It is concluded that with regard the formation of precipitates only copper diamine requires consideration but for operating conditions well below design. In the lead crystallization reactor which is considered the most critical point of the process, because of the prevailing high metal concentrations, following evaporation and cooling, none of the three remaining metals can be expected to form precipitates. It is found, in fact, that the extent of the copper reduction reaction could be reviewed and the economics associated with copper reduction conversion closer to 0.60 assessed, because the need for less leaching agent make-up is apparent together with a reduction in CaO and process water requirements.

The same applies to the assumed 0.10 split fraction for the silver by-pass stream where an increase in revenue associated with an increase in this parameter should be balanced out against the additional investment required by larger equipment capacity for silver cementation.

The possibility of employing denser pulps should be examined but only in conjunction with the performance of the silver cementation by-pass, because the decrease in copper and zinc recovery yields can be expected to offset any gains in make-up feeds, unless the trend in silver recovery can be significantly reversed and approach the one observed for lead.

ACKNOWLEDGMENT

This work was carried out under an EEC research contract, CEC Contract MA1M-0045-C and involved CENIM (Spain) and LNETI (Portugal) under the coordination of Dr. J.L. Limpo (CENIM).
The authors wish to thank Drs. J.L. Limpo and S. Amer (CENIM) and Eng. J.M. Figueiredo (LNETI/DTM) for the advice and information received which were essential for the modelling and simulation work.

REFERENCES

1. The Speedup User Manual, Prosys Technology Ltd, Cambridge, England, 1988.

2. "Study of hydrometallurgical treatment of complex sulphide ores by leaching with ferrous chloride and oxygen in highly concentrated ammonium chloride aqueous solution", Final Report, CEC Contract MA1M-0045-C, June, 1990.

Modelling in a Pachuca tank—flow and mixing phenomena

J. A. Trilleros
M. Martínez
Department of Materials Science and Metallurgical Engineering, Complutense University, Madrid, Spain

SYNOPSIS

A number of experimental devices were used to carry out this work by means of a simulation technique. A pilot plant fitted with two experimental reactors of one and four cubic meters of overall volume, respectively, and the required air-water systems was selected for this study. The liquid circulating flow, mixing times and mixing phenomena of the different Pachuca tanks versions (submerged gas jets with/without draft tubes and bundles) were calculated. A heat zone which provide an upper area of hot liquid, instead of a material tracer, was used to mark the fluid elements.

INTRODUCTION

The gas-type reactors have often been used in industrial processes. A gas dispersal system, with/without a central draft tube, was used, mostly, in these designs.

In the metallurgical industry, Pachuca tanks are used as leaching reactors for the hydrometallurgical reduction of nonferrous metals (gold, uranium, zinc, copper, etc.). All Pachuca tanks are usually classified as an airlift loop reactor, with/without draft tube, when the gas is dispersed in the liquid and a steady circulation flow of the liquid is induced due to the difference between the bulk specific gravity of the aerated liquid in the central zone (with/without the draft tube) and that at the annulus zone.

The gas, (preferably, air) dispersion in the liquid leads to the development of bubbles which are lifted and pushed towards the central zone but return, afterwards, to the starting zone and, at times, if the section of the descent is small, gas might be found entrained in the bubbles, (1) to (3).

The different effects on the liquid circulation due to the draft tube of the reactors should be pointed out. The differences are due: 1) the specific gravities differential which is developed between the gas expansion zone and another one, at the same level, in the liquid return zone, 2) the liquid is pushed, in the upper zone of the reactor, due to the high speed discharge of the two-phase flow which sets in movement the nearby liquid.

The main geometrical parameters to design this type of reactors show a large disparity in their values which are determined by the operational scale and the type of process. In biotechnology, it may be seen that the height/diameter ratio of the reactors fluctuates between 4.8 and 10.7 for scales which range from a laboratory type to those applied in actual manufacturing plants, while the diameters ratio of the draft tubes and the reactors range from 0.20 to 0.87. An 0.70 value is usually selected if settlement problems are noticed in the draft tube, (4) to (7).

Those values are significantly different from the ones found in the Pachuca tanks (8), since, in the latter, the height to diameter ratio ranges from 1.4 to 3.0, and the ratio between the draft tube and the reactor diameter goes from 0.11 to 0.46. It should be kept in mind that the tanks of this type, used in manufacturing plants, have diameters which range from 4 to 10 m, while the height never exceeds 16 m. Due to this design differences as pointed out by Evans, et al. (9), the superficial velocities of the gas in the Pachuca tanks are higher than those in bioreactors. The velocities, in the first mentioned, range from 0.10 to 0.73 m/s.

Although, mathematical simulation processes have often been used to study the metallurgical reduction processes, models of analog systems with prototypes of a pilot plant (10) have been occasionally used. The proper mix of the information gathered by such procedures has been useful to approach, with some assurance, the scaling up problems.

Quite often, therefore, the basic studies of flow properties have been based on models of the air-water system. The kinematic viscosity of water is higher than that of a sizable number of systems used in Extractive Metallurgy. As a result, when a turbulent flow is applied, for many field conditions, the findings from experiments on water can be extrapolated since, in the actual systems, a higher Reynolds number would be applied and both the turbulent flow and the dynamic similarity would be kept.

The liquid flows which circulate in those tanks together with the overall response of the flows (11) to (13), and the time required to achieve blending are significant to pickup knowledge about flows in such tanks, (14) to (16).

As regards the overall flow response in those tanks, the possibility of any existing dead zones, short-circuit and/or cross-over flows

508

should be reported since those conditions can reduce the effective volume of the tank.
The problems of the actual flow are linked to
those of scaling up, and the less-than-ideal conditions in scaling up are often the uncontrolla-
ble factor which affects the flow perfomance and
leads to significant errors either in the design
or in the operation of the tanks.
A knowledge of the actual flow can be drawn from
the residence times distribution in the tank
through the use of different tracer procedures
(17). Of those procedures, the thermal tracer is
preferred since neither the medium nor the system are distorted and the responses are quick
and neat.
The mixing times in the reactors have been determined, experimentally, for airlift loop reactors,
with/without a draft tube. The times tend to decrease as the gas flow increases and the liquid
flows which are circulated follow the same trend.
The mixing times, as established by Nakanishi,
et al. (18), on an empirical basis for a steel
refining gas blowing process; can be correlated
to the energy supplied during this gas expansion
per unit of liquid weight held in the tank, both
for real and analog systems (air-water), can be
expresed as follows:

$$t_m \propto \varepsilon^{-n}$$

Here, the experimental data were grouped within
a range with a slope n = 0.40, which takes in
different scales.
As published in the literature, the exponent of
the above formula ranges from 0.23 to twice this
value for different systems, scales and tracer
techniques. As shown by the experimental work
carried out by our research team, based on different tank designs, based on airlift and liquid
recycling, the exponent -n- ranges from 0.23 to
three times this value.

EXPERIMENTAL APPARATA AND TECHNIQUES

The experimental work was carried out at a pilot
plant scale in Pachuca tanks of a cylindrical
shape, in which the lower conical zone was left
out, since it was decided to determine the effect
of the volume of liquid held in the tank with/
without a draft tube. To that end, an extensive
range of draft tube/tank diameters ratios reaching up to a maximum of 0.40 was selected. A pipes bundle for parallel flow airlift was also
tested.
The conventional techniques were applied to determine the flows of the liquid airlifted through
the draft tube and thermal tracer procedures were
also used to determine the liquid flows which
circulate through the tanks and the mixing times,
and to analyze the flow related to each tank design. A simulation process has been also applied
and the facility was operated with an analog air
water system.

Apparata

The work was carried out in two cylindrical shaped tanks of 1.6 and 4.1 m³ capacity, 1.30 and
1.60 m high, and 1.25 and 2.00 m in diameter,
respectively.
A wealth of information was collected about the

smaller reactor since the research was centered
mainly on that type while, later on, the larger
reactor was operated to study the effects of
scaling up on those situations regarded as more
significant.
A multiple device consisting of 1.0 mm in diameter and 100 mm long capillary stainless steel
tubes, was fitted at up to twelve different locations in both tanks to study the dispersal of
air.
The reactor is fitted with an overflow system
to control the level. This system is continuously operated so that the liquid is fed towards
the bottom. The reactor has been operated, obviously, on a batch process basis.
In figure 1, a simplified flow diagram of the
facility can be seen where both the air circuit,
from the pilot plant, and the hot water circuit
are shown.
In the air line, the flows are checked and controlled up to a maximum of 2.7 kg/h. When a
thermal tracer was used in the tests, the air
flows have ranged from 0.18 kg/h to the mentioned maximum value.
A hot water, zone of 12 cm thick, is developed
by this thermal tracer technique in the upper
zone of the reactor due to the dispersion
through a hot water distributor which is generated by the water heated up to room temperature
by saturated steam at 8 kg/cm².
The tanks were operated with/without draft tubes
of diameters ranging from 28 to 500 mm and intermediate ones of 61.84, 125 and 250 mm. A single
jet tube was used for air dispersal, but increased to 6 or 12, when the draft tube was 84 mm or
more. A tank fitted with tubes bundles was also
used for airlift. A two-phase flow was airlifted
by 6 and 12 pipes bundles. Pipe lengths of 1100
and 550 mm have been tested with all the designs.

Experimental technique

The air flows have been metered by orifice meters and rotameters previously calibrated. The
airlifted water flows have been measured directly by weighing per unit of time, and the circulating water flows were based on the temperature
readings at the different zones of the reactor.
The pressure differences and overpressures were
read on water-colum and mercury pressure-gauges.
The thermal tracer technique, which develops a
hot zone with temperatures ranging from 40 to
60 ºC, has been applied to study the flows in
the Pachuca tanks.
The temperature ranges in the reactor have been
measured by 25 K-type (chromel-alumel) thermo-
couples set at five radial and as many length-
wise positions: five termocouples were arranged
in the hot and twenty in the cold zones, at four
different height levels. A FLUKE unit (HELIOS
Host Computer Interface Data-logger) linked to
a PC was used to log the data.

ANALYSIS OF THE FLOW PHENOMENA

As already mentioned, a 1.6 m³ reactor was used
to start the experimental work and test all the
designs and variables of the operation as shown
in the previous section, and taking into account
a rather broad response of the reactor. The larger reactor was used, afterwards, to study the

scaling up performance under the most favourable flow conditions.

The possible effect of the height/diameter ratio, in the Pachuca tank, has been also taken into account. Therefore, the required studies are already underway to commission another facility with a prototype reactor of up to 4 m³ capacity, 1.25 m diameter and variable height.

Liquid circulation

The liquid flow airlifted through the draft tube has been determined and, later, the liquid flow which goes through the reactor, both the flow contributed by the draft tube and that induced by the level of the liquid movements in the upper zone of the reactor, as the two phase is discharged, have been separately evaluated.

The volume of liquid airlifted through the central draft tube has been shown to grow quickly at the same rate of the air flow due to the pressure differential which is established when air is introduced, but the trend of the gas flow decreases and approaches a steady value, even if the air flows are increased, as may be seen in figure 2. But, if the draft tube length is reduced to half, both flows show an increasing growth so that, for the same air flow, the liquid flows are lower than in the previous case and this trend is increasingly more evident as the expanded air flow is lowered.

The trends are not different from those in the previous paragraph, if the draft tube diameter is increased. The water flows tend to a steady value which remains always within the same range of values. However, for small air flows, the respective water flows decrease, when the draft tube diameter is increased. A different response of the water flows, as shown in Figure 3, is not noticeable when the jets of the dispersion system are increased from one to six or twelve.

The liquid flows which pass through the reactor have been determined from the temperatures distribution curves in the cold zone of the reactor. The starting points of the growing vertical evolution curver of the temperatures have been applied as a reference and the time required by the hot front to shift along 80 cm. in height, as well as the surface speed of the liquid and its flow has been estimated.

The different values of the liquid flows for smaller draft tube, both of the circulation and airlifting types, have been grouped in table I. The circulated and airlifted flows for small volumes of air are seen to be in the same range of values, but the situation is different when the air flow is larger since the turbulent effect, added to the upper expansion of the two-phase flow, is increased by that flow.

The different values of the liquid flow for different reactor designs, have been grouped in table II.

Mixing phenomena in the Pachuca tank

Both the flow and the mixing times for each reactor design have been determined based on the temperatures distribution throughout the reactor. As may be seen, there are no significant differences in the temperature values of the five radial positions. Therefore, the distribution curves, nearest to the reactor wall, will be used in this study. The decreasing trend of the temperatures, in the upper hot zone of the reactor, can be noticed in those curves which show a different shape for each design. Obviously, if a perfect mixing flow could be achieved, the signal would be one step lower towards a balanced temperature and the potential delay, starting at the time interval which is required for the vertical drop of the temperature, could be determined. A similar signal would be pickedup for the temperatures in the cold zones, but here the step would frow towards a thermal balance and the possible delay times could be evaluated for each of them.

When the liquid circulation was rather small, a limit situation would be reached which would lead to an ideal plug flow. The response of the temperatures in the hot and cold zones would show linear profiles. These profiles would tend towards a balance temperature although the time required to reach that condition would be quite long.

Based on the recorded temperature-time readings, determined experimentally, it is evident that the different Pachuca tank designs, with/without draft tube, show a significant effect of the mixing flow in the upper 40% volume of the tanks while the time delays of the progress of that mixing front can be observed as readings approach the reactor bottom.

Generally speaking, the delays in the progress of the mixing front can be said to decrease as the expanded air flow grows. Also, for the same gas flow, the mixture effect is enhanced as the draft tubes diameters increase since the energy spent, due to friction, decreases, fig. 4 to 7. When the length of the draft tubes is reduced in half, the mixing effect, in the upper half of the reactor, increases while the delay times, in the lower half of same, will be reduced, fig. 8 and 9. A higher mixing effect and shorter delay times are found when the air expands freely in the reactors, but this trend is less significant as the expanded gas flow increases, fig. 16 to 18.

As shown earlier by our research team (24), the maximum water flows were airlifted by the draft tubes of diameters from 22 to 28 mm, even for smaller air flows. In view of the foregoing, a pipes bundle was selected for experiments with parallel flow air-lift. As shown by these experiments, the mixing conditions were good and the delay times, at the reactor bottom, were reduced which led to shorter mixing times, fig. 10 and 11. It was also found that, when the draft tube and the twelve pipes bundle have the same cross-section, the flow responses are similar, regardless of the air which is dispersed through one or twelve capillary tubes in the draft tube.

When the length of the draft pipes, in the reactors, is reduced in half, the mixing conditions are boosted even further and the delay times are shortened so that, for the same air flow, the mixing times can be smaller than those in the Pachuca tank less a draft tube, fig. 12 and 13. For the draft tubes of a diameter equal to 84 mm or over and gas dispersal through twelve capillary tubes, the mixing in the flow is increased but the times required for homogeneizing are reduced.

A similar development is noticed in the mixing times. The data from the smaller sized reactor is shown in table III. When high volumes of air are expanded, there are several tank designs which demand a minimum mixing time: the Pachuca tanks with a draft tube of 500, 250 or 125 mm diameter; the bundles of twelve stub pipes, and those designs lacking a draft tube. It can be seen that, as the expanded air flow decreases, the shorter mixing times are achieved by the reactor fitted with a 125 mm in diameter draft tube, regardless of whether it is a full or half length tube. The last mentioned design of a tank is regarded, therefore, to provide the best possible flow conditions.

In the largest volume reactors, the flows pattern is similar to that in the tanks which hold less liquid, fig. 14 and 15. The phenomena of scaling up, limited to the mixing times, is correlated to the specific expansion energies and follows the equation which was stated earlier, fig. 19. The exponents of the specific energy are shown in table IV. There is a general tendency for the exponent to grow as the draft tube's diameter goes from 0.35 to 0.61, this tendency being reversed when the reactor works with half length draft tubes.

FINAL CONSIDERATIONS

The following considerations can be drawn from the experimental study carried out which was summarily described in this text.

The applied thermal tracer technique is both accurate and sensitive and does not upset, in any way, the system's momentum transport. The profiles fo the temperature evolution, within the hot zone, provide a relative knowledge about larger or smaller involvement of the mixing flow in the upper zone of the tank, while the profiles of the temperature evolution, in the colder zones of the tank, are useful to determine the circulating liquid flows and to know the time delays of the turbulent front progress which provides, somehow, an idea of the mixing action at each zone.

The best flow conditions and shorter mixing times are achieved when the tanks are fitted either with draft tubes of 125 mm in diameter and above; or 12 tube bundles and free dispersal are used. As regards the air flows range which were selected for this experiment, the best flow conditions are found for tanks with a diameters (d/D) ratio around 0.1, regardless of the draft tube length.

Concerning the scaling up interval used here, the mixing times can be correlated through the proposed Nakanishi formula and the parameters for the potential correlation should be found experimentally.

The use of tanks with 4 m^3 holding capacity but of larger diameter/height ratios was regarded as more convenient since in this way the delay times of the turbulent fronts can be compared. A more reliable design of the Pachuca tanks would be afforded by this approach to meet best optimum flow requirements.

NOMENCLATURE

D	Reactor diameter (m)
F_g	Gas mass flow rate (Kg/s)
F'_g	Gas mass flow rate (Kg/h)
F_1	Liquid mass flow rate (Kg/s)
F'_1	Liquid mass flow rate (Kg/h)
L	Draft length (mm)
Q_1	Lift liquid mass flow rate (Kg/h)
T	Temperature (ºC)
V	Reactor capacity (m^3)
d	Draft diameter (mm)
t	Time (s)
t_m	Mixing time (s)
ϵ	Mixing power input (W/m^3)

REFERENCES

1. Lamont A.G.W. Can J. Chem. Eng., 1958, pp. 153-60.
2. Clark N.N. Minerals and Metallurgical Processing, 1984, Nov., pp. 226-32.
3. Clark N.N., Flemmer R.C.L. Chem. Eng. Sci., 1984, vol 39, pp. 170-73.
4. Chisti M.Y. Airlift Bioreactors, 1989, Elsevier Applied Sci., London.
5. Chisti M.Y., Moo-Young M. Chem. Eng. Commun. 1987, vol. 60, pp. 195-242.
6. Weiland, P. Ger. Chem. Eng., 1984, vol. 7, pp. 374-85.
7. Merchuk J.C., Siegel M.H. J. Chem. Technol. Biotechnol., 1988, vol. 41, pp. 105-20.
8. Hallett C.J., Monhemius A.J., Robertson, D.G. C. Extraction Metallurgy'81, Symp. Inst. Min. Met., London, 1991, pp. 308-19.
9. Shekhar R., Evans J.W. Metallurgical Transactions B, 1989, vol. 20B, pp. 781-91, and 1990 vol. 21B, pp. 191-203.
10. Szekely J., Evans J.W., Brimacombe J.K., The Mathematical and Physical Modelling of Primary Metals and Processing Operations, 1988, J. Wiley and Sons. Inc., N. York.
11. Bello R.A., Robinson C.W., Moo-Young M. Can J. Chem. Eng., 1984, vol. 62, pp. 573-77.
12. Chisti M.Y., Halard B., Moo-Young M. Chem. Eng. Sci., 1988, vol. 43, pp. 451-57.
13. Clark N.N, Jones A.G. Chem. Eng. Sci., 1987, vol. 42, pp. 378-85.
14. Pandit A.B., Joshi J.B., Chem. Eng. Sci., 1983, vol. 38, pp. 1189-215.
15. Rouseau I., Bu'lock J.D. Biotechnol. Lett., 1980, vol. 2, pp. 475-80.
16. Shah Y.T., Belkar B.G., Godbole S.P., Deckwer W.D. AIChEJ, 1982, vol. 28, pp. 373-79.
17. Ford D.E., Meshelkar R.A., Ulbreet, J. Prac. Tech. Inst., 1971, vol. 17, pp. 781-9.
18. Nakanishi K, Fujii T., Szekely J., Ironmaking Steelmaking, 1975, vol. 3, pp. 1193-7.
19. Leher L.H., IEC. Proc. Des. Dev., 1988, vol. 7, pp. 226-33.
20. Szeleky J., Shener T., Chag C.W. Ironmaking Steelmaking, 1979, vol. 6, pp. 285-91.
21. Mori K., Sano M. The 19th Comittee (Steelmaking). Japan So. for the Promotion of Sci., May, 1980.
22. Kato T. Okamoto T. Denkseiko (Electric Furnace Steel), 1979, vol. 50, pp. 128-34.

23. Trilleros J.A., Recio A. Fluid Mixing 4, I. Chem. E. Symposium Series nº 121. 1990, Bradford, pp. 215-39.

24. Lombardero L., López F., Trilleros J.A., Otero J.L. Extraction Metallurgy'89, Symp. Inst. Min. Met., London, 1989, pp. 361-85.

TABLE I

One Jet

d	L	F'_g	F'_1	Q_1
28	1100	0.86	2592	1573
		0.47	2088	1512
		0.18	1476	1224

TABLE II A. One Jet

	L = 1100		L = 550	
a	F'_1	F_1	F'_g	Q_1
28	0.86	0.72	0.86	1,51
	0.47	0.56	0.50	0,92
	0.18	0.41	0.19	0,60
61	0.86	2.26	0.86	3.27
	0.50	1.85	0.50	2.59
	0,19	1.32	0.19	1.29
84	0.90	3.37	0.90	4.02
	0.54	2.89	0.54	2.83
	0.18	1.79	0.21	1.52
125	0.86	5.20	0.86	3.77
	0.54	4.30	0.54	2.77
	0.21	2.35	0.21	1.39
250	2.78	8.43	----	----
	0.86	3.41	----	----
	0.54	3.22	----	----
	0.19	1.60	----	----
500	2.74	5.84	----	----
	0.86	2.31	----	----
	0.54	1.80	----	----
	0.19	1.22	----	----

TABLE II B

	L = 1100		L = 550	
	F'_g	F_1	F'_g	F_1
Bundle	0.79	4.16	0.76	5.00
12	0.54	3.19	0.50	4.62
Pipes	0.19	1.92	0.19	2.65
Bundle	0.90	2.83	0.94	3.28
6	0.43	2.02	0.43	2.35
Pipes	0.15	1.30	0.18	1.47
Air	0.86	5.66	----	----
Free	0.50	3.77	----	----
Airlift	0.21	1.68	----	----

TABLE III

A) One Jet

	L = 1100		L = 550	
d	F'_g	t_m	F'_g	t_m
28	0.86	1388	0.86	688
	0.47	1760	0.50	940
	0.18	2635	0.19	1448
61	0.86	443	0.86	262
	0.50	551	0.50	354
	0.19	752	0.19	692
84	0.90	303	0.90	255
	0.54	370	0.54	383
	0.18	585	0.21	613
125	0.86	235	0.86	246
	0.54	255	0.54	354
	0.21	440	0.21	638
250	0.86	290	----	----
	0.54	345	----	----
	0.19	560	----	----
	0.86	400	----	----
	0.54	420	----	----
	0.21	621	----	----

TABLE III
B) 12 Jets

L	d	F'_g	t_m
1100	84	0.90	290
		0.40	340
		0.14	580
	125	2.74	110
		1.05	200
		0.68	205
		0.17	330
	250	2.74	116
		0.97	210
		0.41	270
	550	2.78	130
		1.08	205
		0.18	430
550	125	1.00	177
		0.54	186
		0.17	367

TABLE IV

L	1100	550	
d	n	n	
28	-0.41	-0.48	
61	-0.35	-0.67	
84	-0.40	-0.57	
125	-0.49	-0.44	
250	-0.59	----	
500	-0.58	----	
28	-0.53	-0.41	(1)
28	-0.61	-0.42	(2)
28	-0.57	----	(3)

(1) Bundle 12 Pipes
(2) Bundle 6 Pipes
(3) Air Free Airlift

TABLE III
C) Bundles

D = 28 mm

n	L	F'_g	t_m
12	1100	0.79	270
		0.54	350
		0.19	578
	550	0.76	204
		0.50	200
		0.19	370
6	1100	0.90	410
		0.43	568
		0.15	780
	550	0.94	320
		0.43	440
		0.18	645

TABLE III
D) Air Free Airlift

Jet	F'_g	t_m
1	0.86	194
	0.50	315
	0.21	540
12	2.95	120
	0.86	160

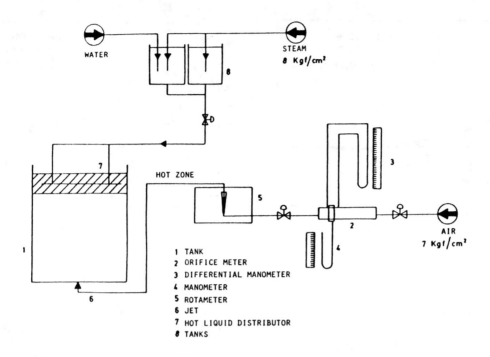

FIG. 1.

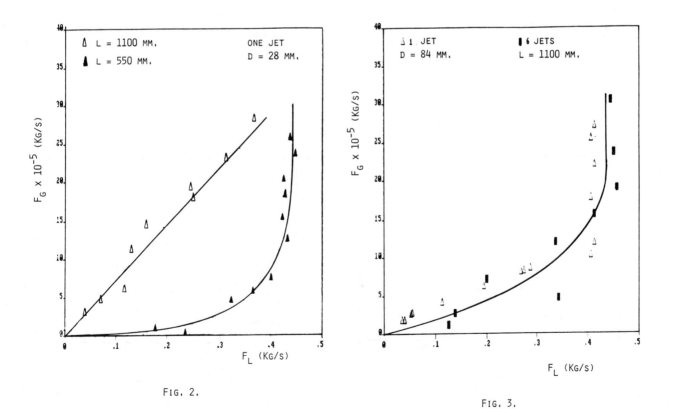

FIG. 2.

FIG. 3.

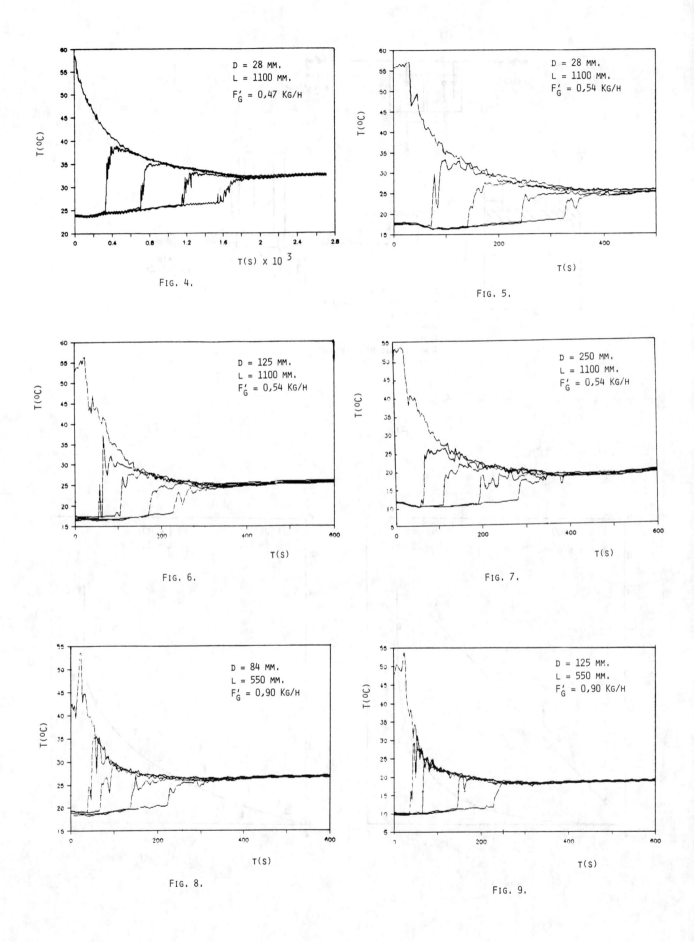

FIG. 4.

FIG. 5.

FIG. 6.

FIG. 7.

FIG. 8.

FIG. 9.

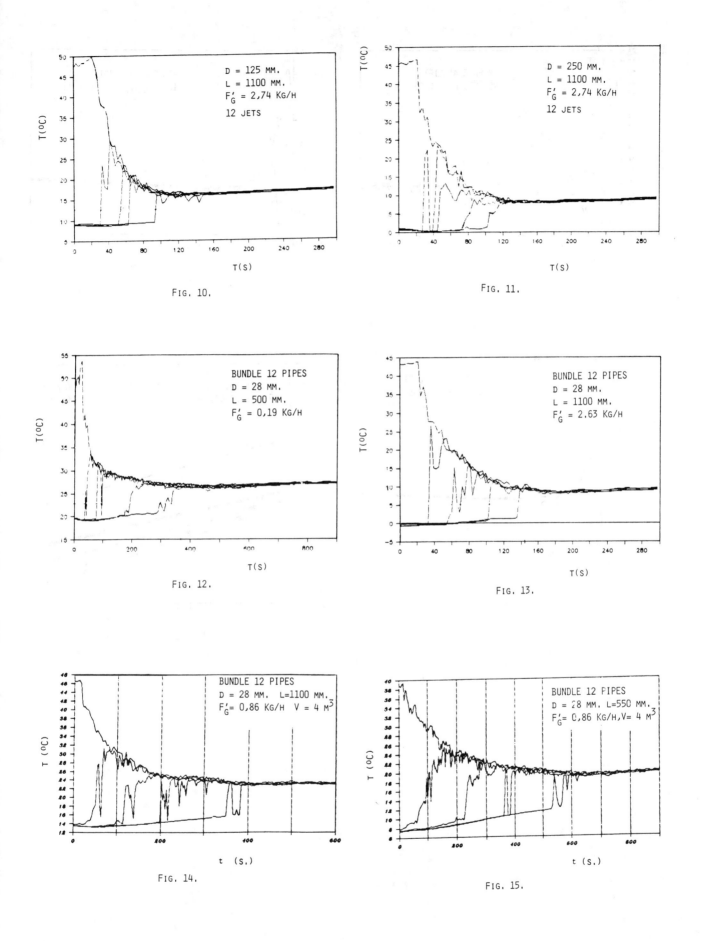

FIG. 10.

FIG. 11.

FIG. 12.

FIG. 13.

FIG. 14.

FIG. 15.

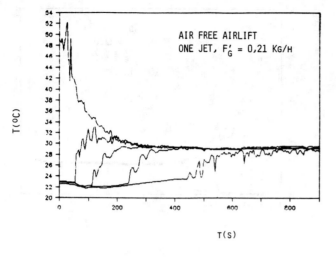

FIG. 16

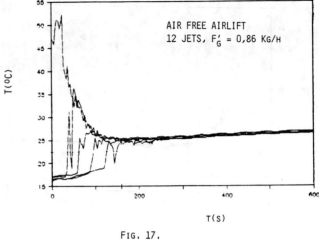

FIG. 17.

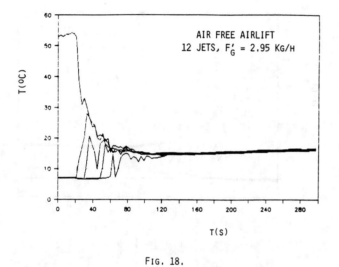

FIG. 18.

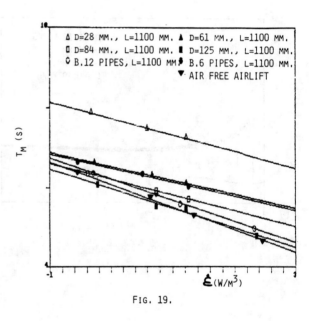

FIG. 19.